Connecting Indian Wisdom and Western Science

Plant Usage for Nutrition and Health

Traditional Herbal Medicines for Modern Times

Each volume in this series provides academia, health sciences, and the herbal medicines industry with in-depth coverage of the herbal remedies for infectious diseases, certain medical conditions, or the plant medicines of a particular country.

Series Editor: Dr. Roland Hardman

Volume 1
Shengmai San, edited by Kam-Ming Ko

Volume 2
Rasayana: Ayurvedic Herbs for Rejuvenation and Longevity, by H.S. Puri

Volume 3
Sho-Saiko-To: (Xiao-Chai-Hu-Tang) Scientific Evaluation and Clinical Applications, by Yukio Ogihara and Masaki Aburada

Volume 4
Traditional Medicinal Plants and Malaria, edited by Merlin Willcox, Gerard Bodeker, and Philippe Rasoanaivo

Volume 5
Juzen-taiho-to (Shi-Quan-Da-Bu-Tang): Scientific Evaluation and Clinical Applications, edited by Haruki Yamada and Ikuo Saiki

Volume 6
Traditional Medicines for Modern Times: Antidiabetic Plants, edited by Amala Soumyanath

Volume 7
Bupleurum *Species: Scientific Evaluation and Clinical Applications,* edited by Sheng-Li Pan

Volume 8
Herbal Principles in Cosmetics: Properties and Mechanisms of Action, by Bruno Burlando, Luisella Verotta, Laura Cornara, and Elisa Bottini-Massa

Volume 9
Figs: The Genus Ficus,
by Ephraim Philip Lansky and Helena Maaria Paavilainen

Volume 10
Phyllanthus Species: Scientific Evaluation and Medicinal Applications edited by Ramadasan Kuttan and K. B. Harikumar

Volume 11
Honey in Traditional and Modern Medicine, edited by Laïd Boukraâ

Volume 12
Caper: The Genus Capparis, Ephraim Philip Lansky, Helena Maaria Paavilainen, and Shifra Lansky

Volume 13
Chamomile: Medicinal, Biochemical, and Agricultural Aspects, Moumita Das

Volume 14
Rhodiola rosea, Alain Cuerrier and Kwesi Ampong-Nyarko

Traditional Herbal Medicines for Modern Times

Connecting Indian Wisdom and Western Science

Plant Usage for Nutrition and Health

Edited by
Luisella Verotta
Maria Pia Macchi
Padma Venkatasubramanian

CRC Press
Taylor & Francis Group
Boca Raton London New York

CRC Press is an imprint of the
Taylor & Francis Group, an **informa** business

CRC Press
Taylor & Francis Group
6000 Broken Sound Parkway NW, Suite 300
Boca Raton, FL 33487-2742

First issued in paperback 2019

ISBN-13: 978-1-4822-9975-5 (hbk)
ISBN-13: 978-0-367-37770-0 (pbk)

Library of Congress Cataloging-in-Publication Data

Connecting Indian wisdom and western science : plant usage for nutrition and health / edited by Luisella Verotta, Maria Pia Macchi, and Padma Venkatasubramanian.
pages cm. -- (Traditional herbal medicines for modern times)
Includes bibliographical references and index.
ISBN 978-1-4822-9975-5 (hardcover : alk. paper) 1. Materia medica, Vegetable. 2. Medicinal plants. 3. Herbs--Therapeutic use. 4. Nature, Healing power of. I. Verotta, Luisella II. Macchi, Maria Pia. III. Venkatasubramanian, Padma.

RS164.C695 2015
615.3'21--dc23 2015004523

Visit the Taylor & Francis Web site at
http://www.taylorandfrancis.com

and the CRC Press Web site at
http://www.crcpress.com

Contents

Foreword

NUTRITIONAL DIMENSIONS OF THE GLOBAL CRISIS IN LIFESTYLE DISEASES

Global concerns about the rise in lifestyle or noncommunicable diseases (NCDs) have highlighted the role that diet plays in obesity, heart disease, diabetes, and cancer, among others.

A March 2013 statement from the World Health Organization (WHO), the leading public health arm of the United Nations, clearly indicates that this issue is now on the WHO agenda: "Of the 57 million global deaths in 2008, 36 million, or 63%, were due to non-communicable diseases (NCDs), principally cardiovascular diseases, diabetes, cancers and chronic respiratory diseases" (World Health Organization, 2011).

The WHO cites obesity as an important example of a disorder that is becoming a serious problem throughout Asia, Latin America, and parts of Africa, despite widespread undernutrition. In some countries, the prevalence of obesity has doubled or tripled over the past decade.

Over the past few years, through the use of various measures, it has become well documented that a modern Western diet is a significant causal factor in both the declining health of the world population and the increase of NCDs.

Popular belief presumes that NCDs afflict mostly high-income populations. However, evidence tells a different story:

- Nearly 80% of NCD deaths occurs in low- and middle-income countries, and NCDs are the most frequent causes of death in most countries, with the exception of those in Africa (World Health Organization, 2012).
- NCDs in Africa are projected to exceed communicable, maternal, perinatal, and nutritional diseases as the most common cause of death by 2030.

It is now well known that chronic diseases are largely preventable. Modifiable NCD risks include diet, physical inactivity, tobacco use, and alcohol consumption.

Diet, known for many years to play a key role as a risk factor for chronic diseases, is now assuming a primary role.

The WHO's 2013 World Health Report notes: "Diet and nutrition are important factors in the promotion and maintenance of good health throughout the entire life course. Their role as determinants of chronic NCDs is well established and they therefore occupy a prominent position in prevention activities."

And *The Journal of the American Medical Association* noted in July 2013 that diet was a more important factor associated with disease than physical inactivity or high body mass index. Of the top 6 of 17 risk factors for death, diet has now surpassed even tobacco as the greatest risk factor for mortality.

THE MEDITERRANEAN DIET

Researchers are turning their attention to traditional patterns of eating, and no tradition has received more research attention than that of the Mediterranean, namely, the Mediterranean diet.

Studies and, in some cases, meta-analyses of studies have shown that

- Greater adherence to a Mediterranean-style diet is associated with lower risk factors for cardiovascular disease.
- Adherence to a Mediterranean diet is associated with a lower risk of age-related disease such as dementia.
- A Mediterranean diet with large amounts of vegetables and fish gives a longer life.

These and related findings have led many in the world of nutrition to conclude that the Mediterranean diet is an approach that has global value and should be swiftly applied across the board to combat the rising tide of NCDs.

What has not been pointed out is that, for all its value, the Mediterranean diet is a Western diet studied by Western scientists on Western populations—with conclusions being drawn for a global population.

At the time of writing (September 2014), the population of Asia is estimated at 4.427 billion, and the population of Africa is estimated at 1.069 billion.

This combined non-Western population of 5.496 billion people represents more than three-quarters (76%) of the global population of 7.211 billion people.

It would thus seem reasonable to ask how a diet drawn from the traditions of less than a quarter of the world's population is being recommended for more than three-quarters of the world's population, which has different cultural and genetic histories—involving different taste preferences and food availability—and different epigenetic histories—involving different metabolomics and different nutrigenomics.

One could also ask how a small section of the Western world (i.e., the Mediterranean) has got it so right and everyone else has not.

If Western cultural dominance is not the explicit driver here, certainly the push toward a Mediterranean approach to solving a global health crisis is good for Western business, as anyone can attest who has seen the growing plethora of advertisements for olive oil in India's news media.

In the global dialogue about the growing problem of lifestyle causes for NCDs, there is a fundamental perspective missing, that is, an appreciation that there is a strong cultural influence on taste and food preference as well as genetic and epigenetic influences on what foods people can benefit from eating.

Research published in the *Journal of Gerontology* in 2014 has finally begun to differentiate the suitability of the Mediterranean diet in producing health benefits in different genetic groupings.

A study by Koyama *et al.* (2014) looked at a population of initially well-functioning older adults and found a significant correlation between strong adherence to the Mediterranean diet and a slower rate of cognitive decline among older

African American, but not white, adults. This study, the researchers noted, is the first to show a possible ethnicity-specific association between the Mediterranean diet and cognitive decline.

More than that, the study may well be the first to underscore that the Mediterranean diet, unsurprisingly, is not a one-size-fits-all approach to nutrition. This new research opens the door for more genetically based approaches to understanding culturally relevant dietary traditions of different genetic groupings in Asia, Africa, and other non-Western contexts.

Clearly, other cultures than those in the Mediterranean region have evolved healthy eating patterns, using the foods available to them and with preparation methods often unique to that culture.

In this context, there has been a growing body of research generated on East Asian traditions, which requires wider global attention in our understanding the influence of diet on NCDs.

TRADITIONAL ASIAN DIETS

- In East Asia, number of influential studies have been conducted showing the cardiovascular and longevity benefits of regional dietary traditions. Some of these are summarized below.
- *Chinese eating traditions.* The China–Oxford–Cornell Study on Dietary, Lifestyle and Disease Mortality Characteristics in 65 Rural Chinese Counties compared the health consequences of diets rich in animal-based foods to diets rich in plant-based foods among people who were genetically similar. The study included a comparison of the prevalence of Western diseases (coronary heart disease, diabetes, leukemia, and cancers of the colon, lung, breast, brain, stomach, and liver) in each county. Diet and lifestyle information from inhabitants of the same counties collected 10 years later found that as blood cholesterol levels rose, so did the prevalence of "Western" diseases.
- *Traditional Korean diet.* The Republic of Korea has lower rates of chronic diseases and a lower-than-expected level of fat intake and obesity prevalence than other industrialized countries with similar economic development. This is notable because the community has largely maintained its traditional high-vegetable diet despite major social and economic change.
- *Traditional Japanese diet.* Japanese people are more likely to reach 100 years old than anyone else in the world, a fact that researchers attribute to their diet. Dr. Craig Willcox, a coprincipal investigator of the Okinawa Centenarian Study, has summarized the benefits of the local diet:

  > The Okinawans have a low risk of arteriosclerosis and stomach cancer, a very low risk of hormone-dependent cancers, such as breast and prostate cancer. They eat three servings of fish a week on average, plenty of whole grains, vegetables and soy products, more tofu and more konbu seaweed than anyone else in the world, as well as squid and octopus, which are rich in taurine—that could lower cholesterol and blood pressure.

These are the perspectives that are needed to drive a new agenda in looking at how the dietary traditions of local cultures may be the best starting point in examining the best nutritional practice for those societies.

While East Asian nutritional traditions have begun to garner research attention, South Asian traditions, with their further spread of influence to Southeast Asia, are to date relatively neglected in the research literature.

INDIA'S TRADITIONS OF NUTRITION

Ayurveda, the natural healthcare tradition of India, is grounded in a sophisticated theory of personalized nutritional, lifestyle, and medicinal guidelines that are consistent with an emerging twenty-first-century prioritization of patient-centered healthcare and personalized medicine. Leading molecular biologists in India have been examining the genetic basis of Ayurvedic body types, which are used to set personalized dietary and medicinal guidelines, and are finding strong genetic evidence in support of the Ayurvedic groupings of mind–body types.

Ayurveda has always placed an emphasis on prevention and on diet. A classic saying of one of the foundational texts on Ayurveda, the *Charaka Samhita*, is as follows: "Without proper diet, medicine is of no use. With proper diet, medicine is of no need."

Accompanied by a deep understanding of chronobiology, Ayurveda gives guidelines on the best foods and exercise both for body types and for seasons of the year and times of the day. What today is called a "wellness lifestyle" has been present in Ayurveda for millennia as *dinacharya* and *ritucharya*—daily and seasonal routines of diet and exercise, respectively.

Like Ayurveda, Unani medicine—the Greco-Arabic system of medicine found throughout the Islamic world—is a naturalistic system with a comprehensive clinical literature and a sound theory of disease origin, patient typologies, and dietary and healthcare guidelines.

PIONEERING A BRIDGE AND SHARED JOURNEY BETWEEN WEST AND EAST

In this groundbreaking book, the Chapters 2 through 5, written largely by Indian authors, trace the early history of Ayurveda as well as the basic principles and dietetics of the Ayurveda, Siddha (the South Indian tradition of medicine that shares theoretical assumptions with Ayurveda), and Unani systems of medicine.

The Chapters 6 through 10, by Western scientists and nutritionists, discuss the current global focus on nutrients, nutraceuticals, and, importantly, the Mediterranean diet.

Chapter 11 begins what will, no doubt, be a long and continuing journey of mutual exchange and dialogue. Here, the authors work to create an integrative perspective for understanding India's health traditions through the lens of modern science. And they look at how India's millennia of experience can enrich the West's notions of optimal nutrition, prevention, and personalized healthcare.

These nutritional traditions, so deeply embedded in India's history, are used in some form or other by more than 1.5 billion people both within South Asia and

Southeast Asia. And, on sheer numbers alone, they certainly warrant at least as much research and policy attention as the Mediterranean diet and the above-mentioned traditions of East Asia. However, there is much more than the sheer democratic principle of ensuring that the preferences of the majority are not overruled by the enthusiasm and priorities of the minority.

What is at the core of this exchange is a frontier of exploration into deep insights into human biology, individual differences and seasonal influences on metabolism, and a conceptual approach to nutrition that can be applied across foods and cultures.

For these reasons, this book is a welcome arrival. It is a book led by Mediterranean researchers and supported by Indian researchers. They have explored together the wisdom of Indian dietary and nutritional knowledge. And the focus has not been just on ingredients and recipes for wellness. Rather, the book begins with the underlying theories of nutrition, the evidence, and personalized application as a basis for expanding an understanding of how nutrition can be culturally based and locally sourced to address health prevention and health promotion needs in a personalized way for vast sectors of the global population.

Accordingly, this will become a landmark book, opening the way to a new generation of research into Indian and South and Southeast Asian nutritional theories and traditions—with implications leading all the way back to the Mediterranean and its global diaspora.

Gerard Bodeker
University of Oxford
and
Columbia University

Introduction

> Charity, longevity, intelligence, happiness, contentment, strength and knowledge are all rooted in food.
>
> **Charaka Samhita**

Plants are partners with animals in the living world. They give animals the energy (as food) to survive and the oxygen to breathe and also contribute to the beauty of the planet with their colors and fragrances. Plant phytochemicals have been the source of important drugs used in degenerative diseases and provided scientific insight into understanding the pathways regulating the basic mechanisms of life (Argyropoulou *et al.*, 2013; Patil *et al.*, 2009; Szychowski *et al.*, 2014). It is noteworthy to mention that plants are renewable and sustainable resources if appropriately managed (science helps in understanding which plants are most favorable to be used as nourishment). They help inhabitants of this planet to successfully combat malnutrition, which is the major cause of diseases in many developing countries. Vegetables are also known to help in counteracting the diseases more common in wealthier countries, like obesity and cardiovascular diseases.

According to the World Health Organization, the use of plants plays a strategic role in healthcare for about 80% of the rural population in developing countries. India is one of the countries in the world where there is great use of medicinal plants; they are referred to as *ratnas*, precious gems of a divine nature. It is estimated that in India the local health traditions alone include about 200,000 herbal preparations, using more than 8000 species of plants. In addition, there is considerable knowledge of hundreds of edible preparations with therapeutic or preventive properties. Indian traditions associated with food are deeply related with all aspects of the cycle of life, starting from the first nutrition provided to newborns, even before breast milk. Plants are part of rituals, ceremonies, and festivals all over India. These traditions are also ecosystem-specific, need based, culturally linked, and autonomous in different village communities. At the household level, housewives still keep alive this rich traditional knowledge, with a number of home remedies made with spices and other food items. In many communities and rural areas, Indian women have a common capacity to evaluate the qualities of food in relation to the individual constitution and suited to different climates and seasons. However, there is a lack of scientific documentation regarding this wealth of knowledge, with a risk of it being lost. This is also the current problem where women have less time to spend on cooking than in the past, thus forgetting traditional recipes and going for fast food.

The Mediterranean basin is home to considerable biodiversity, including 22,500 endemic vascular plant species. It is the Old World region where the olive tree grows, a plant rich in religious meanings and one of the most important sources of nutrients. Scientifically speaking, olive oil is used not only in seasoning but also as a food, as it contains high-quality fats and is rich in lipophilic vitamins and polyphenols, which no other fat contains. The use of Mediterranean plants as medicines engraved the

Mediterranean culture with its *Fitoterapia* and dates back to the ancient Egyptians, Greeks, and Romans. The rich knowledge and practices have provided significant inputs to contemporary biomedicine and nutrition. Physicians were first botanists with a deep knowledge of the potential of plants as poisons or as food.

Vegetables and fruits are nowadays globally recommended by registered physicians as a consistent part of diet, and programs have been launched in developed countries to encourage balanced diets to tackle obesity and cardiovascular diseases.

The potential of food and lifestyle to support health and reduce the risk of diet-related diseases has developed into quite a recent branch of study under nutrition called *nutrigenomics* (Müller and Kersten, 2003; Sales *et al.*, 2014), which aims to explore the role of nutrition in gene expression. As described in this book, the concept that food and lifestyle influence health is in fact an old one. Nutrigenomics is the area of nutrition that uses molecular tools to search, access, and understand the various responses to diet on individuals or populations. It is a discipline involving different professionals (nutritionists, bioinformatists, biologists, physicians) seeking to elucidate how the components of a particular diet (the bioactive compounds) may affect the expression of genes, in terms of up or down-regulation (Kumar *et al.*, 2014). As chemists know very well, "*we cannot understand life if we do not understand molecules*" (Linus Carl Pauling, Nobel Prize in Chemistry in 1954 and for Peace in 1962). The effects of nutrition on health and disease cannot be understood without a profound understanding of how nutrients (and in general all the chemicals we consume daily) act at the molecular level.

It is indeed widely recognized that macro- and micronutrients are potent dietary signals that influence the metabolic programming of cells and have an important role to play in the control of homeostasis. In fact, it is known that diet alters both the genetic composition and the metabolic activity of the resident microorganisms of the human gut (microbiome). There is growing evidence that this alteration not only is the key to chronic diseases like obesity and inflammatory bowel disease but may also establish an environment conducive to lethal viruses and bacteria. We now know that the gut microbiome rapidly responds to altered diet (over 1–2 days), and this offers the opportunity to use changes in diet and lifestyle as a most convenient approach to protecting human health from many diseases (David *et al.*, 2014).

Although advances in chemical and pharmacological evaluation of plant phytochemicals have occurred in recent years, a straightforward relationship between the phytochemical composition and the total activity of the plant has not yet been established. It seems reasonable to assume that the "synergy" between and within a particular class of compounds might be responsible for the bioactivity profile of the plant.

We know that the Western scientific perspective is only one of the ways to help human beings achieve a healthy life. We need to review and recognize the contribution of local health traditions of different cultures to the streams of biology and medicine in a systematic manner. In this sense, the *great medical traditions* that originated in the Chinese, South Asian, and Mediterranean civilizations can contribute to new knowledge when explored scientifically using biomedical tools. India and China have consistent theoretical bases and practical clinical applications that have proven effective for more than 3000 years. At the same time, traditional Hippocratic medicine represents the foundation of modern medicine and has survived relatively unaltered for more than 2000 years. The Greek conception of cure is also quite unchanged in the Unani system of medicine, practiced in the Islamic regions of Asia, especially in India.

This book aims to showcase the perspectives of the traditional medical systems of the South Asian and Mediterranean regions, particularly with respect to food, nutrition, and health. Wherever possible, it has correlated traditional knowledge with contemporary modern science and nutrition.

The book is divided into four sections: the introduction presents an overview of the historical times, which establishes the role of plants as remarkable milestones for nutrition and medicine. It is interesting to note that although different cultures grew in different parts of the world, geographically far from each other, some information on the use of plants and their influence on the body was accepted universally. The ancient healing systems of the East and the West shared a similar cosmology and quite similar basic concepts. Such a common holistic worldview has disappeared in the West, but it is still alive and quite well integrated with the modern lifestyle in India. Indian wisdom emanates from a truthful interaction with Nature. People refer to food in the *Taittiriya Upanishad* (a post-Vedic text dated around the eighth to seventh century BC) as God because food is the life source of all living beings. The ritual offering of food and special dishes in Indian temples is often based on Ayurvedic principles. The use of certain herbal drugs and the selection of food vary, depending upon the season and region. The Ayurvedic knowledge system has a functional understanding of how Nature influences the human body and how one's diet can be modified accordingly to support a healthy life.

Section I describes the Indian wisdom and knowledge about plants and the theoretical basis of the Indian systems of medicine, with particular emphasis on the most known holistic approach to wellness called *Ayurveda*, the science of life. The section covers a substantial range of detail on diagnosis and prescriptions on how to maintain the body in good health conditions and includes suggestions on appropriate dietetic regimens according to individual constitutions.

The Unani system of medicine describes how this system is keeping alive the ancient Greek healing tradition in modern India. Finally, an attempt to show the opportunities of integration between the Indian system and the Western system of medicine is provided.

Section II describes the Western approaches to nutritional science, emphasizing the role of *basic* and *secondary* nutrients, according to recent scientific research. As an example, the beneficial contribution of the Mediterranean diet to cardiovascular, degenerative, and neurocognitive diseases is described and supported by updated studies with an aim of limiting the damage caused by unhealthy daily behaviors such as smoking, lack of exercise, and junk food. The most modern approaches of science applied to nutrition and, more generally, to health result in a new paradigm that moves from the particular to the whole and gives a scientific explanation to folklore and traditional uses. This is true for many natural products that are currently used as drugs (e.g., the antitumor drugs taxol, camptothecin, and vincristine) and a plethora of studies on the properties of phytochemicals present in plants used as food supplements, such as curcumin, resveratrol, and flavonoids (Essa *et al.*, 2012; Kennedy and Wightman, 2011; Khuda-Bukhsh *et al.*, 2014).

Section III is the largest part of the book. The nutritional value and traditional use of 50 plants from across the world are described as cards in monographies. Several of these are wild and lesser known species, rich in nutrients. Cards are organized detailing distribution, botanical features, traditional uses, and bioactive components that could be responsible for the beneficial effects as nutrients or food supplements. The authors deliberately condensed the information on traditional uses in India and Western countries for a better comparison and have described how scientific research supports in most cases the use of plants as food.

It is interesting to note how some plants, particularly spices, are defined traditionally as *food balancers* in the sense that their contribution to the maintenance of health is related to their influence on making other food more digestible or absorbable and to their stimulation of metabolism. In fact, they are used in tiny amounts as flavorings, since their active constituents are sometimes toxic at high concentrations.

Over the centuries, spices have undoubtedly played a very important role in the world economy; spices have been connected to fascinating myths and have been extremely precious at times in history when communication between the East and the West was rather difficult. They have caused conflicts and have marked the economic development of those countries that were able to monopolize their trade. Over 80 spices are grown throughout the world, and 50 are cultivated in India. Many of them possess interesting medicinal properties that deserve serious consideration (Kannappan *et al.*, 2011). The use of spices in India is an ancient form of culinary art: it involves the sophisticated ability to change the quality of food and have a balanced and personalized diet. Spices improve digestion by promoting enzyme action. According to Ayurveda, using spices that suit a constitution and a specific season can greatly improve health. Indians, including illiterate women, are by and large familiar with these principles and practices.

Such knowledge can be easily applied to any kind of cuisine: in different parts of the world, we can learn to use the locally available spices and aromatic herbs to balance the food by adopting the millenary principles of Ayurveda. Spices work at multiple levels, including taste, interaction with food and with other spices, and the way they are cooked. Although the biochemical aspect is very important, it is just one aspect. For this reason, interaction between different approaches is so important, the art of cooking being of course one. Ayurveda pays great attention to the preparation of food, which is regarded as something sacred. When we cook, we are involved in a process of transformation that improves the digestion and assimilation of food. The influence of cooking methods is also the subject of many researches, in an attempt to avoid aggressive cooking treatment, which may affect the functionality of foods (Jiménez-Montreal *et al.*, 2009). Sustained by scientific studies, the traditional food recipes around the world can turn any kitchen into a preventive and healing center. For this reason, some easy recipes selected from different gastronomic traditions are included in the third section of this book.

The scientific literature has been critically reviewed. Our aim is not to be exhaustive on specific topics but to compare traditional uses with modern scientific achievements and, most of the time, to validate the ancient knowledge.

Chapter 11 focuses on *integration*. How can we use the existing data to achieve an integrated approach to a balanced diet or more generally to preserve wellness?

Both Ayurveda and the Mediterranean traditions carry a consistent intellectual heritage based on apparently simple concepts such as harmony and equilibrium in life. Reevaluating and valorizing this wisdom from the perspective of food can play an important role in the decisions on use and control of natural resources. When we become conscious about our diet, we also learn to respect our body and our planet. Feeding ourselves properly is to maintain the earth for the future, since everything is connected. Our medical heritage teaches us a way to respond to the current environmental challenges, such as food safety, security and biodiversity conservation. We need to be able to hand over our biological and cultural heritage to the generations that will come.

> Let food be the medicine and medicine be the food.
>
> **Hippocrates**

The editors, who belong to different research fields, decided to join in preparing this book because they strongly believe that it is a challenge of our times to integrate different systems of culture and knowledge, especially concerning the healing sciences.

Luisella Verotta is a natural products chemist, researcher, and member of the teaching staff at the Department of Chemistry of the University of Milan. Maria Pia Macchi is an anthropologist and a member of the International Society of Ethnobiology working in India for the valorization of traditional medicine in a hot spot of biodiversity. LV and MPM are both consultants to the Italian nongovernmental organization (NGO) Centre for Orientative Education (COE), active for more than 20 years in the valorization of traditional medical systems of different parts of the world. When LV and MPM came to India in 2003, they visited the Foundation for Revitalisation of Local Health Traditions (FRLHT) at Bangalore and met Padma Venkatasubramanian (a biochemist), who heads the Ayurvedic R&D Centre at FRLHT and researches quality assessment of plants and fundamental Ayurvedic principles. This was the first step for a subsequent collaboration. Some experts from the FRLHT were invited to Italy to give lectures at various congresses, and the foundation was selected for training organized by the COE during a project for integration between Ayurveda and biomedicine in Bangladesh. Recently, MPM and LV also started a collaboration with the Tibetan Medical and Astrological Institute at Dharamsala, focusing on the importance of diet as an adjuvant to pulmonary tuberculosis. A booklet on nutrition was published with the contribution of the Italian Ministry of Foreign Affairs and distributed free of cost to schools and monasteries to create awareness about the essential role of diet. In the different projects promoted by the COE, efforts have been made to integrate the biomedical approach with the traditional one, especially in the field of nutrition.

This book publishes during the EXPO 2015 event that is being held in Milan, Italy, and whose theme is *feeding the planet: energy for life*, a six-month exhibition that involves the world on food in all aspects. It was 2012 when we planned to sustain the initiative with a publication as a way to bridge the gap between holistic traditional views and modern scientific approaches in a field so important as plants and nutrition. We have brought in contributions from authors who are experts from different academic backgrounds.

Luisella Verotta
Maria Pia Macchi and
Padma Venkatasubramanian

REFERENCES

Argyropoulou A., Aligiannis N., Ioannis P., Trougakos I.P., Skaltsounis A.-L. Natural compounds with anti-ageing activity. *Nat. Prod. Rep.* **2013**, *30*, 1412–1437.

David L.A., Maurice C.F., Carmody R.N., Gootenberg D.B., Button J.E., Wolfe B.E., Ling A.V., Devlin A.S., Varma Y., Fischbach M.A. *et al.* Diet rapidly and reproducibly alters the human gut microbiome. *Nature* **2014**, *505*, 559–563.

Essa M.M., Vijayan R.K., Castellano-Gonzales G., Memon M.A., Braidy N., Guillemin G.J. Neuroprotective effect of natural products against Alzeheimer's disease. *Neurochem. Res.* **2012**, *37*, 1829–1842.

Jiménez-Montreal A.M., Garcìa-Diz L., Martìnez-Tomé M., Mariscal M., Murcia M.A. Influence of cooking methods on antioxidant activity of vegetables. *J. Food Sci.* **2009**, *74*(3), H97–H103.

Kannappan R., Gupta S.C., Kim J.H., Reuter S., Aggarwal B.B. Neuroprotection by spice-derived nutraceuticals: You are what you eat. *Mol. Neurobiol.* **2011**, *44*(2), 142–159.

Kennedy D.O., Wightman E.L. Herbal extracts and phytochemicals: Plant secondary metabolites and the enhancement of human brain function. *Adv. Nutr.* **2011**, *2*, 32–50.

Khuda-Bukhsh A.R., Das S., Saha S.K. Molecular approaches toward targeted cancer prevention with some food plants and their products: Inflammatory and other signal pathways. *Nutr. Cancer* **2014**, *66*(2), 194–205.

Koyama A., Houston D.K., Simonsick E.M., Lee J.S., Ayonayon H.N., Shahar D.R., Rosano C., Satterfield S., Yaffe K. Association between the mediterranean diet and cognitive decline in a biracial population. *J. Gerontol. A Biol. Sci. Med. Sci.*, **2014**, *70*(3), 352–357. doi:10.1093/gerona/glu097.

Kumar H., Kim I.-S., More S.V., Kim B.-W., Choi D.-K. Natural product-derived pharmacological modulators of Nrf2/ARE pathway for chronic diseases. *Nat. Prod. Rep.* **2014**, *31*, 109–139.

Müller M., Kersten S. Nutrigenomics: Goals and strategies. *Nat. Rev. Genet.* **2003**, *4*, 315–322.

Patil B.S., Jayaprakasha G.K., Chidambara Murthy K.N., Vikram A. Bioactive compounds: Historical perspectives, opportunities, and challenges. *J. Agric. Food Chem.* **2009**, *57*, 8142–8160.

Sales N.M.R., Pelegrini P.B., Goersch M.C. Nutrigenomics: Definitions and advances of this new science. *J. Nutr. Metab.* **2014**, Article ID 202759. http://dx.doi.org/10.1155/2014/202759.

Szychowski J., Truchon J.F., Bennani Y.L. Natural products in medicine: Transformational outcome of synthetic chemistry. *J. Med. Chem.* **2014**, *57*(22), 9292–9308.

World Health Organization. *Global Status Report on Noncommunicable Diseases 2010*, Geneva, Switzerland, **2011**.

World Health Organization, Media Centre, *The Top 10 Causes of Death*. **2012**. http://www.who.int/mediacentre/factsheets/fs310/en/index2.html.

Note from the Editors

DISCLAIMER

This book is not intended to prescribe a treatment. The authors do not recommend self-diagnosis or self-medication. The information herein contained cannot be considered as a substitute for consultation with a physician.

Editors

Luisella Verotta, master degree in chemistry, is a senior assistant professor of organic chemistry and adjunct professor of natural products chemistry at the University of Milan, Italy. Verotta is a member of the Faculty of Science and Technology teaching staff. Her main research studies are in the realm of bioactive natural products, aimed at obtaining lead compounds for the development of new therapeutic agents, particularly noncytotoxic chemopreventive agents and antimalarial compounds. She has pursued parallel lines of investigations, such as the synthesis of compounds of pharmaceutical relevance based on natural scaffolds, the study of their medicinal chemistry, and the mechanism of activity. Recent researches are based on the synthesis of new entities inspired by natural products to produce new functional biomaterials for biomedical and food applications or as nutraceuticals. She has been a visiting scientist in many Third World countries, introducing young researchers to the phytochemical study of medicinal plants. She is an extramural phytochemical expert for the Centre for Orientative Education (COE), an Italian NGO that fosters and supports projects in Third world countries (TWC) for the sustainable use of natural resources. She has authored and coauthored more than 110 research papers (including full papers, book chapters, patents) and more than 110 communications to congresses, including poster presentations, oral communications, and invited and plenary lectures. She holds research contracts with several companies interested in the discovery of new bioactive natural products and acts as a reviewer of several international scientific journals. She is a member of the editorial board of *Fitoterapia* (Elsevier).

For CRC Press/Taylor & Francis Group, she coedited *Herbal Principles in Cosmetics: Properties and Mechanisms of Action* in 2010.

Maria Pia Macchi is an anthropologist member of the International Society of Ethnobiology. After a few years of field work in the Amazon of Ecuador and an academic collaboration with Padua University Human Ecology Workshop, she focused her interest on Indian traditional systems of medicine. During frequent visits to India and Nepal, she took part in different trainings, congresses, and workshops on Ayurveda and Siddha medicine, developing her interest in the traditional use of medicinal plants and food integrators. In 1997, she founded an ethnobotanical garden in the Western Ghats of Tamil Nadu with the objective to actively involve local communities in the conservation and enhancement of knowledge related to local health traditions. Since 2003, she has been the president of the nonprofit organization Magia Verde, Italy, and supports a pilot project on traditional medicine revitalization in Tamil Nadu in partnership with Vivekananda Kendra. She took part in various projects and awareness campaigns focused on environmental conservation, nutrition, and integration between different systems of medicine, promoted by two Italian NGOs, Cooperation for the Development of Emerging Countries (COSPE) and COE. She also has a long-term collaboration with the Agency for Environmental Protection at Riva del Garda in the ethnobotany section.

Padma Venkatasubramanian holds a master's degree in health management from McGill University, Montreal, Québec, Canada, and a doctorate in microbial biochemistry from the University of Cambridge, England. She is a professor at the School of Life Sciences, Institute of Trans-disciplinary Health Sciences and Technology, Bangalore, India. She was the director of the Institute of Ayurveda and Integrative Medicine, Bangalore. Her main research studies are of a transdisciplinary nature, aimed at bridging traditional Indian medical knowledge and modern science. The areas of her interest are to understand the concepts in Ayurveda and use them to develop cost-effective, contemporary applications. She has demonstrated how certain *rasayana* herbs can be used at the household level to promote wellness and manage anemia. Based on the traditional practice of storing drinking water in copper pots, she has designed and developed a copper device that kills diarrheagenic pathogens in drinking water. This is a potential low-cost, point-of-use intervention to prevent diarrheal deaths globally. The other area of her research is to scientifically demonstrate the importance of traditional ways of identification, collection, processing, and storage of medicinal plants. This contributes to resource management and sustainable harvesting that is important for conservationists as well as for the industry. She has a patent on a field diagnostic kit to authenticate medicinal plants and semiquantify phytochemical markers. She has several research publications to her credit and delivered invited lectures at several national and international conferences. She guides PhD scholars and college students on research projects. Her engagement with the industry as a consultant has led to the development of more than 20 herbal products based on traditional knowledge. She is on the scientific advisory board of several government and nongovernment bodies.

Contributors

A.V. Balasubramanian
Centre for Indian Knowledge Systems
Chennai, India

Gerard Bodeker
Nuffield Dept of Primary Health Care
University of Oxford
Oxford, UK

and

Department of Epidemiology
Columbia University
New York, USA

Marina Calloni
Department of Sociology and Social Research
University of Milano-Bicocca
Milan, Italy

Indukuri Kalyan Chakravarthy
Maharishi Ayurveda Privatklinik
Bad Ems, Germany

Cristian Del Bo'
Department of Food, Environmental and Nutritional Sciences
University of Milan
Milan, Italy

Francesca Ghelfi
Department of Experimental Oncology
European Institute of Oncology
Milan, Italy

Maria Pia Macchi
Magia Verde Onlus
Pistoia, Italy

Unnikrishnan Payyappallimana
United Nations University—Institute for the Advanced Study of Sustainability
Tokyo, Japan

Katia Petroni
Department of Bioscience
University of Milan
Milan, Italy

Karin Pirc
Maharishi Ayurveda Privatklinik
Bad Ems, Germany

Marisa Porrini
Department of Food, Environmental and Nutritional Sciences
University of Milan
Milan, Italy

Lucilla Titta
Department of Experimental Oncology
European Institute of Oncology
Milan, Italy

Chiara Tonelli
Department of Bioscience
University of Milan
Milan, Italy

Padma Venkatasubramanian
School of Life Sciences
Institute of Trans-disciplinary Health Sciences and Technology
Bangalore, India

Luisella Verotta
Department of Chemistry
University of Milan
Milan, Italy

Roohi Zaman
Department of Ilmul Saidla
National Institute of Unani Medicine
Bangalore, India

1 Indian and Mediterranean Traditional Dietetics

An Historical Overview

Maria Pia Macchi

CONTENTS

> How the world of nature appears to us depends on how long we take to observe it...
>
> **Collingwood, 1945**

The first contacts between Indian and Western cultures probably date back to the time of Indus Valley Civilization, around the third millennium BC. This civilization was roughly contemporaneous with the Old, Middle, and New Kingdoms of Egypt and the Sumerian and Akkadian period in Mesopotamia. There were two main capitals: Mohenjo-daro in the south and Harappa in the north. The excavations of Mohenjo-daro show a great knowledge in the field of public health and sanitation. *Shilajit*, a drug still in use in Ayurveda, has been found in these excavations (Ranade and Deshpande, 2006). Several seals discovered in both the regions point to a reverence for plants and their products, which are also common to the medical traditions of Egypt and Mesopotamia (Zysk, 2009). The identification of the Harappan seals found in Mesopotamia evokes the possibility of a reciprocal influence between the medical notions belonging to the Indus and Sumerian civilizations.

The great ancient medical traditions were in contact with each other, but maintained their peculiarity. They share a general theoretical holistic feature: health is harmony between man and the ecosystem, and diet plays a primary role. There is a common humoral theory with four humors in the Mediterranean tradition, three humors in the Indian tradition, and six humors in the Chinese tradition (Leslie, 1998). Health is always associated with a condition of equilibrium between these humors. It is very interesting to note that the humoral medicine is a common folk medicine also in other parts of the world, but only in Asia it continues for thousands of years in a sophisticated institutional form. Therefore, "Asian systems of medicine can provide today fascinating opportunities both to observe directly practices that continue ancient scientific modes of thought and to analyze the historical processes that mediate their relationship to modern science and technology" (Leslie, 1998, 1). The creative and practical applications of Ayurveda can build a bridge between the East and the West, especially in the field of nutrition.

1.1 AYURVEDA: HISTORICAL BACKGROUND

> Residing forever in the lap of time, this great science of good health and longevity is revealed from era to era to those who aspire to drink its nectar of wisdom
>
> **Giri and Jain, 2013**

Ayurveda is without doubt the oldest medical system that has been used continuously throughout human history. It is considered a divine science, created by Brahma for the benefit of human beings, carefully memorized, and transmitted orally from generations to generations for thousands of years before being recorded in writing (Ranade and Deshpande, 2006). Ayurveda means "the knowledge or science (*veda*) for longevity (*ayus*)": it is a universal healing art that has been part of India's collective consciousness probably since prehistoric times. Even a child of a small village and his illiterate mother know that yogurt can cause accumulation of phlegm (*kapha*). Indian people daily talk, knowingly or not, in the Ayurvedic idiom (Ranade and Deshpande, 2006).

Zysk sees Ayurveda as a medical tradition emerging from the ascetic milieu existing in the North India in the fifth century BC (Smith *et al.*, 2008) and based on the "*pancha mahabhuta* (five elements) theory" (Wujastyk, 1998). Three biological humors (called *dosha*) control these different elements: *vata*, *pitta*, and *kapha*. In every living being, the composition of these principles is different. By understanding which principle dominates in a person, we can infer the nature of that person and then we can select a proper, balanced diet for him or her (Smith, 2011). Ayurveda has a consistent theoretical basis and practical clinical application, but to understand and fully appreciate it, we have to overcome our linear process of thought and enter an intuitive vision of the whole: there is a deep connection between the physiological and psychological functions of the body. "Cosmic energy and universal matter are at once intertwined in our foods, thoughts and activities. The body is sustained by *prana*, the life breath and *annam*, the earth food. Breath and food influence the tissues of the body and the thought of the mind" (Tiwari, 1999, 16). Food appears as the manifestation of the eternal consciousness: "*Purusha* is all that contains past, present, and future. He is also the governor of *Amrta* (immortality) and that which is manifested by *Anna* (diet)" (Gupta, 2011, 1). All living beings are interrelated through the chain of food production.

The same model of nutrition has been used for more than thousands of years with a practical approach. Many references to the principles of Ayurveda are already in the Veda, the sacred texts dating probably from 1500 to 1000 BC. The theoretical concepts of the Veda were formalized and gradually supplemented by observations based on a more scientific approach during the so-called classical period (600 BC–AD 1000). Unfortunately, many of these compilations no longer exist. Only three works are available today: *Charaka*, *Sushruta*, and *Vagbhata*, which are considered to be the most useful compendia, known as *Samhitas* or compilations. These texts have undergone some modification by various authors in subsequent periods and are all written in the Sanskrit language (Ranade and Deshpande, 2006).

The *Charaka Samhitas*, the first classic text, gives a clear definition of Ayurveda: "It is called 'ayurveda' because it tells us (*vedayati*) which substances, qualities, and actions are life-enhancing (*ayusya*), and which are not." These first texts already focus quite clearly on prevention, by giving practical advice on almost every imaginable aspect of life, including diet (Wujastyk, 1998). In Chapters 25 through 28, we find a critical discussion on dietetics with a detailed classification of compatibility and preparations of different food articles, such as nuts, vegetables, grains, and milk products (Smith *et al.*, 2008). There is also the first description of importance of taste, potency, and post-digestive effect of food. According to the *Charaka Samhitas*, food is one of the three pillars supporting body: wholesome food is the main cause of health, whereas unwholesome food is the main cause of growth of the diseases (Gupta, 2011). *Charaka*, for the first time, stated that medicines are not merely for the treatment of diseases, but for restoring homeostasis (Ranade and Deshpande, 2006).

Therefore, from the beginning of its long history, Ayurveda deals with foodstuffs and their medical properties in an elaborate manner. The relationship between health and the cycle of seasons is also central (Zimmermann, 1975). Even if the *Sushruta Samhitas* is mainly a compilation about surgery, we can find here a detailed description of dietetics. "A healing diet gladdens the heart, nourishes the body and revives memory." The diet has also a direct relationship with the mind and can improve spirituality (Gupta, 2011).

After the conquest of Alexander the Great, in the fourth century BC, the political and cultural relations between the two civilizations were documented by the historians in the retinue of the great leader as well as by Megasthenes who was sent to the court of Chandragupta Maurya in 300 BC. Megasthenes refers that the Indian physicians effected cures by regulating diet rather than by using medicines (Lochan, 2003). Alexander was so impressed by Ayurvedic practitioners, who were able to cure many diseases mostly by dietary methods, that he carried with him some of them when he went back to his country. It is interesting to note that one of the main functions of the royal physician was the supervision of the royal kitchen to ensure a healthy diet to the king and to prevent attempts at poisoning him (Basham, 1998).

During the Buddhist period (323 BC–AD 642), Ayurveda made great academic progress with the setting up of the first universities. Indian culture entered its golden age and started to attract philosophers and scholars from other parts of the world, including China, Rome, Greece, and the Middle East.

Among the earliest medical texts, the so-called Bower Manuscript was a collection of medical and divinatory texts dating from the late fourth or the early fifth century AD, owned by Yasomitra, a senior Buddhist monk. It was buried in a memorial *stupa* dedicated to Yasomitra and laid there for a thousand of years until it was found and bought by the lieutenant Hamilton Bower in 1890. The importance of the manuscript was recognized and was sold to the Bodleian Library in Oxford. The beginning of the first treatise describes the mythical origin of garlic and its medical uses, indicating that Ayurveda was already practiced through Central Asia in the first centuries of the present era (Wujastyk, 1998).

Under the influence of Mahayana Buddhism (from the fourth century BC until the sixth century AD), the medical science previously developed mainly for kings, court leaders, soldiers, and male members of elite castes started to become more popular. Medicinal plants were planted along the sides of streets to be used freely by all, and many public hospitals were founded by the Emperor Ashoka. Along with Buddhist missionaries who traveled in many neighboring countries such as Sri Lanka, Myanmar, Thailand, and Indonesia, Ayurveda spread abroad. During this period, the use of alchemic preparations (*Rasa Shastra*) started. Vagbhata was a Buddhist and his work was translated into the Tibetan language. He wrote the *Asthanga Samgraha* in the seventh century AD as a compilation of the best from *Charaka Samhitas* and *Sushruta Samhitas*. He also introduced more ethical precepts and stressed the importance of a proper diet and physical exercise for promoting health and longevity. There are references to the daily and seasonal regimens and to the six tastes and their right combination (Wujastyk, 1998).

The *Samhitas*, the *Madhava*, and the *Asthanga Samgraha* are mentioned in the Ali bin Rabbani Tabari's *Firdaus al-Hikmat*, the Arabic medical compendium compiled in AD 856 (Askari, 1957). Therefore, the four chief classical Ayurvedic medical texts were known beyond India by the ninth century (Basham, 1998).

During the Muslim invasion (tenth to twelfth century), when the conquerors destroyed the universities and burned the libraries, Ayurveda faced a moment of decline, but when the Arabian system of medicine started to spread throughout India, there was a rich exchange of knowledge between the two medical systems. "Muslims rulers acted as patrons of famous Ayurvedic scholars, and Hindu aristocrats patronized Yunani physicians" (Leslie, 1998, 14). The later texts of Ayurveda are deeply influenced by alchemy, a science prominent in Islam, and are marked by the introduction of new drugs learned by Arabs.

From the point of view of nutrition an interesting text is the *Bhojanakutuhalam*, written in the seventeenth century by Sri Raghunatha, a great scholar supported by the king Ekoji who established his kingdom in Tamil Nadu in 1678. The title of the book is a compound word composed of two Sanskrit words: *bhojana* (meaning food eating or food feeding) and *kutuhala* (meaning wonderful, excellent, praised, celebrated, desire, and curiosity). Therefore, this treatise is about food, the serving of food, and the enjoyment of food that is excellent, praiseworthy, and pleasant, and that excites our curiosity. It is a complete work to learn how to prepare, serve, and consume excellent food that is most beneficial to our health in accordance with Ayurveda tradition. The material comes from various Ayurvedic treatises such as *Sushruta Samhitas*, *Bhavaprakasa*, *Dharmasastra*, and *Manusmrti*. It is a poetic compilation on dietetics and nutrition for all ages. It contains the Ayurvedic properties of most of the ingredients used in the preparation of meals, sweets, and savories. The general dictum is that one should only consume wholesome food in the right quantity and only after the previously ingested food has been completed digested (Venkat *et al.*, 2012).

Ayurveda faced again an even worst decline during the British colonization (eighteenth to nineteenth century). In the early years of the British regime, there was a symbiosis between Western and Indian systems of medicine, but with the spread of English medical schools, a negative attitude toward Ayurveda started: the state patronage stopped and the existing schools were closed. Even if relegated only to the *gurukula* level, the traditional Indian system of medicine continued to serve about 80% of the country (Ranade and Deshpande, 2006). With the birth of Independence movements, from 1920 onward, Ayurveda started again to receive a political support with the establishment of formal colleges and universities. Ayurveda has proven to have been effective for thousands of years, and its importance from a socioeconomic and therapeutic point of view cannot be underestimated. It should not therefore be surprising if, at the Alma Ata conference where the World Health Organization proclaimed "Health for all by 2000" with the decisive contribution of traditional medicine, Ayurveda was mentioned as a reference (Saibene, 2004, 131).

1.2 DIETETICS IN THE GREEK AND ROMAN TRADITIONS

> So in fixing regimen pay attention to age, season, habit, land, and physique, and counteract the prevailing heat or cold. For in this way will the best health be enjoyed
>
> **Hippocrates**
> *(Jhones, 1923–1995, Reg.H. II. 4.49.) (Longrigg, 2013)*

After the conquest of the Indus Valley by Darius I (522–486 BC), the whole of northwestern India was a vital part of the international world that stretched from the Indus to the Mediterranean, with a deep and significant mutual cultural enrichment (Lochan, 2003). The strong connections and similarities between Indian and Greek medical theories can be due to indirect contacts through the Persians, as Darius had Greek physicians (Filliozat, 1964).

In many aspects, the Greeks are as much Eastern as Western people (Frawley, 2010), and their ancient culture has a common root in nutritional science. Diet was very important in their philosophy and medicine, probably in relation to the need of athletes during the trainings for the games (Frutos Gonzàlez and Guerrero Peral, 2011). The schemes forged during those times persisted with great stability in the Western culture for over 2000 years. Early modern physicians attributed the longevity of patriarchs and philosophers of that period to their healthy diet and lifestyle. Greek diet was based on a system of binary oppositions in the universe: hot/cold, dry/moist, and bitter/sweet. Health was considered a state of balance between those principles; the unbalance or predominance of one principle was referred to as one of the main causes of illness and even death.

Already in Homer scripts (eighth century BC), it was mentioned a physician, Podalirius, with a special interest in diet. The pre-Socratic author's observations of nature gave importance to diet. Alcmeones of Croton, a predecessor of Hippocrates, clearly underlined the importance of the diet, the lifestyle, the environment, and the weather to understand and prevent illness. He was the first physician who formulated a doctrine of health as a balance among the fluids of the body. This theory was later developed by Hippocrates of Kos in the "theory of the four humors."

Hippocrates (460–370 BC) was an excellent physician, who is considered the Father of Medicine as he clearly differentiated medicine from magic and religion. He was the leader of the medical school of Kos and promulgated the idea of *vis medicatrix naturae*, or the power of nature to cure itself. This natural tendency could be aided by a clean environment and by a regimen of suitable diet and lifestyle linked to the time of year or the person's age (Bhasme *et al.*, 2007).

In "The Regimen in Health," Hippocrates directly connected dietetics with the cycle of time, a concept similar to the Ayurveda *Ritucharya*: "So in fixing regimen pay attention to age, season, habit, land and physique, and counteract the prevailing heat or cold. For in this way will the best health be enjoyed" (Jhones, 1923–1995, 136). Diet was an important part of every treatment and a detailed dietetic advice was provided to the patients. Therefore, during the winter season more food is needed, whereas during the summer season less food is necessary (QiaoSheng, 2011). "In summer and in autumn food is most difficult to assimilate, easiest in winter, next easiest in spring" (Jhones, 1923–1995, 4, 107, Aph. I.XVIII).

The importance of a balanced diet was a basic concept similar to that of Ayurveda. The Hippocrates motto, "Your Food Is Your Medicine and Your Medicine Is Your Food," is also common in the Indian culture even today. The theory of balance between tastes is also present in the Hippocratic School. The bitter, the salty, the sweet, the tart, the astringent, and the tasteless are not harming if they are mixed and well combined. Only when a specific taste stands alone, it can produce damage.

Hippocratic physicians strongly believed in the natural healing process of rest, good diet, fresh air, and hygiene. Their main support to the patient was to build his or her personal strength and the administration of drugs was only the last drastic resource. Their cosmovision was grounded on the Empedoclean principle of the four elements (earth, air, fire, and water), identified in the body as the four humors: blood, yellow bile, phlegm, and black bile. A relationship between the four humors and the seasonal changes was also described: winter, with its cold and sometimes humid climate, was thought to cause an abundance of phlegm. There was an association between the humors and the signs of zodiac: signs of water were believed to be prone to be phlegmatic. In "The Regimen of Acute Diseases," Hippocrates provides eating advice for people suffering from acute diseases (Lloyd, 1978). The other famous treaty, "On Regimen," is probably the work of a disciple of Kos school. This treaty provides advice on eating habits, sexual habits, bathing, and physical exercise, according to the age, the place of living, and the time of year, and became a reference in the Western world.

Hippocrates' successors developed the theory of Hippocratic temperaments: choleric, sanguine, phlegmatic, and melancholic. The Hippocratic system was strictly based on the clinical observation of the patient, but the following philosophers, Socrates, Plato, and then Aristotle, developed more theoretical aspects, giving less importance to the concrete observation.

Plato's dialogs present a number of dietary concepts close to what we call today the Mediterranean diet (Skiadas and Lascaratos, 2001). In "The Republic" (332C, Skiadas and Lascaratos, 2001), we read that the regulation of diet is one of the physician duties as the medicine is "the art that renders to bodies drugs, food, and drinks." The trainers of gymnastics were also experts in dietetics, as exercises, hygiene, and life style were considered important for health (Protagoras, 314A, Skiadas and Lascaratos, 2001). In the "Timaeus" (89C–D, Skiadas and Lascaratos, 2001; Taylor, 1928), Plato suggests that the therapeutic approach to human diseases should be made through the regulation of diet rather than through medicines. Diet can reinforce not only the body but also the spirit of the patient (Skiadas and Lascaratos, 2001) and should be simple and balanced with a variety of foods selected on the basis of their availability. Plato recommended eating all kinds of nutrients in moderate quantities: cereals, legumes, milk, honey, fish, and fruit. He suggested consuming sweet foods, meat, and wine in moderation. The consumption of high quantity of fats was disapproved and the use of cereals, considered of divine origin, was greatly encouraged. The diet in the ancient Greece was slightly different in different locations, but the basic features were more or less the same.

The relationship between Ayurvedic medicine and Greek medicine of the Hippocratic era is quite interesting. The name of Hippocrates refers to the horse (*hippo* in Greek), and the divine healers of the Veda, the Ashwins, were also closely related to this same animal (*ashwa* means "horse" in Sanskrit). The *tridosha* doctrine is somewhat analogous to the ancient Greek humoral system of Hippocrates and Galen (Wujastyk, 1998). Several Hippocratic authors recommended medications attributed to India and Plato's theory of vision resembled that of Ayurveda (Filliozat, 1964). Indian herbs were already popular in Greece in the fourth century BC as we can see in the Theophrastus history of plants (Hort, 1916). Two Indian remedies based on pepper and cinnamon are reported in the Hippocratic treaty "On the Female Diseases" and are mentioned under their Persian name.

A comparative philology shows that at least part of the mythological sources of Indian and Greek systems of Medicine is probably the same (Lochan, 2003). Although many hypotheses have been made about the analogy between the two theories of humors and the elements of the two cultures, probably the two systems grew in a fundamentally independent way, even if the possible contacts led to reciprocal modifications (Saibene, 2004). About the *tridosha* doctrine, for example, there are many fundamental differences: the Indian theory focuses on five elements, including ether, whereas the Greek theory includes only four elements. Again, in the Indian theory, the *dosha*s themselves cannot produce any disease, but only alterations of *dosha*s vitiate the constituents of the body and produce diseases in them. In the Greek theory, it is the imbalance of the humors that constitutes disease (Lochan, 2003).

After the death of Aristotle and his pupil Alexander the Great, a great medical school was founded in Alexandria, the capital of Egypt, at the court of King Ptolemy. This school played a central role in the development of medical thought and practice from 300 BC through the seventh century BC (Bhasme *et al.*, 2007). The most important investigators were two Greek anatomists, Herophilos of Chalcedon (335–280 BC) and Erasistratos of Iulis (330–250 BC). They both stressed the value of diet and lifestyle. These concepts were transmitted down to Roman authors such as Cato, Plutarch, and Pliny. The Romans adopted many ideas from the Greeks. A key figure was Claudius Galen from Pergamum (AD 129–216), the heir of the school of

Alexandria, considered the most important contributor to medicine after Hippocrates. He was a great teacher with a solid philosophical background combined with a lot of clinical experience. Galen applied the finalism of Aristotelian philosophy to the Hippocrates theory of humors (Frutos Gonzàlez and Guerrero Peral, 2011). The main point of Galen theory was that food, after having got into the body, is transformed by natural warmth in the stomach into different substances. Part of these substances is assimilated, whereas the rest is excreted. The four humors are the main products of this process and are responsible for maintaining the moisture of different organs of the body and also provide nutrition. The four humors combine with the four primary qualities: warmth, cold, dampness, and dryness (Lone *et al.*, 2012). Therefore, the blood is hot and moist, the yellow bile is hot and dry, the phlegm is cold and moist, and the black bile is cold and dry. A man is considered healthy when this combination is well balanced and sick when the balance is disturbed. Therefore, the therapy was based on the principle of *contraria contrariis curantur.* Hot diseases were cured by cold remedies, moist by dry, and vice versa (Burgel, 1998). Drugs and foods were categorized after this scheme, with a detailed work of description of cereals, pulses, fruits, vegetables, and meat. In the Galenian system, we find, like in Ayurveda, a strong ethical element: the physician has a role of instructor who teaches to the patient how to preserve a state of equilibrium in the different spheres of his life.

1.3 THE GREEK AND ROMAN TRADITIONS MEET THE ARABIAN WORLD

As long as you can heal with food, do not heal with medications

Razhes
(Farid Sami Haddad, 1989)

Galen had no effective successor in the West: the ancient medical tradition chiefly survived in the Byzantine Empire, where a great production of medical texts was inspired by the ancient Greek and Roman knowledge. The theological quarrels of the fourth and fifth centuries were the cause of the expulsion of many intellectuals from Constantinople. Many of them were attracted to the court of the Baghdad caliphs where they had the opportunity to protect and transmit the Mediterranean heritage to the Arabian world. Whereas in Europe the medicine was governed by religion and superstition, in Baghdad the House of wisdom (*Bait al-hikma*) was flourishing, with a library closely similar to that of Alexandria. The ancient texts from Greece and Rome (mainly Hippocrates, Plato, Aristotle, Dioscorides, and Galen) were translated into Arabic and studied by Islamic scholars (Bhasme *et al.*, 2007). The backbone of theoretical instructions was the Alexandrian canon, a selection of 16 books written by Galen. This translation work was done mainly during the ninth century and is one of the great cultural achievements in the history of human spirit (Burgel, 1998). Hippocrates' theory of regulation of diet and lifestyle was developed and refined by Arabic, Jewish, and Persian physicians and philosophers, and combined with the Indian knowledge. Key figures such as Razi (Razhes, AD 865–925) and Ibn Sina (Avicenna, AD 980–1037), compiled a number of encyclopedias in "a constant process of reshaping and rearranging the Greek heritage by shortening, broadening, commenting on, and systematizing the ancient source material" (Burgel, 1998, 52). Diet was considered a means to acquire health and prevent diseases, and detailed dietary guidelines were established for different ages or seasons of the year.

Avicenna was called the Prince of Eastern Philosophers and Physicians. Like Rhazes, he challenged the thoughts of Socrates, Aristotle, and Galen regarding the dichotomy of mind and body. He recognized the nutritional and psychological benefits of breastfeeding and noted the close relationship between emotions and the physical condition of patients. He emphasized the importance of music during breastfeeding for the proper development of mind (Modanlou, 2008). His *Canon of Medicine* is the masterpiece of Arabian systematization of Galen legacy (Meyerhof, 1931) extensively used in European medical schools for centuries (Hulse, 2007).

Muslim rule carried the Galenic system to India and Southeast Asia when medical learning was still flourishing in Islam (Burgel, 1998). The practitioners of the two systems seem to have collaborated, because each had much to learn from the other (Basham, 1998): some treatises of Indian medicine were already translated into Arabic and Persian languages, and the Indian doctors occupied the positions of great importance at the court of Baghdad from the seventh century onward (Saibene, 2004). The Muslim world learned much from India, but its medical system was mainly based on the Greek sources. After a golden age of remarkable achievements, Arabic medicine faced a slow decline and is currently widely practiced mainly in South Asia with the name of Unani (which means "Greek medicine" in Arabic language).

1.4 BACK TO THE WEST

The great Greek and Latin heritage was brought back to the West by the Arabs with the translations of Constantinus Africanus (1015–1087), one of the most memorable figures of the School of Salerno, located in the south of Italy. This medical school was the main source of medieval dietetics, founded by a Latin, a Greek, an Arab, and a Jew. This legendary origin sums up the features of medical knowledge of the Middle Ages as a combination of Eastern and Western cultures.

The most relevant book of this school, the *Flos medicine vel regimen sanitatis salernitanus* (*Medicine according to the Salernitan Regimen of Health*) includes the rules of hygiene and dietetics.

Another famous book on dietetics is the *Taqwin as-sihha* (*Table of Contents of Health*) written by Ibn Butlan, a Christian physician trained in Baghdad. It was translated into Latin under the name of *Tacuinum sanitatis* at the Sicilian court of Manfred, the son of the emperor Frederic II. The description of different kinds of fruits, vegetables, spices, meat, fish, and drinks follows the Hippocratic principles. Like the other handbooks that belong to the so-called medieval Galenism, these two *regimina sanitatis* (health regimes) are structured around the six nonnatural things, or *res non naturales*, the factors of prevention for a good health condition: environment, exercise, food, sleep, evacuations, and emotions (Frutos Gonzàlez and Guerrero Peral, 2011). The typical medieval concept is that a good diet and lifestyle can help the body to heal itself. To ensure a proper digestion, a balanced quantity of food is always suggested. "Copious suppers cause stomachs the greatest discomfort/ To sleep well, dine little" (*Flos medicine vel regimen sanitatis salernitanus*, 20–21).

This diet-based literature is very rich and interesting with a number of specific recipes, scientific disquisitions, preparations techniques, and ceremonial service suggestions (Pelner Cosman, 1983). The Arabic influence is evident in the belief that there is a close connection between the body and the soul. Therefore, the *regimina sanitatis* suggests avoiding anger and listening to music to enhance digestion. In the most outstanding *regimina*, there is a significant reference to neurological diseases (Frutos Gonzàlez and Guerrero Peral, 2011). A lot of interesting references are related to the so-called food of love, the edible antidotes to sexual apathy (Pelner Cosman, 1983) with many similarities with ayurvedic descriptions of a number of foods such as spices, garlic, onion, and pomegranate.

Western dietetics until the sixteenth century presented many similarities with the Eastern tradition, a common language and concept of health. By that time, the Galenic dietetic tradition, with its concepts of humors, qualities, and temperament, started to disappear from the Western scenery. From about the middle of the eighteenth century, these notions are less and less used in the official medical writing. In their place, a rather different language emerged, a language that spoke not of qualities, but of constituents of food (Shapin, 1998).

1.5 CONCLUSION

"Current trends in cosmopolitan medicine show a growing awareness that much can be learned from traditional modes of healing including the great Asian medical systems" (Zysk and Wujastyk, foreword to Leslie, 1998). Indian Food Science and Technology has a history of at least 5000 years and it is not limited to a single country. The Greeks, Persians, and many other cultures have added their own contributions during more than 2000 years.

From an historical point of view, it is very interesting to see how Indian systems of medicine maintained their basic principles from over 5000 years in a dynamic way, redefining many of their practices in relation to changing times and mind-sets. This heritage has constantly evolved in a sophisticated way and actually demonstrates a potential capacity to be adapted for various parts of the world. The modern health scene shows a revival of many traditional systems of medicine: among them none is probably as diet based as Ayurveda. The Ayurvedic dietary approach reflects an integral understanding of the elements and vital force (*prana*) within the food. Practical aspects of cooking, spicing, and food combination are deeply described in detail. The treasure of information elaborated in the past necessitates to become more and more relevant nowadays, whereas modern science has concentrated much on the nutritional value of each food, but the food has also a considerable holistic impact on health.

Still today, like in the past, there is a lot to share between the ancient healing systems of Eastern and Western science: the premises for a possible integration can be attributed to common cultural factors. A professional and qualified pluralistic approach to nutrition, in which different medical systems can be integrated and shared with others without losing their particular specificity, has a lot of chances to provide effective solutions to many global health and environmental problems.

REFERENCES

Askari S.H. Medicines and hospitals in Muslim India. *J. Bihar Res. Soc.* **1957**, *43*, 7–21.

Basham A.L. The practice of medicine in ancient and medieval India, in Leslie C. (ed.) *Asian Medical Systems: A Comparative Study*. Motilal Banarsidass Publishers, New Delhi, India, **1998**, 18–43.

Bhasme A., Dsouza R., Smith A. *Medicine. The Voyage through Time*. B. Jain Publishers, Delhi, India, **2007**, 92–132.

Burgel J. Secular and religious features of medieval Arabic medicine, in Leslie C. (ed.) *Asian Medical Systems: A Comparative Study*. Motilal Banarsidass Publishers, New Delhi, India, **1998**, 44–62.

Collingwood R.G. *The Idea of Nature*. Clarendon Press, Oxford, **1945**, 15.

Filliozat J. *The Classical Doctrine of Indian Medicine. Its Origins and Its Greek Parallels*. Munshiram Manoharlal, Delhi, India, **1964**, 229–257.

Frawley D. *Gods, Sages and Kings. Vedic Secrets of Ancient Civilization*. Motilal Banarsidass Publishers, New Delhi, India, **2010**, 290.

Frutos Gonzàlez V. de, Guerrero Peral A.R. Neurology in medieval *regimina sanitatis*. *Neurologia* **2011**, *26*(7), 416–424.

Giri J., Jain P. *Sukham Ayu. Cooking at Home with Ayurvedic Insights*. Wisdom Tree, Delhi, India, **2013**, 5.

Gupta L.P. *Biogenic Secrets of Food in Ayurveda*. Chaukamba Sanskrit Pratishthan, Delhi, India, **2011**, 1, 3, 5, 13.

Haddad F. S. A(l)Razy (Razhes). *J. IMA* **1989**, *18*, 27.

Hulse J. *Sustainable Development at Risk: Ignoring the Past*. Cambridge University Press, New Delhi, India, **2007**, 186.

Jhones W.H.S. (ed. & trans.) *Hippocrates*. Harvard University Press, Cambridge, MA, **1923–1995**, Reg. H.II, Aphorisms I.XVIII.

Leslie C. (ed.) *Asian Medical Systems: A Comparative Study*. Motilal Banarsidass Publishers, New Delhi, India, **1998**, 1–17, 356–367.

Lloyd G.E.R. (ed.) *Hippocratic Writings*. Penguin, Harmondsworth, **1978**, 186–205, 272–276.

Lochan K. *Medicines of Early India*. Chaukamba Sanskrit Pratishthan, Delhi, India, **2003**, 77–79.

Lone A.H., Ahmad T., Anwar M., Imam H., Habib S. Perception of health promotion in Unani herbal medicine. *J. Herbal Med.* **2012**, *2*, 1–5.

Longrigg J. *Greek Medicine: From the Heroic to the Hellenistic Age. A Source Book*. Routledge, London, **2013**, 46.

Meyerhof M. Science and medicine, in Arnold T.W. and Guillaume A. (eds.) *The Legacy of Islam*. University Press, Oxford, **1931**, 311–355.

Modanlou H.D. Avicenna (AD 980 to 1037) and the care of the newborn infant and breastfeeding. *J. Perinatology* **2008**, (28), 3–6.

Ordronaux J. (transl.) *Flos medicine vel regimen sanitatis salernitanus*. J.B. Lippincott & Co, Philadelphia, PA, 1871.

Pelner Cosman M. A feast for Aesculapius. Historical diets for asthma and sexual pleasure. *Annu. Rev. Nutr.* **1983**, (3), 1–34.

QiaoSheng D. Seasonal changes and seasonal regimen in Hippocrates. *J. Cambridge Stud.* **2011**, *6*(4), 128–144.

Ranade S., Deshpande R. *History and Philosophy of Ayurveda*. Chaukhamba Sanskrit Pratishtan, Delhi, India, **2006**, 8, 12, 16, 26, 41, 70.

Saibene A. Ayurveda: Integration between traditional medicine and official medicine, in A.V., *Forest Medicine: Traditional Medicines, a Resource to Discover*. Centre for Orientative Education (COE), Milano, Italy, **2004**, 127–135.

Shapin S. The philosopher and the chicken: On the dietetics of disembodied knowledge, in Lawrence C., Shapin S. (eds.) *Science Incarnate: Historical Embodiments of Natural Knowledge*. University of Chicago Press, Chicago, IL **1998**, 21–50.

Shastri K.N. (ed.) *Charaka Samhita*. Chaukhambha Vidyabhavan, Varanasi, India, **1970**.

Skiadas P.H., Lascaratos J.G. Dietetics in ancient Greek philosophy: Plato's concepts of healthy diet. *Eur. J. Clinical Nutr.* **2001**, *55*(7), 532–537.

Smith A. *Ayurvedic Nutrition*. Motilal Banarsidass Publishers, New Delhi, India, **2011**, 44–45.

Smith F.M., Wujastyck D., Albany N.Y. (ed.) *Modern and Global Ayurveda: Pluralism and Paradigms*. State University of New York Press, Albany, NY, **2008** (introduction).

Taylor A. *A Commentary on Plato's Timaeus*. Oxford University Press, Oxford, **1928**.

Theophrastus. *Enquiry into Plants and Minor Works on Odours and Weather Signs* [trans. Hort A.]. William Heinemann, London; G.P. Putman's Sons, New York, 1916.

Tiwari M. *Ayurveda Secrets of Healing*. Motilal Banarsidass Publishers, New Delhi, India, **1999**, 16, 343, 344, 349.

Venkat P., Gangadharan G.G., Lakshmithathachar M.A., Alwar M.A. (eds.) *Bhojanakutuhalam*. I-AIM, Bangalore, India, **2012**.

Wujastyk D. *The Roots of Ayurveda*. Penguin Books, New Delhi, India, **1998**, 3, 4, 17, 42, 195–206, 236–244.

Zimmermann F. Rtu-satmya, le cycle des saisons et le principe d'appropriation. *Purusartha, recherches de sciences socials sur l'Asie du sud*, **1975**, *2*, 87–105.

Zysk K.G. *Medicine in the Veda*. Motilal Banarsidass Publishers, New Delhi, India, **2009**, 2–4.

Section I

Traditional Systems of Medicine in India

A Holistic Approach to Health

2 Principles of Ayurveda for Food, Nutrition, and Health

Unnikrishnan Payyappallimana and Padma Venkatasubramanian

CONTENTS

2.1 INTRODUCTION

Every civilization in the world has its own culinary culture using locally available items of plant and animal origin. Traditional cuisines evolve over centuries and become an important part of a culture. Local communities possess the knowledge of identifying, collecting, and cooking ecosystem-specific food resources. The communities are also largely aware of the biological effects of the food items on the human body, with different perspectives on the importance of food, nutrition, and their impact on health. This chapter focuses on certain unique Ayurvedic principles that provide an Indian perspective on health and healing through food. It also looks at Ayurvedic classification of foods and related aspects that are essential for healthy living.

India is well endowed with a rich biodiversity and an associated traditional knowledge. Traditional medical knowledge exists in two forms: (1) codified, as in written manuscripts, and (2) noncodified, which is passed down from generation to generation as oral traditions. *Ayurveda*, *Siddha*, *Unani*, and *Gso-wa-rig-pa* (Tibetan medicine) are codified medical systems that are still practiced today. Yoga is a separate discipline, which is primarily a spiritual practice, but certain aspects of Yoga, such as postures and exercises (*asana*, *kriya*) and breathing (*pranayama*) and meditation (*dhyana*) techniques, have been integrated with the health system to streamline the health of the body and mind.

Indian medicine broadly covers a range of healthcare and nutritional perspectives and practices. This ranges from simple household-level practices to the specialized knowledge of health practitioners in local communities and further to the institutionalized medical systems such as Ayurveda, Siddha, Unani, and Tibetan medicine (AYUSH, 2002). The highly decentralized health and nutritional culture among traditional and indigenous communities exemplifies the inseparable link between the diverse ecosystems and the cultural traditions related to health, nutrition, and well-being in the Indian subcontinent. This biocultural relationship between health and nutrition is embedded in proverbs, folk songs, and cultural expressions of the communities across the subcontinent. Traditional medicines are governed formally by acts, rules, and regulations by the Department of Ayurveda, Yoga, Unani, Siddha and Homeopathy (AYUSH), Ministry of Health, Government of India (Figure 2.1).

Apart from local health knowledge and codified medical literature, nutritional percepts are also available in several of the nonmedical literature and related practices. For instance, the *Bhagavad Gita*, a prominent philosophical text of the subcontinent, elaborates on the qualities of food and their influence on the nature of an individual from a spiritual and psychological perspective. For example, a person of *sattvik* (pure) nature likes food that promotes longevity, mental and physical strength, and happiness, whereas a person of *rajas* (excitable) nature likes foods that are pungent, sour, salty, hot, and spicy, that cause dryness, and inflammation in the

The Department of Indian Medicine and Homeopathy was established under the Ministry of Health, Government of India, in 1995 (and renamed as AYUSH in 2003; www.indianmedicine.nic.in).

There are over 700,000 registered medical practitioners who have graduated from over 500 dedicated AYUSH colleges and universities across the country, which are regulated by the Central Council for Indian Medicine.

Medicines prepared in Ayurveda, Siddha, and Unani systems are sourced from natural materials, including plants, animals, metals, and minerals.

The Drugs & Cosmetics Act (1940) and Drugs & Cosmetics Rules (1945) as well as amendments (2010) thereafter govern and regulate traditional medicines prepared by over 9000 manufacturing units in the country.

It is mandatory for all traditional medicine (TM) manufacturing companies to obtain a good manufacturing practice license from AYUSH per Schedule T of Drugs & Cosmetics Act (2000).

There are national institutes dedicated to each of the systems of medicine and an extensive network of research centers under the Central Council of Research in Ayurveda and Siddha.

FIGURE 2.1 Indian traditional medicine regulated by formal systems by the Government of India.

body, leading to unhappiness and ill health. A person of *tamas* (indifferent) nature has an affinity for foods that are stale, tasteless, and putrid (Prabhupada, 2002). Another classical scripture of the subcontinent, *Chandogyopanishad*, says that the purity of food is reflected in the purity of mind and in turn in the intellectual functions of an individual (Jha, 1923).

2.2 AYURVEDA: A HOLISTIC PERSPECTIVE ABOUT LIFE

The term *Ayurveda* denotes knowledge of life and longevity in their various facets. The core objective of Ayurveda is to ensure a happy life (*sukhayu*), sustainable happiness in life (*hitayu*), and long life (*dhirghayu*). Ayurveda indicates the appropriate (*hita*) and inappropriate (*ahita*), happy (*sukha*) and sorrowful (*dukha*) conditions of living. It also recognizes how much of each of the various conditions is appropriate and inappropriate for a long, happy life (Sharma, 2001).

To achieve a happy, long life, certain fundamental principles are outlined in Ayurveda. One such cardinal principle is the inherent relationship between the *outside* and *inside* worlds: the macrocosm (*loka*) and microcosm (*purusha*) share common elements (Figure 2.2). This relationship is further interpreted in terms of the theories of five elements (*panchamahabhuta*), and the three functional principles are roughly correlated as humors (*tridosha*). This unusual theory binds the entire microcosm and macrocosm, be it humans, materials, or Nature (Sharma, 2001). The three *dosha*s are *kapha*, which is the basis for nourishment and structures; *pitta*, which is responsible for all transformations; and *vata*, which regulates movement and destruction. Ayurveda has elaborate descriptions of materials (*dravya*), qualities (*guna*), the action of the

Purushoayam lokasammitah Man is the embodiment of the universe.

FIGURE 2.2 The connection between man and the universe. (Sharma P.V., ed. *Charaka Samhita, Vol. 1.* Chaukhambha Orientalia, Varanasi, India, **2001**, Chaps. 2, 5.)

materials in relation to the context of health or disease (*karma*), an understanding of the materials in their generality (*samanya*) and specificity (*vishesha*), and the holistic or total nature of materials (*samavaya*) (Sharma, 2001).

Achieving health through Ayurveda is based on the premises that substances, qualities, or actions of similar nature have an increasing effect in the system and vice versa. For example, the intake of excess fat increases kapha in turn leading to an increase in adipose tissue and related health problems in the body. This is similar to the *doctrine of signatures*, which recommends, for example, the use of stem bark for skin conditions, *haridra* (yellow color of turmeric) for jaundice, and *laksa* (lac) and *manjista* (*Rubia cordifolia*, red-colored materials) for hemorrhage (Chunekar, 2004). Similarly, *usna virya*, a substance with hot potency, will increase heat in the body and create vitiation of *pitta* and *rakta* (blood). Excessive physical action (*karma*), such as running or hard exercise, can increase *vata* in the body, resulting in joint pains. Thus, the actions are interpreted through their qualities and appropriately used for increasing or decreasing a *dosha* in a disease (Sharma, 2002).

In humans, the *mind–body* constitution or *prakriti* manifestation is interpreted in terms of *dosha*. The *prakriti* of an individual is determined at the time of conception. The lifestyle and intervention, be it diet, exercise, or medicines, will need to be modulated depending on the basic nature or constitution of the individual. *Prakriti* is a unique Ayurvedic way of classifying individuals on the basis of the predominance of *dosha*. Although all individuals have all three *dosha*s, there are seven basic types of *prakriti*, depending on the predominant *dosha*(s), namely, *vata*, *pitta*, *kapha*, *vata-kapha*, *vata-pitta*, *pitta-kapha*, and *sama-dosha*. An individual with *sama-dosha prakriti* has all three *dosha*s in equal proportion (Sharma, 2001). However, this type of individual is rare to find. There is no "good" or "bad" *prakriti*; each has its strengths and weaknesses. Ayurveda recommends an ideal way of living depending on one's own *prakriti*, which includes diet, seasonal foods, and precautions.

Traditionally, the *prakriti* of an individual is identified (Figure 2.3) by a set of observations about his or her physical appearance and physiological and psychological functions (Sharma, 2001). In recent times, there has been much interest generated by research findings that indicate that there could be a genetic basis for *prakriti*, which could open up a whole new way of operationalizing personalized medicine (Mukherji and Prasher, 2011). Patwardhan and Bodeker (2008) refer to this potential area as ayugenomics and discuss the emerging evidence to indicate a correlation between *prakriti* types and genomics and phenomics.

Even though there are broad thumb rules for universal ways of healthy living, Ayurveda mentions that the homeostasis, or equilibrium point, will vary from individual to individual on the basis of his or her *prakriti* and other internal and external determinants. Thus, healthy state is the maintenance of balance of one's

- *Traditional method*: Ayurvedic practitioners determine an individual's *prakriti* through elaborate observations and extensive questioning of the individual/carer.
- *Ayusoft*: The Centre for Development of Advanced Computing (C-DAC), India, has developed and tested a software application called Ayusoft for *prakriti* analysis. Details of this application are available at http://pune.cdac.in/html/hbcg/ayusoft/ayusoft.asp.
- *Do it yourself*: There are several D-I-Y questionnaires that are freely available on the Internet that can also help you determine your *prakriti.*

FIGURE 2.3 Methods of *prakriti* analysis.

own constitution. *Prakriti* is used to interpret and manage health and disease states. For example, a person of *pitta prakriti* is prone to stress and peptic ulcer, which further vitiates the *pitta* in his body. He or she can balance the *dosha* by eating foods and dietary supplements that have cooling or *pittahara* (*pitta*-reducing) properties, for example, barley, oats, and sweet fruits. Food and drinks such as coffee, alcohol, spices, and pungent items that aggravate *pitta* should be avoided.

2.3 AN AYURVEDIC PERSPECTIVE ON HEALTH

Ayurveda defines perfect health as *svasthya*, which means "to be established in oneself or one's own natural state" (Figure 2.4). When the *structural* and *physiological* (*sama-dosha*) factors, metabolic processes (*samagni*), body tissues (*samadhatu*), eliminative system (*samamalakriya*), senses (*prasannendriya*), and mind (*prasannamana*) are in a state of equilibrium, a person achieves pure awareness and a contented self (*prasanna atma*). An individual with this state of equilibrium is considered healthy (Sharma, 2004).

Ayurveda recognizes 10 determinants that affect the health of an individual (*dasa vidha pariksha*) (Murthy, 2001):

1. *Dushya* (body tissues and excretory products)
2. *Desha* (place of residence and body)
3. *Bala* (strength: physical and mental)
4. *Kaala* (seasons)
5. *Anala* (digestion and metabolism)
6. *Prakriti* (genetic and phenetic constitution)

Samadosha samagnischa samadhatu malakriyah
Prasannatmendriya manah svastha ityabhidhiyate

FIGURE 2.4 *Shloka* (verses) from ancient Ayurvedic manuscript on the concept of health (*svasthya*) (Sharma P.V., ed. *Susrutha Samhitha, Vol. 1.* Chaukhambha Visvabharati, Varanasi, India, **2004**, Chaps. 15, 41.)

7. *Vaya* (age)
8. *Sattva* (determination)
9. *Satmya* (habituation)
10. *Ahara* (food)

Though this framework (*dasa vidha pariksha*) is primarily used for disease diagnostics, it is nothing but the determinants of health, portraying the comprehensive and holistic perspective of Ayurveda. Ayurveda mentions the importance of both internal (body tissues, age, digestion, body constitution, etc.) and external factors (time, seasons, food, etc.) as health determinants. An interesting aspect to note is that disposition of the mind (*sattvam*) or mental temperament is also mentioned as an important factor that can keep a person healthy. Thus, the above 10 parameters help a physician in proper diagnosis and accordingly help in choosing the line of treatment (Murthy, 2001).

2.4 AYURVEDA AND NATURAL RESOURCES

Ayurveda has a documented history of more than 3000 years of using plants, animals, metals, and minerals for the purpose of food and medicine. It has a unique way of understanding plants. Right from the earliest Ayurveda treatise, *Charaka Samhita*, exhaustive descriptions about plants and their medicinal uses are available. The information relates to nomenclature, descriptions for identification, biological properties and actions, habitat, regional specifications of substitutes, time and methods of collection, methods of classification, computing and combining plants, processing, application in specific stages of disease conditions, incompatibility, contraindications, recipes, poisons, and so on.

There are a good number of classical Ayurvedic texts and manuscripts that deal with plants. These are largely *Samhitas* (classical Ayurvedic books), *Samgrahas* (compact books written on the basis of classics), *Nighantus* (*Materia Medica*/lexicons), *Vyakhyas* (commentaries), and texts dealing with specific subjects such as *bhaishajya kalpana* (pharmacy). When one screens through the entire evolutionary history of Ayurveda, one would find that Ayurvedic knowledge has been quite dynamic. This knowledge has been updated from time to time depending on the study and inclusion of new materials and sharing between cultures. Materials with unknown identity were omitted and new ones introduced. A number of drugs were introduced from other parts of the world through foreign contacts and trade, and these have also been incorporated into Ayurveda's therapeutic armamentarium. For instance, spices, aromatics, and other drugs were exported from India, whereas drugs such as *hingu* (asafoetida; *Ferula asafoetida*), *kesara* (saffron; *Crocus sativus*), and *yashtimadhu* (liquorice; *Glycyrrhiza glabra*) were imported (Chunekar, 2004). Exotic plants such as brinjal, green chillies, potato, tapioca, tomato, pineapple, tea, and coffee have been studied on Ayurvedic parameters and incorporated in the ayurvedic *Materia Medica*.

Apart from written texts, a major part of plant information is preserved through the oral traditions of Ayurveda, through the living traditions (*parampara*) in various parts of the Indian subcontinent, and through a number of handwritten manuscripts, of which many remain unpublished and unresearched even today.

Yuktivyapasraya is one of three ways of disease management in Ayurveda. It deals with treatment through the rational administration of drugs and diet. This is otherwise called *Sadravya Chikitsa* (treatment with drugs) (Sharma, 2002). *Dravyaguna vijnana* is the speciality in Ayurveda, which elaborates the qualities and biological actions pertaining to food and medicine. It covers the promotive, preventive, and curative applications of drugs. Drugs are classified on the basis of the source as plant (*oudbhida*), animal (*jangama*), and mineral (*parthiva*) (Sharma, 2002, 2006).

The word *oushadhi* denotes herbaceous plants. The word *oushadha* in Sanskrit can be elaborated to mean *oso daho dhiyate atreti va*, which means a burning transformation resulting in a beneficial expression. In other words, plants carry the creative energy that works behind all life processes, especially all transformations. *Oushadhi* is also elaborated as *Osam rujam dhavati iti*, which means the one that removes the pain (Taranatha, 1990). Imbalances in the transformative function of the life process result in disease and any substance that regulates this process is called *oushadha*. Since plants constitute the major source of *oushadha*, the term *oushadhi* for plants becomes appropriate.

Different parts or derivatives from a plant source are used to treat a particular disease condition on the basis of an understanding of their properties. These materials, after application of different methods of processing, become a combination/formulation (*yoga*). The word *oushadha* is used to denote the entity at all three levels, as a plant source, a plant part, or a processed drug. The plant parts used are root bark, heartwood, secretions, stalk, expressed juice, tender leaves, alkali, latex, fruit, flower, ash, oil, thorn, leaves, leaf buds, tubers, sprouts, pedicle, petal, stamen, seed, fruit, and the whole plant (Reddy, 2005).

Dravyaguna vijnana, or the science of substances used as drugs, is divided into *padartha vijnana* and *dravya vijnana*. *Padartha vijnana* is the science that deals with the study of the basic existential principles, that is, substratum (*dravya*), qualities (*guna*), actions (*karma*), generic (*samanya*), specific (*vishesha*), and inherence (*samvaya*). This knowledge provides the basis to understand the dynamics of drug action and *dravya vijnana* is the detailed pharmacology. *Dravya vijnana* is divided into three subtopics: nomenclature (*nama*), identification (*rupa*), and application (*yuktijnana*), which constitute the three-tier understanding of *oushadha* (Sharma, 2006).

A therapeutic *dravya* or material is described in Ayurveda by its qualities (*guna*) and by the actions (*karma*) it causes in the body, that is, pharmacodynamics, as perceived by the body–mind complex, where the five senses and cognitive ability play an important role. The descriptions are on the basis of the taste (*rasa*), quality (*guna*), potency (*virya*), postdigestive effect (*vipaka*), special properties (*prabhava*), action (*karma*), action on *dosha* (*doshakarma*), body tissues (action on *dhatu*), excretory mechanism (*mala*), and so on. The *panchamahabhuta* composition in the material is the most essential indicator of pharmacodynamics (Sastry, 2009). The *bhoutika* composition of a plant drug is evaluated best by its *rasa* (taste). It is also understood that each *rasa* has a different effect on the *tridosha*s in the body. It is important to consume foods of all the tastes at a quantity and quality that corresponds to one's constitution on a daily basis to stay healthy.

2.5 AN AYURVEDIC PERSPECTIVE ON FOOD

Food (*ahara*), as one of the three important pillars (*tristambha*) of life, draws major attention in Ayurveda, the other two being sleep (*nidra*) and regulated sexual life (*brahmacarya*) (Sharma, 2001). Food being an important prerequisite for a healthy life, major classical texts such as *Charaka Samhita* (Sharma, 2001), *Susruta Samhita* (Sharma, 2004), and *Astanga Samgraha*, which were codified *c*. 300 BC–AD 700, have elaborate descriptions and have dedicated sections on food:

- Descriptions of various foods and beverages (*drava dravya vijnaniya*)
- Classification of foods and their qualities (*annasvarupa vijnaniya*)
- Food safety (*annaraksha vidhi*)
- Incompatibilities between different types of food (*viruddhanna vijnaniya*)
- Instructions on consumption (*annapana vidhi*)
- Food quantities and intake (*matrasitiya*)

There is an entire section in the Samhitas dedicated to the quantity of food to be consumed (*matrasitiya*); the section also details the diseases that manifest because of food intake at improper quantities. According to these texts, the quantity of intake is determined mainly by two factors: the digestive capacity of the person who consumes and the nature of food consumed. A simple directive for a healthy meal recommends that half of one's stomach should be filled with solid food, a quarter with liquids (mainly water), and a quarter left empty for movement of *vayu*, or air. This directive should be practiced especially by those who have poor digestive capacity, poor strength, and poor health status, as well as those who have sedentary lifestyles. Such instructions are quite subjective and can only be internalized through regular observation and practice of what is good for each individual (Murthy, 2001).

There are also other sections, including *dravyabhediya* (on materials), *rasabhediya* (on tastes), and *doshabhediya* (on *dosha*s), which have important references to philosophical, physical, and physiological references to food (Sharma, 2002). Apart from the major texts of Ayurveda, there are also texts such as *Bhojana kuthuhala* and the various texts of *paka sastra*, which are dedicated to food, nutrition, health, and well-being (Lakshmithathachar, 2012). Texts such as *kshemakutuhala* further elaborate on various food materials, their qualities, and specialized cooking methods (Thathachar, 2009).

2.5.1 Classification of Foods

Ayurveda classifies food on the basis of various factors.

2.5.1.1 Classification Based on Appropriateness (*Hita Ahara*)

Classification of foods as *appropriate* (*hita ahara*) and *inappropriate* (*ahita ahara*) is an important concept in Ayurveda: if one follows *hita ahara* throughout one's life, then the chances of getting sick are low. Food is not for mere taste or to satiate hunger but also for nutrition. Foods that are tasty may not always be appropriate for the body, and eating to only tackle hunger is not adequate for the body's nourishment (Sharma, 2002). There is growing recognition of the fact that there is a natural tendency in modern societies to select sweet, low-fiber, high-fat foods. Extensive research and development are undertaken to bring out hybrids and genetically modified food plants and animals whose original qualities are many times altered to cater to the palate; for example, the supersweet American corn is genetically selected to be white or yellow (instead of black or other colors) and have more sweetness and less fiber. Along with the colors, important antioxidants and anticancer phytochemicals, such as anthocyanins and β-carotene, are lost in the process (Robinson, 2013).

2.5.1.2 Classification Based on Effect (*Sattva, Rajas,* and *Tamas*)

The introductory verses of the Samhitas suggest a strong link between the five psychological dispositions of lust, anger, greed, attachment, and ego and disease manifestations (Sharma, 2002). A strong connection between food and these psychological temperaments (*manasa prakriti—sattva, rajas, tamas*) is outlined well in Ayurvedic and other non-Ayurvedic literature, such as the *Bhagavad Gita* (Prabhupada, 2002).

2.5.1.3 Classification Based on Properties and Actions (*Dravyaguna Vijnana*)

The most important classification of foods in Ayurveda is based on the preponderance of five elements (*panchamahabhutas*) and the manifestation of the same as different tastes (*rasa*) and qualities (*guna*) that are perceived by the sensory organs. These properties are said to impact the biological action (*karma*).

The Ayurvedic classification of food or drug is based on their biological action on the body rather than the food ingredients per se. Ayurveda uses the composition of six core tastes (*shad rasa*) in foods as a guide to balance the diet. There are six basic tastes: sweet (*madhura*), sour (*amla*), salt (*lavana*), pungent (*katu*), bitter (*tikta*), and astringent (*kashaya*). A well-balanced diet should include all six *rasa*s. An excess of any one taste in the diet leads to health problems. In fact, according to Ayurveda, the taste of a substance is a very important determinant of its biological action (Sharma, 2002).

Ayurveda recognizes that digestion and metabolic processes play an important role in deciding a drug's or substance's biological effect. All food substances go through the digestive cycle, namely, gross digestion at the gastrointestinal tract (*jatharagni paka*), tissue-specific metabolism (*dhatvagnipaka*), and digestion/metabolism at the elemental levels (*bhutagni paka*). The five-element (*bhoutika*) composition of some substances continues as is even after digestion, whereas in some others it may be altered altogether depending on the metabolism, and the change is reflected as *vipaka* (postdigestive effect). When there is more than one taste (*rasa*) for a particular substance (*dravya*), the *vipaka* is a better indicator of its action.

The action of each substance is decided by its inherent potency. These are of two types: hot (*usna*) and cold (*sita*). The potency is attributable to some intrinsic properties called *guna*s. There are 10 opposite qualities (*dvandva guna*), which are different from each other because of the relative preponderance of basic elements (*mahabhutas*). The biological activity of a plant that is not explained through these parameters (i.e., *rasa*, *guna*, *virya*, etc.) is considered as *prabhava*. There are approximately 150 types of pharmacological actions (*karma*) listed in the Ayurvedic literature. These are the specific actions the *dravya* exerts on the body by different combinations of the above-mentioned factors. The actions are also classified in relation to *dosha*, tissues (*dhatu*), and waste materials (*mala*), for example, depletion (*langhana*), nourishing (*brmhana*), emesis (*vamana*), and purgation (*virecana*). According to classical texts, substances do not act solely because of the power of their qualities (*gunas*). Indeed the power of their physical nature (*panchamahabhutic* composition) and the quality (*guna*) are equally important. However, substances do act in a specific time after reaching a specific substratum (*dhatu*), thus achieving a result that is specific to each substance. The drugs act through the following factors: drug substance (*dravya*), quality (*guna*), substratum (*adhikarana*), time (*kala*), application (*yukti*), means (*artha*), intent (*abhipretha*), action (*karma*), stratagem (*upaya*), and effect (*phalam*). What the drug achieves is called the drug/substance action, in what manner it does depends on its potency, where it acts is the target site, when it does is the time of action, how it does is the stratagem, and what it does is the final result (Sharma, 2006).

2.5.1.4 Classification Based on Solid and Liquid Foods (*Dravadravya*)

The classification of food and drinks is also based on the morphological nature, the qualities as well as the physiological actions. The major classifications (*varga*) are grains and pulses (*sali*, *vrihi*, *trina dhanya*, *yavadi*, *simbi*), processed foods (*kritanna*), meat and meat products (*mamsa*), leafy vegetables (*saka*), fruits (*phala varga*), salts (*lavana*), food supplements, various forms of water (*jala varga*), milk and milk products (*ksira varga*), sweets (*madhura varga*), honey (*madhu*) and its products, oils (*taila varga*), and alcoholic drinks (*sura*) (Murthy, 2001; Sharma, 2004). Extensive descriptions of each type of food are available in literature including water. For example, water being the most essential aspect of nutrition, sections from Ayurveda texts on foods and drinks (*dravadravya vijnaniyam*) start with descriptions of different types of water and their qualities (Murthy, 2001). *Astanga Samgraha*, a prominent text of Ayurveda, describes the nature of soil and the tastes and qualities of water found in those regions. The texts also describe the different sources of water and their qualities. Descriptions of the properties and biological action of foods processed using different methods are also featured: for example, there are detailed descriptions of the effect of processing (*samskara*) on the properties (*gunas*) of rice in *Charaka Samhita* (Sharma, 2001). Properties of raw, dried, fried, smoked, grilled, pickled foods, and substances cooked in steam, with various adjuvants, methods of processing, and so on, feature in ancient texts. Food is also classified on the basis of the land it is grown on and its effect at an individual level (Sharma, 2002).

2.6 UNIQUE AYURVEDIC PRINCIPLES AND PRACTICES

To understand Ayurvedic perspectives on food, nutrition, health, and disease, it is important to know some of the underlying principles that are unique to Ayurveda.

2.6.1 *Prakriti* and Diet

Ayurveda has an individual-specific prescription of food. As discussed earlier, every individual is born with a specific constitution, or *prakriti*, which is unchangeable throughout life. The relationship between food and the individual's constitution (*prakriti*) is outlined well in Ayurveda. For instance, individuals with a *vata* constitution have a tendency to like foods that are sweet (*madhura*), sour (*amla*), and salty (*lavana*); *pitta* has affinity for sweet (*madhura*), bitter (*tikta*), and astringent (*kashaya*); and *kapha* has a liking for pungent (*katu*), bitter (*tikta*), and astringent (*kashaya*) tastes (Sharma, 2001).

2.6.2 *Dinacharya* and *Ritucharya*: Daily and Seasonal Regimens

According to Ayurveda, following a daily (*dinacharya*) and a seasonal (*ritucharya*) regimen in diet and lifestyle is of immense importance for a healthy life. There are elaborate descriptions of the type of foods to be consumed in accordance with the diurnal and seasonal cycles as well as the age group to which the individual belongs.

2.6.2.1 Concept of Seasonal Foods

According to Ayurveda, there are six seasons that are divided into two major categories: depletive (*adana*) and replenishing (*visarga*). These concepts are conceived on the basis of the action the sun has on the earth. In the season of depletion (*adana kala*), the sun draws energy out of the earth, whereas in the season of replenishing (*visarga*), it provides energy. This in turn causes certain changes in the body in terms of nourishment and depletion. Further finer changes are described in the discussion of the three seasons in each category. The bodily metabolism according to Ayurveda is closely linked to various seasons. It is said that all tastes should be taken always; however, according to seasons, those tastes that have a stronger positive effect on the body should be consumed. For example, in summer the digestive processes are weak. Therefore, Ayurveda advocates the intake of sweet, light drinks, soups, and so on to pacify the *vata dosha* that gets accumulated in the body, whereas in winter, when the digestive power is good, heavy food items are appropriate. It is also said that the *dosha*s accumulated in winter should be pacified in spring, those accumulated in summer in the rainy season, and those accumulated in the rainy season in autumn (Sharma, 2002).

2.6.3 Incompatible Diet (*Viruddha Ahara*)

A concept that is unique to Ayurveda is that of incompatibility (*viruddha*) of foods. The incompatibilities may be due to place, time, digestive capacity, dose and quantity, processing, and potency, among several others. For example, Ayurveda advocates that fish should not be consumed along with milk. The sweet *rasa* and the post-digestive effect, in such a combination can cause a state of derangement of *kapha* in the body (*abhishyanda*). It is also said that the two have opposite potencies, such as hot and cold, respectively, causing incompatibility. The descriptions go to finer details, such as the variety of fish (e.g., *cilicima*) in particular, which is contraindicated. Another example is the consumption of milk with sour materials such as fruit, which is a common feature in a continental breakfast. Yet another interesting instruction is not to eat radish and turnip along with milk, which can lead to skin diseases. Honey should not be consumed when hot and honey and ghee should not be consumed in equal quantities. There are extensive descriptions on incompatibility and the remedial measures to overcome the effects of consumption of incompatible foods. It has also been mentioned, however, that a person who performs adequate exercise, has good digestive capacity, and is strong may not be affected by such incompatibilities (Sharma, 2001).

2.6.4 Convalescence Foods (*Pathya*)

According to Ayurveda, consumption of the right food by an individual in a diseased state can promote the healing process and improper foods can delay or prevent the healing process. Thus, favorable food (*pathya ahara*) and unfavorable food (*apathya ahara*) in disease state as well as in convalescence period get prominence in Ayurveda. Along with medicines and other disease management methods, food instructions form an important element of Ayurvedic therapy. In the case of *panchakarma*, food restrictions are often required not only during the treatment period but for almost an equal number of days after the completion of treatment. These prescriptions may be disease, patient, and food specific. For example, as a general recommendation, strong tastes (spicy, sour), deep-fried food, certain vegetables such as pumpkin and brinjal, fish, cold food items, yogurt, and alcohol are generally avoided during medication. At the same time, instructions are also given to take yams, gourds, fruits, spices, brown rice, buttermilk, and so on. In certain regions, there is also a tradition of intake of drinking water medicated with certain spices, sandal wood, acacia, and so on. There are also special foods and instructions specific to pregnancy, postnatal period, childhood, old age, and so on (Tripathi, 1998). Food as a rejuvenative (*rasayana*) that can delay ageing is another major topic in Ayurveda, which has been dealt with in detail in chapter 3 in this book.

2.6.5 Digestive Fire (*Agni*)

According to Ayurveda, three components are required for food (*ahara*) to be optimally used by the body: the involvement of "digestive fire" (*agni*), the removal of unmetabolized substances (*ama*), and the nourishment of tissues (*poshana*) when the nutrients are assimilated by the body (Figure 2.5).

Ayurveda gives utmost importance to the digestive process in nutrition. Although foods consumed may be rich in nutritious content such as proteins or micronutrients, but what really matters is their interaction with the body's digestive processes and their actual absorption into the system. Conversion of *ahara* to nourishment is realized when there is "ignition of fire" (*deepana*) and "cooking/transformation" (*pachana*) in the body. This thinking is at the root of the metaphorical understanding of digestion as a process of "burning" and the various digestive "fires" such as *jatharagni*, *dhatvagni*, and *bhutagni*, roughly translated as digestive functions at the levels of the gastrointestinal system, tissue, and the five elements, respectively (Dwarakanatha, 2003).

It has been mentioned that properly ingested food is the route to perfect functioning of sensory and motor functions as well as long life. The essence (*ojas*, *tejas*) of

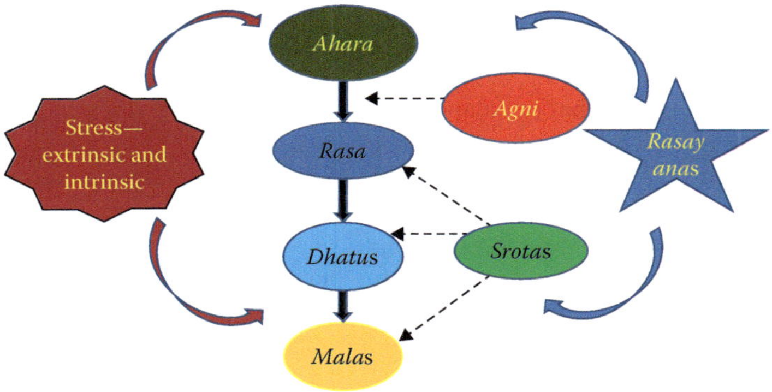

FIGURE 2.5 *Agni* (metabolic fire) at the levels of gastrointestinal system, tissues, and the *panchamahabhuta* plays a crucial role in determining health. Agni (metabolism) vitiation is the root cause for several health problems even before it manifests as a pathological condition.

the seven body tissues, seven types of body tissues (*dhatus*), sense organs, strength, pleasant disposition, nourishment, and overall intellect depend on good nutrition and well-consumed food. Being the transformative metabolic process equated to "fire" in Ayurveda metaphorically, good food is vital for strengthening the "digestive fire" or *agni* (Dwarakanatha, 2003).

The interaction between food and the digestive process is reflected in the dynamic awareness of food tastes (*rasa*), quality and quantity to be consumed, seasonal variations to be observed, constitution, age and strength of a person and his or her relation to food, variation of *dosha*s, mental predispositions (*satva*), habituation (*satmya*), and food in relation to disease states, among several others.

Per modern nutritional science, nourishment of the body depends on the nature of food, digestion and metabolism, uptake of nutrients at the molecular and tissue levels, and the elimination of wastes. Enzymes (e.g., pepsin, pancreatin bile) and certain hormones (such as thyroid-stimulating hormone, insulin, and leptin), supported by micronutrients and vitamins, are involved in the regulation of the digestive system. These play a role in breaking down the food materials and in absorption, distribution, metabolism, and excretion.

Thirteen types of *agni* are narrated by Ayurveda:

- *Jatharagni*, responsible for the digestion of food, located in the stomach and at the beginning of the intestine
- *Dhatvagni*, seven in number, responsible for the formation of the seven tissues
- *Bhutagni*, five in number, responsible for the digestion and assimilation of the five *mahabhuta* fractions (*amsas*) from the food mass

Jatharagni impairment causes indigestion at gross levels, whereas *dhatvagni* and *bhutagni* impairments cause disturbance in the metabolism of specific tissues of the body.

Agni may drastically vary according to person and condition. It may exist in any of the following states:

- *Samagni*: The normal or balanced *agni* that usually gets impaired by irregularity in food habits.
- *Tikshnagni*: The sharp *agni* is capable of digesting any amount of food.
- *Mandagni*: It is dull again which is disturbed by even small amount of irregular food.
- *Vishamagni*: The irregular *agni* is the opposite of *samagni*. It may sometimes gets impaired without any cause.

The *agni* of a person is assessed by his or her capacity to digest foods; for example, a *mandagni* person will purge following consumption of even a small quantity of milk. Maintenance of optimum *agni* is important for health. Variation in *agni* causes the formation of *ama* (unmetabolized substances), which causes several diseases. Correction or maintenance of *agni* is crucial to any Ayurvedic treatment. *Kaya Chikitsa* (general medicine) is considered as the treatment for impaired *agni* (Dwarakanatha, 2003).

2.6.6 *Ama* and Purification Procedures

Ama is an important concept mentioned in classical Ayurveda texts. The food that is not properly digested in the system for long periods of time resides and circulates in the body. *Ama* is mainly caused by deranged or vitiated *agni*, in particular *jataragni* (i.e., *agni* in the *koshtha*, or gastrointestinal system). This causes several acute and chronic disease conditions. How to prevent *ama* and treat *ama*-related conditions is also outlined well in Ayurveda (Sharma, 2002).

Elaborate procedures including *shodhana* (purification), *panchakarma* (the five treatment procedures), and *rasayana* (rejuvenating methods) are administered for "digesting" and "eliminating" *ama*. These techniques also correct systemic dysfunctions and rejuvenate the system at the tissue level. They involve preparatory procedures such as oleation and fomentation of the body; the main procedures of systemic correction such as induced vomiting, purgation, enemas, and nasal purge; and post purificatory procedures such as diet and regimen.

Fasting plays an important role in maintaining health and eliminating undigested food. Methods of fasting are discussed in detail in Ayurvedic texts. It is interesting to note that fasting has been a part of almost every society in the world and has been associated with religious functions; for example, Lent in Christianity, Ramzan in Islam, and *Vrat* in Hinduism have fasting as an important component associated with prayers. Recent evidence demonstrates that intermittent fasting improves health and enhances life span in fruit flies, nematode worms, and rodents (Stipp, 2012).

Although the physiological and other health benefits of *panchakarma* processes are not well understood in terms of contemporary science, they are unique to Ayurveda, and need to be explored (Valiathan and Thatte, 2010). Some forms of purificatory processes, such as purgation, sudation, and oil baths, were commonly followed health practices in India. However, these practices have eroded today.

It is increasingly becoming evident in recent times that the microflora in the gut influences health, diseases, and moods of human beings (O'Hara and Shanahan, 2006). Ayurveda identifies "fire" in the gut (*jataragni*) as a major control point of homeostasis and *panchakarma* as possibly a way of *purifying* and *normalizing* the gut. It is possible that there is a correlation between *prakriti* and/or *koshtha* (gut) types and native gut microflora. Further research is required to understand the specific benefits of *panchakarma* in gut health.

2.6.7 Consumption of Food

There are different codes of conduct for consumption of food. As a general direction, according to Ayurveda, eating what is good and bad together (*samasana*), eating soon after a meal (*adhyasana*), eating inappropriate quantities (*amatrasana*), and irregular eating (*visamasana*) should be avoided.

Another interesting discussion refers to the subtle qualities of food based on factors such as inherent nature (*svabhava*), combining (*samyoga*), processing (*samskara*), quantity/dose (*matra*), place (*desa*), time (*kala*), consumption practices (*upayoga vyavastha*), and state of health of an individual (*svasthya* or *asvasthya*) (Sharma, 2001).

2.6.8 *Rasayana*

Rasayana chikitsa, also called as *jara chikitsa* (literally translated as "age treatment"), is a separate branch of Ayurveda that deals with rejuvenation, health and aging. *Rasa* is the primary product of digestion at the gastroenteric tract (*jatharagnipaka*) and may be correlated with *chyme*, the semifluid mass derived from partial digestion of food in the stomach, and *ayana*, which refers to the channels that transport nutrients to the tissues and excrete wastes from the tissues. *Rasayana* is a set of methods that teaches one how to live a long and healthy life and delay aging. These methods are applied for preventive, promotive, and curative care. They include not just food and dietary supplements but also drugs, exercise, and lifestyle. Regulation of the digestive fire (*agnivardhana*), purification of the channels of nourishment (*srotosodhana*), and nourishment of tissues (*dhatuposhana*) are the three main ways of creating the *rasayana* effect, that is, health maintenance and life span extension (Balasubramani *et al.*, 2011). Disturbance in any of the three processes will affect health and cause diseases.

2.7 BUILDING A BRIDGE BETWEEN AYURVEDA AND BIOMEDICINE

Both modern nutrition and dietetics and Ayurveda insist on the need for diets to be balanced. However, the indicators used to identify the balance are different. A balanced diet is advised by a modern nutritionist for a healthy life, with carbohydrates, proteins, fats, vitamins, and minerals in quantities that are in descending order, which is called the *food pyramid*. Foods are classified on the basis of their chemical content as macronutrients and micronutrients. Carbohydrates, proteins, and fats are macronutrients, and vitamins and minerals are the micronutrients needed for the body. Apart from this, the fiber content and phytochemicals called secondary metabolites are also of importance; for example, tomatoes contain lycopene and green chillies possess capsaicin, which are secondary metabolites that have antioxidant properties. Carbohydrates that include starch, sugars, and polysaccharides provide energy; proteins made up of amino acids help build the body, provide immunity, and repair damage; and fats are important for absorption of certain vitamins and act as a storehouse of energy. Vitamins and minerals are mostly derived from foods and act as catalysts in the metabolism of macronutrients. They are essential for enzymatic reactions, and minerals such as calcium and phosphorus are vital for building bones and teeth.

Systems biology has existed since the 1960s but is an emerging paradigm in bioscience in recent times. Systems biology researchers develop scientific methods to achieve a *holistic* understanding of physiology. It is now believed that physiology and human existence cannot be reduced to physics and chemistry as molecules and genes (Noble, 2008). Various techniques, in particular mathematical modeling and bioinformatics, are employed to integrate the multilevel information about cells, tissues, organs, and entire organisms to get a holistic picture of the functioning of the organism. This essentially is a shift from the conventional reductionist approach of biomedicine that was once based on the *one-gene one-molecule* theory. Yet, systems biology still builds on the bottom-up approach from genes to cells to systems in organisms, whereas Ayurvedic biology has a top-down, holistic understanding that integrates not only the organismal and suborganismal processes but also the dynamic interactions and influences of Nature. However, in Ayurvedic biology the subcellular (genetic, molecular) processes are not dealt with per modern scientific terms. Therefore, a lot can be gained by marrying the top-down understanding of Ayurvedic biology and the bottom-up understanding of systems biology (Patwardhan *et al.*, 2004). According to Ayurveda, the whole universe and its myriad animate and inanimate manifestations are made up of the same fundamental components: earth, water, air, fire, and ether. The homeostatic state of the *tridosha*s is different for different persons, and it is this equilibrium that is crucial for maintaining the health of the individual. Diseases are largely treated by understanding the *dosha* vitiation and bringing them back to equilibrium.

2.7.1 Exploring Ayurveda for Managing Anemia—A Case Study

According to the World Health Organization (WHO), almost 30% of the world population is anemic, and in developing countries every other pregnant woman is anemic, silently contributing to infant and maternal morbidity and mortality. Iron deficiency anemia (IDA) is one of the predominant causes of anemia. In this section, the authors compare Ayurvedic and biomedical perspectives and line of treatment of anemia. One of the authors (PV) of this chapter is also leading a research program to develop Ayurveda-based low-cost herbal supplements for iron deficiency anaemia. Some of the work-in-progress on anemia has been shared under this section.

2.7.1.1 Biomedical and Ayurvedic Perspectives on Anemia: A Comparison

The biomedical system looks at IDA essentially as a nutritional deficiency and a deficiency in iron. Therefore, iron–folic acid tablets are used as supplements for managing anemia. IDA can be correlated with a condition called *pandu* in

Ayurveda, though *pandu* is a broader condition. Ayurveda defines *pandu* as a disease in which the skin gets discolored to mild yellowish or pale white. *Pandu* is a pathological entity and is basically a *pitta* predominant disorder. It is a symptom seen in the derangement of *rasa* (chyme)-carrying channel (Sharma, 2001). A comparative understanding of anemia in biomedicine and Ayurveda is presented in Table 2.1.

TABLE 2.1
Comparison of the Understanding of Anemia per Biomedicine and Ayurveda

S. No.	Criteria	Biomedical Understanding	Ayurvedic Understanding
1	Etiology	Increased blood loss, for example, peptic ulcer, hemorrhoids, hematuria, hemoptysis, and excessive menstruation; increased requirements, for example, pregnancy and lactation, childhood, adolescence, and inadequate dietary intake; decreased absorption, for example, intestinal malabsorption and partial or total gastrectomy	Excess intake of alkaline, sour, salty, spicy foods, incompatible foods, unwholesome foods, alcohol; sleeping during day time; lack of exercises; faulty management of purificatory measures and seasons; suppression of natural urges; anxiety, fear, anger, sorrow, and so on
2	Pathology	IDA develops when the supply of iron is inadequate for the hemoglobin biosynthesis. When the tissue stores of iron are exhausted, the supply of iron to the marrow becomes insufficient for the formation of Hb; hence, a state of IDA develops. This depends on increased blood loss, increased requirements, inadequate dietary intake, and decreased intestinal absorption, which vary with the age and sex of the patient.	Owing to the above cause, the *dhatvagni* (metabolic activity at the tissue level) becomes weak and the *rasavaha srotas* (*rasa*-carrying channels) get obstructed and the formation of *rakta dhatu* (blood tissue) is hampered, which leads to *alparaktata* (decreased formation of *rakta dhatu*).
3	Clinical features	Weakness, fatigue, dyspnea on exertion, palpitations, pallor of the skin, mucous membrane, sclera, and older patients may develop angina and congestive cardiac failure. Long-standing chronic IDA causes epithelial tissue changes in some patients, such as changes occurring in nails—spoon-shaped nails, atrophic glossitis, angular stomatitis, and so on.	Diminution of complexion, strength, palpitation, and exhaustion Advanced symptoms such as tinnitus, loss of digestion, debility, malaise, giddiness, pallor of skin, fever, dyspnea, anorexia, irritability, aversion to cold, exertion on climbing stairs, cramps in shanks, pain in the waist, thighs, legs, and lassitude
4	Line of treatment	Correction of the disorder after ruling out the underlying cause by medical or surgical or preventive measure Correction of iron deficiency by oral therapy or parental therapy	Purification by emesis and purgation Pacification by metabolic correction using certain herbs like ginger Promotive by supplying nutrition

2.7.1.2 Management of Anemia through Ayurveda

Following is the line of management of anemia according to Ayurveda depending on the type, stage, severity of the condition and the dosha involvement.

- Purification (*shodhana*)—by emesis and purgation
- Pacification (*shamana*)—by metabolic correction
- Promotion (*poshana*)—by nourishment

The *pandu* patient is subjected to unction at first with *sneha*s (unctuous herbal clarified butter containing formulations) such as *panchagavya*, *maha tikta*, or *kalyanaka ghrita*, depending on his or her strength. He or she is then subjected to strong emesis and purgation. Pacifying the *dosha*s by metabolic correction then follows the promotive therapy (Sharma, 2001; Udupa, 2004).

2.7.1.3 Properties of Plants Used in Treating *Pandu*

The common properties of anemia drugs are as follows:

- Drugs used to treat anemia as per ayurveda possess bitter (*tikta*), pungent (*katu*), sweet (*madhura*) taste and light (*laghu*), dry (*ruksa*), penetrating (*tiksna*), and unctuous (*snigdha*) *guna*; they are hot in potency (*usna virya*); they possess hot postmetabolic activity (*katu vipaka*) and pacifies all three doshas (*tridoshahara*).
- They have effects such as the following: appetizer (*dipana*), digestive (*pacana*), strengthening (*balya*), rejuvenating (*rasayana*), good for heart (*hridya*), reducing edema (*sothahara*), anthelmintic (*krimighna*), reducing fever (*jvaraghna*), good against skin diseases (*kusthaghna*), purifying the blood (*raktasodhaka*) (Mishra, 2007).
- These drugs are commonly used in treating not just anemia (*pandu*) but also jaundice (*kamala*), edema (*sotha*), diseases of the duodenum (*grahani*), fever (*jvara*), spleen diseases (*pliha roga*), hemorrhoids (*arsas*), dyspnea (*svasa*), respiratory diseases (*kasa*), diseases of the heart (*hridroga*), and so on.

Formulations to manage anaemia are designed using these properties and actions (Chunekar, 2004; Mishra, 2007).

2.7.1.4 Role of Food in *Pandu*

Other than medicines, nutritionally rich foods and dietary supplies also play an important role in the management of anemia. A wide variety of grains, fruits, and vegetables are used to manage anaemia. Grains include old rice, wheat, barley, and ragi; pulses include green gram, Bengal gram, black gram, soybeans, and pepper; vegetables include tomato, ash gourd, snake gourd, and bitter gourd; green leafy vegetates include agase, spinach, drumstick, and *amaranthus*; fruits include papaya, pomegranate, sweet lemon, oranges, lemon, grapes, mango, *amla* (Indian gooseberry), and banana; and herbal supplements include ginger and pepper. Others such as buttermilk, milk, dates, dry grapes, almonds, groundnuts, eggs, chicken, and mutton are also good supplements. Among liquids, barley water, coconut water, buttermilk, lemon juice, sugar solution, *amla* and grape juice, *khadira* (Acacia catechu) water, and coriander water are said to be beneficial against anemia. Some of these formulations, which have one or two ingredients, can be easily made at home. Some of these herbs, vegetables, and fruits can be grown in home gardens, which can be a self-help way to balanced nutrition, for example, *dadima* (pomegranate—*Punica granatum*), *shigru* (drumstick—*Moringa pterygosperma*) leaves, *kaidarya* (curry leaves—*Murraya koenigii*), *punarnava* (*Boerhavia diffusa*), *agathi/agastya* (*Sesbania grandiflora* syn. *Agathi grandiflora*), and *matsyaksi* (*Alternanthera sessilis*) (Tripathi, 1998).

2.7.2 Use of Transdisciplinary Methodologies

The principles and practices of Ayurveda are based on epistemologies that are different from those of biomedicine. Hence the exercise of bridging the two knowledge systems has to be completed with caution, so the meaning is not lost or distorted. It is akin to comparing the whole and the part. Although the whole possesses the parts, it is not equal to the parts or even the sum of the parts. "Whole systems" clinical trial approaches and protocols are being adopted by groups researching Ayurveda across the world to compare the clinical efficacy of Ayurvedic and biomedical treatments (Furst *et al.*, 2011; Roy *et al.*, 2007; Witt *et al.*, 2013). This technique uses clinical observations and predetermined end points of treatment, without interfering with the physician's ways of diagnosis or line of treatment. Although the whole systems clinical trial approach is ideal, use of appropriate bioassays and preclinical models can also provide insights into and proof of Ayurvedic concepts.

The research laboratories at the Institute of Trans-Disciplinary Health Sciences and Technology (ITD-HST), Bangalore, India, are pioneering the process of bridging Ayurveda with biomedicine. One aspect that is being studied is to understand Ayurvedic concepts including *rasayana*, *agni*, and *srotas*, using in vitro and in vivo bioassay models. Using the fruit fly (*Drosophila melanogaster*) model, the ITD-HST demonstrated that pomegranate, a well-known *rasayana*, extends life span (*vyasthapana*) (Balasubramani *et al.*, 2014). Pomegranate also protects the fruit flies from infection and promotes fecundity and general activity.

Agni (metabolism) is one of the important factors involved in transformation of food to absorption and nourishment of body tissues. In biomedicine too, there is recognition of the process of catabolism, anabolism, and excretion of wastes, which has been succinctly recognized as absorption, distribution, metabolism, and excretion. Bioavailability of a material in the blood is crucial for its absorption, and there are in vitro models through which bioavailability of a substance can be tested. *Amla* (Indian gooseberry, *Phyllanthus emblica*) A common fruit that has been mentioned in Ayurveda in the management of anemia (*pandu*) because it is said to have *agni*-promoting property. Hence the effect of amla juice was tested on iron dialysability and uptake, using in vitro cell-free and cell-based models (Caco2 intestinal cells and HepG2 liver cell lines). *Amla* increased iron bioavailability more than threefold. Such substances are affordable dietary supplements that can be scientifically tested and promoted for anemia management (Venkatasubramanian *et al.*, 2014).

2.8 CONCLUSION: NEED FOR CONVERGENT APPROACHES TO NUTRITION AND HEALTH INTEGRATING TRADITIONAL KNOWLEDGE ABOUT FOODS

On the one hand, 50% of people from developing countries suffer from malnourishment and deficiency diseases, and on the other, chronic diseases contribute to 60% of the 56 million deaths in a year globally, with unhealthy diets being the major cause (WHO, 2002). Thus, societies are facing a double burden of diseases of the haves and the have-nots. Noncommunicable diseases, including diabetes, cardiovascular disease, hypertension, and obesity, are on the increase the world over. High stress levels compounded by an altered food culture (including consumption of fast, processed, fusion foods) and lack of physical exercise are important reasons cited for the growing metabolic syndrome, which is emerging as a health epidemic and a major public health concern (NIN, 2010). Inadequate, unwholesome, and imbalanced diets affect the health of an individual. What we eat, how much we eat, when we eat, and how we live has a huge impact on our health. En masse shifts in food intake and lifestyle can make personal health conditions such as obesity and diabetes a public health epidemic. Policies and practices in agriculture and trade, science and technology, environment and healthcare, education and public awareness, mass media promotions, and so on greatly influence trends in community behavior (Rayner *et al.*, 2006). For example, the European food policy to promote the intake of fruits and vegetables in schools has had a positive impact on reducing obesity among school children (de Sa and Lock, 2008). Thus, personal and public health are interdependent in contemporary times and can affect each other (Arah, 2009). Appropriate integration of traditional medical systems such as Ayurveda and biomedicine can provide the world with affordable preventive and promotive care solutions through self-management or home-based management. This approach will help achieve universal health coverage. It will also generate new knowledge providing new insights into the human body and mind and the dynamic interactions with Nature.

REFERENCES

Arah O.A., On the relationship between individual and population health. *Med. Healthc. Philos*. **2009**, 12, 235–244.

AYUSH. National Health Policy on AYUSH—2002. New Delhi, India, **2002**. (Accessed from http://indianmedicine.nic.in, on June 28, 2014).

Balasubramani S.P., Mohan J., Chatterjee A., Patnaik E., Kukkupuni S.K., Nongthomba U., Venkatasubramanian P. Pomegranate juice enhances healthy lifespan in *Drosophila melanogaster*: An exploratory study. *Front. Public Health*, **2014**, 2, 245.

Balasubramani S.P., Venkatasubramanian P., Kukkupuni S.K., Patwardhan B. Plant-based rasayana drugs from Ayurveda. *Chin. J. Integr. Med.* **2011**, 17(2), 88–94.

Chandogyopanishad. Gita Press, Gorakhpur, India, 742.

Chunekar K.C. (ed.) Bhavaprakasa Nighantu. Chaukhambha Visvabharati, Varanasi, India, **2004**, 41, 64, 110, 233.

De Sa J., Lock K. Will European agricultural policy for school fruit and vegetables improve public health? A review of school fruit and vegetable programmes. *Eur. J. Public Health* **2008**, 18(6), 558–568.

Dwarakanatha C. *Digestion and Metabolism in Ayurveda*. Chaukhambha Krishnadas Academy, Varanasi, India, **2003**, 5.

Furst D.E., Venkatraman M.M., Krishna Swamy B.G., McGann M., Booth-Laforce C., Manohar P.R., Sarin R., Mahapatra A., Kumar P.R.K. Well controlled, double-blind, placebo controlled trials of classical Ayurvedic treatment are possible in rheumatoid arthritis. *Ann. Rheum. Dis.* **2011**, 70(2), 392–393.

Jha G. (translator). The Chhandogya Upanishad and Sri Sankara's commentary, Part 2. VC Seshacharri, Madras, India, **1923**, 218.

Lakshmithathachar M.A., Alwar M.A., Shankar R. *Bhojanakutuhalam.* Institute of Ayurveda and Integrative Medicine, Bangalore, India, **2012**.

Mishra S.N. (ed.) *Bhaishajya Ratnavali.* Chaukhamba Surabharati Prakashan, Varanasi, India, **2007**, 375–388.

Mukherji M., Prasher B. Ayurgenomics: A new approach in personalized and preventive medicine. *Sci. Cult.* **2011**, 77(1/2), 10–17.

Murthy K.R.S. (ed.) *Vagbhata's Astanga Hrdayam, Vol. 1.* Chaukhambha Orientalia, Varanasi, India, **2001**, 53–57, 124, 179.

National Institute of Nutrition (NIN). *Dietary Guidelines for Indians*, ICMR Publications. New Delhi, India, **2010**.

Noble D. Claude Bernard, the first systems biologist, and the future of physiology. *Exp. Physiol.* **2008**, 93(1), 16–26.

O'Hara A.M., Shanahan F. The gut flora as a forgotten organ. *EMBO Reports*, **2007**, 7(7), 389–340.

Patwardhan B., Bodeker G. Ayurvedic genomics: Establishing a genetic basis for mind–body typologies. *J. Altern. Complement. Med.* **2008**, 14(5), 571–576.

Patwardhan B., Vaidya A.D.B., Chorghade M. Ayurveda and natural products drug discovery. *Curr. Sci.* **2004**, 86(6), 789–799.

Prabhupada B.S. *Bhagavadgita Yatharupa.* ISKCON, Bangalore, India, **2002**, 736–738.

Rayner G., Hakes C., Lang T., Bell W. Globalisation for health—Trade liberalization and the diet transition: a public health response. *Health Promot. Int.* **2007**, 21, S1.

Reddy K.R.C. *Bhaishajya Kalpana Vijnanam.* Chaukhambha Sanskrit Bhavan, Varanasi, India, **2005**, 9–13, 111, 114.

Robinson J. Breeding the nutrition out of our food. *The New York Times*, **2013**.

Roy C.R., Thatte U., Liu J. Clinical trial methodology. In *Traditional, Complementary and Alternative Medicine: Policy and Public Health Perspectives.* Bodeker G., Burford G. (eds.), Oxford University, Oxford, **2007**, 389–340.

Sastry J.L.N. *Dravyaguna Vijnana, Part 1.* Chaukhambha Orientalia, Varanasi, India, **2009**, 1–2.

Sharma P.V. (ed.) *Astanga Samgraha of Vagbhata, Vol. 1.* Chaukhambha Orientalia, Varanasi, India, **2002**, 1, 10, 58, 165, 258–259, 319, 335, 367.

Sharma P.V. (ed.) *Charaka Samhita, Vol. 1.* Chaukhambha Orientalia, Varanasi, India, **2001**, 5–9, 75, 189, 228, 304–308, 375–376, 440.

Sharma P.V. (ed.) *Susrutha Samhitha, Vol. 1.* Chaukhambha Visvabharati, Varanasi, India, **2004**, 173, 462.

Sharma P.V. *Dravyaguna Vijnana, Vol. 1.* Chaukhambha Bharati Academy, Varanasi, India, **2006**, 7, 28.

Stipp D. How intermittent fasting might help you live a longer and healthier life. *Sci. Am.*, December **2012**. (Accessed from http://www.scientificamerican.com/article/how-intermittent-fasting-might-help-you-live-longer-healthier-life/, on July 1, 2014).

Taranatha. *Vachaspathyam, Vol. 2.* Choukhambha Sanskrit Series, Varanasi, India, **1990**, 1563.

Thathachar M.A.L., Alwar M.A., Shankar R. *Kshemakutuhalam.* Institute of Ayurveda and Integrative Medicine, Bangalore, India, **2009**.

Tripathi B. *Pathyapathyanirnayah.* Chaukhambha Sanskrit Pratishtan, Delhi, India, **1998**, 2.

Udupa M.H. *Comprahensive Kayachikitsa and Principles of Ayurveda.* Laveena Publications, Bangalore, India, **2004**, 797.

Valiathan M.S., Thatte U. Ayurveda: The time to experiment. *Int. J. Ayurveda Res.* **2010**, 1(1), 3.

Venkatasubramanian P., Kaul I.B., Varghese R. K., Koyyala S., Shivakumar A. Amla (*Phyllanthus emblica* L.) enhances iron dialysability and uptake in in vitro models. *Curr. Sc.* 2014, 107, 11, 1859–1866.

Venkatasubramanian P. *Annual Report of the Institute of Ayurveda & Integrative Medicine.* FRLHT, Bangalore, India, **2013**.

Witt C.M., Michalsen A., Roll S., Morandi A., Gupta S., Rosenberg M., Kronpaß L., Stapelfeldt E., Hissar S., Müller M., Kessler C. Comparative effectiveness of a complex Ayurvedic treatment and conventional standard care in osteoarthritis of the knee—Study protocol for a randomized controlled trial. *Trials*, **2013**, 14, 149.

World Health Organisation (WHO). *The World Health Report 2002: Reducing Risks, Promoting Healthy Life, Geneva*, Switzerland, **2002**.

3 The Quest for Longevity in Ayurveda

Dietetic and Lifestyle Advice

Karin Pirc and Indukuri Kalyan Chakravarthy

CONTENTS

3.1 HOLISTIC NUTRITION IN AYURVEDA

Ayurveda contains a treasure house of possible approaches to help fight illnesses by restoring the balance of mind and body and uses nutrition to obtain positive and specific effects on the body and mind, including many aspects that are not even considered in modern dietary therapies. Examining the roots of the word *ahara* gives us a clue to the depth and scope of this science: "to get," "to aim at," "to cause."

The science of nutrition as it is practiced in Western medicine analyzes primarily the content of food, which is evaluated as either *good* or *not good*. It emphasizes that the food needs to contain a significant amount of crucial elements, such as vitamins and minerals, and should also be a rich source of fiber. The taste of the food is considered more or less inconsequential and to be of no therapeutic value.

The assumption is that food has the same effect on every human being, and as a consequence no consideration is given to differences between individuals. In the comprehensive science of *ahara*, the Ayurvedic science of nutrition, a completely different approach is taken. In this body of knowledge, you will find different advice given to different individuals. In addition, a person's current state of health, age, and strength of digestion, or *digestive power*, are taken into consideration; on the basis of these factors, an appropriate nutritional plan is laid out specifically for that person. Of course, the nutritional ingredients of food are of great importance, but the right choice and preparation of food are just as important. Moreover, the science of *ahara* offers additional insights into the subtle effects of each food on the human physiology. Food is classified according to its effect on the three *dosha*s. This type of classification ensures use of all elements of the food to precisely balance the *dosha*s of an individual and to help in preventing diseases and supporting the healing process. Another essential factor that is considered is *agni*, the digestive power, which can vary tremendously from person to person. A strong *agni* allows the mind and body to produce useful elements following the digestion of the food; it also ensures that the body cells readily absorb food constituents.

Only if all these different aspects work in complete harmony is it possible for a person's physiology to assimilate the nutrients completely. When the *agni* is totally strong and healthy, and the nutrients are completely assimilated, a highly refined substance that is deeply nourishing to both body and mind is created.

Those who eat right don't need medicine. For those who eat wrong, medicine is of no use. This Ayurvedic slogan underlines the principle that Ayurveda does not make any distinction between the value of food and the healing effect of a medicinal herb or plant. With Ayurvedic knowledge, one is able to precisely analyze each nutritional element on the basis of specific criteria. If one examines each healing plant with regard to its effect—as one would evaluate all elements of one's daily food and beverages—food can also serve as a means to promote healing.

3.2 NUTRIENTS AS CARRIERS OF INFORMATION

The Vedic Science of Maharishi Mahesh Yogi (1997) provides some insights into the interconnection of the fundamental structuring dynamics of life and the three *dosha*s, in other words, the interconnection between the unmanifest levels of life and the manifest levels of life. From the unified level of life—physicists call it the unified field; the Maharishi's Vedic Science calls it pure, silent, unbounded consciousness—the whole variety of creation comes out. For those whose consciousness is clear (*rishi*s, or "seers"), the internal fluctuations of the more subtle levels of material expressions of life are easily cognized. The most subtle, unmanifest impulses or vibrations of the *unified field of natural law* are identified by the Maharishi's Vedic Science as the knower (*rishi*), the process of knowing (*devata*), and the known (*chhandas*). These fundamental structuring dynamics interact within themselves to express the whole variety of creation. All aspects of physical reality sequentially unfold from the underlying unity (*samhita*) of the unmanifest impulses, including the three *dosha*s (*vata*, *pitta*, and *kapha*).

The organizing and structuring power of the three *dosha*s is found operating throughout the whole universe. On the one hand, the *dosha*s govern the harmonious interaction of all mental and physical processes of the human being, and on the other hand, they are found in all our food sources. Because every grain and every kind of vegetable is an expression of a combination of certain *dosha*s, each meal provides the human body with their influence. This has an effect on the *dosha*s of the mind, as well as the body. Since time immemorial, Ayurvedic practices have made use of this knowledge. Therefore, we see that our daily meals build and nourish our body—when correctly used, the food we eat can also help to heal it.

There are many variables on either side—the wide range of combination of the *dosha*s in different foods and the wide range of conditions in each person's physiology. Knowledge of Ayurveda can assist in the cure of various diseases by specific means, because of the highly differentiated possibilities of effects of different foods and food combinations.

3.3 BALANCING THE *DOSHAS*

All complaints of body and mind in Ayurveda are first considered from the viewpoint of the underlying, subtle aspects of the specific symptoms seen on the surface. This is why every disorder is analyzed precisely with respect to the imbalance of *dosha*s and sub-*dosha*s. In order that a person may regain psychological and physiological balance, Ayurvedic therapy has the following approach: increase the influence of a specific *dosha* that the body is lacking, or reduce the influence of a *dosha* that exists in excess (the excess disrupts the normal functioning of the mind–body).

Generally, the approach is to treat the person with the addition of the quality that is absent. For example, if a person feels cold (an indication of an excess of *vata dosha*), the recommendation is to have warm foods and beverages (*pitta* strengthening). If the person is nervous (*vata* aggravation), Ayurveda balances the situation with a recommendation for more silence, such as deep sleep before midnight or meditation (strengthening of *kapha dosha*). Ayurvedic therapies follow this basic principle: the healing power of food in prevention of disease and promotion of healing is likewise accomplished with the building up of *dosha*s that are weak and the reduction of *dosha*s that exist in excess.

3.4 *RASA*—THE "TASTE"

We know that Ayurveda identifies six different tastes: sweet (*madhura*), sour (*amla*), salty (*lavana*), pungent (*tikshna*), bitter (*tikta*), and astringent (*kashaya*). Because each of these tastes either strengthens or reduces a *dosha*, the taste of food itself is an important instrument in balancing the *dosha*s or disrupting their proper functioning.

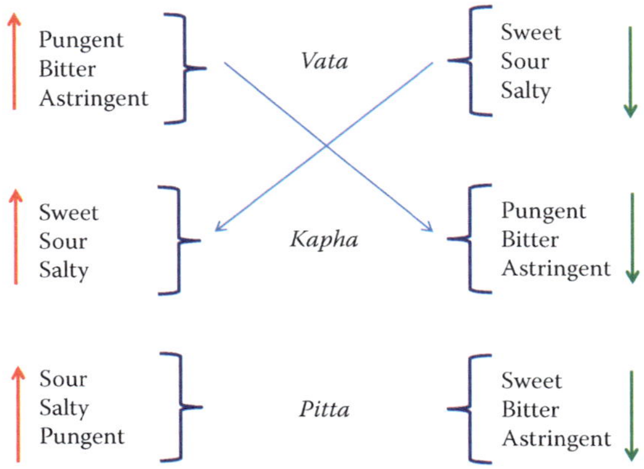

FIGURE 3.1 The influence of the six tastes on the three *dosha*s.

This influence is completely independent of the positive healing effect of any particular nutritional food source (Figure 3.1).

Of course, most foods do not contain just one flavor; they are a combination of several flavors. If you want to influence some imbalance in the body, you could use the various tastes in food to reestablish balance.

A simple example from medical practice is the common cold. A cold indicates a pronounced dysfunction of *kapha dosha*, because of the production of mucus. A proven Ayurvedic remedy to alleviate a cold is ginger tea or hot water with a mixture of powdered ginger, cloves, and black pepper. Here, the hot taste is part of the Ayurvedic therapy because it effectively balances *kapha*. It takes away the heaviness from the head, it is anti-inflammatory, and it reduces the production of mucus in the respiratory system.

The knowledge about taste is also used for prevention of disorders in Ayurvedic therapy. To keep the balance of the *dosha*s, all six flavors should be present in a meal, if at all possible. For instance, if you serve the following items at a meal, all six tastes would be available: some sweet fruit, a little bit of lemon, some salt, some spicy herbs, some bitter herbs, salad or vegetables, and some astringent beans or legumes.

3.5 *GUNAS*—THE QUALITIES OF NUTRITION

Ayurvedic experts also consider the *guna*s (fundamental qualities) that are found in various food items and remedies. In the chart, you will see a list of *guna*s (*Charaka Samhita*), which, you will note, exist in pairs of opposite values.

Let us go back to the example of the acute infection of the respiratory tract. *Kapha* is cold by nature. It also displays stability (*sthira*), heaviness (*guru*), and viscousness (*picchila*). The ginger water or the spiced tea works as a remedy not only because of its hot flavor, but because it reduces the cold *guna* (*shita*) found in *kapha*. The heat of the beverage also provides the following qualities (*guna*s): mobile (*sara*), clear, nonslimy (*vishada*), and light (*laghu*)—all opposite *guna*s of *kapha* (Table 3.1).

TABLE 3.1
***Gunas*—The Fundamental Qualities in Ayurveda**

1	Heavy (*guru*)	Light (*laghu*)
2	Cold (*shita*)	Hot (*ushna*)
3	Oily (*snigda*)	Dry (*ruskha*)
4	Slow (*mandala*)	Quick, hot (*tikshna*)
5	Stable (*sthira*)	Mobile (*sara*)
6	Soft (*mridu*)	Hard (*kathina*)
7	Clear, nonslimy (*vishada*)	Viscous, slimy (*picchila*)
8	Rough (*khara*)	Smooth (*sh/akshna*)
9	Gross (*sthula*)	Subtle (*sukshma*)
10	Subtle (*sukshma*)	Fluid (*drava*)

3.5.1 *Virya*, *Vipaka*, and *Prabhava*

Virya (literarily "power"), the potency of a nutritional item, is evaluated according to eight different qualities: hot, cold, light, heavy, oily, nonoily, dull, and hot. In this case, the Ayurvedic expert looks at the effect on the human metabolism, rather than considering the nature of the item itself, as it is done with the *guna*s. For instance, *ghee* is an ideal remedy for *vata* or *pitta* excess when you consider it according to the *guna* qualities, because it is heavy, oily, and soft. It therefore balances the lightness of *vata* and *pitta*. Its sweet *rasa* (taste) pacifies both of them. Beyond these considerations, when it comes to its *virya*, its effect within the body–mind system, it is cooling and—apart from the pacifying effect—it can balance *pitta* aggravations, such as burning and heat.

Vipaka (literally "to take apart" and "to cook") describes the delayed effect within the body, which is released through digestion—independent from the original taste (*rasa*). Let us consider ginger: Its *rasa* is quite hot, but it turns sweet in the process of digestion, which everybody can experience by noticing its taste in the mouth after chewing it for a longer time.

By this process, ginger turns into an herb that perfectly activates *agni*, the power of digestion. Because of its sweet *vipaka*, it does not cause *pitta* disorders—despite its heating influence. This is in marked contrast to chilli and white pepper that are very hot by *rasa* and also stay hot during their digestion (*vipaka*). Therefore, it is likely that they will cause *pitta* disorders.

Prabhava (the "specific quality") is the specific healing effect of a food item or a medicinal plant that cannot be explained by *rasa* (taste), its *guna*s (qualities), *virya* (potency), or *vipaka* (delayed effect).

3.5.2 *Sattva*, *Rajas*, and *Tamas*—Mental *Doshas* in Food

For several thousand years, Ayurvedic science has known about the deeper connections that explain how our food influences our positive and negative emotions.

The *doshas*—*vata*, *pitta*, and *kapha*—correspond to an even more subtle level to the mental *doshas* of *sattva*, *rajas*, and *tamas*, and these qualities are also expressed in various food items.

Sattva: *Sattvic* food strengthens the positive mental capacities of a human being, such as his power to enjoy, to be humble, to be friendly, and to have broad comprehension. It consists of fresh food with soft, cooling *virya*. It is sweet, light, oily, and fresh, as in the case of cooked rice or wheat, fresh fruits, cow milk, *ghee*, and so on.

Rajas: Food that increases *rajas*, such as spicy or salty food, tends to produce unhappiness, worries, and illnesses. Food that is too hot or too dry also has a *rajasic* quality. *Rajasic* food has a taste that is too intense—such as chilli, aged cheese, garlic, meat, fish, and alcohol.

Tamas: *Tamasic* food is heavy, cold, and dry. Spoiled, reheated, or stale dishes increase *tamas*. It is tasteless, spoiled, or rotten. It increases the body's decay and makes the mind dull, slack, and lazy.

Food in Maharishi Ayurveda is used not only as a healing remedy for the body, but also directly as a means to balance mental qualities. The latest scientific research has given credence to this ancient wisdom that has been passed down through the ages. Scientists have found serotonin, the hormone that indicates happiness; melatonin, the hormone that promotes good sleep and that is also a powerful antioxidant; and tryptamine in significant quantities in plants. Until recently, it has been believed that these substances were only synthesized in animals (Badria, 2002).

It is interesting to note that the highest melatonin amounts were found in rice (Hattori *et al.*, 1995), the food that Ayurveda asserts is the most sattvic, vitalizing food item that can harmonize our body functions and positively influence our mental disposition.

3.6 COMBINING AND PREPARING FOOD

Of course, in most meals that people prepare, there is a combination of foods that increase or decrease the effects on the *doshas*. However, it is important to note that in addition to this, the different ways of preparing food can intensify or balance the effects on the *doshas*.

Ayurveda points out that there is a wide range of effects, depending on how you prepare the food. You could eat a raw carrot, boil it in water, or heat it with *ghee*. You could squeeze it and drink the fresh juice. You could drink it with or without any spice. You could have it cold or warm.

Some scientific studies showed the value of the tried-and-true Ayurvedic recommendations. If, for instance, one group was given cold, freshly squeezed juice from raw carrots and a control group the same juice boiled for a short time, after 1 hour the latter had a significantly higher absorption of carotenoids than the former that drank the cold, raw juice. Ayurveda knows that the cold juice increases *vata* and *kapha* and decreases *pitta*, whereas the hot juice balances *vata* and *kapha* and strengthens *pitta*—one of the reasons why the body can make use of the vitamin content more easily.

In the following list, you will find some recommendations for a healthy diet in accordance with maintaining the balance of the three *dosha*s and the individual's power of digestion:

Prakriti: Selection of food with regard to its ease of digestion (light/easy or heavy/difficult) and its targeted effect on the balancing of the *dosha*s.
Karana: To develop other qualities of the dishes through cooking and flavoring. Better digestion is guaranteed through cooking the food. By means of appropriate combinations, the right flavor of a meal is intensified.
Samyoga: Through the right combination of various food items, the negative elements are minimized and the positive ones are strengthened.
Rashi: The right amount of food differs depending upon the individuals and their digestive power. *Rashi* also takes into account the right proportion of the various food items to each other.
Desha: Food that is cultivated in the country where one lives is normally more easily digested than imported foods.
Kala: Food should be eaten during its normal growing season.
Upayoga sanstha: Regular intake of meals is considered here as well as, respecting one's hunger level and noticing of the point of satiety—which means only eating after the previous meal has been completely digested.
Upayokta: Subjective tolerance that is discovered through individual experience.

If you want to learn about *ahara*, the Ayurvedic science of nutrition, you will find many recommendations and rules for each of these criteria. It will take time to understand and deal with them correctly. One should not forget that the *rishi*s who experienced these laws in detail and passed down these instructions to younger generations could do so only because of their very refined state of consciousness. Even today, every human being has the experience of using his or her "inner intelligence" or "inner voice," which is considered an important tool for providing the right direction for day-to-day living.

Again and again, it is a great joy to see that people who have not paid much attention to their body signals in the past have begun to become enthusiastic about Ayurveda even after hearing a brief explanation. They intuitively realize that this kind of nutrition is good for them. After a short period of practice, they learn more and more about what is good for them and what is not. At the same time, one should still respect one's own impulses (even if they differ from the Ayurvedic recommendation), because these impulses are an expression of what the body and mind need under certain conditions to reestablish his or her own balance. Everybody knows firsthand that sometimes desires that are not life-supporting crop up due to old, wrong habits or due to an existing disorder. That is why it is very helpful to know as many details as possible about Ayurvedic cuisine. That way, we can give our body and mind the possibility of choosing those food items that balance and ultimately satisfy them.

3.6.1 Rules for Nutrition for Daily Life

A good first step is to follow these "golden rules" that can be accomplished without a huge amount of background knowledge. It only takes a short time to see positive results in one's body–mind system and improved feelings of well-being in day-to-day life.

- If possible, eat meals regularly, that is, at the same time each day.
- Lunch should be the main meal of the day.
- The food should be balanced and should contain all six tastes. The individual constitution and the power of digestion should be taken into account when preparing the amount and the composition of the particular dish.
- Eat while sitting in pleasant surroundings and in silence. Avoid reading, watching TV, as well as disagreeable discussions, while eating.
- Chew well and fill your stomach only three quarters full. Be aware of your point of satiety.
- Avoid frozen food or drinks, which decrease the digestive power.
- Take hot water in sips to increase *agni.* (Milk does not go well with vegetables, fruits, or salty dishes.)
- Sit quietly for some minutes after the meal and before you start working again.
- Avoid eating before the previous meal has been digested. Wait until a distinct feeling of hunger comes. Allow 3–6 hours between meals—according to how you feel.
- Between meals, sip a little bit of hot water because it increases the digestive power and purifies your body.
- Dinner should be taken early and should be light. Avoid meat and fish and also sour milk products such as hard cheese, cream cheese, or yogurt.

Each of these points takes into account the digestive power of each individual at different times of the day. All these recommendations help strengthen *agni.*

3.7 *AGNI*—THE INDIVIDUAL DIGESTIVE POWER

In our body, the *devata* value is expressed as a transformational value, as the *pitta dosha*, directing our metabolism and governing the multiple transformational processes in our physiology. The first and therefore very basic and most powerful impulse of *pitta* in our body is *agni*, the digestive fire.

In the Vedic literature, *agni*, the "bright fire," plays an important role. The Veda begins with the word *agni.* It is the first impulse from which the whole Veda and the whole Vedic literature unfold themselves step by step according to their own underlying structure. From this first, very subtle impulse, the whole creation comes out. Many people are familiar with the phrase: "And God said: 'Let there be light!' And there was light!" And so from this first, subtle vibration, our inner intelligence creates and maintains our body day by day.

Within the human body, this "cosmic fire" that exists throughout creation is expressed as the digestive energy that functions throughout our mind–body system. That is why *agni* not only governs the digestive process within the intestine, but also regulates the whole metabolism with regard to the transformation of different food elements inside the *dhatu*s (the body tissues) as well as in each cell through enzymatic transformation processes. If the digestive energy works powerfully and without any disturbance, it has positive effects on the whole organism. A human being with good assimilation and metabolic strength is able to digest food very well and to ward off germs from outside with great efficiency. One is strong, healthy, and full of energy, and remains vital until old age. If *agni* is working powerfully in the physiology, the "combustion" process is complete; therefore, it automatically produces fewer metabolic waste products.

3.8 *AMA*—TOXINS THAT DISRUPT OUR HEALTH

Inseparably connected with *ahara*, the Ayurvedic science of nutrition, it is the knowledge of detoxification. Only the body whose cells are almost free from metabolic deposits is capable of absorbing and using new substances from any given food in an ideal way.

Ama is the collective term in Ayurveda for all substances that burden the body. When *ama* accumulates, it prevents the unrestricted flow of one's inner intelligence, no matter where it comes from. The literal translation of *ama* is "undigested."

Generally, Ayurveda distinguishes three kinds of *ama*:

- Undigested food (endotoxins)
- Toxins that are absorbed from outside (ektotoxins)
- Mental *ama*

3.8.1 Undigested Food

Most of the *ama* in our bodies is generated by ourselves—especially when the individual's digestive energy (*agni*) is not powerful enough to completely digest any given food. The whole human digestive tract contains enzymes, chemical substances that are able to break down food elements into tiny little molecules. If the cooperation between the enzymes from the saliva, the gastric juice, the pancreas, and the intestine is out of balance, the digestion processes cannot proceed correctly. One consequence is as follows: instead of the food being broken down into small constituent parts that can be used by the cells of various tissues, larger parts are left behind. Because the body cannot use them any more, it leaves them deposited somewhere.

3.8.2 Toxins

Other harmful substances that get into our body are also a kind of *ama*: pesticides, herbicides, heavy metal or other environmental toxins, chemical preservatives, or artificial colors that are inherently dangerous to every human body, because they cannot be broken down by the body.

3.8.3 Mental *Ama*

It is interesting to note that according to an old Ayurvedic concept, metabolic toxins can also have mental origin. This plays an important role for those of us who are rushed and stressed in today's busy world. For the *rishis*, the body and mind were an inseparable unit. They knew that each impression changes the body chemistry—each impression has *to be digested*. Modern medicine has known about *mental ama* for decades. They have observed the effects of too much hydrochloric acid in the stomach and the stress hormone adrenaline. They have seen mental *ama* in the form of various imbalanced neuropeptides or in the production of free radicals when someone is under emotional stress. If a person is worrying all the time or if he or she gets angry or is furious, he or she does in fact produce corresponding molecules that are a burden on his or her bodily system (Goldstein and Kopin, 2007). Moreover, negative emotions reduce the digestive power and this causes even more unwanted digestive deposits (Naliboff *et al*., 2004).

Due to a combination of these three basic sources for the production of *ama*, a layer of waste products gets created in the body. This hinders the frictionless transport of substances and influences the free flow of inner intelligence. In this way, deposits end up right where the body has its weak points. These deposited useless molecules are like grains of sand in an engine (like throwing a monkey wrench into the works) because they block other chemical reactions within the cells or they destabilize them in some way.

3.8.4 Avoiding and Reducing *Ama*

Ama with other factors plays a causative role in the development of nearly every disease. Ayurveda takes into account the choice of food, how foods are combined, and how to eat meals properly. It is only logical that Ayurveda pays a lot of attention to avoid the production of *ama* on one side and to systematically help to decrease the *ama* that is already there on the other side.

One of these simple recommendations is the regular intake of hot water—water that has been boiled for about 5–10 minutes. Due to its temperature, it enhances *agni* directly. That is why the food items can be processed better and absorbed while eating. When sipped between meals, moreover, it can efficiently support the elimination of water-soluble toxins from the *dhatus* (tissues). This is very important because only tissues that are not blocked by food deposits can absorb the new molecules that are important for the metabolic processes of the *dhatus* and the cells. Only then they can work properly.

To reduce *ama* on a regular basis, Maharishi Charaka (one of the most famous Ayurveda *rishis*) recommends having a liquid diet for one day during the week. All sorts of freshly pressed fruit or vegetable juices may be consumed, as well as freshly cooked grain or vegetable soups. It is important to eat a sufficient amount of those so that the fasting person does not suffer from a *vata* imbalance, such as weakness, freezing, or restlessness. Sipping hot water between meals enhances *agni* and supports the elimination of toxins. Through this intentional "easing up" on the digestive system, *pitta* is released, which the body can then use to mobilize metabolic deposits and to eliminate them.

Of course, there are many other recommendations to reduce the amount of food and therefore increase the elimination capacity of the body. Maharishi Ayurveda also uses classical procedures that have been passed down through the ages. For instance, Maharishi Panchakarma therapy is used to mobilize and eliminate *ama* systematically.

3.9 THREE PILLARS FOR THE HEALTH OF BODY AND MIND

The three main aspects of the Ayurvedic nutritional science, *ahara*, are as follows:

Shamana (literarily "to calm down" and "to smooth"). All measures that balance the various *doshas* are called *shamana*. The recommendation for food to balance the *doshas* is based on the doshic constitution of a person, as well as taking into account the imbalances.

Langhanam (literarily "to relieve"). To gain a healthy body, both the body and the soul need regular "time-outs," because even if one lives a healthy lifestyle, metabolic toxins are produced. With regard to *langhanam*, one takes into account all measures that reduce the already existing toxins in the body by relieving the digestive tract, for example, by means of fasting, taking liquid food, taking less food, and allowing a longer time between meals. By these means, the self-healing properties of the body are strengthened so that the purification of the cells is supported.

Pachana (literarily "to cook" and "to digest"). This is the third main pillar of nourishment. Here one considers methods that activate *agni*: for instance, herbal teas, spicier and hotter meals, and all herbs that stimulate the digestive power in such a way that the elimination of toxins is supported.

3.10 *OJAS*—THE SUBSTANCE OF CONSCIOUSNESS

Shamana, *langhanam*, and *pachana* are important to nourish the physiology in an ideal way. The more the body is freed from metabolic toxins, the better the inner intelligence can flow within the whole system and the more *ojas* is produced by the mind–body system.

Ojas is the most refined end product of digestion. When the food molecules have undergone all transformational processes in the body through all the *dhatus* and in this way have been refined to the maximum, *ojas* is the result.

This "substance" has such a refined structure that it is both material and consciousness (nonmaterial) at the same time. To date, scientists have not found any molecule that identifies *ojas*, although its qualities are described in great detail in the classical Ayurvedic texts (Priyavrat Sharma). It is possible to perceive it clearly in a person—it is reflected by radiant skin and bright eyes. *Ojas* is important because it connects the body and the mind at every point in the physiology, and it makes sure that the structuring and healing energies of the wholeness (*Samhita*—the unified field) of the person permeate the body. That is why it is a crucial element for any healing process.

Effects of Ojas:

- It brings about vitality and strength.
- It enhances immunity (*ojas* protects from bacteria, viruses, and allergens).
- It keeps digestion and metabolism running.
- It builds body tissues and regulates transformation.
- It maintains youthfulness and slows down the aging process.

- It is responsible for the balance of the three doshas.
- It prevents psychological and psychosomatic disorders.
- It improves concentration (focus), positive thinking, and a proper perspective on problems.
- It enables the experience of bliss.

These are only some of the positive effects that come about when the fundamental source of energy and intelligence—the qualities of the unified value of consciousness, *samhita*—expresses itself without any impediments in the human physiology. The ultimate aim of the Maharishi Vedic Approach to Health is to help to refine the physical system by eliminating all impurities that have been created by poor nutrition, weak digestion, and improper preparation of food. This is so important because *ama* is generally responsible for the separation of consciousness and matter. It overshadows the frictionless flow of natural laws that are responsible for all physical and mental homeostatic mechanisms and the self-healing processes. However, good assimilation of nutritional elements is the basis for a healthy body and mind, which naturally produces an abundance of *ojas* and a free flow of inner intelligence.

3.11 BALANCED DIET

The term "balanced" simply means that a diet meets your nutritional needs while not providing too much of any nutrient. A balanced diet is to get the right type and amounts of foods and drinks to supply nutrition and energy to maintain body cells, tissues, and organs, and to support normal growth and development.

Though it sounds very easy to maintain a balanced diet in our daily life, it is not so. Every day there is a new emerging theory in diet that can create confusion, whereas the ancient wisdom of India, Ayurveda, has well-defined concepts of diet, which aim at an individual's wellness. It should be noted that each person's needs are different because they depend on a universal average diet chart; thus, it is difficult to achieve a balanced diet for each.

Balanced diet, according to Ayurveda, is what is in proper quantity and quality according to each individual. The quantity of food depends on the capacity of the individual's *agni*. It is *agni* that determines the quantity of food one should take to be properly digested. The amount of food that does not disturb the equilibrium of the internal physiology and is digested and metabolized in proper time is to be regarded as the proper quantity of food (*Agnivesa*, *Charaka Samhita*). *Sarvagraha* refers to the whole quantity of food and *parigraha* refers to the quantity of each item.

Sarvagraha: the whole quantity. Ayurvedic thumb rule of quantity is that one-third of the stomach should be filled with solids and one-third with liquids, and the rest should be left empty. If you can take 300 g of pasta for your lunch, take only 100 g of it and take 150 ml of liquids and finish your lunch. Indeed it is a simple calculation and individual specific.

Therefore, what should be included in a diet and how much? This is the most important aspect of food. There is no clear picture available, as the diet is different in different regions and varies because every person's need is different. It can be difficult to come to a conclusion. Again Ayurveda comes to the rescue with its unique, yet universal, approach.

Charaka, the oracle of Ayurveda, has clearly mentioned in his *magnum opus—Charaka Samhita*—that a meal should contain food items with all the six tastes. This alone constitutes the balanced diet of Ayurveda. Entering into details, the six tastes are *madhura* (sweet), *amla* (sour), *lavana* (saline), *katu* (pungent), *tikta* (bitter), and *kashaya* (astringent). The five elements (known as *Panchamahabhutas*) are giving to each taste its chemical and physical properties.

Sarvarasaabhyaso Balakaranaam Srestham means that a regular intake of all the six tastes in a diet is the best way to promote *bala* or health. Every food item, fruits, vegetables, cereals, or else that we consume has a taste and that taste determines its qualities. The foods that should be taken regularly are listed by *Charaka*: rice, red rice, barley, green gram, rock salt, gooseberry, rain water, cow's milk, cow's *ghee* (clarified butter), and multifloral wild honey. The list, a guideline, is simply a balanced diet (Table 3.2).

Ayurveda classified food into different categories (*varga*):

1. Cereals (*sookadhanya varga*)
2. Pulses (*sameedhanya varga*)
3. Meat (*mamsa varga*)
4. Vegetables (*saaka varga*)
5. Fruits (*phala varga*)
6. Vegetable spices (*harita varga*)
7. Fermented beverages (*madya varga*)
8. Water (*jala varga*)
9. Milk and milk products (*gorasa varga*)
10. Sugarcane and related products (*ikshu varga*)
11. Cooked food (*krutanna varga*)

Each *varga* or category includes many food items based on their qualities. A diet should include all the above categories according to the season, habitat, and nature of the person. According to Ayurveda, we should first eat something sweet, followed by sour, salt, pungent, and bitter taste. Finally, we should eat something astringent.

TABLE 3.2
Food for Regular Intake and Its Assessment

Food Item	Group	Class
Rice and red rice	Cereals	Carbohydrates
Green gram	Pulses	Proteins
Meat	Meat and poultry	Proteins
Ghee—clarified butter	Fat	Lipids
Gooseberry	Fruit	Vitamin
Milk	Liquids	Minerals
Rock salt	Salts	Electrolytes

Therefore, we can start our meal with a fruit such as pomegranate or gooseberry, followed by a soup of lentils or green gram spiced with garlic and ginger. The main meal can follow, ending with buttermilk, lassi, or something equivalent with the astringent taste.

The time of the meal also plays a vital role for our health. Ayurveda suggests having two meals a day. A meal in the morning and a meal at the end of the day is the perfect choice for a healthy life. However, it should be noted that without having appetite one should not take meals.

The quality of the food is determined by its nature (*prakruti*), place of growth, time of collection, storage method, and method of preparation. Ayurveda has dealt with each and every aspect in detail. The *prakruthi* of the food is essential and basically can be of two types: heavy (*guru*) or light (*laghu*). Based on these simple principles, the quantity should be adjusted.

Heavy or light property of food is related to the time taken for its digestion. For example, though green gram and black gram are pulses, the former is light or easy to digest and the latter is heavy or difficult to digest. Thus, green gram is considered best and is included in the list of daily consumable foods. During winter, heavy foods such as cheese, butter, and meat can be consumed, and during summer, light or easily digestible foods should be used.

In the same way, it is stated that though some of the food items have similar tastes, their properties can be different. For example, rice and apple fall under *madhura rasa varga*, that is, class of foods having sweet taste, and thus they should increase body weight. However, often we see people on fruit diet to lose body weight, whereas people on regular diet do not. Why some foods having the same taste act in different ways? The reason is the heavy or light nature of the food item. We know that sweet taste is formed by two elements: earth (*prthvi*) and water (*jala*). The combination of these two elements is the key in deciding how much heavy is the item among the sweet group. Rice is having more *prthvi* or solid *bhoota* and less *jala* or liquid *bhoota*, whereas apple is having more *jala* or liquid and less *prthvi* or solid *bhoota*. More solid *bhoota* induces more heaviness and, because rice is more solid, it causes more heaviness or *guru* property to the food.

A balanced diet of an individual depends on the season and place. A person should always give a great importance to his or her food (*Agnivesa*, *Charaka Samhita*, Sutrasthana, 7, P-54). Food is the basis of life, health, as well as diseases (*Agnivesa*, *Charaka Samhita*, Sutrasthana, 28, P-181). An individual should always indulge in good food and regimen, and keep away from worldly pleasures as to attain a peaceful and disease-free life (*AS. Hrudayam*, Sutrasthana 4) (Table 3.3).

There are certain food items restricted in Ayurveda for daily use (Sharma and Bhagvandash, 2009). These food items should not be used on a regular basis. They are as follows:

1. Dried meat
2. Dried leafy vegetables
3. Lotus stem

4. Meat of animals that are diseased, not well fed
5. Spoiled milk
6. Pork
7. Beef
8. Fish
9. Curd/yogurt
10. Black gram

Also it should be noted that food prepared from flour, such as pizza, burger, and *paratha*, should not be used on a regular basis and should be taken in medium or low quantity (Table 3.4).

TABLE 3.3
Ayurvedic Food List—Best and Worst

Group	Best	Worst
Saali dhanya—cereals	Red rice (hand pound, unrefined)	*Yavaka* (unripe barley)
Sameedhanya—pulses	Green gram	Black gram
Lavana—salts	Rock salt	*Ooshara* (from earth) salt
Saaka—leafy vegetable	*Jeevanti* (*Leptadenia reticulata*)	*Sarshapa*/mustard leaf
Ghruta—*ghee* (clarified butter)	Cow's *ghee*	Sheep's *ghee*
Ksheera—milk	Cow's milk	Sheep's milk
Taila—oil	Sesame oil	*Kusumbha* oil (oil from safflower seeds)
Kanda—tubers	Yam, ginger	Potato
Phala—fruit	Grapes (raisins)	*Nikucha* (*Artocarpus nikucha*)
Ikshu varga—sugarcane product	Sugar—unrefined pure sugar	Molasses or fermented sugarcane juice
Jala varga—water	Rain water	River water during the rainy season

TABLE 3.4
Balanced Diet Chart

***Prakruti* (Nature)**	**Rasa**	**Food**
Vata prakruti	Sweet+++, sour++, salt++, pungent+, bitter+, astringent+	Red rice, wheat, green gram, black gram, sesame oil, *ghee*, pomegranate, banana, milk
Pitta prakruti	Sweet+++, bitter++, astringent++, salt+, pungent+, sour+	Red rice, wheat, green gram, ghee, pomegranate, milk
Kapha prakruti	Pungent+++, bitter++, astringent++, sour+, salt+, honey+	Red rice, green gram, honey, bitter gourd

+++ means more, ++ means medium, + means less consumption.

3.11.1 Fruits

Fruits have been a common item of diet since very earliest times. They were the favorite foods of the sages living in the forests as they are one of the most *sattvic* foods. Apart from their normal uses, many fruits have some useful medicinal properties for curing diseases and disorders. In the Purana and the great epic, fruits are offered to gods, guests, and forefathers. Especially some fruits such as banana, coconut, mango and pomegranate are sacred in Hindu tradition and are part of many religious rituals.

Charaka has included a number of fruits in the nutritional items mentioned in a balanced diet. The grape was reputed the best one. *Sushruta* has described 84 fruits in his *Samhita* (Shastri, 1970, Sutrasthana 46): for him the best was the pomegranate. There is a copious mention of fruits in Buddhist and Jain works: during that time mango became very popular and there are references of mango grooves donated to Lord Buddha. Mango gardens were also developed by Ashoka who greatly encouraged plantation of trees, preferably fruit-bearing ones, on avenues as a religious and social service (Sharma, 2009).

Fruits are mostly made up of water and secondarily of ether (Nibodhi, 2010). They are the lightest food group, generally fast to digest, so Ayurveda recommends to eat them alone or before other foods. They should never be eaten after meals. Generally, they are sour or sweet, with a cooling energy and a sweet postdigestive effect. They have a refrigerant, cleaning, and nurturing action. They are a suitable food for *pitta* type, second to *vata*, and third to *kapha*. *Pitta* type should avoid very sour fruits. Fresh, sweet, and ripe fruits are suggested. For *vata* type, cooked fruits are the best choice as *vata* have low enzymatic function and raw fruits can cause gas and bloating. Dry fruits are not suitable for *vata*. The *kapha* type can eat mildly sour or sweet fruits. It will be aggravated by too sweet or sour fruits and by fruits with high water content. Dry fruits are recommended to *kapha* in moderate quantity.

It is always better to eat only seasonal fruits when fully ripe. The amount of fruits in the diet should be reduced during the cold season. An option can be to "warm" the fruits cooking them with spices such as cinnamon, cardamom, nutmeg, or ginger. Sour fruits are better in the morning and sweet fruits are better in the afternoon. The quality of fruits is very important: only organic farming fruits are good for health (Smith, 2011).

3.11.2 Dairy Products (*Dugdha Varga*)

Milk is sweet and has a cold potency and a heavy nature. It is considered a *rasayana*, as it promotes tissue building, strength, and intelligence. It pacifies *vata* and *pitta*, and is beneficial in case of psychological disorders, blood impurities, breathing disorders, depletion of tissues, and hemorrhoids. It is especially useful for children, aged, and emaciated, and for those who repeatedly indulge in sex. It creates a balance between *dosha*, *dhatu*, and *mala*, and specifically enhances the *ojas*.

Cow milk has a sweet taste; it is cold, heavy, and unctuous. It pacifies *vata* and *pitta*, it is rejuvenating, it improves complexion, and it gives strength.

Goat milk has similar properties; it is sweet, light, and cold, and pacifies *vata* and *pitta*. It is an appetizer, useful for depletion of tissue, hemorrhoids, diarrhea, menorrhagia, blood disorders, and fever. It is recommended almost for all diseases because goats consume mostly beneficial bitter substances and little water and are subjected to sufficient exercise.

Buffalo milk has certain similarities with cow milk; it is sweet, cold, and heavy, but it is sleep promoting, it improves sperm cells and semen, it has a clogging nature, and it reduces appetite.

Human milk has a sweet taste; it is light and cold, and pacifies *vata* and *pitta*. It is an appetizer and useful in case of eye pain, injuries, and wounds. It is considered the best milk for *nasya* (nasal treatment) and *ascyotana* (eye treatment).

Boiled and cooled milk subsides *pitta*, whereas boiled hot milk subsides *kapha–vata*. Similarly, overboiled milk possesses heavy and unctuous properties, is rejuvenating, and promotes strength. Generally, raw milk produces *ama* as it clogs the channels and is heavy in nature. Only breast milk has to be consumed in raw state: once boiled, human milk will be spoiled and aggravates all *dosha*s.

Milk obtained from the early morning milking is usually heavy and promotes fat accumulation due to the effect of moon and due to lack of exercise in the night. Milk obtained from the animal during the evening time is generally good for eyes and reduces fatigue, due to the effect of sun rays and the exercise done by the animal during the day.

The dry layer of *cream* on the repeatedly boiled milk is heavy, cold, and aphrodisiac, and pacifies *pitta* and *vata*.

Curd is considered hot and unctuous, with a sweet or sour taste. The after digestion taste is sour. It is an appetizer and promotes strength and fat. It is indicated for the management of painful urination, cold, chronic fevers, diarrhea, loss of appetite, and debility. Sweet curd reduces *vata–pitta*, whereas sour curd increases *pitta*, *kapha*, and *rakta* disorders.

Cow milk curd is sacred and is the best among all types of curd. Goat milk curd is also considered as one of the best curd types. It is light, reduces *tridosha*, and increases appetite. It is indicated for bronchitis, cough, hemorrhoids, and emaciation. Use of curd is forbidden during spring, autumn, and summer. However, its use is encouraged or considered best during winter and rainy seasons. Generally, curd consumption is forbidden during evenings or nighttime. Curd with water reduces fatigue, promotes strength and adds taste to food, cleanses the channels (*srota*s), and reduces *kapha* and *vata*. A blend of curd and water is called *lassi*: it is sweet, increases *kapha*, and pacifies *vata* and *pitta*.

Butter has a sweet taste; it is light and cold in nature. Fresh butter is astringent and has sour taste; pacifies *pitta* and *vata*; reduces burning sensation; improves *agni*; is good for eyes, hemorrhoids, and cough; and heals wounds. Very old butter is heavy and aphrodisiac, increases fat and *kapha*, reduces swelling, promotes strength, and is especially considered an elixir for the children.

Ghee (*clarified butter*) is considered as *rasayana*; it is sweet, heavy, unctuous, cold, good for eyes, and an appetizer; and it reduces *vata* and *pitta*, emaciation, herpes and *raktha* disorders. It improves *kapha*, complexion, *ojas*, beauty, and intelligence. Old *ghee* is always best; reduces *tridosha*, especially *vata* and *pitta*; and pacifies burning sensation, blood-related disorders, toxicity, and giddiness. It is indicated for ear problems, eye disorders, headache, skin disorders, swelling, vaginal problems, fever, bronchitis, hemorrhoids, and rhinitis. It is an appetizer and is useful for enemas (*basti*) as well as *nasya*. The good and bad qualities of *ghee* are similar to those of milk. *Ghee* obtained from cow milk is the best, whereas that obtained from sheep milk is the worst.

3.12 VEDIC REGIMENS FOR HEALTHY LIVING

Time is transformation. It is the healer and it moves on. The continuous transformation of time brings about many changes in nature. Ayurveda teaches that human has to establish himself or herself (*svastha*) as a healthy individual with harmony of the internal nature (*microcosm*) in respect of external nature (*macrocosm*). The effect of time, day and night, different seasons etc., in the individuals is well understood by the great scholars of Ayurveda and they have proposed regimens to be followed accordingly to maintain health and longevity. These principles or code of conduct are based on the nature and individual and thus they are highly effective to maintain health not only for the present generation but also for the generations to come.

> *Kaalo hi naama bhagavaan*. ("Time is God and is the factor responsible for transformation.")

Time is one of the causative factors of the universe, known as *kaarana dravya*. Vedic system of measurement of time is fascinating and highly advanced. The measure of time from a wink of eye—known as *nimesha*—to the 4.32 billion solar years or *kalpa* is just an example of the inscrutable intelligence of Vedic seers. Human physiology has a deep relation to the changes in the outside world. The diurnal rhythm has impact on every system of the body such as hormones, enzymes, and immune system. These patterns are even accepted in the present day by scientists, who are working hard on this part of physiology, which still is in its budding stage. One such advancement in medical field is chronomedication, that is, the time of administration of medicines, when they act more effectively. However, Ayurveda has developed chronomedication with the name of *bhaishajya kala* (time of administration of herbs) thousands of years ago.

The human body works in a systemic way. The production, active state, and inhibition of hormones, enzymes, and other neural transmitters are time specific and change according to the outer atmosphere. These ups and downs of the human functions are natural and required for the body to cope with the external atmosphere. Sometimes, we need to adopt new measures to cope with the changing seasons. People living in cold countries try to eat more fat to provide more energy for the body to produce more heat to keep the body temperature up. In the same way, people living in hot/temperate regions eat more liquid and less fat food to cope with the scorching sun, which causes poor digestive capacity and more dehydration. Thus, we see different types of cuisines and lifestyle in different living conditions.

3.12.1 *Dina Charya* (Daily Regimens) and *Rutu Charya* (Seasonal Regimens)

Daily regimens deal with the regular day-to-day activities of life such as time to wake up, brushing, bath, time to eat, quantity of food, and time to bed. Seasonal regimens deals with dos and don'ts of daily regimens according to each season.

Ayurveda deals very elaborately with the lifestyle. It is the only system of medicine dealing with lifestyle aspects. Every aspect of lifestyle, from getting up from the bed to retire to bed, was dealt with very clearly.

3.13 *RASAYANA*—THE QUEST FOR LONGEVITY

An accurate exploration of the medical doctrines of the ancient cultures came across many references on how to prolong life. In fact, from immemorial time, man is trying to achieve the most impossible task of becoming immortal or at least to live many hundreds of years.

Many civilizations around the world had different kinds of rituals and medicines to attain the elixir of long life. One such ancient and highly advanced civilization is that of the Indus valley and Ayurveda is its medical system. It is the most emphatic, advanced, and enigmatic medical system of the world.

Ayurveda is the subbranch of *Atharva Veda*, which is one among the four *Veda*s. Vedas are the oldest testaments available on the earth. They are pragmatic and divine as they were revealed to the philosophers (the *rishi*s), who were in deep meditation to discover the truth of the universe. The inner aim of Ayurveda is to prolong life to achieve the ultimate goal of humans: salvation (*Charaka Samhita*, Sutrasthana 1/1). Ayurveda is something more than a medical system, as it is intended to maintain good health and to recover from diseases to achieve longevity.

Ayurveda literally means the wisdom of life. It is a complete healthcare system because it is the only and unique science to deal with every aspect of life. The alarming increase of diseases related to change in lifestyle is one of the most talked subjects of the present day, but no medical system deals with lifestyle, that is, what to eat, when to eat, when to bath, how much to exercise, what to wear, when to sleep, and so on. Ayurveda offers some opportunities to live healthily.

The science of rejuvenation and long life is known as *rasayana tantra*, a separate branch in Ayurveda created to restore youth and bestow longevity. *Rasayana* is a technical term that is a combination of two words *rasa* and *ayana*, and means rejuvenation (literally the pathway of the primordial tissue). *Charaka*, the oracle of Ayurveda, defines *rasayana* as the therapy that allows the body to develop healthiest tissues, with a considerable physical and psychic improvement.

The procedures involved in *rasayana* therapy are as follows:

1. *Nitya rasayana*—Physical, moral, social, and spiritual conduct to make self in tune with nature and ready for the therapy.
2. *Panchakarma*—Purification of the body. It clears the *srota*s (channels) from wastes and makes them prone to the proper flow of the nourishments.

It removes the *mala* or the waste from even the most micro cells and channels of the body.

3. *Kutipravesa*—Residence in the specified hut for a defined period of time. The essential part of the therapy is to keep the patient in a relaxed condition, away from all kinds of connections or links with the external world.
4. *Rasayana* herb/herbs—Prescribed herb/herbs are given to the patient under medical control to induce the desired effect to entirely rejuvenate. Herbs such as *amalaki* (*Emblica officinalis*, gooseberry), *vidanga* (*Embelia ribes*), and other very common herbs have been described for the purpose of prolonging life for hundreds of years with astonishing results. Milk and red rice are given to sustain the body along with the above *rasayana* herbs.

The Vedic principles and code of conduct to achieve a healthy life are described under the topic *nitya rasayana*, which means daily *rasayana*. In the broader understanding, *nitya rasayana* brings peace and harmony with the ecosystem, by giving a guideline for a personal, social, moral, and spiritual behavior. The suggested personal conduct includes celibacy, early wakeup and sleep to avoid exhaustion, a regular intake of cow milk and *ghee*, and following a rational, good, and simple lifestyle. The social conduct includes avoiding anger, violence, and use of alcohol. The moral conduct includes being peaceful, sincere, and kind with everybody, and not being egoistic. The spiritual conduct includes chanting Vedic *mantra*s; following *dharma saastra*; praying to gods; worshipping the cows, *Brahman*s, and teachers; following the spiritual path; doing meditation; and regularly contributing to charity. These codes of conduct are aimed to implement the individual's health and longevity.

A brief interpretation of the Ayurvedic physiology will make it easier to understand how *rasayana* therapy works.

Body is an amalgam of *bhootas* (physical aspect), *manas* and *indriya* (sensory aspect), and *atma* (spiritual aspect). The physical and sensory bodies are supplied with nourishment from the food. *Rasa* (the essence of food) circulates in the body through innumerable *srota*s or channels and nourishes physical body 24 hours a day, whereas the sensory aspect of the body is nourished only when the *hrudaya* (the abode of consciousness) is active. Therefore, if good food is taken, it creates healthy conditions at both physical and sensory levels. After digestion, the food is converted into *rasa* or the essence and *mala* or the waste. Food requirement directly depends on the amount of work, both physical and mental, so more work means more food is required. In this sense, the quantity of *rasa* and *mala* is also directly dependent on the food quantity.

The metabolism depends on *agni*, the factor responsible for digestion and assimilation: an optimum level of *agni* indicates a good metabolism. In this sense, a proper *agni* will clear *srota*s and will aid in good nourishment. Reducing body exertion is also important, as the body will require less food, and thus there will be less amount of *mala* (waste) to be thrown out. The accumulation of *mala* over time leads to block in the *srota*s or channels, which further leads to diseases. This is the reason why Ayurveda advises for seasonal cleansing of the body from the accumulated wastes through *panchakarma*.

If the metabolism is reduced gradually close to zero, as in *samadhi* state of yoga, the intake of food and excretion of wastes will also come to almost zero. When this state is attained, the body will start to purify from in-depth with the power of *agni*. This is the most difficult state to achieve, unless the person is mentally and physically strong. Once the person is able to withstand this state, the addition of *rasayana* herbs will enhance the body cells to regenerate. Common people cannot emulate the diet of yogis who can survive only on *prana*, but they can follow the *nitya rasayana* conducts along with a proper diet. A correct, nourishing diet, not too heavy and supplemented with specific herbs, is the fundamental for body and mind rejuvenation.

In many yogic texts, the description of the *sattvic* diet based on vegetarian food is reported, including dairy products, whole grains, beans, vegetables, fruits, and nuts. This diet supports a spiritually oriented life. The rejuvenative diet suggested by Ayurveda is similar, with more emphasis on the nutritive properties of food, with special care to reduce *vata dosha* (Frawley, 2013). Excess in eating, particularly oily, heavy, and very sweet, foods promotes the aging process. Fast food and junk food should be avoided. Only natural and light foods can promote longevity if the quantity of food is adequate. Vegetarian food is recommended as it holds the power of *prana*, in which immortality resides. The main items suggested for rejuvenation are as follows: fruits, seeds and nuts, whole grains and heavy roots, and vegetables such as potatoes, beetroot, and carrot (Frawley, 2013). Mild spices are usually included in the diet to promote *agni*.

The sweet taste in right quantity promotes rejuvenation, whereas an excess quantity promotes diseases and aging. The quality is very important: refined sugar and high fructose corn syrup build up *ama* and should be avoided. Only honey, cane sugars, fruit sugars, milk sugars, and special fruit jellies are suggested. Sugar is more rejuvenative if balanced with the right spices (especially with ginger and cardamom). The sour taste is not rejuvenative, with the exception of gooseberry and yogurt.

The salty taste can accumulate in the body and clog the channels, but a small amount is necessary. The pungent taste is not rejuvenating, but it stimulates digestion and provides a support to the diet. Generally, the use of spices aids the digestion. The best spices for a rejuvenative diet are ginger, turmeric, and cardamom. Cinnamon, fennel, cumin, coriander, basil, fenugreek, mint, saffron, and rosemary are also good.

Fruits are the strongest rejuvenative items, especially tropical fruits and berries. The intake of freshly made fruit juices is highly recommended.

Vegetables should be well cooked and spiced. Raw vegetables and sprouts should be used in small amount. The best vegetables are asparagus, sweet potatoes, yams, squash (particularly winter squash), green beans, green peas, carrots, beets, and celery. Greens are helpful in small quantity.

The best grains for rejuvenation are whole wheat and rice (especially basmati).

Beans disturb *vata dosha*, so only split yellow mung is recommended. The *kicharee*, made of an equal part of basmati rice and split yellow mung, helps to restore weakened digestive functions and is one of the best rejuvenative foods. It is cooked with *ghee*, salt, turmeric, and ginger.

Nuts and seeds are excellent if taken in small amounts.

Oils are a good rejuvenative for the nerves. The rejuvenative therapy also includes external application of oils. The oils should be fresh and of good quality. The *ghee* is considered the best of all, excellent for *vata* and *pitta dosha*s.

Dairy products form the basis of many rejuvenation diets. Cow milk increases *ojas* and should be taken warm, with ginger, cardamom, and cinnamon. *Ghee* and yogurt are also considered good rejuvenators.

The preparation of food has also an important role. Food that is overcooked or cooked more than once should be avoided, along with dead food (processed and canned or frozen too long). The traditional way of cooking, with love and personal feeling and involvement, is emotionally rejuvenative. It is also suggested to take the major meal around twelve o'clock when the digestive fire is strongest. Heavy foods should be avoided after sunset. Heavy, sweet, and mucus-forming food should be avoided for breakfast. In the early morning, a mild spiced tea with cinnamon, ginger, or basil will clean any residual of *kapha* or *ama*. It is suggested to follow a light and detoxifying diet during late spring and summer. A more nutritive and rejuvenative diet is better in autumn and winter.

It is noteworthy that the diet always needs to be adjusted according to an individual's constitution, climate, season, and location.

REFERENCES

Badria F.A. Melatonin, serotonin, and tryptamine in some Egyptian food and medicinal plants. *J. Med. Food*, Chowkhamba Sanskrit Series office, Varanasi, India, **2002**, *5*(3), 153–157.

Frawley D. *Soma in Yoga and Ayurveda. The Power of Rejuvenation and Immortality*. Motilal Banarsidas Publishers, Delhi, India, **2013**.

Goldstein D.S., Kopin I.J. Evolution of concepts of stress. *Stress* **2007**, *10*(2), 109–120.

Hattori A., Migitaka H. *et al.* Identification of melatonin in plants and its effects on plasma melatonin levels and binding to melatonin receptors in vertebrates. *Biochem. Mol. Biol. Int.* **1995**, *35*(3), 627–634.

Mahesh Yogi M. *Perfection in Education*. Maharishi Vedic University Press, Fairfield, IA, **1997**.

Naliboff B.D. The effect of life stress on symptoms of heartburn. *Psychosom. Med.* **2004**, *66*, 426–434.

Nibodhi. *Annapurna's Prashad*. Mata Amritanandamayi Mission Trust, Amritapuri, India, **2010** (2nd edn.).

Sharma P.V. *Fruits and Vegetables in Ancient India*. Chaukhamba Orientalia, Varanasi, India, **2009**, III–XIII.

Sharma R.K., Bhagvandash (trans.) *Agnivesa, Charaka Samhita,* Sutrasthana (Text with English translation). Krishnadas Academy, Varanasi, India, **2009**, Chapter 5, 36, 38; Chapter 7, 54; Chapter 25, 130, 132; Chapter 28, 181.

Shastri K.N. (ed.) *Charaka Samhita*. Chaukhambha Vidyabhavan, Varanasi, India, **1970**, Sutrasthana 46.

Smith A. *Ayurvedic Nutrition*. Motilal Banarsidas Publishers, Delhi, India, **2011**, 115–118.

4 The Principles of Diet Therapy in the Unani System of Medicine

Roohi Zaman

CONTENTS

Your food shall be your medicine.

Hippocrates

4.1 INTRODUCTION

The Unani system of medicine is a great healing art as well as a science. It treats a person as a whole and not as a group of individual parts. It is aimed at treating the body, mind, and soul. It is a science from which we learn the various states of the body in health and ill-health as well as the means by which health is likely to be lost and when lost how it may be restored (Avicenna, 2007).

Unani (pronounced *Yūnānī* in Arabic, Bengali, Hindi, Persian, and Urdu) means "Anatoli" and has its origins in the Greek word, the name of a place given to a Greek-populated coastal region of Anatolia. This major traditional holistic healing system, one of oldest forms of medicine, covers all aspects of health, encouraging physical, mental, emotional, and spiritual well-being. It was Hippocrates (*Buqrat*, 460–377 BC), a great philosopher and physician, who freed medicine from the realm of superstition and magic and gave it the status of science and presented the basic theoretical framework of Unani medicine.

A number of other Greek scholars enriched this system considerably; of them Jalinus (Galen AD 131–200), who drew inspiration from Hippocrates, stood out as the one who stabilized its foundation on which Arab physicians such as Rhazes (AD 850–923), Avicenna (AD 980–1037), Al Zahravi (Albucasis), the surgeon, and Ibn Nafis constructed an imposing edifice. Hence, it is also known as *Greeko-Arab* medicine.

This system was also enriched by imbibing what was best in the contemporary systems of medicine in Egypt, Syria, Iraq, Persia, China, India, and various parts of Central Asia. At that time, Greek medical knowledge was known to be the best medical science. In India, the Unani system was introduced by Arabs, and soon it took firm roots in Indian soil and has been serving vast sections of the people in the country.

The Unani physician believes that the healthy state of the human body is maintained by a power known as *tabiat* or *quwwat-e-mudabbira* (*medicatrix naturae*), which controls, regulates, and restores the physiological mechanisms of the body and helps in potentiating the immunity of the body and its resistance against various ailments. Suppression of this gifted power leads to disease. Therefore, the duty of the physician is to use such methods/treatments that encourage the body's own innate healing response (*tabiat*); this can be achieved by stimulating the *hararate ghariziya* (vital force of the body), which is decreased in a diseased person making that person vulnerable to environment and pathological challenges.

4.2 BASIC FUNDAMENTALS AND PRINCIPLES

In the Unani system of medicine, the human body is considered as a single unit, made of seven components known as *umoor-e-tabiya*. These seven components are *arkan* (elements), *mizaj* (temperament), *akhlaat* (humors), *aaza* (organs), *arwaah* (life force/oxygen), *quwa* (faculties), and *afa'al* (functions). *Arwaah* and *hararate ghariziya* are not the same: they have different functions.

According to Unani philosophy, the body is made up of four basic elements, namely, earth, air, water, and fire, which have different temperaments, that is, cold, hot, wet, and dry, respectively. Different combinations of these four elements lead to a new *mizaj* (temperament), that is, hot and wet, hot and dry, cold and wet, cold and dry.

The body has simple and compound organs, which receive their nourishment through four humors (*akhlaat*), that is, *dam* (blood), *balgham* (phlegm), *safra* (yellow bile), and *sauda* (black bile). Each humor has its own temperament; blood is hot and moist, phlegm is cold and moist, yellow bile is hot and dry, and back bile is cold and dry. Every person attains a temperament according to the preponderance of the humors in his or her body and it represents the person's healthy state. The temperament of a person may be sanguine (dominance of blood), phlegmatic (dominance of phlegm), choleric (dominance of yellow bile), or melancholic (dominance of black bile).

Unani physicians believe that health is a state of body in which there is equilibrium in the humors and functions of the body. The humoral balance is maintained or balanced by the power of self-preservation or adjustment called *quwwat-e-mudabbira* (*medicatrix naturae*) in the body. When this power weakens, the equilibrium of the humors is disturbed quantitatively or qualitatively or both, and the physiological functions of the body are deranged due to the abnormal temperament of the affected organ or system resulting in a disease. Therefore, the aim of the Unani physician is to find out the cause of the underlying disruption of humors, so that it can be corrected and disease can be cured. An imbalance of humors may be due to external factors such as an injury, incorrect diet, and environment factors or internal factors such as improper digestion or both.

The state of health (*sehat*) is the end product of specific proportion and appropriate ratio of the above *al-umar-al-tabi'yah* (basic principles of physiology) under the influence of six factors (*Asbab-e-Sitta Zarooriya*): *Al-Hawa al muhit* (atmospheric air), *al-makul wa'l-mashrub* (food and drinks), *al-harkat wa'l-sukun al badaniya* (physical or bodily movement and repose), *al-harkat wa'l-sukun al nafsaniya* (mental or psychic movement and repose), *al-naum wa'l-yaqzah* (sleep and wakefulness), and *al-Istifragh wa'l-ihtibas* (evacuation and retention). These six factors essentially influence each and every human body; therefore, these are called as *Asbab-e-Sitta Zarooriya* and any alteration/change in their specific proportion or appropriate ratio of *Asbab-e-Sitta Zarooriya* predisposes disease (*marz*).

4.3 TREATMENT MODALITIES IN THE UNANI SYSTEM

4.3.1 *Ilaj-Bil-Ghiza* (Diet Therapy/Nutritional Therapy)

Treating certain ailments by controlling the intake of food, the Unani physician believes that most of the diseases can be cured by administrating specific diets or by regulating the quantity and quality of the food; for example, cold water is better for the person possessing a hot temperament, whereas it is bad for the person having a cold and wet temperament. If a patient has excessive thirst, which is an indication of an ill temperament of hot type, a diet having a cold temperament such as *khurfa* (*Portulaca oleracea*) should be used, but if a patient is complaining of cough (*su'al*) due to cold, these cold diets shall aggravate the condition and a hot diet will be beneficial for the patient. The details of the importance of diet therapy, its general principles, and some modified diets recommended are mentioned in Chapter 5.

4.3.2 *Ilaj-Bid-Dawa* (Pharmacotherapy)

If diet therapy alone is not sufficient, pharmacotherapy is advised in addition. These drugs may be derived from plants, minerals, or animals. Unani physicians have mentioned some principles related with pharmacotherapy, which are as follows.

Mizaj-e-advia (drug temperament): According to Unani medicine, every drug has its own temperament due to its specific phytoconstituents. The temperament of drugs is classified into five types: hot, cold, dry, wet, and sometimes neutral. A drug may be hot and dry, hot and wet, cold and dry, or cold and wet. Drugs used in a particular disease should have a temperament opposite to the prevailing temperament of the target organ responsible for the disease.

4.3.3 *Darjaat-e-Advia* (Potency of Drugs)

The four temperaments may be classified into four grades: first, second, third, and fourth. The first grade has the lowest potency and fourth grade has the highest potency; for example, the latex of *Papaver somniferum* is cold 4° and dry 4°, whereas the fruit of this plant known, as *post-e-khashkhaash*, is cold 2° and dry 2°. The selection of potency of the drug depends on the severity of diseases. If the disease is severe, drugs having higher potency are selected.

4.3.4 *Ilaj-Bid-Tadbeer* (Regimental Therapy)

These are the various rejuvenate and detoxification regimens/therapies. They are not only curative but also widely used for the prevention of diseases. Unani physicians attempt to use simple physical means to cure a disease. However, for some specific and complicated diseases, special techniques are adapted by using certain regimes using formulations/equipments/guidance known as regimental therapies (*llaj-bid-tadbeer*). These may be used in adjuvant to dietotherapy and pharmacotherapy.

Some of commonly used regimental therapies are as follows:

Hammaam—Turkish bath and various types of medicated baths
Dalak—Massage and physiotherapy
Riyazat—Physical exercise
Takmeed—Fomentation
Hijamat—Cupping
Fasad—Venesection
Ishaal—Purgation
Qhai—Emesis
Idrar-e-Baul—Diuresis
Kai—Cauterization
Taleeq—Leeching

4.4 PRINCIPLES OF DIET THERAPY

Diet can be defined as a food intake pattern in which substances that are used for the purpose of providing nutrition to the body through metabolism (Hamdani and Hussain, 1980). Diet therapy is a therapy by which health is attained by modulation of dietary habits. It seeks to restore the imbalance in the body caused due to some genetic infection or errant lifestyles. Diet is an important component of these essential factors. The Unani system of medicine believes that a physician is not the healer but an assistant to natural *tabiat* (physique) of the body, which is the true healer; hence, diet should be in accordance with it. Diets have been mentioned according to various stages of life to maintain the digestive system at its best (Hamdani and Hussain, 1980).

4.5 BENEFITS OF HEALTHY FOODS

"Nutrition is the science of food, the nutrient and the other substances there in, their action, interaction and balance in relation to health and disease and the processes by which the organism ingests, digests, absorbs, transports, utilizes and excrete food substances" (Williams, 1995). Nutrition, which depends on food, is also of utmost importance in the cure of disease. The primary cause of disease is a weakened organism or lowered resistance in the body, arising from the adoption of a faulty nutritional pattern. There is an elaborate healing mechanism within the body, but it can perform its function only if it is abundantly supplied with all the essential nutritional factors. A diet that contains liberal quantities of seeds, nuts, grains vegetables, and fruits would provide adequate amounts of all the essential nutrients. These foods are therefore called basic food groups and a diet containing these food groups is an optimum diet for vigor and vitality.

We live in a world with rapidly changing elements in our environment, food supply, population, and scientific knowledge. On exposure to various environments, our physical bodies and our personalities change and with them our personal needs and goals. These constant changes of lives must be in some kind of positive balance to produce healthy living. Thus, to be realistic within these life concepts of change and balance, our study of food, nutrition, and healthcare must focus on health promotion. Although we may view and define health and disease in different ways, a primary basis for promoting health and preventing disease must always be good food and the sound nutrition it provides (Williams, 1995).

Diet plays a vital role in the maintenance of good health and in the prevention and cure of disease. In Nutrition in health and disease, the words of Sir Robert McCarrison (1936), one of the best known nutritionists, "the right kind of food is the most important single factor in the promotion of health; and the wrong kind of food is the most important single factor in the promotion of disease."

The word *diet* is taken from the Latin word *dieta*, which means a way of living in Greek. In the Unani system of medicine, abundant literature regarding diet therapy has been discussed. Diet therapy *(ilaj-bil-ghiza)* is a unique nonmedicinal therapy by which the patients are treated and health is attained through modulation in dietary habits, that is, fasting and use of foodstuffs in more quantity that have less nutritional value or less quantity that have more nutrients. It is concerned with those receiving a normal diet as well as those for whom a modified diet has been prescribed. Modified diets are the principal therapeutic agents in some metabolic diseases and chronic diseases. Diet therapy serves as a preventive measure as well as a therapeutic aid.

According to Unani, foods are broken down by the body's "heat." When *khamrae* (enzymes) come into contact with food, the enzymes supposedly release heat that breaks it down. How nutritious a particular food is depends on various factors, including how much of each humor it creates, what season it is, and the person's *temperament*. Unani medicine deals with different types of diet such as *dawae ghizaee* (a substance with more medicinal value than nutritional value), *ghizae dawai* (a substance with more nutritional value than medicinal value), *kaseef ghiza* (heavy foods), *lateef ghiza* (light foods), and *motadil ghiza* (moderate foods).

4.6 HISTORICAL BACKGROUND

Diet therapy has a long history; it stems from the Zhou dynasty (1000 BC). Zhang Ji, a distinguished physician in the Han dynasty, realized the action of diet therapy during the rehabilitation from the disease (Qureshi, 2006).

Nutrition as a discrete or separate discipline has existed since Vedic times (8000 BC). *Charaka Samhita* and *Sushrutha Samhita* described all major branches of medicine, including dietetics. Egyptians, Romans, and Greeks such as Hippocrates (460–327 BC), Aristotle (384–377 BC), Celus (53 BC–AD 7), and Galen (AD 130–200) were great pioneers in dietetics. They gave much importance to diet during health and sickness; however, dietetics gained more popularity during the time of Mohammed the Prophet (PBUH) (AD 570–632) (Ibn Al Qaim Al Jouzi, 2008).

The Holy Quran speaks very highly of the health and medicinal value of pomegranate, figs, dates, olives, milk, honey, and ginger, and forbids certain harmful foods such as blood and putrefied and unslaughtered meat. Later on, Arabian physicians such as Rhazes (AD 850–923), the Galen of the Arabs, improved dietetics and advised "when you can heal by diet prescribe no remedy."

After Rhazes, another Persian named Avicenna (AD 989–1036), "the Persian Galen, gave much more importance to diet in sickness and health. His noted book *Canon of Medicine* has remained a classical medical textbook for many centuries all over Europe and Arabia (Siddiqui and Khan 2008).

Diet therapy is classified according to the following:

1. Intake of diet
2. Temperament of food
3. Heaviness to digest
4. Nature of chime formed
5. Abstinence/fasting (*tarke ghiza*)
6. Reduction in intake of diet/quantity (*taqleele ghiza*)
7. Modification in quality of diet

Fasting is the process of purification and an effective and quick method of cure. It assists nature in its continuous effort to expel waste matter and disease-producing noxious substance from the body, thereby correcting the faults of improper diet and wrong lifestyle. It also leads to repair and regeneration of various tissues of the body. Fasting affords a physiological rest to the digestive, assimilative, and protective organs. As a result, the digestion of food and the utilization of nutrients is greatly improved after fasting.

Food may be withheld, reduced, or given in a moderate quantity or deliberately increased. Food is withheld when the physician wants the body to be occupied with maturing humors and morbid matter. It is reduced when desired to maintain the strength and provide sufficient time to the body to deal efficiently with the morbid condition rather than involve itself in the digestion of a large meal. When the problem is dual, the most important aspect should be given priority. If strength is low, it should receive prior attention, whereas if disease is severe, it should be attended to first; therefore, diet is reduced in quantity or quality or both. Quality is lowered by giving food that is heavy and bulky but of low nutritional value; for example, a diet mainly of vegetables and fruits is bulky but of fewer calories. Food is generally stopped or reduced in quantity in the case of acute diseases. Food serves to replace the loss from the wear and tear of body and provides nutrition. When the body is too weak to digest heavy food and the need is urgent to maintain the strength and restore the loss of vitality, easily assimilable food should be given. Highly nutritious but slowly assimilable food is given when strengthening of the body is desired, which prepares it for vigorous exercise and activity (Hamdani and Hussain, 1980; Ibn Al Qaim Al Jouzi, 2008; Qureshi, 2006).

In the Unani system, the normal health of a person depends on the proper combination and physiologic effects of the four humors: blood, yellow bile, phlegm, and black bile, and their temperament (Ibn-e-Sina, 1998; Qureshi, 2006; Siddiqui, 2006). The constitution of food (*mizaj*/temperament) refers to its effect in the healthy body (pharmacodynamics). Every food/drug produces certain perceptible and imperceptible symptoms when the same is administered in various concentrations. This effect is described in terms of one's feelings in different intensities (Ibn Rushd, 1987). If food is consumed in a certain quantity and after its assimilation one does not notice any effect or change in one's body, it is considered as *motadil* or moderate (Williams, 1995). If it produces a slight effect of heat/cold/dryness and the same passes off without leaving any feeling after 4 h, the food is considered as having the constitution of hot/cold/dry/wet in the first degree. If the intensity of the effect is slightly more and remains in the body for more than 8 h and does not harm the body, the food is considered to belong to the second degree. If the effect of a given substance produces a remarkable but tolerable effect and remains in the system for more than 18 h but does not endanger one's life, it is considered to belong to food having constitution of hot/cold/dry/wet in the third degree. Finally, if the consumed substance produces an intense effect immediately after 1/2 h to 1 h and endangers one's life, it is known as having a fourth degree constitution. The intensity of the effects of foods depends on one's own constitution/temperament. According to production of calories, diet is classified as light, heavy, or moderate. A light food is that which produces thin blood, whereas a heavy food is that which forms thick blood, and a moderate food produces blood of average consistency.

Food may be of high calories or low calories, and in this respect, it is classified as follows:

1. *Lateef Ghizae*: It is an easily digestible food. It is of two types:
   a. *Ghiza-e-lateef qalil-ul-taghzia*, easily digestible, low caloric food, such as fruit juices and vegetable soups
   b. *Ghiza-e-lateef kaseer-ul-taghzia*, easily digestible, high caloric food, such as half boiled egg, meat distillate, and wine/grape extract
2. *Kaseef Ghizae*: It is not an easily digested food, for example, meat, pulses, and fruits such as mango and jackfruit. This is also of two types:
   a. *Ghiza-e-kaseef qalil-ul-taghzia*, difficult digestible, low caloric food, such as dry animal meats/fish
   b. *Ghiza-e-kaseef kaseer-ul-taghzia*, difficult digestible, high caloric food, such as beef
3. *Mutawasit/Motadil Ghiza* (moderate) in between the preceding two group, such as meat soup, cow's milk, and vegetables such as pumpkin and fresh fruits

Chyme (*chymoos*) is the liquid substance found in the stomach before it passes through the pyloric valve and entering the duodenum. It results from the mechanical and chemical breakdown of a bolus and consists of partially digested food, water, hydrochloric acid, and various digestive enzymes. Chyme slowly passes through the pyloric sphincter and into the duodenum, where the extraction of nutrients begins. Depending on the quantity and contents of the meal, the stomach will digest the food into chyme anywhere between 40 min and a few hours.

According to the production of chime, diet is classified as follows (Hamdani and Hussain, 1980; Ibn-e-Sina, 1998; Ibn Rushd, 1987; Qureshi, 2006):

1. *Ghiza jayyadal kaimoos*: Good chyme is formed, for example, *ma al-laham* (meat distillate), *aab-e-anar* (pomegranate juice), and *aab-e-seb* (apple juice).
2. *Ghiza radi al-kaimoos*: Bad chyme is formed, for example, *kalijee* (liver), *Laham al-bat* (duck meat), and dry salted meat.

As diet therapy is the use of food as an agent in effecting recovery from illness, it is concerned with those receiving normal diet as well as those for whom modified diet has been prescribed. Modified diets, as mentioned earlier, are the principal therapeutic agents in some metabolic diseases; they serve as a preventive measure as well as a therapeutic aid. The purpose of diet therapy is to maintain good nutritional status, to correct deficiencies that may have occurred, to afford rest to the whole body or to certain organs that may have been affected, to adjust food intake to the body's ability to metabolize the nutrients, and to bring about changes in the body weight whenever required.

The Unani system recommends certain modified or specified diets in specific diseases; some of these are described below (Avicenna, 2007; Hamdani and Hussain, 1980; Ibn Rushd, 1987; Kabir, 2002; Qureshi, 2006):

- *Mau sha'eer* (barley water): It is useful in acute conditions such as fever and burning micturition; it is cool and moist in nature.
- *Jubn* (milk): It is moderate in temperament with little influence of cold and moist. It possesses three major qualities: it is light with excess water content and helps eliminate morbid matter; it maintains body heat and nourishes and energizes body moderately; and it cleanses the system.
- *Maul jubn* (milk water): It is considered as very good purgative and is used in bilious conditions such as jaundice; it is dilute and light, easily absorbable and digestible. Oil content keeps the organs soft, prevents irritation, and forms very good chyme.
- *Mau'llahm* (meat distillate): this is prepared by the process of distillation. It is rich in calories and easily digestible and is advised in malnourishment and weak patients.
- *Maurraib* (buttermilk): It is cold and moist in temperament, easily absorbable, and cool; it quenches thirst.
- *Maul asl* (honey water): Honey is boiled with water or sometimes with herbal decoction. It is useful in fever, irritation, thirst, and dry cough.
- *Maul fawakh* (fruit juices): It is rich in essential nutrients and easily digestible; it is useful in general weakness and debility.

- *Sikanjabeen* (oxymel): It is prepared by boiling one part of vinegar and two parts of honey. It is useful in purging out thick chime; it is a soothing medicine; it removes yellow bile; and it acts as an emetic in hot water and as an antiemetic in cold water.
- *Gulkhand*: Rose petals are preserved in sugar syrup or honey to form a jam, which is useful in reducing fever. It is a good laxative, a liver and heart tonic, and a purgative when consumed in large quantities.

4.7 CONCLUSION

A well-balanced and correct diet is thus of utmost importance for the maintenance of good health and the healing of diseases. Diet therapy plays an important role in treating and preventing diseases. Diet therapy minimizes the duration of treatment and improves the drug tolerance by slight modification in diet or proper management of diet. In chronic illnesses such as diabetes mellitus, hypertension, and renal diseases, modified diets help in preventing complications and improve the health condition.

In the Unani system, diet therapy gives importance to the formation of quality and quantity of humors, which are believed to be the end product of digestion and which provide nutrition to the organs, whereas the Western system believes in nutritional importance.

REFERENCES

Avicenna. *The General Principles of Avicenna: The Canon of Medicine*. CCRUM, MOHFW GOI, Delhi, India, **2007**. 182–187, 359–361.

Hamdani S.M., Hussain K. *Ussole tib*. Muslim University Press, Ali Manzil, Aligarh, Uttar Pradesh, India, **1980**. 405–409.

Ibn Al Qaim Al Jouzi. *Tibbe Nabvvi*, Darul Salfia, Mumbai, India, **2008**. 229–235.

Ibn-e-Sina. *Al Qanoon Fil-Tibb*, vol. 2. Idarae Kitabusshifa, Delhi, India, **1998**, 1st edn. 194–195.

Ibn Rushd A.M. *Kitab ul Kulliyath*. CCRUM, MOHFW GOI, Delhi, India, **1987**, 2nd edn. 246–256.

Kabir H. Introduction to *Ilmul Advia*. Shamsher Publisher and Distributors, Aligarh, Uttar Pradesh, India, **2002**, 1st edn.

Qureshi E.H. *Ilaj ba Zariye Ghiza*. CCRUM, MOHFW GOI, Delhi, India, **2006**, 2nd edn.

McCarrison R. Nutrition in health and disease, *Br. Med. J.* **1936** 2(3951): 611–615.

Siddiqui M.A. *Usoole Ilaj*. Jamia Hamdard, Delhi, India, **2006**, 1st edn. 80–96.

Siddiqui M.M.H., Khan M.S. Dietotherapy and its significance with special reference to the management of *Dique* (tuberculosis). *Indian J. Trad. Know.* **2008**, *7*(3), 397–400.

Williams S.R. *Basic Nutrition and Diet Therapy*. Mosby-Year Book, St. Louis, MO, **1995**, 10th edn. 319–477.

5 Traditional Indian Concepts of Nutrition

Toward an Integrative Approach

A.V. Balasubramanian

CONTENTS

5.1 THE SOCIAL ORGANIZATION OF INDIGENOUS HEALTH TRADITIONS

The Indian subcontinent abounds, as it were, with a variety and diversity of health traditions. It has perhaps the longest unbroken health tradition, which has not only a stream of practitioners but also textual and theoretical backing in terms of the Ayurvedic and Siddha systems of medicine (Balasubramanian and Radhika, 1989).

What is most remarkable about the Indian medical tradition is that it prevails at two different levels: the classical system (*sastric*) and the folk system (*lok parampara*). By the classical system, I refer to the codified systems such as the Ayurveda, Siddha, and Unani traditions. They are characterized by institutionally trained practitioners, a body of texts, and highly developed theories to support their practices. India also has a folk tradition (or what may be termed as the *lok parampara*), which is an oral tradition passed on from father to son or mother to daughter (or daughter-in-law) or from *guru* to *sishya* in tens and thousands of villages through the ages.

The relationship between folk and classical traditions is found to be symbiotic. There is a strong commonality of underlying theory and worldview expressed at the level of *pancha mahabutha* (theory of composition of matter) and *tridosha* (theory of causation of disease). There is also a striking common ground between the technical terms that are used by expert practitioners and what is known to the folk practitioners. Technical vocabulary such as *vata*, *pitta*, *kapha*, *ushna*, *sheeta*, *laghu*, *guru*, *guna*, and *virya* is a consistent part of the knowledge of folk practitioners and common households.

Traditionally, in India there has been a vast body of knowledge and literature on food and nutrition based on an entirely different worldview and understanding that is linked with and embedded in traditional medicine concepts such as those of Ayurveda. In this approach, food is considered to be one of the three pillars for health and for the treatment of diseases (the other two being the use of medicines and appropriate lifestyle/behavior). There is a vast body of literature that describes the properties of foods based on Ayurvedic parameters such as the following:

- What are the *dosha*s (the body humors) that are aggravated or balanced by the food?
- What is their impact on the body tissues such as the blood and the skin?
- What is taste? It is recognized in traditional medicine that taste has a bearing on nutrition and a balanced meal is expected to have a suitable blend of the six tastes, that is, sweet, sour, pungent, salt, astringent, and bitter tasting substances
- The functions of specific foods are described, such as groups of foods that have common properties: *chakshushya* (wholesome for eyesight), *garbha poshaka* (nourishing the fetus), and *mutrala* (diuretic)
- The effect that foods have on the mind is also recognized and the foods are assigned properties such as *nidrakara* (inducing sleep), *madhya* (intoxicant), and *mohakara* (causing delusion).
- Foods that are considered as *pathya* or *apathya* (indicated or contraindicated) for specific conditions or illness (e.g., fever, cold) are also described

In Section 5.2, an overview of the manner in which traditional medicine looks at the properties of materials is provided—in the context of both medicine and food (Balasubramanian and Radhika, 1990; Nanal and Nanal, 1991).

5.2 BASIC TERMINOLOGIES

Although nutritional properties in modern medicine are described using terms such as protein, caloric content, and vitamins, traditional medicine uses entirely different terms. Among the basic terms used, some of the key concepts are as follows:

1. *Rasa*—This literally refers to the taste of a substance. It is believed that different tastes impart or correspond to different nutritional qualities. For example, a substance that is bitter in taste is likely to have properties that are *krimigna* (it controls *krimi*, "deworming") and a substance that is sour in taste can increase the intensity of *agni* (capacity to digest).
2. *Guna*—This refers to a large number of specific qualities.
3. *Virya*—This refers to the potency of a substance, which may be *ushna* (hot) or *sita* (cold). This indeed refers to the overall impact or feeling in the human after ingesting/digesting the food—a later part of this chapter contains the description of a study on this subject.
4. *Ganas*—*Charaka*, another classic Ayurvedic authority, classified drugs into 50 groups or *ganas* according to their action. Various other authors have added more terms to describe the action of all substances, be it food or medicine.

5.2.1 DEFINITION OF SOME *GANAS*

There are rigorous technical definitions given for various *ganas* in the Sanskrit language, which carry rich meanings and connotations for the expert practitioner. Some of them are listed below to give a flavor of these terms and examples of substances that are the members of the *gana*:

1. *Sandhaaneeyam*: *Sandhaaneeyam bhagna sandhaana kaarakam. Sandhaana* actually means to join or to connect; what helps in rejoining something that has broken is called *sandhaaneeyam*, for example, wheat, garlic, *priyangu* (*Callicarpa macrophylla*).
2. *Deepaneeyam*: *Deepaneeyam vahnekaddipanaaya hitham, deepanam antharagneh, sandhukshanam thasmai hitham deepaneeyam.* What increases the *agni* (*jataraagni*) is called *deepana*, for example, ginger, garlic, pepper, and asafetida.
3. *Keshyam*: *Keshaaya hitham keshyam.* What is good for the hair is *keshyam*, for example, sesame and garlic.
4. *Chakshushyam*: *Chakshushe hitham chakshushyam.* What is beneficial to the eyes is *chakshushyam*, for example, *shigru* (*Moringa oleifera*).
5. *Moothralam*: *Moothram laathi moothralam.* What brings about more (increases formation of) urine, for example, rice, black gram, sesame, water melon, brinjal, and *shata pushpa* (*Pimpinella anisum*).
6. *Vranaropana*: *Vranam ropayathi ithi vranaropanam.* What heals (fills) the wound is *vranaropanam*, for example, wheat, honey, and ghee (clarified butter).

A comprehensive list of such terms is found in *Nighantus* and other medical literature.

5.2.2 *Nighantus*

There is an abundance of valuable knowledge on *dravyaguna* (literally "properties of substances") in the form of *Nighantus*. *Nighantu* literally means a vocabulary or glossary of words. Initially, *Nighantus* contained only the synonymous words that had implicit meaning. This was the ancient method of conveying the idea in a concise form. The Ayurvedic *Nighantus* also followed this pattern and the earlier *Nighantus* described the drugs by way of synonyms. Later on, the properties and actions were added on and a new pattern of *Nighantus* was evolved, where properties and synonyms were described.

5.2.2.1 How Is a *Nighantu* Used?

I can take as an example the Bhaavamisra's *Bhavaprakasha Nighantu*. In the field of *materia medica*, Bhaavamisra made valuable contributions. If I want to know the *guna* (properties) and *karma* (action) of any substance, for example, a fruit such as plantain, I will have to look up the appropriate group under the texts. Bhaavamisra's *Bhavaprakasha* gives the synonyms, types, and qualities of plantain as follows (Sh. 33, 34):

> The synonyms of plantain are *kadali*, *vaaranaa*, *mocha*, *anubusaara*, and *amshumatiphala*.
>
> The quality changes if the fruit is raw or ripe, that is, raw plantains are tasty (sweet), cooling, anticarminative, they increase *kapha*, heavy, unctuous, they control *pitta*, *raktha*, thirst, burning sensation, injury, emaciation, and *vata*. *Mochaphalam swaadu sheetam vishtambho kapha krut guru snigdham pittaasra truddaaha kshatakshaya sameerajit* (Balasubramanian and Radhika, 1990). Ripe fruits are sweet (in both taste and *vipaka*), cooling, aphrodisiac, and bulk promoting; they increase taste and *mamsa* (flesh); and they cure hunger, thirst, eye diseases, and *prameha* (diabetes). *Pakwam swaadu himam paake swaadu vrushyamcha bruhmanam kshut trushna, netra gada hrunmehaghnam ruchi, maamsakrut* (Balasubramanian and Radhika, 1990).

Also the different varieties of plantains and their individual properties have been described. In general, the qualities of plantain (both raw and ripe) and their action on our body can be understood by the discussion above. In a similar way, the *guna/karma* of vegetables, flowers, drugs, and so on can be looked into and used accordingly.

5.2.2.2 Today's *Nighantus*

Some of the *Nighantus* that are available today are as follows:

1. *Dhanvantari Nighantu*
2. *Madanapaala Nighantu*
3. *Raja Nighantu*
4. *Kaiyadeva Nighantu*
5. *Bhavaprakasha Nighantu*
6. *Amarakosha*
7. *Shabdakalpadruma*

8. *Vaachaspathyaabhidhaana*
9. *Medinikosha*
10. *Dravyagunashathaka*

There are also other works in local languages in different regions.

5.3 FOLK KNOWLEDGE

The body of knowledge and practices related to nutrition exists both in the classical and the folk forms. Although this is a subject that merits a detailed investigation, a brief overview is provided below. There is a vast body of textual literature in Sanskrit and various other Indian languages (such as Tamil, Telugu, and Kannada), which not only have detailed and rich descriptions of the properties of foods but also provide a theoretical framework for this knowledge. Linked to this understanding, there is a rich folklore and traditional knowledge that still strongly influences and governs the choice of food by the people of India. This includes the understanding of what is healthy food, what is a balanced diet, which foods are appropriate or inappropriate during various seasons, incidence of illness, and various stages of life and growth such as infancy, women attaining puberty, pregnant and lactating mothers, and old age (Balasubramanian and Radhika, 1989, 1990; Nanal and Nanal, 1991).

An effort has always been made to give a complete and scientific solution for every problem that emerges at each time. This can be understood when one reads the *Nighantu* from the beginning till the end. *Charaka Samhita* says, *Pareekshya karino hi kushala bavanthi*, that is, "It is only those who test the knowledge obtained from texts with their practical experience who benefit from the *shastraas*." One might wonder if this kind of knowledge is fossilized—is this just a matter of referring to Ayurvedic sources that have compiled such information already? The real test for this as a knowledge system would be to apply the theoretical framework and principles of Ayurveda to evaluate new substances. If one takes a look at *Nighantus* as they have evolved through the last few centuries, I observe that the more recent *Nighantus* make note of the properties of substances that have been introduced into India from outside in more recent times, such as tobacco and new vegetables such as cabbage and cauliflower. Such efforts are continuing even in today's context. It is possible to evaluate any "new" substance (i.e., a substance of recent origin or new in the sense of being unknown to ancient physicians who compiled the *Nighantus*) and to assess its food value or medicinal properties based on the *pancha mahabutha siddhantha*. Analysis of one such food, namely, custard, has been done by Vaidya Ramesh Nanal (1989) from Mumbai. Per this analysis, custard controls *vata* and *pitta*; aggravates *kapha*; increases *rasa dhatu*; purifies *raktha*; and increases *mamsa*, *medas*, and *sukra*. Similarly, its action on diseases and contraindications have also been given.

5.4 APPROACH TO INTEGRATION

Having seen that there are two distinct and parallel traditions of knowledge and practices in nutrition, namely, the modern Western and the traditional, I now address the question of whether and how they can be integrated. To begin with, I would like to suggest that the integration that I speak about is quite different from other kinds of integration from which I differentiate this. For example, in modern physics, the theory and understanding relating to two different areas, namely, electricity and magnetism, were integrated under a single *electromagnetic theory* of matter and they were subsumed under *Maxwell's equations*. In a similar manner, today one often hears about working toward a *unified field theory*, by which scientists generally mean that electromagnetic and gravitational forces can be described in a unified manner by the same overarching set of equations (Feynman *et al.*, 1995, 1998).

I, however, feel that integration in our context would have a different kind of meaning and significance. Here I have a situation where solutions to the same problem can be derived from two different approaches to nutrition: traditional and modern. These approaches may be integrated at the level of practical application. In India, there are a wide range of efforts to carry out integration using several medical systems simultaneously or in parallel largely driven by the user group, namely, the patients. It is of great interest to examine this effort for the lessons that it has for possible integration of knowledge systems related to nutrition. A detailed description of the manner in which many users are integrating at the user level has been provided in this author's earlier paper entitled "Seeing with Two Eyes" (Balasubramanian, 2010). A large number of patients are involved in attempts to integrate inputs and solutions from different medical systems to take care of immediate health problems.

The systems included modern medicine, Ayurveda, Siddha, yoga, naturopathy, and homeopathy. Also on the scene or in the background were astrology and religious practices. Generally, one system was chosen for the primary line of intervention, with one or more further systems supplementing it. The choice of primary and supplementary systems was based on a number of factors, which are as follows:

- Exposure to various medical systems based on family background and previous experience
- Extent of exposure to modern/Western ideas
- Nature of the specific condition

However, a brief description of the salient features of this approach is given below.

5.4.1 "Seeing with Two Eyes"

It appears that in solving medical problems, seeing from the perspective of two different medical systems may be compared to the physiology of *seeing with two eyes*. There are clear advantages in seeing with two eyes, which are as follows:

- Improved resolution at edges, increased contrast, and better ability to read when the print is small or illumination is poor.
- Information obtained on the depth of field gaining *perspective*.
- Information provided by two eyes is thus information of a different logical type.

Various issues and problems that are encountered when such integration is attempted are discussed below.

5.4.2 Specific Aspects of Integration

Integration has to take place bearing in mind various aspects, including fundamental differences in approaches between modern and traditional knowledge. It would also be incomplete and distorted if it does not take into consideration insights that come in from the expert's viewpoint and if it is totally *ad hoc* and only at the operational level. In this section, the following four specific aspects relating to integration and its pros and cons are discussed.

1. Methodology and approach to scientific theory
2. Risks of *ad hoc* usage
3. Modern studies on traditional knowledge, for example, hot and cold
4. Noncommensurate concepts

5.4.2.1 Methodology and Approach to Scientific Theory: Contrasting Traditional and Modern Approaches

One has to be open to the possibility that the methods and approaches of different scientific systems may vary widely to the extent that to someone who is educated in one system, the first encounter with elements of a different system may lead to suspicion, skepticism, and a feeling of "this does not make any sense." A few years ago, at the Centre for Indian Knowledge Systems, I had a fascinating experience on the basis of which I had to delve into this matter in some detail. Per the insight and understanding that I have from *vrkshayurveda* (traditional Indian plant science), plant products that have a particular *rasa* (taste) such as a bitter taste or specific types of smells are excellent candidates for having biological control properties that may be effective against pests. I put this to practice in a very effective manner when I found that the leaves and stems of the plant *Andrographis paniculata* (known in Sanskrit as *bhunimba*), which are extremely bitter in taste, were found to have excellent pest control properties. However, I realized that a statement such as how the smell or taste of a plant can give a clue to its biological properties is viewed with a great deal of discomfort and probably considered as mumbo jumbo or nonscientific to those who are trained in modern scientific methods. It was necessary for us to delve into the roots of such thinking—from where did I derive the idea that scientific theories or discourses cannot be carried out using parameters such as smell or taste? The answer to this question took us as far back as Aristotle in the history of Western scientific method. I have paraphrased in this section extensively from a fascinating paper by Singh (2000) who looked into this.

As a founding act of the scientific method, Aristotle made an epoch-making distinction between trustworthy and untrustworthy qualities. Trustworthy qualities are those that can be grasped and cross-corroborated through more than one sense organ such as shape, size, and number, which, for instance, can be grasped by the sense organ of vision as well as that of touch. Untrustworthy qualities are those that can be grasped by only one sense organ and cannot be corroborated by another sense organ. Warding off "untrustworthy" qualities in the quest for knowledge is one of the cornerstones of the scientific method (Aristotle, 1961). Galileo Galilee gave finishing touches to the founding principles of the scientific method and its consequent power. Galilee opened another new argument in favor of the distinction between primary and secondary qualities when he voiced, somewhat dramatically, a contemporaneously shocking sentiment, by saying that there is more in common between the weight of an apple and the weight of the moon than between the weight of an apple and the color of the apple. What he noticed was a new clinching argument, which was based on the relationship between measurement and primary and secondary qualities. Secondary qualities cannot be measured simply and they cannot be determinably related to each other, whereas primary qualities can be determinably related to each other. The primary qualities of objects can all be measured, whereas their secondary qualities cannot be measured (Galilee, year unknown).

How was this viewed in Indian tradition? Let us take a look at how the logical schools of thought in India addressed this issue. Prasastapada had distinguished three classes of qualities that are grasped by the sense organs (Bhatti, 1997), which are as follows:

1. Those that can be acquired through two sense organs—number, magnitude, distinctness, conjunction, separation, farness, nearness, instrumental fluidity, attraction, and speed.
2. Those that can only be acquired through one sense organ—sound, touch, color, taste, and smell.
3. Those that can only be acquired through the internal sense organ (*manas*)—cognition, contentment, discontentment, desire, aversion, and effort.

All these three classes of qualities are trustworthy qualities that present themselves to elementary experience. Error and falsehood regarding the knowledge of qualities are important issues, and according to Prasastapada, "they need to be understood independently and not in relation with the basic sensory mode of acquisition of qualities." Aristotle not only overlooks the third class of qualities altogether but also relates trustworthiness with corroboration through more than one sense organ, a feature of the first class of qualities.

The overall lesson from this is the understanding that I cannot and should not dismiss discussions on nutritional or therapeutic properties based on parameters such as smell or taste, no matter how strange or unscientific it may appear based on Western scientific learning. This may leave us with the question of the grounds on which some understanding may be accepted or rejected. In my view, these grounds may only be the testing out of the implications for such theories in practice. In the example that I have cited earlier, I received a lead or a clue based on *vrkshayurveda* to the effect that plants that have a bitter taste may be good biopesticides, and on these grounds, the leaves and stems of the plant *Andrographis paniculata* were an excellent candidate. However, we did put this lead to practical testing and found that the leaves and stems of the plant were indeed excellent as biopesticides to control fruit and shoot borers in some vegetables. In this sense, the understanding from *vrkshayurveda* was validated through our practical experience in the field.

5.4.2.2 Risks of *Ad Hoc* Usage

Integration of systems cannot mean the picking up and use of bits and pieces of one tradition with disregard for the wisdom or knowledge accompanying it in terms of the context of its use. As an illustration, consider this example of how modern research on a specific aspect of tradition may merely increase confusion when modern practitioners utilize pieces of it *ad hoc* (Tiwari, 2007). A recent review of antidiabetic properties of bitter gourd states in its review of scientific literature, "Modern scientific analysis of [bitter gourd's] antidiabetic properties reveal that it has the capacity to regulate vitiated carbohydrate digestion, glucose metabolism and utilization, possesses insulin mimetic and secretagogue activities, and corrects the impaired antioxidant defence in diabetes" (Tiwari, 2007, 1697). Based on this kind of thinking, some modern medical practitioners might suggest that their patients consume bitter gourd, or drink its juice, in quantity, based on the underlying thought that even if it does not help, it can do no harm. However, Ayurveda cautions against such generalized practice, pointing out that although bitter gourd is beneficial to diabetics who are overweight and obese (*kapha* type), it may not have such benefits and could even harm the very different kind of diabetics who are thin and emaciated (*vata* type) (Gangadharan, 2010).

5.4.2.3 Modern Studies on Traditional Knowledge—Hot and Cold

In India, choices of food among the people at large are still derived from the Ayurvedic worldview and understanding. This is an area where the local communities abound in various do's and don'ts, which often appear to be not comprehensible to the modern investigator. A detailed study in Andhra Pradesh of 144 *dai*s (traditional birth attendants) and 640 rural women has recorded a number of do's and don'ts in the matter of food during pregnancy (Krishna, 1984). Among the foods to be avoided were pumpkin, banana, brinjal, gongura (*Hibiscus sabdariffa*), guava, and papaya. In general, foods considered "hot" were avoided during pregnancy. Yet another study in Andhra Pradesh lists that eggs, jaggery, and papaya, which are considered to be hot foods, are forbidden for fear of abortion (Parvathi, 1968). The same study also states that citrus fruits, buttermilk, and curd are forbidden for fear of cold. The fact is that many of such practices are supported by Ayurvedic theory and would make sense if I work to understand it by analyzing them through Ayurvedic worldview and criteria. In fact, concepts such as hot and cold are not only not understood but also dismissed as superstition. Consider the following question, which appeared in a journal published by the National Institute of Nutrition (Reader's Question, 1983).

> It is true that there is a widespread belief that some foods are "hot" and others "cold." However there appears to be no scientific basis for such classification. It is not desirable to limit the choice of food based on such beliefs.

There has been an attempt to directly understand and assess the concepts of traditional medicine such as hot or cold employing modern nutritional laboratory techniques. At the outset, it may appear that the concepts of traditional medicine such as hot and cold are completely outside the scope of understanding of modern medicine, and few would think that there is any kind of dialog possible. However, it is interesting to note that in the late 1960s an attempt was made by a scientist from the National Institute of Nutrition, Hyderabad, to try and understand the biochemical basis for the Ayurvedic understanding of hot and cold foods (Ramanamurthi, 1969). A list of food items that are believed to produce either heat or cold when consumed was prepared, and from among these, two diets were formulated. The first consisted mainly of hot foods and the second cold foods. Hot foods were generally those that, when consumed, produced a subjective feeling of burning eyes, burning micturition, and a feeling of warmth all over the body. Four normal healthy adult subjects from the laboratory staff were selected for the metabolic study. Protein, fat, and calorie intakes were adjusted according to the Nutrition Advisory Committee recommendations. The subjects were first placed on a diet comprising mostly of the hot foods for a period of 10 days followed by the second period of 10 days when they were given a diet consisting mainly of cold foods. The period of intake of each of these types of foods was 10 days. When the subjects were consuming hot foods, the acidity of the urine was much higher than when they were consuming cold foods. The high acidity of the urine may be the possible cause for the burning sensation during micturition. Urinary excretion of sulfur was also found to be higher in hot foods than in cold foods, indicating a possibility that the high sulfur content of the urine may be the cause for the high acidity of the urine. Nitrogen retention was found to be lower in hot foods than in cold foods, although the total nitrogen content was the same in both the diets. Surprisingly, this remains the only research report in India where such an effort has been made to comprehend Ayurvedic parameters through a laboratory exercise.

5.4.2.4 Noncommensurate Concepts

It may not always be possible to use multiple medical systems simultaneously. Using more than one medical system has interesting advantages and should be taken forward. However, I may foresee instances where two such approaches may be so different as to be almost incompatible, for example, the comprehension of morning sickness (termed *nausea gravidarum*) experienced by women in pregnancy. Modern science understands this phenomenon in terms of levels of certain hormones such as estrogen, human chorionic gonadotropin, or hypoglycemia. The traditional Ayurvedic view stands in stark contrast. Pregnant women are termed *douhrdini*, literally meaning "endowed with two *hridayas*." In this context, *hridaya* is not merely the physiological heart but a seat of emotion. The pregnant woman is seen as one who has a second center and, in the fetus nurtured by her, a second seat of emotion growing within her! This gives Ayurveda a completely different insight into the mental state of a pregnant woman, her likes and dislikes, inexplicable reversals of tastes and preferences, and so on. What is significant is that this is not merely a philosophical understanding, but Ayurveda offers various time-tested practical solutions in terms of how the pregnant woman can be helped through her morning sickness. In such case, the gap between the two approaches is so large that it is not feasible to integrate them in any simple manner. These approaches would be rich areas for research, which would highlight foundational differences between the two systems.

5.5 SYNERGY BETWEEN MULTIPLE TRADITIONS

During the last 100 years or so, the phenomenal success and progress of modern medicine has led to a situation where in India all modern education on medicine, health, and nutrition draws only on modern nutritional concepts based on biomedicine.

For example, this has meant that, in terms of understanding the properties of food substances, it has been exclusively based on protein, carbohydrates, calories, vitamins, minerals, and so on. This has no doubt been reinforced by the significant achievements and successes in addressing diseases caused due to deficiency of specific nutrients such as vitamin A, vitamin C, and iron. The National Policy on Nutrition adopted by the Government of India in 1993 makes no mention about the concepts drawn from traditional Indian nutrition and its frame of thinking is entirely modern Western (National Nutrition Policy, 1993). Similarly, although the Government of India has departments of Indian systems of medicine with research councils in charge of Ayurveda, Siddha, and Unani systems of medicine, they normally run parallel educational courses with very little integration with the biomedical stream.

In India, mainstream health professionals tend to treat practices drawn from traditional nutrition to be almost the same as superstition. In fact, it would not be an exaggeration to state that most of them are of the opinion that these practices have no theoretical basis or foundation, and these are considered to be the "obstacles in the part of educating the public in general about nutrition." The illustration that I provide below drives home this point.

In retrospect, I can see that through the ages, every geographical location of the world has nurtured and produced sciences and technologies that bear the distinct stamp and character of its own people and civilization. However, during the last few hundred years and increasingly for a hundred years or so, an opinion has developed that the Western tradition of science and technology is unique and universal based on scientific instrumental evaluations (Tiwari, 2007). This viewpoint has been not only propagated by the mainstream scientists in the West but also internalized by professionals in various other parts of the world, particularly developing countries.

Today, I see that the public health experts and professionals are often at a loss to comprehend old practices. This is because their training is entirely drawn from the principles of nutrition based on modern Western practices and they have had no appropriate education of the traditional nutrition theory and practices. Hence, there is an urgent need for a cross-cultural dialog on this matter and to build bridges of understanding. This is necessary to ensure that the enormous corpus of traditional literature on this matter plays a proper role in achieving good health for all in today's context.

In this chapter, I have discussed a possible approach to integration of traditional Indian concepts of nutrition with the modern understanding by taking as a starting point an approach that is widely prevalent in India with respect to integration of medical systems for treatment. Some salient features of this approach have been outlined and some specific aspects related to integration and its pros and cons have been discussed. These range from examining the aspects of methodology and scientific theory to taking note of some interesting past efforts and looking at the risks of *ad hoc* incorporation of bits and pieces of knowledge and practice from the traditional system to a modern framework. I hope that this would stimulate discussion about how integration can take place. In our view, a basic starting point for any such effort would be a dialog between the experts who are well versed in each of the systems with an attitude of mutual respect and a conviction that in today's context no system is complete and each system can indeed learn from and be enriched by such an interaction.

REFERENCES

Aristotle. *De Anima* (N.D. Rose d.). Oxford University, Oxford, **1961**.

Balasubramanian A.V. Seeing with two eyes: How professionals can help patients trying to integrate medical systems. *J. Ayurveda Integr. Med.* **2010**, *1*(3), 177–182.

Balasubramanian A.V., Radhika M. *Ayurvedic Principles of Food and Nutrition: Part I.* Lok Swasthya Parampara Samvardhan Samithi, Chennai, Tamil Nadu, India, **1990**.

Balasubramanian A.V., Radhika M. *Local Health Traditions: An Introduction.* Lok Swasthya Parampara Samvardhan Samithi, Chennai, Tamil Nadu, India, **1989**.

Bhatti, S. (ed.) *Padarthadharma Samgraha of Prasastapadacharya* with the commentary *Nyayakandali*. Durgadhar Jha, Varanasi, India, **1997**.

Feynman R.P., Leighton R.B., Sands M. *The Feynman Lectures on Physics—Vol. II: Mainly Electromagnetism and Matter.* Narosa Publishing House, New Delhi, India, **1995** (9th reprint).

Feynman R.P., Leighton R.B., Sands M. *The Feynman Lectures on Physics—Vol. III Quantum Mechanics.* Narosa Publishing House, New Delhi, India, **1998** (13th reprint).

Galilee. G. *Dialogue Concerning Two New Sciences* (Text with English translation). Year unknown.

Gangadharan G.G. Bengaluru, India, Personal communication, **2010**.

Krishna T.P. Traditional birth attendants (*dais*) and rural woman, their concepts and attitude towards nutrition and child health in Andhra Pradesh. *India Pediatr.* **1984**, *21*(1), 29–34.

Nanal R.M. *Custard: Ayurvedeeya drishtikon se gun karmaadi ka nischay* (Determination of properties from an Ayurvedic perspective). Ayurvedeeya Mahasammelan Patrika, New Delhi, India, **1989**, 17–19.

Nanal V.M., Nanal R.M. *Ayurvedic Principles of Food and Nutrition: Part II.* Lok Swasthya Parampara Samvardhan Samithi, Chennai, Tamil Nadu, India, **1991**.

National Nutrition Policy. Dept. of Women and Child Development, Ministry of Human Resource Development, Govt. of India, New Delhi, India, **1993**.

Parvathi K.R. Socio cultural factors and malnutrition in Telangana and Andhra Pradesh. *Proc. Nutr. Soc. of India* **1968**, 32–43.

Ramanamurthi P.S.V. Physiological effects of "hot" and "cold" foods in human subjects. *J. Nutr. Diet.* **1969**, *6*, 187–191.

Reader's Questions. *Nutrition.* **1983**, *17*(4), 29.

Singh N. What is scientific method and how is it related to Sastra? In D. Prahaladacharya (ed.), *Vijnana Vimarsha*, Rashtrya Samskrita Samsthan, New Delhi, India, **2000**.

Tiwari A.K. Karela. A promising antidiabetic vegetable therapy. *Curr. Sci.* **2007**, *92*(12), 1697–1701.

Section II

The Pillars of Western Nutritional Science

6 Nutrients
The Essence of Life

Cristian Del Bo' and Marisa Porrini

CONTENTS

6.1 NUTRIENTS: DEFINITION AND CLASSIFICATION

The approach to defining the nutritional adequacy of diet has progressed according to scientific understanding of the biochemical and physiological basis of human nutrition.

The primary role of diet is to provide an adequate mix of foods and beverages in appropriate amounts, to sustain growth and development of the body and to optimize its numerous functions. This may be achieved by means of nutrients and other substances of nutritional interest, widely distributed in foods that, following absorption in the gastrointestinal track, intervene in many physiological processes and metabolic pathways essential for life. Nutrients include macronutrients, such as carbohydrates, proteins, fats, and water that the body needs in relatively large amounts, and micronutrients, such as vitamins and minerals provided in minor amounts. Moreover, there are other compounds in food that, although not considered as nutrients (e.g., polyphenols, carotenoids, etc.), are involved in human metabolism and functions. Some nutrients can be also synthesized by the organism starting from other substances [e.g., saturated fatty acids (SFAs) are synthesized from glucose, glucose from some amino acids]; other ones must be derived from the diet and are consequently considered essential. The categories of essential nutrients include vitamins, minerals, essential fatty acids (EFAs), and essential amino acids (EEAs).

6.2 MACRONUTRIENTS

Macronutrients (carbohydrates, proteins, fats) supply the energy necessary to the organism: fats yield 9 kcal/g (37.8 kJ/g) and proteins and carbohydrates yield 4 kcal/g (16.8 kJ/g). Water does not supply energy and might not be considered a macronutrient; however, it is needed in large amounts for all body functions, representing approximately 60% of the weight of an adult individual.

6.2.1 CARBOHYDRATES

Dietary carbohydrates are the bulk of the diet. They are classified as simple (monosaccharides and disaccharides) or complex (soluble and insoluble polysaccharides) depending on their chemical structure and the rate of digestion and absorption (Cummings and Stephen, 2007). The major providers of food carbohydrates are cereals and their derivatives, representing about 50% of all the carbohydrates consumed worldwide, followed by legumes, fruits, vegetables, and milk products. Carbohydrates have a role to play in several important roles, including structural components of cells; components of RNA and DNA (ribose and deoxyribose, respectively); and integral features of many proteins and lipids (glycoproteins and glycolipids, respectively), especially in cell membranes, where they are essential for cell–cell recognition and molecular targeting (Cumming *et al.*, 1997; Elia and Cummings, 2007). Above all, carbohydrates represent the first source of fuel and energy of the human body, especially for the brain and the nervous system (Galgani and Ravissin, 2008). According to different recommended daily allowances in use in numerous countries, carbohydrates should supply approximately 45%–60% of the total energy intake, mainly as complex carbohydrates (EFSA, 2010b; FAO/WHO, 2007).

In this regard, it is important to consider that the amount and type of carbohydrates consumed have an effect on insulin secretion and postprandial glycemia. In general, fully digestible carbohydrates are rapidly absorbed in the small intestine and induce a rapid glycemic and insulinemic response, whereas slow-digestible carbohydrates are slowly or incompletely absorbed in the small intestine, inducing lower blood glucose concentration, and partially fermented by bacteria in the large intestine (Wong and Jenkins, 2007). The glycemic response to different kinds of carbohydrates can be evaluated by the glycemic index (GI), which can be defined as the indexing of the glycemic response of a fixed amount of available carbohydrate from a test food to the same amount of available carbohydrate from a standard food consumed by the same subject (the standard food can be glucose or, better, white bread) (Monro and Shaw, 2008). In general, the insulin responses relate well to glycemic responses. From GI, it is possible to calculate the glycemic load, which is the product of the GI of the food and the amount of carbohydrates, as a global indicator of the glucose response and insulin demand induced by the amount of food consumed (Monro and Shaw, 2008).

The glycemic response to carbohydrates is influenced by intrinsic and extrinsic factors that alter the rate of gastrointestinal motility, digestion, and absorption, as well as by the nature of carbohydrates; particle size; cooking method; food matrix; and the presence of fiber, fats, and proteins. This implies that GI is not the same for all kinds of starches as well as of simple sugars, but varies a lot among different food sources (Björck and Elmståhl, 2003). The starchy staples of traditional cultures were often foods that had low GIs, such as pasta, whole-grain pumpernickel breads, cracked wheat or barley, dried peas, beans, and lentils; also, sugars coming from milk (lactose) and fruit (fructose) have low GI (Björck *et al.*, 2000). However, over time the new dietary trends resulted in the selection of foods rich in rapidly digested carbohydrates and hence with high GI. Numerous findings have emerged to suggest that the dietary GI is of potential importance in the treatment and prevention of chronic diseases (Brand-Miller and Buyken, 2012). For example, low-GI diets have been shown to lower urinary C-peptide excretion in healthy subjects, improve glycemic control in diabetic subjects, reduce serum lipids in hyperlipidemic subjects, increase high-density lipoprotein cholesterol concentrations, and, in large cohort studies, decrease the risk of developing diabetes and cardiovascular disease (CVD) (Mirrahimi *et al.*, 2014; Riccardi *et al.*, 2008).

The hypothesized metabolic effects relate to the fact that a reduced rate of glucose absorption after the consumption of low-GI carbohydrate foods will reduce the postprandial rise in gut hormones (e.g., incretins) and insulin (Jenkins *et al.*, 2002).

Notwithstanding some issues that are still being debated, the GI has proven to be a more useful nutritional concept than the chemical classification of carbohydrates (as simple or complex, sugars or starches) allowing new insights into the relation between the physiological effects of carbohydrate-rich foods and health.

6.2.2 Lipids

There is a wide variety of lipids, including fatty acids, triglycerides, phospholipids, and sterols (e.g., cholesterol). These compounds are almost ubiquitous in foods, spacing from vegetables to meat products. Some of them are only present in some foods. For example, plant oils from seeds or nuts are rich in triglycerides. From a qualitative point of view, the length of the carbon chain, the presence of the double bonds, their location, and their geometric isomerism affect the physical, biological, and nutritional properties of lipids.

The main function of lipids is to provide energy for many tissues and organs, in particular for the cardiac and skeletal muscles (Burlingame *et al.*, 2009). Lipids have the highest energy content (1 g provides about 9 kcal). In addition, they are important signaling molecules and key components of cell membranes in all tissues, as well as precursors of regulatory substances of the cardiovascular system, blood coagulation, renal function, and immune system (Burlingame *et al.*, 2009). Other important functions of lipids include enzymatic activation and molecular transportation (i.e., they act as carriers for fat-soluble vitamins and other compounds such as carotenoids, by helping their absorption). Linoleic acid (18:2n-6) and α-linolenic acid (18:3n-3) are

EFAs; after absorption, they are converted into other polyunsaturated fatty acids (PUFAs) of the series *n*-6 [e.g., arachidonic acid (AA)] and *n*-3 [e.g., eicosapentaenoic acid; docosahexaenoic acid (DHA)], respectively. The *n*-6 and *n*-3 PUFAs are important structural components of cell membranes; they can affect cell membrane properties such as fluidity, flexibility, permeability, and the activity of membrane-bound enzymes (Russo, 2009). DHA is selectively incorporated into retinal cell membranes and postsynaptic neuronal cell membranes, suggesting an important role in vision and nervous system function (Bradbury, 2011). Moreover, several studies indicate that *n*-6 and *n*-3 fatty acids can modulate the expression of a number of genes, including those involved in fatty acid metabolism and inflammation such as peroxisome proliferator-activated receptors (PPARs) and liver X receptors (LXRα and LXRβ) (Bordoni *et al.*, 2006). Although *n*-3 and *n*-6 fatty acids share some metabolic pathways, their pathophysiological effects are conflicting (Schmitz and Ecker, 2008). The *n*-3 PUFAs help reduce inflammation, whereas most *n*-6 PUFAs tend to promote inflammation (Schmitz and Ecker, 2008). Moreover, the *n*-6 PUFAs, especially AA, promote platelet aggregation and vasoconstriction, whereas the *n*-3 PUFAs have a minor pro-inflammatory and pro-aggregating effect and a major vasodilator effect (Wijendran and Hayes, 2004). Thus, their balance in the diet is essential to maintain the physiological functions with a view in improving general health and reducing the risk of diseases (Wijendran and Hayes, 2004).

An important metabolic effect of *n*-3 PUFAs is the reduction of circulating levels of triglycerides, whereas *n*-6 PUFAs reduce low-density lipoprotein (LDL) cholesterol. At the molecular level, *n*-3 fatty acids act by suppressing lipogenesis and upregulating fatty acid oxidation; these effects are mainly exploited through the regulation of mechanisms involving the transcription factors LXRα, sterol regulatory element binding protein-1 (SREBP-1), and PPARs (Bordoni *et al.*, 2006; Kersten, 2002; Wijendran and Hayes, 2004).

Concerning monounsaturated fatty acids (MUFAs), several studies have shown to favorably affect a number of CVD risk factors, including plasma lipids, lipoproteins and cholesterol, factors related to thrombogenesis, LDL oxidative susceptibility, and insulin sensitivity (Gillingham *et al.*, 2011). However, evidence suggests that diverse sources of MUFAs (animal fats vs. vegetable oils) may have different effects in the modulation of CVD risk factors (Degirolamo and Rudel, 2010; Truswell and Choudhury, 1998).

In a balanced diet, the contribution in energy provided by fats (mainly from triglycerides) varies according to the age of the individuals. According to data from numerous longitudinal and clinical studies, the reference fat intake is 20%–35% of the total energy for adults, depending also on the amount of carbohydrates in the diet: there is in fact a range within which the organism is able to maintain a good physiological and functional balance (EFSA, 2010a). Moreover, there is a strong scientific debate among nutritionists about the rate in which SFAs, MUFAs, and PUFAs should be introduced through the diet. Generally, it is assumed that the relative amount of SFAs should not exceed 10% of the total energy for the prevention of CVD risk factors (EFSA, 2010b; FAO/WHO, 2010). On the contrary, no agreement has been reached regarding the amount of MUFAs and PUFAs, and the best ratio *n*-6/*n*-3 for the population.

The minimum requirement for the adult comes from studies on the basis of which it appears that 250 mg/day of PUFAs is enough to maintain the healthy status of individuals (EFSA, 2010a).

Evidence from epidemiologic and clinical studies indicates that the quality of fat is more important than its total amount in promoting health. For example, it was estimated that replacement of 5% of energy from SFAs with energy from unsaturated fatty acids would reduce the risk of cardiovascular events by 42%, whereas the replacement of 2% of energy from *trans*-fatty acids with energy from *cis*-fatty acids would reduce risk by 53% (Hu *et al.*, 1997). Therefore, these findings suggest that replacing SFAs and *trans*-fatty acids with *cis*-monounsaturated and *cis*-polyunsaturated fats is more effective than reducing overall fat intake in preventing cardiovascular events (Hu *et al.*, 1997; Lawrence, 2013).

6.2.3 Proteins

Proteins are important biomolecules required for the structure, function, and regulation of body tissues and organs. In particular, they carry out several functions as antibodies, enzymes, messengers, structural components, transporters, and storage components (Matthews, 2006). Proteins are also a good source of energy (4 kcal/g). However, they must be the last source of energy to be used of all macronutrients. In case of extreme starvation, the body muscles are used to provide energy through a process called muscle wasting. Proteins are formed by chains of amino acids linked by peptide bond and characterized by the proportion and the sequence in which these amino acids bind to form the peptide chain (Matthews, 2006). Proteins have the peculiar characteristic of being continually hydrolyzed and synthesized to allow the organism to modulate the synthesis of its own proteins depending on the evolution of its needs: this is the basis of its ability to adapt (Matthews, 2006). Although at the cellular level all the 20 amino acids used for protein synthesis must be present simultaneously, only 9, the so-called EEAs, are to be introduced preformed with food because the body is unable to synthesize them (WHO/FAO/UNU, 2007). According to the different amino acid composition, the biological value of a protein reflects the percentage of its amino acids that are available for incorporation into endogenous proteins (Millward and Rivers, 1988). The biological value varies widely among animal and vegetable proteins. For example, animal proteins in milk, eggs, meat, and fish have a high biological value; on the contrary, proteins in cereals and vegetables have a lower biological value (Schaafsma, 2005). In fact, legumes and cereals provide only some of the EAAs (e.g., legumes lack methionine, whereas cereals lack lysine), but it has been recognized that mixing them in the diet may improve the value so much until the protein quality of meat, eggs, and dairy products is matched. The daily dietary protein requirement varies from 1.1 g/kg in infants to 0.9 g/kg in children and to 0.8 g/kg in adults (EFSA, 2012; WHO/FAO/UNU, 2007).

6.2.4 Water

Water is the predominant component of the human body and represents approximately 60% of the weight of an adult individual. This percentage is higher in childhood (at birth it is about 80% of body weight) and decreases progressively with age (by the age of 60 years it has decreased to only 50% of body weight) and/or with the increase of fat depots. In adults, up to 67% of the total body water is distributed within cells and, under physiological conditions, represents an index of body cell mass (Oh and Uribirri, 2006). The remaining 33% is external to cells and constitutes the extracellular fluid, which includes the interstitial fluid (23%), plasma (7%), lymph (2%), and transcellular fluid (1%). Water is involved in several functions; for example, it helps the digestive processes, it is a good source of minerals, it is the solvent of metabolic reactions, it regulates cell volume and body temperature, and it allows the transport of nutrients and removal of metabolic waste (Oh and Uribirri, 2006). The amount of water produced with the metabolism (about 350 ml/day) is not sufficient to cover the daily requirement. Thus, water must be supplied by foods and beverages. The requirement varies greatly from individual to individual and depends on the composition of the diet, climate, and physical activity. In physiological conditions, the water daily turnover corresponds to 15% of the body weight in the first few months of life and 6%–10% of body weight in the adults. The physiological loss of water is mainly due to respiration and perspiration (about 1250 ml/day) and the production of urine (800–1500 ml/day) and feces (100–150 ml/day) (Oh and Uribirri, 2006). High water loss can also occur through sweat, vomiting, and diarrhea.

6.3 MICRONUTRIENTS

Micronutrients include vitamins and minerals that should be introduced through the diet in small amounts (grams, milligrams, micrograms). They do not represent an energy source, but most of them are constituents or precursors of enzymes, hormones, and other substances, and are consequently essential for the normal functioning of the human body (WHO/FAO, 2001). Micronutrients are differently distributed in foods, even in relatively large amounts; however, the extent to which they are available for absorption and utilization varies greatly depending on food composition and structure, but also physiological aspects and health status.

6.3.1 Vitamins

Vitamins are a group of organic compounds that are essential for normal metabolism, growth, development, and regulation of cell function (WHO/FAO, 2001). They are divided into fat-soluble (vitamins A, D, E, and K) and water-soluble (vitamins of B group and C) vitamins. Fat-soluble vitamins are provided by green leafy vegetables, milk and dairy products, and plant oils. They share the same absorption mechanisms of lipids and can be stored in the fatty tissues of the body when they are in excess, so they are not excreted easily (WHO/FAO, 2001). This means that humans do not need to introduce them daily. Water-soluble vitamins are widely distributed in foods. Usually they are not stored in the body in large amounts and must be replaced regularly with the diet (vitamins B_6 and B_{12} are exceptions). Vitamins of B group function as coenzymes that intervene in energy and nutrient metabolism, and are also important for the nervous system, red blood cell formation, and a healthy skin (WHO/FAO, 2001).

Most vitamins are well absorbed from foods; thus, their bioavailability is generally high. Each vitamin exerts a specific function in the body and, when their level is low, a specific deficiency disease may develop. However, an excessive intake of vitamins can cause toxic effects; this is theoretically possible for any vitamin, even if vitamin A toxicity is the most frequently observed. Therefore, vitamin supplementation in large amounts must be carefully considered. Except some specific situations, a well-balanced diet remains the most advisable way to fulfill physiological needs. In Table 6.1, the principal functions carried out by each vitamin and the recommended dietary allowance according to the National Recommended Energy and Nutrient Intake Levels are reported (LARN, 2012).

6.3.2 Minerals

Essential minerals, including trace elements, are inorganic elements that have a physiological function within the body. Approximately 4% of body mass consists of minerals. These must be supplied in the diet through foods and fluids, and their intake varies from grams per day for the major (macro) minerals through milligrams to micrograms per day for the microminerals (or trace elements). Macrominerals

TABLE 6.1
Vitamins, Functions, Food Sources, and Recommended Dietary Allowance for the Population

Vitamins	Functions	Dietary Sources	Recommended Dietary Allowance in Males and Females
		Fat-Soluble Vitamins	
Vitamin A (retinol)	Helps maintain healthy skin, eyes, and hair Required to resist infection and maintain integrity of cell membranes	Milk, eggs, meat, cod liver oils, green leafy vegetables, yellow and orange vegetables, and fruits	400–700 μg[a]
Vitamin D (calciferol)	Cooperative bone making and bone/teeth maintenance Needed for calcium and phosphorus absorption and metabolism	Sunshine, dairy products, egg yolk, fish, and cod liver oils	15–20 μg
Vitamin E (tocopherol)	Helps maintain cell membrane integrity Acts as antioxidants protecting vitamin A and fatty acids from oxidation	Plant oils, vegetables, wheat germ, whole grains, egg yolk, nuts, and seeds	5–13 mg[b]
Vitamin K (phylloquinone)	Helps make factors that promote blood clotting and helps bone mineralization	Green leafy vegetables and milk	60–170 μg

(*Continued*)

TABLE 6.1 (*Continued*)
Vitamins, Functions, Food Sources, and Recommended Dietary Allowance for the Population

Vitamins	Functions	Dietary Sources	Recommended Dietary Allowance in Males and Females
	Water-Soluble Vitamins		
Vitamin C (ascorbic acid)	Involved in collagen production Required for healthy skin, bones, cartilage, teeth, and blood vessels	Fresh fruits, in particular citrus, and vegetables	40–105 mg
Vitamin B_1 (thiamine)	Supports normal appetite, digestion, and nervous system function Coenzyme in energy metabolism	Whole grain cereals, legumes, nuts, and meats (pork, ham, and liver)	0.5–1.2 mg
Vitamin B_2 (riboflavin)	Supports normal vision and healthy skin Coenzyme in energy metabolism	Dairy products, meat, green leafy vegetables, and whole grains	0.6–1.6 mg
Vitamin B_3 (niacin)	Supports healthy skin, nervous system, and digestive system Coenzyme used in energy metabolism.	Tuna, dairy products, meat, whole grains, and nuts	7–18 mg[c]
Vitamin B_5 (pantothenic acid)	Essential for metabolism of fats, carbohydrates, and proteins, and for making fatty acids and cholesterol	Liver and kidney, yeast, egg yolk, fish, shellfish, chicken, milk, yogurt, broccoli, legumes, mushrooms, avocado, sweet potatoes, and whole grains	2.0–5.0 mg
Vitamin B_6 (pyridoxine)	Essential for amino acid and carbohydrate metabolism Helps the normal function of nervous system	Green leafy vegetables, meat, fish, poultry, shellfish, legumes, fruits, and whole grains	0.5–1.7 mg
Vitamin B_{12} (cobalamin)	Forms and generates red blood cells Needed for DNA synthesis Maintains a healthy nervous system	Meat and fish products, eggs, and dairy products	0.9–2.4 μg
Biotin	Coenzyme used in energy metabolism, fat synthesis, amino acid metabolism, and glycogen synthesis Required for healthy skin and hair	Egg yolk, liver, and yeast	10–30 μg

(*Continued*)

TABLE 6.1 (*Continued*)
Vitamins, Functions, Food Sources, and Recommended Dietary Allowance for the Population

Vitamins	Functions	Dietary Sources	Recommended Dietary Allowance in Males and Females
Folic acid	Essential for blood cell formation, protein metabolism, RNA and DNA production, and prevention of neural tube defects	Green leafy vegetables, legumes, and seeds	150–400 μg

Source: *LARN: Livelli di Assunzione di Riferimento di Nutrienti ed energia per la popolazione italiana*, Revisione 2012, Documento di sintesi per il XXXV Congresso Nazionale SINU, Bologna, Italy, 34–45, http://www.sinu.it/documenti/20121016_LARN_bologna_sintesi_prefinale.pdf, 2012.

[a] Retinol equivalents.

[b] α-Tocopherol equivalents.

[c] Niacin equivalents.

include calcium, magnesium, and phosphorous, particularly important for body growth and development, and electrolytes include sodium, potassium, and chloride, which are involved primarily in body fluid balance. Microminerals include iron, copper, zinc, cobalt, chromium, and fluoride. They are mostly cofactors and are necessary for the function of enzymes in the body, but the knowledge of their roles in numerous body functions is a rapidly expanding area of knowledge in nutrition (WHO/FAO, 2001). Most essential minerals are not highly available from food sources and this aspect must be taken into consideration when planning dietary protocols. In Table 6.2, the principal functions carried out by macro- and microminerals, and the recommended dietary allowances according to LARN are reported (LARN, 2012).

6.4 SUBSTANCES OF NUTRITIONAL INTEREST

In addition to nutrients, foods are also rich in other compounds that do not have the characteristics of a nutrient, but can exert a protective effect on human health and well-being. These substances of nutritional interest include dietary fiber and bioactive components.

6.4.1 Dietary Fiber

Nonstarch polysaccharides [cellulose, hemicelluloses, pectins, hydrocolloids (i.e., gums, mucilages, β-glucans), resistant oligosaccharides (fructo-oligosaccharides, galacto-oligosaccharides, and other resistant oligosaccharides), resistant starch (consisting of physically enclosed starch, some types of raw starch granules, retrograded

TABLE 6.2
Minerals, Functions, Food Sources, and Recommended Dietary Allowance for the Population

Minerals	Functions	Dietary Sources	Recommended Dietary Allowance in Males and Females
Calcium (Ca)	Aids in bone and teeth mineralization, muscle contraction and relaxation, heart function, and nerve function	Dairy products, fortified soy and rice milk, fish bones	700–1300 mg
Phosphorus (P)	Supports in formation of bones and teeth Regulates release and use of body energy and maintains normal acid/base balance in the body Helps carry fat in the body as part of phospholipids	Meat, fish, poultry, eggs, milk, cereals, and cereal products	460–1250 mg
Magnesium (Mg)	Aids in bone mineralization Necessary for muscle contraction and nerve function Active in many chemical reactions in the body	Whole grains, nuts, green vegetables, and legumes	80–320 mg
Sodium (Na)	Important component of body fluids, mostly outside cells Works with potassium to regulate fluid and acid/alkali balance	Table salts, shellfish, seafood, meat, milk, cheese, bread, bakery products, processed food, and vegetables	700–1500 mg
Potassium (K)	Important component of body fluids, mostly inside cells Works with sodium to regulate fluid and acid/alkali balance	Fruits and vegetables, meat, milk, and legumes	1900–3900 mg
Iron (Fe)	Makes hemoglobin in the blood, which carries oxygen Makes myoglobin, which stops oxygen in the muscles Makes energy in the mitochondria	Meat, legumes, tofu, green leafy vegetables, and breakfast cereals	8–18 mg
Iodine (I)	Part of thyroid hormones (thyroxin and triiodothyronine)	Iodized salts and seafood	90–150 μg
Selenium (Se)	Antioxidant, immunity, possible cancer prevention, and viral infection prevention	Whole grains from selenium-rich soils, poultry, meat, and dairy products	20–55 μg
Zinc (Zn)	Growth, immunity, wound healing, taste, sperm production, antioxidant, and prostate health	Oysters, meat, poultry, and fish	3–11 mg

(Continued)

TABLE 6.2 (*Continued*)
Minerals, Functions, Food Sources, and Recommended Dietary Allowance for the Population

Minerals	Functions	Dietary Sources	Recommended Dietary Allowance in Males and Females
Chromium (Cr)	Works closely with insulin to regulate blood sugar (glucose) levels	Liver, brewer's yeast, whole grains, nuts, and cheese	12–35 μg
Copper (Cu)	Constituent of many enzymes	Legumes, grains, nuts, seeds, and organ meat	400–1000 μg
Fluoride (F)	Constituent of the teeth and bone Useful to prevent cavities in the teeth	Fluoridated drinking water, fish, and tea	1400–4000 μg
Chloride (Cl)	Needed for proper fluid balance and stomach acid	Table salt, large amounts of processed foods, milk, meats, bread, and vegetables	1100–2300 mg
Manganese (Mn)	Constituent of many enzymes	Widespread in foods, especially plant foods	1000–2700 μg
Molybdenum (Mo)	Constituent of some enzymes	Legumes, bread, grains, green leafy vegetables, green vegetables, milk, and liver	17–45 μg

Source: *LARN: Livelli di Assunzione di Riferimento di Nutrienti ed energia per la popolazione italiana*, Revisione 2012, Documento di sintesi per il XXXV Congresso Nazionale SINU, Bologna, Italy, 34–45, http://www.sinu.it/documenti/20121016_LARN_bologna_sintesi_prefinale.pdf, 2012.

amylose, chemically and/or physically modified starches), and lignin (associated with the dietary fiber polysaccharides) are usually defined by the general term of dietary fiber (EFSA, 2010b).

They are complex carbohydrates that cannot be absorbed in the small intestine and do not provide energy. They enter the large bowel where they are partially or completely broken down by bacteria in the colon by a process called fermentation. Although dietary fiber cannot be considered a nutrient, it exerts functional and metabolic effects, which suggests that it is an important component of the human diet (King, 2005; Slavin, 2013). In addition to the increase in satiety and the improvement of bowel function–associated disorders such as constipation and diverticulosis, the introduction of fiber into one's diet has been linked to reduced risk for major chronic diseases, in particular cancers of the colon and rectum (Zeng *et al.*, 2014). Nowadays, it is very difficult to ascertain the beneficial effects deriving from fiber or a specific food supplement with respect to that obtained by consuming high-fiber foods (e.g., whole grains, vegetables, fruits, and legumes). However, there is a strong evidence of benefit to health associated with consumption of diets rich in fiber-containing foods at dietary fiber intakes greater than 25 g/day (EFSA, 2010b).

6.4.2 Bioactive Components

Bioactive compounds are extra-nutritional constituents that typically occur in very small quantities in foods, mainly from the plant kingdom. The principal bioactive components are polyphenols, carotenoids, tocopherols and tocotrienols, glucosinolates, phytosterols, and phytoestrogens (Liu, 2013). Most of them have antioxidant, anti-inflammatory, and antidiabetic properties; they affect the cardiovascular system by reducing triglycerides and LDL cholesterol, and by protecting endothelium from dysfunction (Hooper and Cassidy, 2006; Kris-Etherton *et al.*, 2002; Liu, 2013). Despite their potential beneficial effects and the importance to consume food sources rich in these bioactives, further scientific research is needed to provide evidence of their role in the daily diet. For this reason, there are not dietary recommendations for the population yet.

6.5 THE CONCEPT OF BALANCED AND SUSTAINABLE DIET

The combination, amount, and interaction of nutrients consumed with foods play an important role in human health. Food is not only the *fuel* of the organism and consequently the core element to sustain life, but also a major modifiable and powerful factor in promoting health, preventing and treating disease, and improving the quality of life. It is well known that noncommunicable diseases are currently the main cause of mortality worldwide and that CVD, cancers, diabetes, and chronic lung diseases are the top four (WHO, 2008). Unhealthy diet, tobacco use, physical inactivity, and harmful use of alcohol increase the risk of these diseases through a few key metabolic and physiological changes, such as raised blood pressure, overweight/obesity, hyperglycemia, and hyperlipidemia (WHO, 2004). The dietetic approach to control these disease risk factors can be consequently at the base of a balanced nutrition. In this regard, public institutions and scientific organizations have created, in the major countries of the world, specific directives aimed to synthesize all the knowledge available on the physiological nutritional needs and on the relations between diet and health to give reference values for each nutrient. These values must then be converted in food intake or dietary guidelines. This is an important challenge, as many food patterns result that are able to fulfill the nutrient allowances. This challenge is even more intriguing as nowadays dietary goals go even beyond the prevention of noncommunicable diseases and must take into consideration also social, environmental, and economic aspects.

In the recent decades, the globalization of food production and consumption has exposed populations worldwide to a wide variety of foods pushing to change and homogenize their eating and lifestyle habits and upsetting local realities as well as agriculture and industrial production capacity. Thus, the improvement of dietary habits is above all a societal, not just an individual problem, and the identification of the best food pattern to fulfill the nutrient needs is a complex task that requires a multifaceted approach. Dietary patterns represent the best descriptor of physiological, cultural, and ecological issues. The development of a global food system able to guarantee everyone a balanced diet requires health professional awareness and a commitment to education increasingly complex. The FAO recommends the development of a code of conduct for sustainable diets based on low environmental impacts, which contribute to food and nutrition security and to healthy life for present and future generations (Fatati and Poli, 2013; FAO, 2011). Low-input agro-ecological staple food production, limited animal husbandry, short-distance production–consumption nets, minimal food processing and refining, and information and education about the appropriate food choices could represent some of the key concepts for a sustainable food production and food consumption (Burlingame *et al.*, 2010; FAO, 2011). In this context, the Mediterranean diet (MD) is confirmed as a valid model even in modern times. It succeeds in concealing traditional food productions and way of life with sounded food choices to fulfill nutrient requirements and fit with low-energy use and environment and biodiversity protection.

6.6 MD: AN EXAMPLE OF BALANCED AND SUSTAINABLE DIET

Since the first data from the study of seven countries (Keys *et al.*, 1986), numerous surveys have shown that adherence to an MD can significantly decrease the risk of mortality from CVDs and cancer, as well as the incidence of age-associated cognitive decline.

The MD was identified as the traditional dietary pattern found in Crete, Greece, and southern Italy in the late 1950s and early 1960s, characterized by high consumption of whole grains, legumes, nuts, fruits, vegetables and olive oil, moderate consumption of fish and alcohol, usually in the form of red wine, and low-to-moderate consumption of poultry and dairy products (usually as yogurt or cheese).

As already mentioned, the protective effects of the MD fit well into the current paradigm of studying the overall dietary patterns instead of simply assessing the isolated nutrients in nutritional epidemiology. The rationale is that food items and nutrients may have synergistic or antagonistic effects when they are consumed in combination. Furthermore, the combination of healthy food choice with healthy weight, physical activity, a well-ordered life, and a proper dose of sunlight may increase the beneficial effects.

A recent meta-analysis of cohort prospective studies investigated the association between the adherence to the MD and adverse clinical outcomes (Sofi *et al.*, 2010), and showed that an increase in adherence to the MD was associated with a significant reduction of overall mortality (8%), cardiovascular incidence or mortality (10%), and neurodegenerative diseases (13%). Also considering some limits, for example, considerable variations existing among studies, these results are impressive. Very recently, the results of the PREDIMED study, the only large-scale clinical trial on primary prevention of CVD with an MD, were published (Estruch *et al.*, 2013). In the trial, about 7500 people over 55 (men) or 60 (women) years of age with no CVD but at high cardiovascular risk were randomly assigned to one of the following three diets: an MD enriched with extra virgin olive oil, an MD enriched with nuts, and a low-fat diet (control diet). After about 5 years, the two MDs resulted in a 30% reduction in major cardiovascular events compared to the control low-fat diet. Positive results were also highlighted studying the anti-inflammatory effect of these diets (Estruch, 2010; Urpi-Sarda *et al.*, 2012).

Over the years, many researchers investigated whether the advantages of the MD were transferable to other populations and positive results were obtained in studies conducted in the Unites States (Mitrou *et al.*, 2007), Australia (Kouris-Blazos *et al.*, 1999), the Netherlands (Hoevenaar-Blom *et al.*, 2012), and other countries, confirming, overall, that the key features of the MD appear to be transplantable to other dietary cultures.

Obviously, there are other diets associated with a reduced risk of chronic diseases, such as the traditional Okinawan diet, the modern DASH diet (Dietary Approaches to Stop Hypertension), or the New Nordic Diet, as well as dietary patterns identified by a statistical approach (e.g., factor or cluster analysis), such as the prudent pattern (Poulsen *et al.*, 2014; Sacks *et al.*, 2001; Willcox *et al.*, 2007).

The MD should be also seen as an extremely environmentally sustainable food model characterized by (1) a variety of food practices and food preparation techniques that ensure nutritional quality of diet; (2) a strong commitment to culture and traditions; (3) a respect for seasonality of foods and a less environmental impact, due to low consumption of animal products; and (4) a diversity of landscapes that contribute to the welfare (Burlingame *et al.*, 2010).

6.7 THE ROLE OF FUNCTIONAL FOODS IN A BALANCED DIET

Although many traditional food products, including fruits, vegetables, and whole grains, contain bioactive components with potential health benefits, in the last few years, demand for foods and beverages that sustain and improve health has increased in many parts of the world. These new foods developed to exert a desirable physiological effect are named *functional foods* (Yamada *et al.*, 2008) and are generally obtained modifying their original composition or increasing the content of specific components. The concept of functional food was developed in Japan in the mid-1980s when the Government of Japan decided that greater consumption of certain food products fortified with special compounds could help to limit or reduce the impact of a number of disease risk factors, and therefore assist in controlling the rising health costs associated with an aging population. Later, the concept of functional food was accepted in countries such as the United States, Canada, and Europe, as well as other parts of the world. There is no unitary accepted definition of functional foods globally; thus, the term *functional food* has varied meaning across countries. However, the definition proposed by the European Commission Concerted Action on Functional Food Science in Europe, a large EU project on functional food science in Europe, has been widely accepted: "a food can be regarded as 'functional' if it is satisfactorily demonstrated to affect beneficially one or more target functions in the body, beyond adequate nutritional effects in a way that is relevant to either an improved state of health and well-being and/or reduction of risk of disease" (Diplock *et al.*, 1999, S6). They are not pills or capsules, but just foods that have to demonstrate their effect in amounts that can normally be expected to be consumed in the diet (Doyon and Labrecque, 2008). The scientific validation of the protective effect exerted by a food or a component remains one of the most important and demanding objectives. Functional foods are not regulated in the European Union, but there is a strict regulation of all kinds of claims related to nutrition or health used on food labels and advertising.

6.8 CONCLUSIONS

In the last few decades, food and nutrition science is moving from identifying and correcting nutritional deficiencies to designing foods that could promote optimal health and reduce the risk of disease. The scientific approach exploited in this process has brought the role of whole dietary patterns, as well as the value of traditional models, into even sharper relief.

MD can be considered a sustainable model to maintain health by reconciling the traditional foods with the nutritional need of the population. However, lifestyles and food habits have undergone many changes, and we still need to find the best answers to optimize an individual's health and well-being meeting at the same time the physiological requirements and the social demands of the population.

REFERENCES

Björck I., Elmståhl H.L. The glycaemic index: Importance of dietary fibre and other food properties. *Proc. Nutr. Soc.* **2003**, *62*, 201–206.

Björck I., Liljeberg H., Ostman E. Low glycaemic-index foods. *Br. J. Nutr.* **2000**, *83*, S149–S155.

Bordoni A., Di Nunzio M., Danesi F., Biagi P.L. Polyunsaturated fatty acids: From diet to binding to PPARs and other nuclear receptors. *Genes Nutr.* **2006**, *1*, 95–106.

Bradbury J. Docosahexaenoic acid (DHA): An ancient nutrient for the modern human brain. *Nutrients* **2011**, *3*, 529–554.

Brand-Miller J., Buyken A.E. The glycemic index issue. *Curr. Opin. Lipidol.* **2012**, *23*, 62–67.

Burlingame B., Dernini S. Nutrition and Consumer Protection Division; Food and Agriculture Organization (eds.) *Sustainable Diets and Biodiversity. Directions and Solutions for Policy, Research and Action.* FAO, Rome, Italy, **2010**. 222–294 (Accessed from http://www.fao.org/docrep/016/i3004e/i3004e00.htm, on June 20, 2014).

Burlingame B., Nishida C., Uauy R., Weisell R. Fats and fatty acids in human nutrition: Introduction. *Ann. Nutr. Metab.* **2009**, *55*, 1–308.

Cummings J.H., Roberfroid M.B., Andersson H., Barth C., Ferro-Luzzi A., Ghoos Y., Gibney M. *et al.* A new look at dietary carbohydrate: Chemistry, physiology and health. *Eur. J. Clin. Nutr.* **1997**, *51*, 417–423.

Cummings J.H., Stephen A.M. Carbohydrate terminology and classification. *Eur. J. Clin. Nutr.* **2007**, *61*, S5–S18.

Degirolamo C., Rudel L.L. Dietary monounsaturated fatty acids appear not to provide cardioprotection. *Curr. Atheroscler. Rep.* **2010**, *12*(6), 391–396.

Diplock A.T. Scientific concepts of functional food in Europe consensus document. *Br. J. Nutr.* **1999**, *81*(Suppl 1), S1–S27.

Doyon M., Labracque J. Functional foods: A conceptual definition. *Brit. Food J.* **2008**, *110*, 1133–1149.

EFSA (European Food Safety Authority). Scientific opinion on dietary reference values for fats, including saturated fatty acids, polyunsaturated fatty acids, monounsaturated fatty acids, trans-fatty acids, and cholesterol. *EFSA J.* **2010a**, *8*, 1461–1568.

EFSA. Scientific opinion on dietary reference values for carbohydrates and dietary fibre. *EFSA J.* **2010b**, *8*, 1462.

EFSA. Scientific opinion on dietary reference values for protein. *EFSA J.* **2012**, *10*, 2557.

Elia M., Cummings J.H. Physiological aspects of energy metabolism and gastrointestinal effects of carbohydrates. *Eur. J. Clin. Nutr.* **2007**, *61*(Suppl 1), S40–S74.

Estruch R. Anti-inflammatory effects of the Mediterranean diet: The experience of the PREDIMED study. *Proc. Nutr. Soc.* **2010**, *69*(3), 333–340.

Estruch R., Ros E., Salas-Salvadó J., Covas M.I., Corella D., Arós F., Gómez-Gracia E. *et al.* Primary prevention of cardiovascular disease with a Mediterranean diet. *N. Engl. J. Med.* **2013**, *368*(14), 1279–1290.

FAO (Food and Agriculture Organization). *Ensuring Food Security in a Changing World.* 47–51. Food and Agriculture Organization, Rome, Italy, **2011** (Accessed from http://www.fao.org/docrep/015/i2307e/i2307e.pdf, on June 20, 2014).

FAO/WHO *Fats and Fatty Acids in Human Nutrition.* FAO Food Nutrition Paper 91. World Health Organization Expert Consultation Report/Food and Agriculture Organization of the United Nations, Rome, Italy, **2010**. 1–159 (Accessed from http://www.who.int/nutrition/publications/nutrientrequirements/fatsandfattyacids_humannutrition/en/, on June 16, 2014).

FAO/WHO. *Scientific Update on Carbohydrates in Human Nutrition.* World Health Organization Expert Consultation Report/Food and Agriculture Organization of the United Nations. *Eur. J. Clin. Nutr.* **2007**, *61*(Suppl 1), 137 (Accessed from http://www.who.int/nutrition/publications/nutrientrequirements/scientific_update_carbohydrates/en/, on June 16, 2014).

Fatati G., Poli A. Sustainability of food choices: The position of the Italian experts in nutrition. *Recenti Prog. Med.* **2013**, *104*(12), 609–614.

Galgani J., Ravissin E. Energy metabolism, fuel selection and body weight regulation. *Int. J. Obes. (Lond)* **2008**, *32*(Suppl 7), S109–S119.

Gillingham L.G., Harris-Janz S., Jones P.J. 2011. Dietary monounsaturated fatty acids are protective against metabolic syndrome and cardiovascular disease risk factors. *Lipids* **2011**, *46*(3), 209–228.

Hoevenaar-Blom M.P., Nooyens A.C.J., Kromhout D., Spijkerman A.M. Beulens J.W., van der Schouw Y.T., Bueno-de-Mesquita B., Verschuren W.M. Mediterranean style diet and 12-year incidence of cardiovascular diseases: The EPIC-NL cohort study. *PLoS One* **2012**, *7*(9), e45458.

Hooper L., Cassidy A. A review of the healthcare potential of bioactive compounds. *J. Sci. Food Agric.* **2006**, *86*(12), 1805–1813.

Hu F.B., Stampher M.J., Manson J.E., Rimm E., Coldtiz G.A., Rosner B.A., Hennekens C.H., Willett W.C. Dietary fat intake and the risk of coronary heart disease in women. *N. Engl. J. Med.* **1997**, *337*(21), 1491–1499.

Jenkins D.J., Kendall C.W., Augustin L.S., Franceschi S., Hamidi M., Marchie A., Jenkins A.L., Axelsen M. Glycemic index: Overview of implications in health and disease. *Am. J. Clin. Nutr.* **2002**, *76*(1), 266S–273S.

Kersten S. Effects of fatty acids on gene expression: Role of peroxisome proliferator-activated receptor α, liver X receptor α and sterol regulatory element-binding protein-1c. *Proc. Nutr. Soc.* **2002**, *61*(3), 371–374.

Keys A., Menotti A., Karvonen M.J., Aravanis C., Blackburn H., Buzina R., Djordjevic B.S., Dontas A.S., Fidanza F., Keys M.H. The diet and 15-year death rate in the seven countries study. *Am. J. Epidemiol.* **1986**, *124*(4), 903–915.

King D.E. Dietary fiber, inflammation, and cardiovascular disease. *Mol. Nutr. Food Res.* **2005**, *49*(6), 594–600.

Kouris-Blazos A., Gnardellis C., Wahlqvist M.L., Trichopoulos D., Lukito W., Trichopoulou A. Are the advantages of the Mediterranean diet transferable to other populations? A cohort study in Melbourne, Australia. *Br. J. Nutr.* **1999**, *82*(1), 57–61.

Kris-Etherton P.M., Hecker K.D., Bonanome A., Coval S.M., Binkoski A.E., Hilpert K.F., Griel A.E., Etherton T.D. Bioactive compounds in foods: Their role in the prevention of cardiovascular disease and cancer. *Am. J. Med.* **2002**, *113*(Suppl 9B), 71S–88S.

LARN: Livelli di Assunzione di Riferimento di Nutrienti ed energia per la popolazione italiana. Revisione 2012. Documento di sintesi per il XXXV Congresso Nazionale SINU, Bologna, Italy, **2012**. 34–45 (Accessed from http://www.sinu.it/documenti/20121016_LARN_bologna_sintesi_prefinale.pdf, on June 16, 2014).

Lawrence G.D. Dietary fats and health: Dietary recommendations in the context of scientific evidence. *Adv. Nutr.* **2013**, *4*(3), 294–302.

Liu R.H. Dietary bioactive compounds and their health implications. *J. Food Sci.* **2013**, *78*(Suppl 1), A18–A25.

Matthews D.E. Proteins and amino acids. In Shils M.E., Shike M., Ross A.C., Caballero B., Cousins R.J. (eds.), *Modern Nutrition in Health and Disease.* Linppincott Williams & Wilkins, Baltimore, MD, **2006** (10th edn.), 23–61.

Millward D.J., Rivers J. The nutritional role of indispensible amino acids and the metabolic basis for their requirements. *Eur. J. Clin. Nutr.* **1988**, *42*(5), 67–393.

Mirrahimi A., Chiavaroli L., Srichaikul K., Augustin L.S., Sievenpiper J.L., Kendall C.W., Jenkins D.J. The role of glycemic index and glycemic load in cardiovascular disease and its risk factors: A review of the recent literature. *Curr. Atheroscler. Rep.* **2014**, *16*(1), 381.

Mitrou P.N., Kipnis V., Thiébaut A.C., Reedy J., Subar A.F., Wirfält E., Flood A. *et al.* Mediterranean dietary pattern and prediction of all-cause mortality in a US population: Results from the NIH-AARP Diet and Health Study. *Arch. Intern. Med.* **2007**, *167*(22), 2461–2468.

Monro J.A., Shaw M. Glycemic impact, glycemic glucose equivalents, glycemic index load: Definition, distinctions, and implications. *Am. J. Clin. Nutr.* **2008**, *87*(1), 237S–243S.

Oh M.S., Uribirri J. Electrolytes, water and acid-base balance. In Shils M.E., Shike M., Ross A.C., Caballero B., Cousins R.J. (eds.), *Modern Nutrition in Health and Disease.* Linppincott Williams & Wilkins, Baltimore, MD, **2006** (10th edn.), 149–193.

Poulsen S.K., Crone C., Astrup A., Larsen T.M. Long-term adherence to the New Nordic Diet and the effects on body weight, anthropometry and blood pressure: A 12-month follow-up study. *Eur. J. Nutr.* **2014**. doi:10.1007/s00394-014-0686-z.

Riccardi G., Rivellese A.A., Giacco R. Role of glycemic index and glycemic load in the healthy state, in prediabetes, and in diabetes. *Am. J. Clin. Nutr.* **2008**, *87*(1), 269S–274S.

Russo G.L. 2009. Dietary n-6 and n-3 polyunsaturated fatty acids: From biochemistry to clinical implications in cardiovascular prevention. *Biochem. Pharmacol.* **2009**, *77*(6), 937–946.

Sacks F.M., Syetkey L.P., Vollmer W.M., Appel L.J., Bray G.A., Harsha D., Obarzanek E. *et al.* Effects on blood pressure of reduced dietary sodium and the dietary approaches to stop hypertension (DASH) diet. DASH-Sodium Collaborative Research Group. *N. Engl. J. Med.* **2001**, *344*(1), 3–10.

Schaafsma G. The protein digestibility-corrected amino acid score (PDCAAS)—A concept for describing protein quality in foods and food ingredients: A critical review. *J. AOAC Int.* **2005**, *88*(3), 988–994.

Schmitz G., Ecker J. The opposing effects of n-3 and n-6 fatty acids. *Prog. Lipid Res.* **2008**, *47*(2), 147–155.

Slavin J. Fiber and prebiotics: Mechanisms and health benefits. *Nutrients* **2013**, *5*(4), 1417–1435.

Sofi F., Abbate R., Gensini G.F., Casini A. Accruing evidence on benefits of adherence to the Mediterranean diet on health: An updated systematic review and meta-analysis. *Am. J. Clin. Nutr.* **2010**, *92*(5), 1189–1196.

Truswell A.S., Choudhury N. Monounsaturated oils do not all have the same effect on plasma cholesterol. *Eur. J. Clin. Nutr.* **1998**, *52*(5), 312–315.

Urpi-Sarda M., Casas R., Chiva-Blanch G., Romero-Mamani E.S., Valderas-Martínez P., Salas-Salvadó J., Covas K.I. *et al.* The Mediterranean diet pattern and its main components are associated with lower plasma concentrations of tumor necrosis factor receptor 60 in patients at high risk for cardiovascular disease. *J. Nutr.* **2012**, *142*(6), 1019–1025.

Willcox B.J., Willcox D.C., Todoriki H., Fujiyoshi A., Yano K., He Q., Curb J.D., Suzuki M. Caloric restriction, the traditional Okinawan diet, and healthy aging: The diet of the world's longest-lived people and its potential impact on morbidity and life span. *Ann. N. Y. Acad. Sci.* **2007**, *1114*, 434–435.

WHO (World Health Organization). World Health Assembly 57.17: Global strategy on diet and physical activity. WHO Report, Geneva, Switzerland, **2004** (Accessed from http://www.who.int/nmh/nmh_dpas_wha58_update.pdf?ua = 1, on June 15, 2014).

WHO. World Health Assembly Document A61/8. 2008–2013 Action plan for the global strategy for the prevention and control of noncommunicable diseases. WHO Report, Geneva, Switzerland, **2008** (Accessed from http://whqlibdoc.who.int/publications/2009/9789241597418_eng.pdf, on June 15, 2014).

WHO/FAO. *Human Vitamin and Mineral Requirements*. WHO/FAO Expert Consultation Report, Bangkok, Thailand, **2001** (Accessed from http://www.fao.org/docrep/004/y2809e/y2809e00.htm, on May 5, 2014).

WHO/FAO/UNU. *Protein and Amino Acid Requirements in Human Nutrition Report*. Technical Report Series no. 935. WHO/FAO/UNU Expert Consultation Report, Geneva, Switzerland, **2007** (Accessed from http://whqlibdoc.who.int/trs/who_trs_935_eng.pdf, on June 15, 2014).

Wijendran V., Hayes K.C. Dietary n-6 and n-3 fatty acid balance and cardiovascular health. *Annu. Rev. Nutr.* **2004**, *24*, 597–615.

Wong J.M., Jenkins D.J. Carbohydrate digestibility and metabolic effects. *J. Nutr.* **2007**, *137*(11 Suppl), 2539S–2546S.

Yamada K., Sato-Mito N., Nagata N., Umegaki K. Health claim evidence requirements in Japan. *J. Nutr.* **2008**, *138*(6), 1192S–1198S.

Zeng H., Lazarova D.L., Bordonaro M. Mechanisms linking dietary fiber, gut microbiota and colon cancer prevention. *World J. Gastrointest. Oncol.* **2014**, *6*(2), 41–51.

7 The Role of Nutraceuticals

Katia Petroni and Chiara Tonelli

CONTENTS

7.1 INTRODUCTION

One of the most pressing challenges in the current century is reduction of the impact of chronic disease, including cardiovascular disease (CVD), cancers, and age-related neurodegenerative diseases. Chronic diseases are responsible for at least three out of four deaths in Western countries. In 2005, the World Health Organization predicted that mortality from chronic disease would increase by 17% over the following decade. Adequate preventive measures for chronic diseases would lower the cost of public healthcare; they would be beneficial from a socioeconomic perspective as well, because most of these diseases—some of which are strongly debilitating—affect subjects who are still in the productive years of their life (Daar *et al.*, 2007).

The risk of chronic disease is strongly associated with three sociobehavioral risk factors: smoking, lack of physical activity, and unhealthy diets, which are the causes of predisposing conditions, such as obesity, diabetes, metabolic syndrome, and dyslipidemia. Evidence is accumulating that ischemic CVD, several types of tumors, and neurodegenerative disorders share common pathogenic mechanisms. A large epidemiological study that collected data from 52 countries over the world (INTERHEART; Yusuf *et al.*, 2004) indicated that up to 90% of the risk of developing acute myocardial infarction could be reduced by modifiable risk factors (such as physical activity and fruit and vegetable consumption).

The role of healthy diet in preventing chronic diseases is supported by epidemiological studies, which indicate that the *Mediterranean diet* is associated with a reduced risk not only of cardiovascular events but also of cancer and neurodegenerative disorders. The Mediterranean diet, typical of southern Europe, includes several daily servings of fruit and vegetables, unrefined cereals, and legumes, accompanied by wine in moderation and less frequently meat, cheese, and sweets. Unfortunately, however, the Mediterranean diet is progressively losing its impact because of the global "Westernization" of dietary habits; changes in lifestyles and working conditions, with drastic reduction in physical activity and energy expenditure; and the increases in the price of fruits, vegetables, and fish. The unhealthy diet of the younger generations is becoming the principal cause of increase in obesity and associated complications.

Approaches to prevent most cases of myocardial infarction and other ischemic CVDs could be based on the reintegration of phytonutrients derived from fruits and vegetables in the diet (Martin *et al.*, 2013).

Recently, nutritional studies have been focusing on the evaluation of foods for their protective potential and have unraveled *nutraceuticals* as naturally derived bioactive compounds that have health-promoting, disease-preventing, or medicinal properties (Martin *et al.*, 2011). Most nutraceuticals derive from plants and correspond to several classes of phytonutrients, such as dietary fiber, polyphenols, and carotenoids, whereas a minority of them are of animal origin, such as marine ω-3 polyunsaturated fatty acids (PUFAs).

7.2 NUTRACEUTICALS

The term *nutraceutical* was coined in 1989 by Stephen DeFelice by combining the terms *nutrition* and *pharmaceutical*. According to DeFelice, *nutraceutical* may be defined as "a food (or part of a food) that provides medical or health benefits, including the prevention and/or treatment of a disease" (Brower, 1998). Today, *nutraceutical* is recognized internationally as the singular word that describes the emerging industry of food and food-related substances with potential medical and health benefits. Nutraceuticals can be distinguished into two main categories: (1) nutraceutical products, such as isolated nutrients, dietary supplements, and herbal products, and (2) nutraceutical or functional foods, that is, foods enriched in micronutrients and/or bioactive compounds (i.e., phytonutrients) that confer a health benefit beyond basic nutrition. Functional foods can be considered traditional foods with functional properties (e.g., fresh fruits or vegetables), foods/beverages fortified with nutraceuticals (e.g., milk enriched with ω-3), or innovative functional foods, obtained by breeding strategies that exploit the existing natural biodiversity of some crops or by using plant metabolic engineering.

To develop innovative nutraceuticals, the expertise of plant geneticists, food technologists, nutritionists, medical doctors, and food chemists must be combined. An effective methodology would be to identify on a scientific basis the contribution of bioactive compounds toward improving health through diet and the quantity required

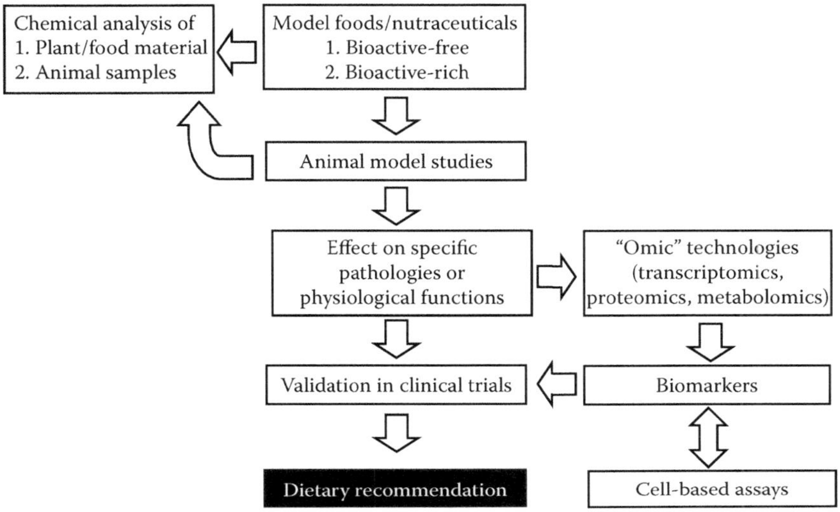

FIGURE 7.1 Schematic representation of nutrigenomics studies for the development of nutraceuticals.

for beneficial effects using a nutrigenomic approach (Figure 7.1). When considering functional foods, bioactive compounds need to be studied in their whole-plant context, that is, in the chemical form in which they are available in the diet and in the context of other metabolites, enzymes, fibers, and so on, in which they are normally ingested, because all these external factors may modify the availability and efficacy of nutraceuticals (Martin *et al.*, 2011). This can be achieved through the analysis of the dietary effects of *model foods*, using a well-defined quality and quantity of bioactive compounds on model mice with chronic diseases, completed by comparing bioactive-rich and bioactive-free diets (Figure 7.1). Such model foods should be designed to be used for both animal feeding studies and human intervention trials, for assessing the role of plant bioactives in the diet. Similarly, to validate the effect of nutraceutical products, the effect of formulated mixtures of bioactive compounds (separately or in combination) have to be tested *in vivo* by their administration in drinking water and using appropriate nutraceutical-free controls. The bioavailability of bioactive compounds have to be determined by chemical analysis of blood/urine samples from the mice under dietary treatment using high-performance liquid chromatography and liquid chromatography coupled to mass spectrometry (Figure 7.1).

Functional model foods can then be used to ascertain the preventive effect of bioactives against chronic diseases and the molecular mechanisms underlying the observed effects by using "omic" tools (i.e., transcriptomics, proteomics, metabolomics, and epigenomics), to determine potential gene and protein biomarkers predictive of a preclinical onset of a diet-related disease. The robustness of such biomarkers should then be verified in human intervention studies using the same functional foods/nutraceutical products, to develop nutraceuticals with a scientifically demonstrated physiological benefit or provide protection against a specific disease. Finally, the molecular actions of these molecules can also be characterized on cellular models, to define their molecular targets and mechanism of action (Figure 7.1).

Therefore, health and medical benefits can be claimed only after well-designed clinical trials have demonstrated that a specific functional food or nutraceutical product is able to enhance short-term well-being and performance ability or to prevent incidence and progression of certain diseases (Lampe *et al.*, 1999). This should be done using suitable biomarkers, which are scientifically verified as linked to the physiological effects and measurable using solid, well-known, and globally accepted procedures. Biomarkers that have been widely accepted as effective in monitoring the efficacy of nutraceuticals to reduce the risk of CVD are, for example, plasma levels of cholesterol and low-density lipoproteins (LDLs), or blood pressure, but robust biomarkers to verify the ability of nutraceuticals to improve mental performance are yet to be developed. Most of the available nutraceutical products lack clinical data and can still only claim a "potential benefit" on the basis of the health-promoting properties of their components.

7.3 HEALTH AND MEDICAL BENEFITS OF SOME NUTRACEUTICALS

Nutraceuticals are commonly defined on the basis of food sources, their chemical nature, and mechanism of action. The most important categories of nutraceuticals are dietary fiber, probiotics, prebiotics, PUFAs, antioxidants (e.g., polyphenols, carotenoids, tocopherols), and other phytonutrients.

7.3.1 Dietary Fiber

Dietary fiber is the indigestible part of plant food and can be distinguished into two main components: (1) soluble dietary fiber, including β-glucans, pectins, gums, mucilages, and hemicelluloses, which can be fermented in colon, and (2) insoluble dietary fiber, including celluloses, some hemicelluloses, and lignins, normally poorly fermented in colon. Vegetables rich in dietary fiber are artichoke, chicory, egg plant, asparagus, garlic, leeks, and onion (Wanders *et al.*, 2011). Dietary fiber can reduce appetite and thus contribute to weight loss, due to the viscosity of its soluble components, which retards emptying of the stomach. When fermented in the colon, dietary fibers form short-chain fatty acids (SCFAs), such as acetate, butyrate, and propionate, which are important in signaling satiety. Viscous fiber lowers the glycemic index of foods and has a beneficial impact on obesity, type 2 diabetes, and risk factors for CVD (Kendall *et al.*, 2010; Meyer *et al.*, 2000).

Nondigestible starch, or resistant starch, commonly found in unripe bananas and partially milled grains, has properties similar to those of dietary fiber, because it is fermented in the colon to produce SCFAs and impact insulin signaling, modulate immune function, and prevent colon cancer. Resistant starch can also induce beneficial changes in the composition of the colonic microbiota. Because of these beneficial effects on gut health and chronic disease, resistant starch and dietary fiber are being promoted as prebiotics (Fuentes-Zaragoza *et al.*, 2011).

7.3.1.1 Probiotics

Probiotics are defined as "living microorganisms which upon ingestion in certain numbers exert health benefits beyond inherent basic nutrition" (de Vrese *et al.*, 2008). Substances that are classified as probiotics possess the following properties: nonpathogenic, nontoxic, resistance to gastric acid, adherence to gut epithelial tissues, temporary colonization of the intestine, beneficial to host, produce antibacterial substances, and inhibit pathogenic strains (Suvarna and Boby, 2005). They generally include lactobacilli (e.g., *Lactobacillus acidophilus, L. casei, L. delbrueckii* subsp. *bulgaricus, L. brevis, L. cellobiosus*), some Gram-positive enterococci (e.g., *Lactococcus lactis, Streptococcus salivarius* subsp. *thermophilus, Enterococcus faecium*), and bifidobacteria (e.g., *B. bifidum, B. adolescentis, B. infantis, B. longum, B. thermophilum*) (Das *et al.*, 2012).

Specific probiotics have been proven to be effective in the prevention and treatment of acute or rotavirus diarrhea and antibiotic-associated gastrointestinal side effects. Interesting results that deserve further study have been published regarding food allergies, atopic eczema, and acute respiratory infections in children and prevention of vaginitis and travelers' diarrhea in adults. Promising results have been reported in patients with inflammatory and irritable bowel disease, cystic fibrosis, and dental caries in children (for a review, see Gorbach [2002]).

7.3.1.2 Prebiotics

A prebiotic is defined as "a selectively fermented ingredient, or a fibre that allows specific changes, both in the composition and/or activity of the gastrointestinal microflora, conferring benefits on the well being and health of host" (de Vrese *et al.*, 2008, 34). Prebiotics are generally nondigestible short-chain polysaccharides, in particular fructose-based oligosaccharides normally present or added in food, which promote the growth of *Lactobacillus* and *Bifidobacteria* in the gut (Hord, 2008). Vegetables rich in fructose-based oligosaccharides are chicory roots, banana, tomato, and alliums. Beneficial effects of prebiotics on health are prevention of diarrhea or obstipation, modulation of the metabolism of the intestinal flora, prevention of cancer, effects on lipid metabolism, and stimulation of mineral adsorption and immunomodulatory properties (Douglas and Sanders, 2008).

7.3.2 Polyunsaturated Fatty Acids

PUFAs may be distinguished into ω-3 (*n*-3) fatty acids and ω-6 (*n*-6) fatty acids. Both are considered "essential fatty acids," because in animals and humans they are produced inefficiently or not at all and must be ingested through the diet (Martin *et al.*, 2013). Animals and humans metabolize dietary linoleic acid (LA; ω-6 PUFA) to form arachidonic acid (AA; ω-6 PUFA) and α-linolenic acid (ALA; ω-3 PUFA) to form eicosapentaenoic acid (EPA; ω-3 PUFA) and docosahexaenoic acid (DHA; ω-3 PUFA).

LA is abundant in vegetable oil and seed oil (especially safflower, sunflower, corn, cottonseed, and soybean oils), whereas ALA is abundant in flaxseed oil, canola oil, and walnuts (Ramsden *et al.*, 2009). However, the conversion of essential fatty acids to AA, EPA, and DHA is inefficient (<5%), so that the balance between ω-3 and ω-6 PUFAs in plasma is determined largely by the relative levels of EPA and DHA to AA assumed in the diet (Martin *et al.*, 2013). EPA and DHA are especially abundant in marine fish (mackerel, herring, salmon, anchovies, sardines, tuna, and oysters), whereas AA is mostly abundant in eggs, poultry, beef, pork, liver, and farmed fish (Ramsden *et al.*, 2009).

Generally, ω-3 PUFAs are associated with anti-inflammatory effects and combat the effects of ω-6 PUFAs from which signaling molecules, particularly prostaglandins, are synthesized and which promote inflammation and induce autoimmune responses. This is thought to be so because ω-6 PUFAs are used to form pro-inflammatory eicosanoids, whereas ω-3 PUFAs form anti-inflammatory eicosanoids (Simopoulos, 2002).

Several studies on cellular and animal models as well as epidemiological and intervention studies showed that higher *n*-3:*n*-6 ratios (optimal would be 4:1) prevent or reduce different types of chronic diseases through the promotion of anti-inflammatory activity by ω-3 PUFAs, although they may also influence the properties of cell membranes and modulate gene expression (Erkkila *et al.*, 2003; He *et al.*, 2002).

7.3.3 Antioxidants

Reactive oxygen and nitrogen species (RONS) are considered as major determinants of many degenerative pathologies, including CVD, neurodegenerative diseases, chronic inflammation, certain cancers, and tissue aging. Essentially, RONS can act by damaging biological structures and molecules, but they can also influence cell signaling either indirectly by changing the redox cellular status or directly by participating in intracellular signaling (Virgili and Marino, 2008). Although the antioxidant capacities of many phytonutrients have been claimed to explain their health-promoting properties (Rice-Evans *et al.*, 1997), many dietary phytonutrients have low bioavailability, making it highly unlikely that, once absorbed, they operate directly as antioxidants to promote health. As an example, flavonoids have been shown to have antispasmodic, antisecretory, antidiarrheal, antiulcer, and bactericidal properties in the gastrointestinal tract probably because of their strong antioxidant capacities (La Casa *et al.*, 2000; Lima *et al.*, 2005; Sunairi *et al.*, 1994). However, there is an emerging view that phytonutrients may act not only by scavenging reactive oxygen species (ROS) or suppressing their production, but also by enhancing the endogenous antioxidant capacity of cells/tissues (e.g., glutathione synthesis) or by influencing signaling pathways through interaction with proteins, enzymes, and nuclear receptors (Virgili and Marino, 2008).

As an example, the chemopreventive activity of phytonutrients has been suggested to be associated with the ability to block the progression of cancer through four mechanisms (Figure 7.2) (Beliveau and Gingras, 2007).

First, several phytonutrients exert their anticancer activity through modulation of enzymes detoxifying from carcinogens or excreting them, such as isothiocyanates (the active metabolites in Brassicaceae species), which protect against cancer by modulating carcinogen metabolism (Talalay and Fahey, 2001).

Second, many phytonutrients have been shown to inhibit tumor growth by direct or indirect induction of apoptosis and cell death of tumor cells.

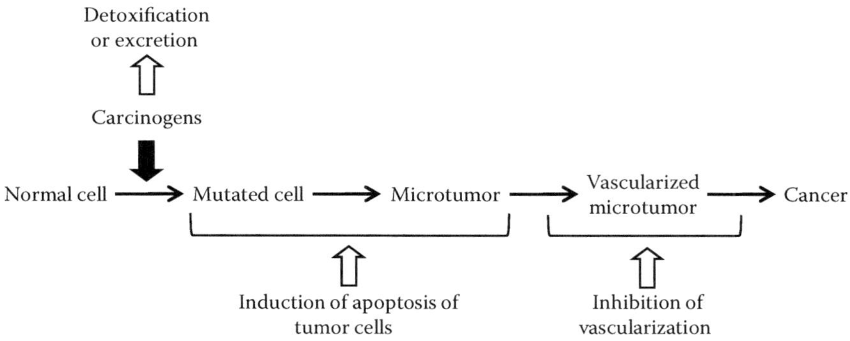

FIGURE 7.2 Mechanisms of the chemopreventive activity of phytonutrients.

Isothiocyanates, curcumin (the yellow pigment in turmeric), and resveratrol (a polyphenol present in grape) have shown a direct strong pro-apoptotic activity against many cancer cell types *in vitro*. However, curcumin and other phytonutrients indirectly promote apoptosis by sensitizing tumor cells to respond to normal pro-apoptotic stimuli (Aluyen *et al.*, 2012; James *et al.*, 2012; Park *et al.*, 2013).

Third, there are several studies suggesting that phytonutrients exert their chemopreventive properties by inhibiting vascularization of latent microtumors, which is necessary to sustain their growth by providing oxygen and nutrients and to invade the host tissues. Such an anti-angiogenic effect is achieved by blocking the vascular endothelial growth factor-2 (VEGF2) receptor, in the case of epigallocatechin gallate (EGCG) from green tea, or by inhibiting both the VEGF2 receptor and the platelet-derived growth factor (PDGF) receptor in the case of ellagic acid (abundant in raspberries and strawberries) and delphinidin (present in blueberries). The inhibition of both receptors appears to be particularly efficacious, because they are both essential proteins for angiogenesis (Labrecque *et al.*, 2005; Lamy *et al.*, 2002, 2006).

Fourth, there is now considerable evidence that chronic inflammation contributes to the progression of different types of cancers, including colon and breast cancers. Several phytonutrients, such as EGCG, resveratrol, and curcumin, exert their reduction of inflammation through inhibition of cyclooxygenase-2 (COX-2), which is a mediator of inflammation, thus reducing the expression of the NF-kB transcription factor (Aggarwal *et al.*, 2004; Surh *et al.*, 2003).

Overall, these chemopreventive mechanisms appear to be important because they are related to the ability of phytonutrients to block the progression of precancerous cells and latent microtumors in tissues and organs of many healthy people. A study that conducted an autopsy of people who have died of causes unrelated to cancer estimated that 30%–50% of middle-aged women had latent breast microtumors, 40% of aged men had prostate microscopic tumors, and 98% of people had small latent asymptomatic microtumors in the thyroid (Black and Welch, 1993).

Among the phytonutrients that are believed to be beneficial for health are many bioactives with strong antioxidant activity. These include the water-soluble polyphenols and the fat-soluble carotenoids and tocotrienols. Dietary advice is to include both types of antioxidants to reduce the risk of chronic diseases (Eastwood, 1999).

7.3.3.1 Polyphenols

Among phytonutrients, polyphenols are synthesized by all plants and used for their own defense against biotic and abiotic challenges. Although endowed with diverse chemical composition (Figure 7.3), polyphenols constitute collectively the most abundant antioxidants in the diet with 10-fold higher levels of polyphenols than vitamin C and 100-fold higher levels than vitamin E and carotenoids. Most polyphenols have higher intrinsic antioxidant activities than vitamins C and E (Rice-Evans *et al.*, 1997). Their role in the protection particularly against CVD was launched by the publication, about 20 years ago, of the epidemiological observation called the *French paradox*, which led to the conclusion that, despite consuming as much dairy fats as people in other northern European countries, the French had a lower mortality due to coronary heart disease, because of regular moderate consumption of red wine (Renaud and de Lorgeril, 1992). More recently, several studies have showed that polyphenols also protect against cancers and neurodegenerative diseases and can act as anti-inflammatories (Surh, 2003; Jennings *et al.*, 2014; Joseph, 1999; Singh *et al.*, 2013). Red wine contains varying levels of resveratrol, flavonols, anthocyanins, and catechins (EGCG), which are all effective antioxidants.

Overall, polyphenols improve endothelial function, reduce the uptake of LDLs and their oxidation and aggregation by direct protein binding to LDL oxidable sites, reduce blood pressure, and inhibit platelet aggregation (Machha and Mustafa, 2005; Miura *et al.*, 2001; Wang *et al.*, 2002).

These beneficial effects are in part due to flavonols, such as quercetin and kaempferol, which can be found in many plants, and at higher levels in onions and apples. Flavonols have been demonstrated to reduce atherosclerosis, stroke, and thrombosis,

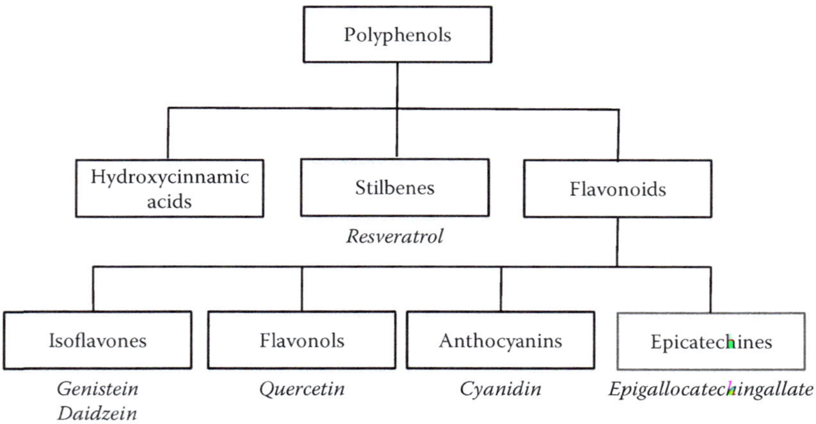

FIGURE 7.3 Polyphenols. The main classes of polyphenols are indicated in white boxes, whereas the most widespread polyphenols present in foods are indicated in italics.

by inhibiting LDL oxidation and platelet aggregation (Viña *et al.*, 2007). Flavonol-rich diets also reduce the incidence of breast cancer and other age-related degenerative diseases by both stimulating the antioxidant defense and inhibiting growth and proliferation of breast cancer cells (Fink *et al.*, 2007).

Among polyphenols, anthocyanins are water-soluble pigments that color the fruit and flowers of many plants and can be found in higher quantity in berries, blood orange, and blue and purple corn and cherries. Evidence that consumption of anthocyanin-rich foods promotes health is supported by recent studies of anthocyanin-rich fruits such as blueberry and cranberry (Cassidy *et al.*, 2013). Preclinical studies with animal models fed with anthocyanin-rich diets demonstrated that they confer cardioprotection (Toufektsian *et al.*, 2008, 2011), that they inhibit adipocyte development and weight gain on a high fat diet (Titta *et al.*, 2010; Tsuda *et al.*, 2003), and that they can extend the life span of cancer-prone mice (Butelli *et al.*, 2008). Dietary anthocyanins have anti-inflammatory effects and reduce oxidation in nervous tissues, so protecting against neurodegeneration (Jennings *et al.*, 2014; Joseph, 1999), and promote visual acuity (Matsumoto *et al.*, 2003).

Interestingly, cardioprotection from blue corn was associated with increased cardiac glutathione levels and increased marine ω-3 levels in the blood, suggesting that dietary anthocyanins modulate cardiac antioxidant defense and synthesis of ω-3 fatty acids (Toufektsian *et al.*, 2008, 2011). Dietary anthocyanins may alter the endogenous PUFA metabolism by promoting the conversion of ALA present in corn into the ω-3 PUFAs, EPA, and DHA. In fact, EPA synthesis involves as the first step Δ6 desaturase, a rate-limiting enzyme in humans, and two condensation steps, whereas EPA conversion to DHA requires three additional steps (Burdge and Calder, 2005). Recent epidemiological studies have revealed that EPA and DHA levels were increased in the blood and cells of red wine drinkers and that components of red wine, other than alcohol, probably flavonoids, were responsible for the fish-like effect observed (de Lorgeril *et al.*, 2008). It is possible that the diverse effects of anthocyanins on different chronic diseases may result from their modulation of plasma ω-3 PUFA levels as the primary mechanism of action. Alternatively, the impact of anthocyanins on ω-3 PUFA metabolism might just be a part of their physiological mode of action.

Catechins, epicatechins, and condensed tannins accumulate in the seed coat of seeds and are present in many fruits, such as grape and blueberry. In particular, EGCG is the major polyphenolic compound in green tea, and many epidemiological, clinical, and preclinical studies have shown that green tea/epicatechin consumption reduces the incidence of breast and prostate cancers and CVD (Arts *et al.*, 2002; Brausi *et al.*, 2008). These phytonutrients are able to modulate gene expression through epigenetic modifications. Epigenetics describes phenomena in which genetically identical cells or organisms have different gene expression and phenotypes without a change in DNA sequence. Epigenetic phenomena are due to DNA methylation, histone modifications (methylation, acetylation, phosphorylation, etc.), and posttranscriptional gene regulation by noncoding RNAs. These epigenetic modifications, collectively defined as epigenome, can be modulated by environmental factors, including diet, mostly during fetal and neonatal development, but also during adult age (Jiménez-Chillarón *et al.*, 2012). EGCG has been shown to directly inhibit DNA methyltransferase 1 (DNMT1) by covalent binding with its catalytic pocket, but also to be methylated by catechol-*O*-methyltransferase, thus releasing *S*-adenosyl-L-homocysteine, which is a potent feedback inhibitor of DNMT (Lee *et al.*, 2005). The dual inhibitory mechanism of EGCG may explain its potent demethylating activity and chemopreventive properties against breast cancer, prostate cancer, and many other tumors. Treatment of different types of cancer cell lines with EGCG has been associated with demethylation and consequent reactivation of tumor suppressor genes (Lee *et al.*, 2005). Among phytonutrients, EGCG also proved to be a potent histone acetyltransferase (HAT) inhibitor against different HAT enzymes (Choi *et al.*, 2009). Furthermore, in liver cancer HepG2 cells, EGCG was found to induce apoptosis and upregulate the expression of miR-16, which is known to suppress the expression of the anti-apoptotic Bcl-2 (Tsang *et al.*, 2010).

Isoflavonoids are polyphenols produced almost exclusively in legumes, such as soybeans, beans, and pea. They are metabolized by gut microbiota, which produce a significantly more bioactive form, equol, which is however produced only by 30% of the population (Setchell *et al.*, 2002). Isoflavonoids (daidzein, genistein, and equol) as well as other polyphenols, such as resveratrol, act as phytoestrogens by interacting with estrogen receptors and regulating estrogen-responsive genes. Phytoestrogens may have estrogen-like effects and protect against osteoporosis, neurodegeneration, and CVD, or they can act as estrogen antagonists by preventing breast and endometrial cancers (McDonnell *et al.*, 1999). Asian populations consuming a high soy diet (rich in genistein and daidzein) show a significantly lower incidence of steroid hormone-responsive cancers (25% less for prostate cancer and 10% less for breast cancer). Possible mechanisms by which phytoestrogens may prevent breast cancer are inhibiting the enzyme aromatase, involved in estrogen synthesis (Shufelt *et al.*, 2012), and downregulating the oncogenic Wnt signaling pathway in the mammary gland (Kim *et al.*, 2006). Interestingly, *in vivo* dietary maternal supplementation of genistein to A^{vy} dams with genistein resulted in increased DNA methylation of offspring and reduced incidence of obesity and tumorigenesis (Dolinoy *et al.*, 2006).

Resveratrol is thought to be one of the principal agents in the health-promoting effects of red wine, but it has also been reported to have anticancer, neuroprotective, antiaging, and anti-inflammatory properties. The neuroprotective effect of resveratrol against stroke has been linked to both its direct antioxidant capacity as a free radical scavenger and the activation of endogenous antioxidant defense mechanisms through transcriptional activation of the Nrf2 transcription factor, which in turn regulates antioxidant enzymes. However, resveratrol directly inhibits the activation of microglia, the resident immune cells of the brain, which secret pro-inflammatory factors (e.g., cytokines, ROS) in response to insults and thus trigger neuroinflammation, an important contributor to the pathogenesis of neurological disorders. Finally, resveratrol exerts its neuroprotective effect by modulating anti-apoptotic pathways, thus reducing the oxidative stress-mediated cell death and apoptosis resulting from ischemic attacks (Singh *et al.*, 2013). Several studies have demonstrated the extraordinary antiaging properties of resveratrol in different organisms ranging from yeast to mammals (Baur and Sinclair, 2006). This antiaging activity has been attributed to the modulation of multiple pathways, such as those related to insulin signaling (insulin/insulin-like growth factor/Akt/FOXO), energy metabolism [AMP-activated protein kinase (AMPK)/SIRT1/PGC1α], and the protein synthesis/

autophagy pathway (mTOR/S6K) (Pan *et al.*, 2012). The chemopreventive activity of grape seed extract and resveratrol in particular has been suggested for several types of cancers, including skin, prostate, breast, lung, and colon cancers, and due to the pro-apoptotic, anti-proliferation, and anti-inflammatory properties of resveratrol. However, only a limited number of human studies on resveratrol have been performed in the past 10 years (Aluyen *et al.*, 2012).

Among polyphenols, curcumin from turmeric has preventive and therapeutic properties associated with its antioxidant, anti-inflammatory, and anticancer properties. Curcumin modulates several molecular targets by altering their gene expression and signaling pathways or through direct interaction (for a review, see Gupta *et al.* [2011] and Shishodia [2013]). As an example, the potent antitumor property of curcumin is exerted by directly inducing or indirectly promoting apoptosis and cell death of tumor cells, by scavenging ROS, and by reducing the inflammatory cancer microenvironment (Park *et al.*, 2013). In addition, curcumin reduces inflammation through direct binding and consequent inhibition of COX1 and COX-2, as well as tumor necrosis factor-α, and regulates the expression of inflammatory cytokines (Gupta *et al.*, 2011; Shishodia, 2013). Several studies have revealed that curcumin affects both DNA methylation and histone modification. Similar to ECGC, curcumin has been shown to directly inhibit DNMT1 by covalent binding with its catalytic pocket, but it is also an inhibitor of HATs and several histone deacetylases (HDACs), suppressing proliferation and inducing apoptosis of cancer cells (Gupta *et al.*, 2011). Recently, *in vitro* and *in vivo* studies, as well as a few human and epidemiological studies, have highlighted the interaction of curcumin with molecular pathways implicated in insulin resistance, hyperglycemia, hyperlipidemia, and other inflammatory symptoms associated with obesity and metabolic diseases, contributing to the unraveling of the molecular mechanisms of the hypolipidemic effects of curcumin (Shehzad *et al.*, 2011; Zingg *et al.*, 2013). Owing to its ability to affect multiple targets, curcumin has the potential for the prevention and treatment of various chronic diseases, including, among others, cancers, atherosclerosis, neurodegenerative diseases, hepatic disorders, obesity, and diabetes (Shishodia, 2013).

Finally, polyphenols have been found to modulate energy metabolism and, as a consequence, reduce insulin resistance, metabolic syndrome, and type 2 diabetes (Hanhineva *et al.*, 2010). Recent studies demonstrated that many dietary polyphenols can act as mimetics of caloric restriction and activate AMPK, an evolutionary conserved sensor of cellular energy, which in turn activates SIRT1 (HDAC) and influences glucose/lipid metabolism and age-related disease (Park *et al.*, 2012). Resveratrol, quercetin, and anthocyanins also extend life span in many species (Pan *et al.*, 2012). The direct target of resveratrol is a phosphodiesterase, which is inhibited by resveratrol, leading to an increase of cyclic AMP and subsequent activation of AMPK (Park *et al.*, 2012).

7.3.3.2 Carotenoids and Tocopherols

Lycopene is a potent lipophilic antioxidant, with a higher antioxidant activity than other carotenoids, such as α-tocopherol, β-carotene, and lutein. It is mainly abundant in tomato, and also in fruits such as grapefruit, papaya, guava, and watermelon.

Dietary lycopene confers protection against CVD by reducing LDL oxidation (Napolitano *et al.*, 2007) and against different types of cancer (i.e., prostate, breast, lung, and stomach cancers) through its ability to impact cell growth regulation (Seren *et al.*, 2008).

Dietary carotenes (α, β, and γ) from plants are precursors of retinol (vitamin A), which are in turn precursors of retinoic acid, an important regulatory factor, and retinal, needed for vision by the retina. Vegetables rich in β-carotene are carrot, tomato, and pumpkin (Asensi-Fabado and Munnè-Bosch, 2010). Both clinical and epidemiological studies have shown that dietary β-carotene protects against LDL oxidation and atherosclerosis, thus reducing CVD, including cerebral infarction (Hirvonen *et al.*, 2000; Kristenson *et al.*, 1997). No detrimental effects have been reported from β-carotene consumption in food, whereas adverse effects have been reported for β-carotene supplementation (Bjelakovic *et al.*, 2012).

Another carotenoid synthesized only by plants is lutein, which is found in high quantities in green leafy vegetables such as spinach and black cabbage, but also in corn. Lutein is concentrated in the macula of the retina, where it protects against oxidative stress and the high-energy photons of blue light. Several studies have shown that an increase in macula pigmentation decreases the risk of age-related macular degeneration (Richer *et al.*, 2004).

Tocopherols are a group of molecules collectively known as vitamin E. The most biologically active form of vitamin E is α-tocopherol, which is a fat-soluble antioxidant that counteracts the production of ROS in adipose tissue and protects against LDL and PUFA oxidation (Packer *et al.*, 2001). The recommended daily intake of vitamin E is 15 mg per day, but it is not generally achieved by diet, apart from a minority of people. High intake of vitamin E has been associated with reduced cardiovascular risk (Stampfer *et al.*, 1993) and risk of breast and prostate cancers (Bostick *et al.*, 1993; Heinonen *et al.*, 1998).

7.3.3.3 Other Phytonutrients

Other phytonutrients with important health-promoting properties are phytosterols and isothiocyanates. Phytosterols prevalent in plants are β-sitosterol, campesterol, and stigmasterol. High levels of phytosterols can be found in many vegetables, such as broccoli, carrot, and cauliflower (Piironen and Lampi, 2003). Clinical trials indicate that 0.8 g of phytosterols per day leads to a significant reduction in LDL cholesterol and total cholesterol in the blood, by competing for cholesterol uptake in the gut. Their action is therefore complementary to anti-cholesterol statins (Normén *et al.*, 2005). Some studies also indicate that phytosterols are protective against breast cancers (Ju *et al.*, 2004). Isothiocyanates are the biologically active metabolites of glucosinolates, which are mostly abundant in broccoli sprouts. Glucoraphanin is the glucosinolate found in high concentrations in broccoli and other members of the Brassicaceae family. It is converted into the biologically active sulforaphane by the enzyme myrosinase, which is released from plant cells during mastication or food processing and is partially inactivated upon cooking. Gut bacteria also possess myrosinase and further contribute to the conversion of residual glucoraphanin into sulforaphane (Moreno *et al.*, 2006). Several studies indicated sulforaphane as associated with reduced incidence and progression of cancer. Sulforaphane modulates phase 1 cytochrome P450 (CYP) enzymes and consequently reduces the activation of pro-carcinogens, thus reducing the accumulation of DNA damage responsible for the initial stages of tumorigenesis. However, it is best known as the most potent inducer of phase 2 enzymes, involved in xenobiotic detoxification. In addition, sulforaphane possesses anti-inflammatory activity by inhibiting cytokine production. Recent human studies suggest the potential beneficial effect of sulforaphane against CVD and upper airway immunity (James *et al.*, 2012).

ACKNOWLEDGMENTS

This work was supported by the EU FP7 ATHENA collaborative project (Grant Agreement 245121) to CT and by Fondazione Umberto Veronesi Grant 2012 to KP.

REFERENCES

Aggarwal B.B., Shishodia S. Suppression of the nuclear factor-kappa B activation pathway by spice-derived phytochemicals: Reasoning for seasoning. *Ann. N.Y. Acad. Sci.* **2004**, *1030*, 434–441.

Aluyen J.K., Ton Q.N., Tran T., Yang A.E., Gottlieb H.B., Bellanger R.A. Resveratrol: Potential as anticancer agent. *J. Diet Suppl.* **2012**, *9*(1), 45–56.

Arts I.C., Jacobs D.R., Jr, Gross M., Harnack L.J., Folsom A.R. Dietary catechins and cancer incidence among postmenopausal women: The Iowa Women's Health Study (United States). *Cancer Causes Control* **2002**, *13*(4), 373–382.

Asensi-Fabado M.A., Munné-Bosch S. Vitamins in plants: Occurrence, biosynthesis and antioxidant function. *Trends Plant Sci.* **2010**, *15*(10), 582–592.

Baur J.A., Sinclair D.A. Therapeutic potential of resveratrol: The in vivo evidence. *Nat. Rev.* **2006**, *5*(6), 493–506.

Béliveau R., Gingras D. Role of nutrition in preventing cancer. *Can. Fam. Physician* **2007**, *53*(11), 1905–1911.

Bjelakovic G., Nikolova D., Gluud L.L., Simonetti R.G., Gluud C. Antioxidant supplements for prevention of mortality in healthy participants and patients with various diseases. *Cochrane Database Syst. Rev.* **2012**, *3*, CD007176. doi:10.1002/14651858.

Black W.C., Welch H.G. Advances in diagnostic imaging and overestimations of disease prevalence and the benefits of therapy. *N. Engl. J. Med.* **1993**, *328*(17), 1237–1243.

Bostick R.M., Potter J.D., McKenzie D.R., Sellers T.A., Kushi L.H., Steinmetz K.A., Folsom A.R. Reduced risk of colon cancer with high intake of vitamin E: The Iowa Women's Health Study. *Cancer Res.* **1993**, *53*(18), 4230–4237.

Brausi M., Rizzi F., Bettuzzi S. Chemoprevention of human prostate cancer by green tea catechins: Two years later. A follow-up update. *Eur. Urol.* **2008**, *54*(2), 472–473.

Brower V. Nutraceuticals: Poised for a healthy slice of the healthcare market? *Nat. Biotechnol.* **1998**, *16*(8), 728–731.

Burdge G.C., Calder P.C. Conversion of α-linolenic acid to longer-chain polyunsaturated fatty acids in human adults. *Reprod. Nutr. Dev.* **2005**, *45*(5), 581–597.

Butelli E., Titta L., Giorgio M., Mock H.P., Matros A., Peterek S., Schijlen E.G.W.M. *et al.* Enrichment of tomato fruit with health-promoting anthocyanins by expression of select transcription factors. *Nat. Biotechnol.* **2008**, *26*(11), 1301–1308.

Cassidy A., Mukamal K.J., Liu L., Franz M., Eliassen A.H., Rimm E.B. High anthocyanin intake is associated with a reduced risk of myocardial infarction in young and middle-aged women. *Circulation* **2013**, *127*(2), 188–196.

Choi K.C., Jung M.G., Lee Y.H., Yoon J.C., Kwon S.H., Kang H.B., Kim M.J. *et al.* Epigallocatechin-3-gallate, a histone acetyltransferase inhibitor, inhibits EBV-induced B lymphocyte transformation via suppression of RelA acetylation. *Cancer Res.* **2009**, *69*(2), 583–592.

Daar A.S., Singer P.A., Persad D.L., Pramming S.K., Matthews D.R., Beaglehole R., Bernstein A. *et al.* Grand challenges in chronic non-communicable disease. *Nature* **2007**, *450*(7169), 494–496.

Das L., Bhaumik E., Raychaudhuri U., Chakraborty R. Role of nutraceuticals in human health. *J. Food Sci. Technol.* **2012**, *49*(2), 173–183.

de Lorgeril M., Salen P., Martin J.L., Boucher F., de Leiris J. Interactions of wine drinking with omega-3 fatty acids in patients with coronary heart disease: A fish-like effect of moderate wine drinking. *Am. Heart J.* **2008**, *155*(1), 175–181.

de Vrese M., Schrezenmeir J. Probiotics, prebiotics, and synbiotics. *Adv. Biochem. Eng. Biotechnol.* **2008**, *111*, 1–66.

Dolinoy D.C., Weidman J.R., Waterland R.A., Jirtle R.L. Maternal genistein alters coat color and protects A^{vy} mouse offspring from obesity by modifying the fetal epigenome. *Environ. Health Perspect.* **2006**, *114*(4), 567–572.

Douglas L.C., Sanders M.E. Probiotics and prebiotics in dietetics practice. *J. Am. Diet. Assoc.* **2008**, *108*(8), 510–521.

Eastwood M.A. Interaction of dietary antioxidants in vivo: How fruit and vegetables prevent disease. *Q. J. Med.* **1999**, *92*(9), 527–530.

Erkkila A.T., Lehto S., Pyorala K., Uusitupa M.I.J. n-3 Fatty acids and 5-y risks of death and cardiovascular disease events in patients with coronary artery disease. *Am. J. Clin. Nutr.* **2003**, *78*(1), 65–71.

Fink B.N., Steck S.E., Wolff M.S., Britton J.A., Kabat G.C., Schroeder J.C., Teitelbaum S.L., Neugut A.I., Gammon M.D. Dietary flavonoid intake and breast cancer risk among women on Long Island. *Am. J. Epidemiol.* **2007**, *165*(5), 514–523.

Fuentes-Zaragoza E., Sanchez-Zapata E., Sendra E., Sayas E., Navarro C., Fernandez-Lopez J., Perez-Alvarez J.A. Resistant starch as a prebiotic: A review. *Starch/Starke* **2011**, *63*, 406–415.

Gorbach S.L. Probiotics in the third millennium. *Dig. Liver Dis.* **2002**, *34*(Suppl. 2):S2–S7.

Gupta S.C., Prasad S., Kim J.H., Patchva S., Webb L.J., Priyadarsini I.K., Aggarwal B.B. Multitargeting by curcumin as revealed by molecular interaction studies. *Nat Prod. Rep.* **2011**, *28*(12), 1937–1955.

Hanhineva K., Törrönen R., Bondia-Pons I., Pekkinen J., Kolehmainen M., Mykkänen H., Poutanen K. Impact of dietary polyphenols on carbohydrate metabolism. *Int. J. Mol. Sci.* **2011**, *11*(4), 1365–1402.

He K., Rimm E.B., Merchant A., Rosner B.A., Stampfer M.J., Willett W.C., Ascherio A. Fish consumption and risk of stroke in men. *JAMA* **2002**, *288*(24), 3130–3136.

Heinonen O.P., Albanes D., Virtamo J., Taylor P.R., Huttunen J.K., Hartman A.M., Haapakoski J. *et al.* Prostate cancer and supplementation with alpha-tocopherol and beta-carotene: Incidence and mortality in a controlled trial. *J. Natl. Cancer Inst.* **1998**, *90*(6): 440–446.

Hirvonen T., Virtamo J., Korhonen P., Albanes D., Pietinen P. Intake of flavonoids, carotenoids, vitamins C and E, and risk of stroke in male smokers. *Stroke* **2000**, *31*(10), 2301–2306.

Hord N.G. Eukaryotic microbiotic crosstalk: Potential mechanisms for health benefits of prebiotics and probiotics. *Annu. Rev. Nutr.* **2008**, *28*, 215–231.

James D., Devaraj S., Bellur P., Lakkanna S., Vicini J., Boddupalli S. Novel concepts of broccoli sulforaphanes and disease: Induction of phase II antioxidant and detoxification enzymes by enhanced-glucoraphanin broccoli. *Nutr. Rev.* **2012**, *70*(11), 654–665.

Jennings A., Welch A.A., Spector T., Macgregor A., Cassidy A. Intakes of anthocyanins and flavones are associated with biomarkers of insulin resistance and inflammation in women. *J. Nutr.* **2014**, 144(2), 202–208.

Jiménez-Chillarón J.C., Díaz R., Martínez D., Pentinat T., Ramón-Krauel M., Ribó S., Plösch T. The role of nutrition on epigenetic modifications and their implications on health. Biochimie. **2012**, 94(11), 2242–2263.
Joseph J.A., Shukitt-Hale B., Denisova N.A., Bielinski D., Martin A., McEwen J.J., Bickford P.C. Reversals of age-related declines in neuronal signal transduction, cognitive, and motor behavioral deficits with blueberry, spinach, or strawberry dietary supplementation. *J. Neurosci.* **1999**, *19*(18), 8114–8121.
Ju Y.H., Clausen L.M., Allred K.F., Almada A.L., Helferich W.G. β-Sitosterol, β-sitosterol glucoside, and a mixture of β-sitosterol and β-sitosterol glucoside modulate the growth of estrogen-responsive breast cancer cells in vitro and in ovariectomized athymic mice. *J. Nutr.* **2004**, *134*(5), 1145–1151.
Kendall C.W.C., Esfahani A., Jenkins D.J.A. The link between dietary fibre and human health. *Food Hydrocoll.* **2010**, *24*, 42–48.
Kim J., Zhang X., Rieger-Christ K.M., Summerhayes I.C., Wazer D.E., Paulson K.E., Yee A.S. Suppression of Wnt signaling by the green tea compound (–)-epigallocatechin 3-gallate (EGCG) in invasive breast cancer cells. *J. Biol. Chem.* **2006**, *281*(16), 10865–10875.
Kristenson M., Ziedén B., Kucinskienë Z., Elinder L.S., Bergdahl B., Elwing B., Abaravicius A., Razinkovienë L., Calkauskas H., Olsson A.G. Antioxidant state and mortality from coronary heart disease in Lithuanian and Swedish men: Concomitant cross sectional study of men aged 50. *Br. Med. J.* **1997**, *314*(7081), 629–633.
Labrecque L., Lamy S., Chapus A., Mihoubi S., Durocher Y., Cass B., Bojanowski M.W., Gingras D., Beliveau R. Combined inhibition of PDGF and VEGF receptors by ellagic acid, a dietary derived phenolic compound. *Carcinog.* **2005**, *26*(4), 821–826.
La Casa C., Villegas I., de la Lastra C.A., Motilva V., Calero M.J.M. Evidence for protective and antioxidant properties of rutin, a natural flavone, against ethanol induced gastric lesions. *J. Ethnopharm.* **2000**, *71*(1/2), 45–53.
Lampe J.W. Health effects of vegetable and fruit: Assessing mechanism of action in human experimental studies. *Am. J. Clin. Nutr.* **1999**, *70*(3 Suppl), S475–S490.
Lamy S., Blanchette M., Michaud-Levesque J., Lafleur R., Durocher Y., Moghrabi A., Barrette S., Gingras D., Beliveau R. Delphinidin, a dietary anthocyanidin, inhibits vascular endothelial growth factor receptor-2 phosphorylation. *Carcinog.* **2006**, *27*(5), 989–996.
Lamy S., Gingras D., Beliveau R. Green tea catechins inhibit vascular endothelial growth factor receptor phosphorylation. *Cancer Res.* **2002**, *62*(2), 381–385.
Lee W.J., Shim J.Y., Zhu B.T. Mechanisms for the inhibition of DNA methyltransferases by tea catechins and bioflavonoids. *Mol. Pharmacol.* **2005**, *68*(4), 1018–1030.
Lima J.T., Almeida J.R.G.S., Barbosa-Filho J.M., Assis T.S., Silva M.S., da-Cunha E.V.L., Braz-Filho R., Silva B.A. Spasmolytic action of diplotropin, a furanoflavan from *Diplotropis ferruginea Benth.*, involves calcium blockade in guinea-pig ileum. *J. Chem. Sci.* **2005**, *60b*, 1093–1100.
Machha A., Mustafa M.R. Chronic treatment with flavonoids prevents endothelial dysfunction in spontaneously hypertensive rat aorta. *J. Cardiovasc. Pharmacol.* **2005**, *46*(1), 36–40.
Martin C., Butelli E., Petroni K., Tonelli C. How can research on plants contribute to promoting human health? *Plant Cell* **2011**, *23*(5), 1685–1699.
Martin C., Zhang Y., Tonelli C., Petroni K. Plants, diet, and health. *Annu. Rev. Plant Biol.* **2013**, *64*, 19–46.
Matsumoto H., Nakamura Y., Tachibanaki S., Kawamura S., Hirayama M. Stimulatory effect of cyanidin 3-glycosides on the regeneration of rhodopsin. *J. Agric. Food Chem.* **2003**, *51*(12), 3560–3563.
McDonnell D.P. The molecular pharmacology of SERMs. *Trends Endocrinol. Metab.* **1999**, *10*(8), 301–311.
Meyer K.A., Kushi L.H., Jacobs D.R., Slavin J., Sellers T.A., Folsom A.R. Carbohydrates, dietary fiber, and incident type 2 diabetes in older women. *Am. J. Clin. Nutr.* **2000**, *71*(4), 921–930.

Miura Y., Chiba T., Tomita I., Koizumi H., Miura S., Umegaki K., Hara Y., Ikeda M., Tomita T. Tea catechins prevent the development of atherosclerosis in apoprotein E-deficient mice. *J. Nutr.* **2001**, *131*(1), 27–32.

Moreno D.A., Carvajal M., López-Berenguer C., García-Viguera C. Chemical and biological characterisation of nutraceutical compounds of broccoli. *J. Pharm. Biomed. Anal.* **2006**, *41*(5), 1508–22.

Napolitano M., De Pascale C., Wheeler-Jones C., Botham K.M., Bravo E. Effects of lycopene on the induction of foam cell formation by modified LDL. *Am. J. Physiol. Endocrinol. Metab.* **2007**, *293*(6), E1820–E1827.

Normén L., Holmes D., Frohlich J. Plant sterols and their role in combined use with statins for lipid lowering. *Curr. Opin. Investig. Drugs* **2005**, *6*(3), 307–316.

Packer L., Weber S.U., Rimbach G. Molecular aspects of α-tocotrienol antioxidant action and cell signalling. *J. Nutr.* **2001**, *131*(2), 369S–373S.

Pan M.H., Lai C.S., Tsai M.L., Wu J.C., Ho C.T. Molecular mechanisms for anti-aging by natural dietary compounds. *Mol. Nutr. Food Res.* **2012**, *56*(1), 88–115.

Park S.J., Ahmad F., Philp A., Baar K., Williams T., Ke H., Rehmann H. *et al.* Resveratrol ameliorates aging-related metabolic phenotypes by inhibiting cAMP phosphodiesterase. *Cell* **2012**, *148*(3), 421–433.

Park W., Amin A.R., Chen Z.G., Shin D.M. New perspectives of curcumin in cancer prevention. *Cancer Prev. Res. (Phila)* **2013**, *6*(5), 387–400.

Piironen V., Lampi A.-J. Occurrence and levels of phytosterols in food. Chapter 1, in Dutta P.C. (ed.), *Phytosterols as Functional Food Components and Nutraceuticals.* Boca Raton, FL: CRC Press, **2003**, pp. 1–27.

Ramsden C.E., Faurot K.R., Carrera-Bastos P., Cordain L., de Lorgeril M., Sperling L.S. Dietary fat quality and coronary heart disease prevention: A unified theory based on evolutionary, historical, global and modern perspectives. *Curr. Treat. Options Cardiovasc. Med.* **2009**, *11*(4), 289–301.

Renaud S., de Lorgeril M. Wine, alcohol, platelets, and the French paradox for coronary heart disease. *Lancet* **1992**, *339*(8808): 1523–1526.

Reuter S., Gupta S.C., Park B., Goel A., Aggarwal B.B. Epigenetic changes induced by curcumin and other natural compounds. *Genes Nutr.* **2011**, *6*(2), 93–108.

Rice-Evans C., Miller N., Paganga G. Antioxidant properties of phenolic compounds. *Trends Plant Sci.* **1997**, *2*(4), 152–159.

Richer S., Stiles W., Statkute L., Pulido J., Frankowski J., Rudy D., Pei K., Tsipursky M., Nyland J. Double-masked, placebo-controlled, randomized trial of lutein and antioxidant supplementation in the intervention of atrophic age-related macular degeneration: The Veterans LAST study (Lutein Antioxidant Supplementation Trial). *Optometry* **2004**, *75*(4), 216–230.

Seren S., Lieberman R., Bayraktar U.D., Heath E., Sahin K., Andic F., Kucuk O. Lycopene in cancer prevention and treatment. *Am. J. Ther.* **2008**, *15*(1), 66–81.

Setchell K.D.R., Brown N.M., Lydeking-Olsen E. The clinical importance of the metabolite equol-a clue to the effectiveness of soy and its isoflavones. *J. Nutr.* **2002**, *132*(12), 3577–3584.

Shehzad A., Ha T., Subhan F., Lee Y.S. New mechanisms and the anti-inflammatory role of curcumin in obesity and obesity-related metabolic diseases. *Eur. J. Nutr.* **2011**, *50*(3), 151–161.

Shishodia S. Molecular mechanisms of curcumin action: Gene expression. *Biofactors* **2013**, *39*(1), 37–55.

Shufelt C., Merz C.N.B., Yang Y.C. Red versus white wine as a nutritional aromatase inhibitor in premenopausal women: A pilot study. *J. Women Health* **2012**, *21*(3), 281–284.

Simopoulos A.P. The importance of the ratio of omega-6/omega-3 essential fatty acids. *Biomed. Pharmacother.* **2002**, *56*(8), 365–379.

Singh N., Agrawal M., Doré S. Neuroprotective properties and mechanisms of resveratrol in in vitro and in vivo experimental cerebral stroke models. *ACS Chem. Neurosci.* **2013**, *4*(8), 1151–1162.
Stampfer M.J., Hennekens C.H., Manson J.E., Colditz G.A., Rosner B., Willett W.C. Vitamin E consumption and the risk of coronary disease in women. *N. Engl. J. Med.* **1993**, *328*(20), 1444–1449.
Sunairi M., Watanabe K., Suzuki T., Tanaka N., Kuwayama H., Nakajima M. Effects of antiulcer agents on antibiotic-activity against *Helicobacter pylori. Eur. J. Gastroent. Hepatol.* **1994**, *6*(Suppl 1), S121–S124.
Surh Y.J. Cancer chemoprevention with dietary phytochemicals. *Nature Rev. Cancer* **2003**, *3*(10), 768–780.
Suvarna V.C., Boby V.U. Probiotics in human health: a current assessment. *Curr. Sci.* **2005**, *88*(11), 1744–1748.
Talalay P., Fahey J.W. Phytochemicals from cruciferous plants protect against cancer by modulating carcinogen metabolism. *J. Nutr.* **2001**, *131*(11 Suppl), 3027S–3033S.
Titta L., Trinei M., Stendardo M., Berniakovich I., Petroni K., Tonelli C., Riso P. *et al.* Blood orange juice inhibits fat accumulation in mice. *Int. J. Obes.* **2010**, *34*(3), 578–588.
Toufektsian M.C., de Lorgeril M., Salen P., Nagy N., Donati M.B., Giordano L., Mock H.-P. *et al.* Chronic dietary intake of plant-derived anthocyanins protects the rat heart against ischemia-reperfusion injury. *J. Nutr.* **2008**, *138*(4), 747–752.
Toufektsian M.C., Salen P., Laporte F., Tonelli C., de Lorgeril M. Dietary flavonoids increase plasma very long-chain (n-3) fatty acids in rats. *J. Nutr.* **2011**, *141*(1), 37–41.
Tsang W.P., Kwok T.T. Epigallocatechin gallate up-regulation of miR-16 and induction of apoptosis in human cancer cells. *J. Nutr. Biochem.* **2010**, *21*(2), 140–146.
Tsuda T., Horio F., Uchida K., Aoki H., Osawa T. Dietary cyanidin 3-O-beta-D-glucoside-rich purple corn color prevents obesity and ameliorates hyperglycemia in mice. *J. Nutr.* **2003**, *133*(7), 2125–2130.
Viña J., Gomez-Cabrera M.C., Borras C. Fostering antioxidant defences: Up-regulation of antioxidant genes or antioxidant supplementation? *Brit. J. Nutr.* **2007**, *98*(Suppl 1), S36–S40.
Virgili F., Marino M. Regulation of cellular signals from nutritional molecules: A specific role for phytochemicals, beyond antioxidant activity. *Free Rad. Biol. Med.* **2008**, *45*(9), 1205–1216.
Wanders A.J., van den Borne J.J.G.C., de Graaf C., Hulshof T., Jonathan M.C., Kristiansen M., Mars M., Schols H.A., Feskens E.J.M. Effects of dietary fibre on subjective appetite, energy intake and body weight: A systematic review of randomized controlled trials. *Obesity Reviews* **2011**, *12*(9), 724–739.
Wang Z.R., Huang Y.Z., Zou J.G., Cao K.J., Xu Y.N., Wu J.M. Effects of red wine and wine polyphenol resveratrol on platelet aggregation *in vivo* and *in vitro*. *Int. J. Mol. Med.* **2002**, *9*(1), 77–79.
Yusuf S., Hawken S., Ounpuu S., Dans T., Avezum A., Lanas F., McQueen M. *et al.* INTERHEART Study Investigators. Effect of potentially modifiable risk factors associated with myocardial infarction in 52 countries (the INTERHEART study): Case-control study. *Lancet* **2004**, *364*(9438), 937–952.
Zingg J.M., Hasan S.T., Meydani M. Molecular mechanisms of hypolipidemic effects of curcumin. *Biofactors* **2013**, *39*(1), 101–121.

8 The Mediterranean Diet from Ancel Keys to the Present

Epidemiological Studies on Selected Populations and Practical Applications in the Modern Cultural Context

Francesca Ghelfi and Lucilla Titta

CONTENTS

8.1 HISTORY

8.1.1 The Relationship between Mediterranean Lifestyle and Health Highlighted by Ancel Keys

The Mediterranean diet should not be considered as a dietary pattern but rather reflective of an integrated approach to living in which foods represent only one of the relevant cultural components, along with community beliefs, traditions, and the sense of pleasure associated with food preparation and intake. Ancel Keys's sabbatical year, and related travels in 1951–1952, opened his eyes to cultural differences in diet, behavior, and the risk of diseases. In particular, he perceived the difference between cardiovascular mortality rates in Mediterranean countries and the United States (Keys, 1995). His first hypothesis was that the traditional Mediterranean diet could influence the health status of the population that practices it. The Mediterranean diet of that time can be described by the following characteristics: an abundance of plant foods (fruit, vegetables, cereals, beans, nuts, and seeds); minimally processed, seasonally fresh, and locally grown foods; fresh fruit as the typical daily dessert, with sweets containing added sugar or honey consumed a few times per week; olive oil as the main source of fat; low amounts of dairy products; fish and poultry consumed in moderate amounts and red meat occasionally; and wine consumed in low-to-moderate amounts normally within meals. Thus, a Mediterranean diet was defined as the dietary pattern found in the olive-growing areas of the Mediterranean region in the late 1950s and the early 1960s when the ill effects of World War II had been overcome and the fast food culture had not yet invaded lifestyle preferences.

The term *Mediterranean diet* was coined, paradoxically, by Americans in the early 1960s, and today, only 50 years from the "discovery" of this characteristic, traditional food model, the United Nations Educational, Scientific and Cultural Organization (UNESCO) declares that this century-long cookery culture, belonging to populations overlooking the Mediterranean basin, represents a "cultural heritage of humanity." However, the debate continues among the scientific community with regard to its defining features because of the enormous differences between food patterns in the region and the potential development of these patterns over time.

8.1.2 Connection between Cholesterol Levels and Risk of Cardiovascular Events

Before the 1950s, most physicians considered atherosclerosis to be an inevitable consequence of aging, about which nothing much could be done. At that time, a lot of controversy affected the hypothesis that high blood cholesterol levels contribute causally to atherosclerosis and coronary heart disease (CHD) (the *lipid hypothesis*) (Steinberg, 2006). Over the next four decades, the case against hypercholesterolemia would become ever stronger, based on experimental pathology, clinical observations, genetic studies, and epidemiology. In 1948, the Framingham Heart Study, under the direction of the National Heart Institute (now known as the National Heart, Lung, and Blood Institute, or NHLBI), embarked on an ambitious project in health research (Dawber *et al.*, 1951). At that time, little was known about the general causes of heart disease and stroke, but the death rates for cardiovascular diseases (CVDs) had been increasing steadily since the beginning of the century and had become an epidemic. The objective of the Framingham Heart Study was to identify the common factors or characteristics that contribute to CVDs by following their development over a long period of time in a large group of participants who had not yet developed overt symptoms of CVD or suffered a heart attack or stroke. The researchers recruited 5209 men and women between 30 and 62 years from the town of Framingham, Massachusetts, and began the first round of extensive physical examinations and lifestyle interviews that they would later analyze for common patterns related to CVD development. The year 1961 marked a turning point, with the publication by the Framingham Heart Study researchers, who demonstrated that cholesterol level, blood pressure, and electrocardiogram abnormalities increase the risk of heart disease (Epstein, 1996). In this year, the American Heart Association accepted the causal relationship and recommended that people at high risk modify their diets to prevent heart attacks.

8.1.3 Relationship between Diet and Cardiovascular Risk

Based on the *lipid hypothesis*, it was presumed that a low-cholesterol diet could decrease blood cholesterol levels (Steinberg, 2006). Several studies were conducted with cholesterol-fed animals; for example, Anitschkow and colleagues showed that simply feeding rabbits with purified cholesterol dissolved in sunflower oil induced vascular lesions closely resembling those of human atherosclerosis, both grossly and microscopically (Anitschkow, 1933). For years there have been doubts about the relevance of these animal experiments to humans because the focus was on dietary cholesterol and humans are much less sensitive to dietary cholesterol than the rabbits and chickens studied in the laboratory. In this perspective, several experimental trials were performed to test the effectiveness of a low-fat diet or a diet rich in different kinds of fats on blood cholesterol.

Already in the 1950s, the evidence obtained in epidemiologic studies indicated that populations subsisting on high-fat diets tended to have high cholesterol concentrations in their blood and a high incidence of atherosclerotic diseases (Keys and Grande, 1957). The emphasis at that time was on total dietary fat and it has been realized that in humans the amount of cholesterol in the blood is affected by ordinary food fats much more than by dietary cholesterol. In the ensuing years, it became increasingly apparent that different fats varied in their effect on serum cholesterol levels.

8.1.4 Seven Countries Study Formally Started in Fall 1958

The Seven Countries Study was the first to systematically examine the relationship among lifestyle, diet, and the rates of heart attack and stroke in contrasting populations. Ancel Keys posed the hypothesis that differences in the frequency of heart attacks and stroke among populations would occur in some orderly relation to physical characteristics and lifestyle, particularly composition of the diet, and especially fats in the diet (Blackburn, 1995). To test this idea, Keys and his colleagues carried out surveys between 1958 and 1970 in populations of men aged 40–59, in 16 selected cohorts in Finland, Greece, Italy, Japan, The Netherlands, the United States, and Yugoslavia (Kromhout *et al.*, 1989). Most of the areas were stable and rural, and had wide contrasts in habitual diet. For example, nine cohorts were made up of rural Europeans, including farmers and men in other occupations common to villages in Europe. Five cohorts were Mediterranean and they were alike in that olive oil provided from 15% to 30% of the total dietary energy, with wine, an important part of main meals, supplying an average of 8%–20% of calories. By contrast, there were four cohorts in rural Finland and Yugoslavia for whom milk and meat fats were the dietary fats and the alcoholic beverages were beer and distilled liquors, often not taken with meals. The Mediterranean people consumed more fruits and fresh vegetables than people in the other areas.

The 5- and 10-year experience of the Seven Countries Study showed marked differences in CHD incidence and mortality among the cohorts. In this study, dietary surveys showed a high correlation between the average percentage of dietary energy from saturated fatty acids (SFAs) and the 10-year CHD incidence and death rate. Completion of the 15-year follow-up confirmed these general relationships between coronary mortality rates and mean characteristics of the cohorts for serum cholesterol, blood pressure, and percentage of the total energy from SFAs (Keys *et al.*, 1986).

Italy presented Ancel Keys and the Seven Countries Study collaborators with the prototypical Mediterranean lifestyle and its apparently healthy diet of grains, pasta, legumes, vegetables, fruits, olive oil, bread, and wine.

8.1.4.1 Italy

Italy, in contrast to other countries researched by the Seven Countries Study, also has one of the greatest traditional cuisines of the world. Moreover, for the Italian investigator concerned with both regional and individual differences, Italy is not just one country or culture. Nicotera in Calabria, Porto San Giorgio on the Adriatic coast, Montegiorgio in the Marche hills above the coast, and Crevalcore in the rich delta of the Po River provide abundant contrasts between northern and southern Italian eating patterns, heavier on meats in the north and on legumes, pasta, and olive oil in the south. The Seven Countries Study was clearly the first epidemiological enterprise of any sort on population-based samples in Italy and the first to show the independent significant predictive power of conventional risk factors within the Italian culture with regard to death from CHD and CVD as well as other causes of death. As much as any other study, the coordination of the dietary and medical surveys in the Italian component of the Seven Countries Study resulted in a universal understanding of the healthy nature of the so-called *Mediterranean diet*. It also revealed the importance of dietary patterns, over and above single foods or nutrients, in determining the population-wide causes of disease and death.

8.1.4.2 Greece

The Seven Countries Study findings in the Greek islands have had a profound influence on research and thinking in CVD epidemiology and on eating pattern and lifestyle approaches to prevention and health promotion. They not only confirmed the salubrious nature of the Mediterranean diet, but they provided a major lesson that an habitual eating pattern relatively high in fats is compatible with good health, low cardiovascular and cancer risk, and substantial longevity, under the particular conditions of the Greek diet. The theory that SFAs are the principal cause of mass elevation of blood cholesterol and, thus, of CHD was therefore addressed in the Greek islands under the special condition of a high-fat, low-saturated fat intake. The Greek studies have also made special contributions to understand the effects of mass change in population risk characteristics and their predictive importance. The data on trends may contribute to explanations of the rapid acceleration in CVD deaths reported recently in the World Health Organization (WHO, 2013) statistics for Greece.

8.1.5 In the 1950s, the American (Western) Lifestyle Was Already Sedentary, Rich in Junk Food and Big-Size Meals

The sources of total calories in the U.S. diet have changed since the beginning of the twentieth century. Food consumption data for households are available from a series of surveys conducted approximately every 5–10 years since 1936 by the United States Department of Agriculture (Slattery and Randall, 1988). A decrease in the percentage contribution of calories from grains and cereals occurred coincident with an increase in the use of sugars and syrups and animal foods. The percentage of total calories from fats and oils increased after 1900. The contribution of total calories from the meat, fish, and poultry group increased until the late 1960s when it started to level off. Total consumption of dairy foods increased after early in the century; in particular, there was a marked increase in consumption of whole milk and a dramatic increase in the use of low-fat milk and cheese after the 1950s. An increase in consumption of cheese means an increase of SFAs in the diet. Despite the consumption of butter has decreased, the consumption of margarine increased significantly. Trends in animal and vegetable fat consumption obtained from food balance sheets also showed a constant increase in vegetable fat and a constant decrease in animal fat after 1961. Changes within these food groups suggest lower amount of SFAs being consumed, but the fact is that new margarines are mostly produced with coconut and palm oil, rich in saturated fats. Significant shifts in food use within major food groups are evident.

Using a variety of data sources and statistical modeling, many studies have suggested that increased food intake is largely, but not completely, responsible for the obesity epidemic. Importantly, occupational physical activity has an even greater potential to have a significant impact on the total caloric expenditure. Using nationally representative data sources, a recent study examined trends in occupational physical activity during the past five decades, in particular the prevalence of service occupations, goods producing occupations, and agricultural occupations for the United States from 1960 to 2008 (Church *et al.*, 2011). This study shows that over the last 50 years in the United States, there has been a progressive decrease in the percentage of individuals employed in goods producing and agricultural occupations, whereas there has been an increase in the percentage of individuals employed in service occupations. This has resulted in a shift away from occupations that require moderately intense physical activity to occupations that are largely composed of sitting and sedentary behavior.

8.1.6 The Mediterranean Lifestyle was Gradually Influenced by Western Diet

Total energy availability per person per day was lower in the Mediterranean countries in the 1960s than in the other European areas (Balanza *et al.*, 2007). Over the last 40 years, there has been a major increase in energy availability in all the European areas, though the increase was quantitatively greater in the Mediterranean countries. Also, energy availability from fats was significantly lower in the Mediterranean and eastern countries than in the north in the period 1961–1963, but during the last 40 years, energy from fat has considerably increased in all three European regions. However, energy availability from carbohydrates was significantly higher in the Mediterranean and Eastern Europe than in the north, and after 40 years, a general reduction in the availability of carbohydrates was observed in all three European regions. In particular, cereals were staple foods in the 1960s both in the Mediterranean countries and in the east, and provided greater energy availability than in the northern countries; cereals were the main source of energy in these two areas. The availability of cereals, especially in Mediterranean Europe, has declined sharply over the last 40 years. Energy availability from meat and dairy products was lower in Mediterranean and Eastern Europe in the 1960s, but there are now no major differences between the three European regions. Energy availability in the form of legumes, nuts, eggs, seafood, sweets, olive oil, fruits, and vegetables showed little variation over time.

In the mid-1960s, Greece had a national average apparent food consumption yielding 2900 kcal/person/day, with 29% coming from fats (13% from olive oil) and 43% from cereals. Meat consumption was a mere 33 kg/person/year and that of sugar 18 kg (raw sugar equivalent) (Alexandratos, 2006). Three and a half decades later, the kcal/person/day had risen to 3700 and fats from 92 to 152 g/day, accounting for 36% of total calories (12% from olive oil). Meat consumption had risen to 88 kg/person/year and that of sugar to 32 kg. No wonder that the incidence of obesity in Greece is among the highest in Europe and the world, with Crete itself holding place of pride in this transformation (International Obesity Task Force, 2005). The evolution of diets in the other Mediterranean countries on the European side tells a similar story. The Italian diet, for example, was fairly close to the Mediterranean model until the 1950s; thereafter, the consumption of vegetable proteins and cereals decreased with an increasing consumption of fat, particularly animal fat (INRAN, 2012). These data show the consequences of globalization and increasing availability of precious foods, such as meat, dairy, and processed food.

In sum, food habits in the Mediterranean countries have deviated enormously from the traditional pattern of the 1960s, which are generally considered to be the Mediterranean diet, and have tended to move toward the food pattern typical of the northern countries. Among the most significant changes observed in the Mediterranean countries, there is a fall in the availability of carbohydrates and an increase in the availability of animal proteins and fats.

8.1.7 Ancel Keys, Med Diet, and Parameters Setting: The Incredible Long View

In 1959, Ancel Keys, with his wife Margaret, wrote a book titled *Eat Well and Stay Well*, dedicated to "everyone who loves to eat and wants to know how the diet affects the heart, arteries, and blood cholesterol" (Keys and Keys, 1959). In this book, one would find several analogies between Keys's statement and current health recommendations.

Keys's *best advice list* starts with the recommendation "see your doctor regularly" and this sentence brings to mind current early detection campaigns. After that, 10 recommendations are listed:

1. If you are fat, reduce.
2. Restrict use of fats in dairy products and in ordinary meats.
3. Use vegetable oils in cookery, not as medicine.
4. Avoid hydrogenated products but if you insist on a spread use the newer margarines.
5. Avoid heavy use of salt and sugars.
6. Favour fish, shellfish, poultry, fresh vegetables, fruits, nonfat milk products.
7. Do not depend on the drugstore or vitamin peddler for a good diet.
8. Be sensible about cigarettes, alcohol, excitement, business strain.
9. Get plenty of exercise, with recreation, and work out-of-doors.
10. Learn more about cookery and have fun without saturation.

Of note is a clear preference for all lifestyle aspects, in step with the Mediterranean way of life. Similar to current standards, the association between body weight, physical activity, and health is emphasized. Also the importance to reduce salt and sugar consumption is a very discussed matter today and the abuse of these additives has a clear association with poor health.

In the 1960s, the use of supplements appeared to be healthy behavior, and since then, the supplement business has increased. However, the most recent health recommendations, such as Keys's advice, discourage against the use of supplements to replace foods or to protect health.

Always in the same book, Keys wrote:

> The purpose of this book is not merely to provide a guide to good eating and sound nutrition with control of calories and total fats; almost equally important is the proper selection of the kinds of fats to eat. All food fats and oils have about the same caloric

> value but they differ in their effects on the blood and on the arteries.... The fats that are most liquid at ordinary temperatures are usually most acceptable. Most, but not all, natural fats of vegetable origin are in this class.... Among the fats to be used with discretion in the diet are dairy fats (in butter, milk, cheese, etc.), beef, pork and lamb fats, lard, coconut oil, margarines, solid shortenings, and the fat in chocolate.... The oils in nuts range from "intermediate" to "good" in respect to the effect on the blood cholesterol.

Nowadays, reduction in SFA consumption is traditionally a major focus of dietary recommendations to reduce CHD risk. Furthermore, consumption of polyunsaturated fatty acids (PUFAs) lowers the total cholesterol:high-density lipoprotein cholesterol (TC:HDL-C) ratio, perhaps the best single lipid predictor of CHD risk (Lewington *et al.*, 2007). These effects on risk factors suggest that PUFA may be an ideal replacement for SFA in the population. To investigate and quantify this effect, a systematic review and meta-analysis of randomized controlled clinical trials was performed that assessed the impact of increased PUFA consumption, as replacement for SFA, on CHD end points (Mozaffarian *et al.*, 2010). In this meta-analysis, increasing PUFA consumption as a replacement for SFA reduced the occurrence of CHD events by 19%; each 5% energy greater PUFA consumption reduced CHD risk by 10%. Whereas nearly all these trials were insufficiently powered to detect a significant effect individually, the pooled results demonstrate a significant benefit of replacing PUFA for SFA on clinical CHD events.

Another anticipation derives from pulses:

> Peas, beans, and lentils have special nutritional properties, and their cooking characteristics are different from other vegetables, especially when prepared from the dried state. They are often called poor man's meat, and they are actually good meat substitutes because of their high content of protein. The complex nutrients in them have a small but fairly consistent cholesterol-depressing effect in controlled experiments on man.

In the 1970s, a protective link was proposed between dietary fiber (in the form of whole grain foods) and ischemic heart disease (Trowell, 1972). Many observational and experimental studies have since examined the relationship between dietary fiber or fiber-rich foods and total cardiovascular risk (CVD) or cardiovascular risk factors, such as hypertension, central obesity, insulin sensitivity, and elevated plasma cholesterol. For example, a recent study showed a significantly lower risk of 9% for both CVD and CHD with every additional 7 g/day of total fiber consumed (Threapleton *et al.*, 2013). Several studies have also shown that legume consumption has been associated with lower risks of CHD in observational epidemiological studies (Bazzano *et al.*, 2001). In support of this evidence, a meta-analysis of randomized controlled trials shows that non-soy legume consumption has a significant beneficial effect on serum cholesterol levels (Bazzano *et al.*, 2011). Interestingly, it was assessed that both TC and LDL cholesterol decreased, whereas HDL-C did not change significantly. Again, the incredible long-term view arising from Ancel Keys is clear.

8.2 ACTUAL SCIENTIFIC EVIDENCE BETWEEN MEDITERRANEAN DIET AND HEALTH

8.2.1 Scientific Literature in Support of the Relationship among Nutrition, Physical Activity, Tradition, Other Lifestyles, and Health: Epidemiological and Intervention Studies on Mediterranean Lifestyle and Diseases Prevention

The relationship between nutrition and health has been studied intensively for more than half a century. Following the publication of the results of the Seven Countries Study, the effect of nutrition on human health has been evaluated by many case control, prospective cohort, and randomized clinical trials, all of which have provided evidence that diet may significantly affect the health state of a population (Serra-Majem *et al.*, 2006). An optimal diet, as a fundamental element of an array of lifestyle issues, may significantly contribute to the maintenance of a healthy status. Conversely, poor-quality diets may favor the development of adverse health states, including obesity, dyslipidemia, hypertension (the combination of which may lead to the *metabolic syndrome*), and the emergence of chronic diseases such as diabetes, coronary artery disease, neoplastic disorders, and neurodegenerative disorders.

The Mediterranean diet is not a homogeneous model within the Mediterranean area. It presents regional variations derived from the same dietary pattern, influenced by various factors, such as sociocultural, religious, and economic determinants, to name a few. In the last few decades, numerous associations have been postulated between health status and the Mediterranean diet and some of its components. Most of the scientific knowledge, used as evidence for the creation of food guidelines, is based on the association between foods or nutrients and the incidence of certain diseases. However, human diet is composed of several food groups that can be consumed in different amounts and in the twenty-first century scientific studies started exploring food patterns in health and diseases. There are two approximations in the analysis of food patterns: developing food indexes, that is, food scores according to the intake of certain foods, or deriving patterns via multivariant analysis by means of a factorial analysis, principal component analysis, or cluster analysis (Bach *et al.*, 2006). Several epidemiological studies have used different methods to evaluate the adherence of a population to the Mediterranean dietary pattern (MDP), based on different principles, including Mediterranean diet score (MDS) and Mediterranean diet quality index. Reaching a consensus is complex because the pattern may be related to a specific Mediterranean region and to a selected period that are adopted as references. Even though it is difficult to quantify the adherence to a diet and despite the fact that indexes require operational definitions, grouping foods to obtain complex scores is a very useful method to evaluate epidemiological associations.

8.2.2 Nutrition and Cancer

Historically, the incidence of overall cancer was lower in Mediterranean countries compared to northern European countries, the United Kingdom, and the United States (Trichopoulou *et al.*, 2000). This difference has been linked to the

Mediterranean food pattern consumed in the Mediterranean countries, but there are variations in the components of the traditional Mediterranean diet between and within countries. Moreover, no isolated study can provide evidence that any factor is a cause of, or is protective against, any disease. Therefore, the nutritional etiology of various forms of cancer has been evaluated in an extensive and detailed report in 2007 by the World Cancer Research Fund (WCRF) and the American Institute for Cancer Research (AICR): *Food, Nutrition, Physical Activity, and the Prevention of Cancer: A Global Perspective* (WCRF-AICR, 2007). An international team evaluated the relationships between foods and drinks, body weight, physical activity, and cancers. The conclusions have been reached on the basis of several systematic reviews of the literature available. The criteria considered to assign the level of evidence were the type of study (with a significant weight given to prospective studies), the heterogeneity within and between studies, their good quality, the presence of a dose–response relationship, and, ultimately, the biological plausibility of the association. In sum, the panel ranked the available evidence into four levels: "convincing," "probable," "limited-suggestive," and a fourth level that collects the effects for which the evidence of association with the tumor is highly "unlikely."

Regarding body composition, the evidence that greater body fatness is a cause of cancers of the esophagus, the pancreas, the colorectum, the breast (postmenopause), the endometrium, and the kidney is convincing. Greater body fatness is probably a cause of ovarian cancer and tumor of the gallbladder. There is also limited evidence suggesting that greater body fatness is a cause of liver cancer. The evidence that greater abdominal (central) fatness is a cause of colorectal cancer is convincing, whereas it is probably a cause of cancers of the pancreas, the breast (postmenopause), and the endometrium. By contrast, greater body fatness probably protects against premenopausal breast cancer. Body fatness directly affects levels of many circulating hormones, such as insulin, insulin-like growth factors, and estrogens, creating an environment that encourages carcinogenesis and discourages apoptosis (programmed cell death). It also stimulates the body's inflammatory response, which may contribute to the initiation and progression of several cancers. Adjusting for serum levels of estradiol diminishes or destroys the association with body mass index (BMI), suggesting that hormones are a predominant mechanism. There is no single well-established mechanism through which body fatness could prevent premenopausal breast cancer. According to the estrogen plus progesterone theory, overweight premenopausal women would be protected because they would be more frequently anovulatory, and therefore less likely to be exposed to endogenous progesterone. However, this theory is not well supported by recent studies, which suggest that natural progesterone could be protective. Normal levels of natural progesterone are likely to be protective, and well nourished, or perhaps over-nourished women, who may become slightly overweight in adulthood, may be protected by their natural fertile condition. Another possible mechanism is that the increased adipose tissue–derived estrogen levels in overweight children could induce early breast differentiation and eliminate some targets for malignant transformation (Hilakivi-Clarke *et al.*, 2001). Anovulation and abnormal hormone profiles are commonly associated with obesity. The age-specific pattern of association of breast cancer with BMI, therefore, is largely explained by its relationship with endogenous sex hormone levels.

There is no evidence that sugar directly causes any type of cancer, but the excessive consumption of sugar is one of the leading causes of overweight and related diseases (Ludwig *et al.*, 2001; Schulze *et al.*, 2004). For this reason, the WCRF recommended to limit consumption of energy-dense foods and avoid sugary drinks. Energy-dense foods are supplies that are mainly made up of processed foods, which often contain substantial amounts of fat and sugar.

The evidence of physical activity and the risk of cancer, which has continued to accumulate since the early 1990s, shows that regular, sustained physical activity protects against cancers of some sites. These include convincing evidence about the protection of colon cancer (Wolin *et al.*, 2009) and a probable evidence for female hormone-related cancers, independently of other factors such as body fatness. In addition, the evidence that physical activity protects against weight gain, overweight, and obesity is convincing.

Even before the discovery of vitamins as essential nutrients beginning in the early twentieth century, vegetables and fruits have been recommended as *protective foods*. Evidence that vegetables and fruits might be protective against some cancers emerged in the 1990s. At least five portions (or at least 400 g) of vegetables and fruits per day has been the common recommendation. It is difficult to attribute to each plant food a specific protective effect, because the real power is the variety of the diet, but it has been possible to classify fruits and vegetables on the basis of their content of micronutrients and phytochemicals and assign them probable evidence. Thus, foods containing carotenoids probably protect against cancers of the mouth, pharynx, larynx, and lung; foods containing β-carotene and also vitamin C probably protect against esophageal cancer; foods containing selenium and also lycopene probably protect against prostate cancer; and foods containing folate probably protect against pancreatic cancer (Table 8.1). Plant foods also contain dietary fiber, convincing in the risk reduction of colon cancer. Considerable effort has been dedicated to characterize the dietary components of what has come to be called dietary fiber

TABLE 8.1
Food Sources and Cancer Prevention

Components	Foods	Cancer Sites
Carotenoids	Tomatoes, carrots, peppers, pumpkin, apricots, herbs	Mouth, pharynx, larynx, lung
β-Carotene	Carrots, pumpkin, apricots, spinach, peppers	Esophagus
Vitamin C	Citrus peel and juice, kiwi, strawberry, peppers	
Selenium	Fish, shellfish, other seafoods	Prostate
Lycopene	Tomatoes, pink grapefruit, watermelon	
Folate	Wheat germ, green leafy vegetables	Pancreas
Organosulfur compounds	Garlic, onion, shallot, leek	Stomach

Source: World Cancer Research Fund (WCRF) and American Insitute for Cancer Research (AICR), *Food, Nutrition, Physical Activity, and the Prevention of Cancer: A Global Perspective*, WCRF/AICR, Washington, DC, 2007.

that might confer health benefit. Naturally occurring dietary fiber is only derived from plant foods. Pulses (legumes) and minimally processed cereals are particularly concentrated sources, but vegetables and fruits also contain significant amounts.

In case of animal foods, there is conflicting evidence for milk and dairy products, given that their impact on the risk of cancer varies in different tissues. Milk probably protects against colorectal cancer, but at the same time diets high in calcium are a probable cause of prostate cancer. Certainly, some type of dairy products like aged cheese, cream and butter can be high in saturated fats, and such dairy products should be consumed occasionally. Milk is not the only food rich in calcium, but represents the major source of calcium in the population.

For red meat, fish, poultry, and eggs, there is neither convincing nor probable evidence that attributes to them any protective effect against cancer risk, whereas the evidence that red meat and processed meat consumption increases the risk of colorectal cancer is convincing. In particular, the panel declared that consuming up to 500 g (cooked weight) of red meat per week does not significantly raise cancer risk. Eating more than this, however, does increase your risk of bowel cancer.

Therefore, in the field of cancer prevention, international institutions seem to agree about the importance of a plant-based diet mainly composed of fruit, vegetables, whole grains, and pulses, with the right amount of animal products and a small amount of processed foods, the same food pattern that is the base of the Mediterranean diet.

8.2.3 Nutrition and Diabetes

The link between Mediterranean diet and the risk of developing type 2 diabetes has been explored in a few studies. A recent perspective investigation study [European Prospective Investigation into Cancer and Nutrition (EPIC) study] assessed the association between adherence to the MDP, using an a priori defined score, and incidence of chronic diseases among a large European population, including Mediterranean and non-Mediterranean countries, with diversity of dietary patterns (Romaguera *et al.*, 2011). The EPIC project is a multicenter prospective cohort based on healthy, middle-aged subjects who agree, following an active invitation, to participate in the study and to have their health status followed up for the rest of their lives (Riboli and Kaaks, 1997). Europe represents an ideal natural laboratory, thanks to the heterogeneous dietary habits still to be found in different European populations, ranging from the Mediterranean diets of Greece, southern Italy, Spain, and the south of France through the central European food patterns of Germany, the Netherlands, and the north of France to the Nordic diet of Sweden and Denmark. The study is based in 22 collaborating centers in 10 European countries and includes populations characterized by large variations in dietary habits and cancer risk. Data are collected on diet, physical activity, sexual maturation and reproductive history, lifetime consumption of alcohol and tobacco, previous and current illnesses, and current medication. The EPIC has collected a large amount of information about the life of 500,000 middle-aged men and women. Thanks to the collected data, it was possible to detect an association between a Mediterranean diet and the risk of type 2 diabetes establishing a case–cohort study, including 11,994 incident type 2 diabetic case subjects and a stratified subcohort of 15,798 participants (Romaguera *et al.*, 2011). The relative MDS (rMED) (score range 0–18) was used to assess the adherence to the MDP on the basis of reported consumption of nine dietary components characteristic of the Mediterranean diet, some potentially beneficial foods (vegetables, legumes, fruits and nuts, cereals, fish and seafood, olive oil, and moderate alcohol consumption) and other potentially detrimental foods (meat and meat products and dairy products). Individuals with a high rMED score range (11–18 points) were 12% less likely to develop diabetes than those with low rMED scores (0–6 points). Overall, in the multiple adjusted model, a two-point increment in rMED was associated with a 4% reduction in the risk of type 2 diabetes. There was evidence of effect modification by age group and no association between rMED and diabetes was observed among the youngest participants (<50 years of age). Another observational study that analyzed data from the Greek cohort of the EPIC population has found an inverse association between MDS and diabetes risk (Rossi *et al.*, 2013). From the EPIC food-frequency questionnaire, the adherence to traditional Mediterranean diet was calculated, as well as the dietary glycemic load (GL) in relation to occurrence of diabetes. As expected, a higher MDS was inversely associated with the risk of diabetes and GL was positively associated with diabetes. A significant protection of about 20% was found for a diet with a high MDS and a low GL. Only few nutritional intervention trials were performed to investigate the role of a Mediterranean diet in diabetes prevention. More relevant is the Prevención con Dieta Mediterránea (PREDIMED) study, a multicenter, randomized, parallel group primary prevention trial conducted in Spain to assess the effects of two Mediterranean diets, supplemented with either extra virgin olive oil (EVOO) or mixed nuts versus a low-fat control diet on cardiovascular and other chronic disease outcomes in individuals at high cardiovascular risk (Garcia-Lopez *et al.*, 2014). Participants are men aged 55–80 years and women aged 60–80 years without prior CVD but having at least three cardiovascular risk factors, namely, smoking, hypertension, dyslipidemia, overweight (BMI > 25 kg/m^2), and family history of premature CVD. On the basis of the initial assessment of individual scores of adherence using a 14-item questionnaire, dieticians gave personalized dietary advice to participants randomly assigned to one of three diet groups. After a median follow-up of 4 years, the incidence of diabetes was 10.1%, 11%, and 17.9% in the Mediterranean diet with olive oil group, the Mediterranean diet with nuts group, and the control group, respectively. When the two Mediterranean diet groups were pooled and compared with the control group, the incidence of diabetes was reduced by 52% (Salas-Salvado *et al.*, 2011). In sum, the results show that a non-energy-restricted traditional Mediterranean diet high in unsaturated fat can be a useful tool for preventing diabetes.

8.2.4 Nutrition and Neurodegenerative Diseases

Alzheimer's disease (AD) is the most common form of dementia in the elderly and one of the biggest fears in later life. It is well established now that AD starts with a prolonged preclinical stage of pathological changes without symptoms and progresses through a prodromal stage with subjective memory complaints and mild cognitive impairment (MCI) before the development of clinical dementia. Over the last few years, only few studies attempted to investigate the association between the adherence to Mediterranean diet and either AD or cognitive decline. AD and MCI have been reported to be more prevalent among subjects with low adherence to a Mediterranean diet. The possible relation between Mediterranean diet and AD has been interestingly showed for the first time by Scarmeas and colleagues in 2006, through a study that comprised a total of 2258 nondemented individuals who were prospectively evaluated for a median time of 4 years of follow-up. In this study, the authors investigated the link between a greater adherence to the traditional Mediterranean diet and the risk of incident cases of AD (Scarmeas *et al.*, 2006). After 4 years of follow-up, there were 262 incident cases of AD and it has been reported that a higher adherence, that is, being in the third tertile of the distribution of the adherence score, was associated with a lower risk of AD, which was found to be reduced by 40%, after adjustment for potential confounders. This result was confirmed by some other studies reported in a recent review that shows an association between higher adherence to a Mediterranean diet and decreased AD risk (Sofi *et al.*, 2010). Interestingly, in two of the studies reported in the same review, the risk of MCI and MCI conversion to AD was found to be lower in subjects with better adherence to this dietary pattern, suggesting that the MDP could be used as a preventive strategy in early prevention programs. In addition, a research group in the United Kingdom has recently published a study with the objective to systematically review the literature to synthesize and evaluate available evidence on the association between the adherence to a Mediterranean diet and the cognitive function or dementia (Lourida *et al.*, 2013). This review identifies 12 papers reporting an association between the adherence to a Mediterranean diet and the cognitive function or dementia. Adherence to a Mediterranean diet may be associated with lower rates of cognitive decline. The benefits of adherence to a Mediterranean diet were particularly evident in studies with mean baseline ages above 75 years, possibly because the higher risk of incident cognitive problems in this age group made it easier to detect differences in outcomes between those who consumed a Mediterranean diet and those who did not. The first data regarding experimental trials have been gathered from the PREDIMED trial, a multicenter program conducted in Spain from May 2005 to December 2010, involving 7500 persons randomized in three groups: a control group advised to follow a low-fat diet and two experimental groups advised to follow a Mediterranean diet supplemented with either EVOO or nuts. In particular, an article with a subsample from 1 of the 11 recruitment centers (PREDIMED-NAVARRA), which assessed the effect on cognition of a nutritional intervention using a Mediterranean diet in comparison with a low-fat control diet, was published (Martinez-Lapiscina *et al.*, 2013). This trial suggests that nutritional intervention with a Mediterranean diet supplemented with either EVOO or nuts is associated with improved global cognition. The benefit of a Mediterranean diet was independent of potential confounders such as age, family history of cognitive impairment or dementia, *apolipoprotein E* genotype, education, physical activity, vascular risk factors, and energy intake. These longitudinal results concur with the recent findings from a cross-sectional analysis at baseline in another PREDIMED subgroup (Barcelona, Spain), whereby increased consumption of EVOO and walnuts was independently related to better cognition (Valls-Pedret *et al.*, 2012). They are also consistent with a positive but not significant association observed for olive oil, monounsaturated fatty acids, and the Mediterranean diet in the EPIC-Greek study (Psaltopoulou *et al.*, 2008), with a similar but stronger association observed in a large Italian cohort (Solfrizzi *et al.*, 2006) and with a slightly improved cognitive performance in the Three-City French cohort (Feart *et al.*, 2009).

8.2.5 Mediterranean Diet and Longevity: The Blue Zones

The number of older people is growing rapidly worldwide. With the increase in life expectancy, the leading causes of death have shifted dramatically from infectious diseases to noncommunicable diseases and from younger to older individuals (WHO–FAO, 2003). In industrialized countries, about 75% of deaths in persons older than the age of 65 years are now from CVDs and cancer. Regardless of predisposing factors, diet and lifestyle influence morbidity and mortality during the course of life. Clearly, longevity is a complex attribute determined by factors such as exposure to disease, variability in sleeping patterns, smoking habits, physical activity, and diet, which have a direct effect, in addition to their indirect emotional and cognitive influence on physiological pathways. Longevity gain marks a significant achievement in human history, which yet also comes with challenges. The Mediterranean diet has been widely reported to be a model of healthy eating for its contribution to a favorable health status and a better quality of life (Sofi *et al.*, 2008). Few epidemiological studies have addressed the question of which factors have the greatest effect on longevity. Trichopoulou *et al.*, (2003) examined, in a cohort study, the effects of the traditional Mediterranean diet (low in saturated fat, high in monounsaturated fat, high in complex carbohydrates, and high in fiber) on health and survival of 182 elderly residents from three Greek villages. They found that a 1 unit increase in diet score (an a priori–defined nutritional pattern reflecting the traditional Greek Mediterranean diet) was significantly associated with a 17% reduction in overall mortality. A recently published longitudinal study has investigated the single and combined effect of a Mediterranean diet and other lifestyle factors on all-cause and cause-specific mortality in European elderly individuals (Knoops *et al.*, 2004). During the 10-year of follow-up, men and women between the ages of 70 and 90 years who had adhered to a Mediterranean diet, were nonsmokers or had stopped smoking more than 15 years ago, were physically active, used alcohol moderately, and had less than half the mortality rate from all causes, CHD, CVD, and cancer, and mortality from other causes than those who did not.

Although it is widely recognized that life expectancy is influenced by a combination of genetic and environmental factors, it is still unknown whether specific characteristics of the natural or cultural environment may facilitate the emergence of extreme longevity. There are places around the world where people live longer and, most importantly, are physically active even after the age of 100 years. Specifically, anthropologists observed that people living in Sardinia (Italy), Okinawa (Japan), Loma Linda (California), Nicoya Peninsula (Costa Rica), and Ikaria (Greece) have very high life expectancy, with the percentage of people over the age of 90 being amazingly high compared with the average age in the developed world. These places have been defined as the *blue zones* and are a part of a large anthropologic and demographic project. Blue zones are places that have the greatest life expectancy and where more people reach age 100 than anywhere else. It has been observed that people living in these areas share common behavioral and lifestyle characteristics, despite the different race, nationality, and regional characteristics they have. Particularly, the investigators of the blue zones reported that "some lifestyle characteristics, like family coherence, smoking avoidance, plant-based diet, moderate and daily physical activity, social engagement, where people of all ages are socially active and integrated into the community, are common in all people enrolled in the surveys" (Panagiotakos *et al.*, 2011). Longevity also seems to be a product of the interaction between physical, social, behavioral, and demographic factors. Tom Perls, founding director of the New England Centenarian Study, and his colleagues published the first health survey on centenarians, reporting on 32 people between the ages of 110 and 119 years (Schoenhofen *et al.*, 2006). CVDs, the leading killer in developed countries, were rare among super-centenarians; only 6% had suffered heart attacks and 13% reported strokes. Diabetes and Parkinson's diseases were also uncommon in the group, striking only 3% of the subjects. In this view, it is reasonable to think that the Mediterranean lifestyle plays a key role in the promotion of healthy aging. In support of that statement, a recent meta-analysis showed that adherence to a Mediterranean diet can significantly decrease the risk of overall mortality, mortality from CVDs, incidence of or mortality from cancer, and incidence of Parkinson's disease and AD (Sofi *et al.*, 2008). These results seem to be clinically relevant in terms of public health, particularly for reducing the risk of premature death in the general population.

8.3 PRACTICAL APPLICATIONS OF A MEDITERRANEAN DIET IN THE MODERN CULTURAL CONTEXT

8.3.1 The Mediterranean Lifestyle: Is It Possible These Days?

The traditional Mediterranean diet is the heritage of millennia of exchanges of people, cultures, and foods of all countries around the Mediterranean basin. It has been the basis of food habits during the twentieth century in all countries of the region, originally based on Mediterranean agricultural and rural models. However, the traditional Mediterranean diet is now progressively eroding due to the widespread dissemination of the Western-type economy, urban and technology-driven culture, as well as the globalization of food production and consumption, related to the homogenization of food behaviors in the modern era (Bach-Faig *et al.*, 2011). A healthy Mediterranean diet pattern has been popularized since 1995 (Willett *et al*, 1995) using the world famous pyramid representation that graphically highlights the food groups to be consumed daily, weekly, or less frequently. The pyramid reflects the process of change that the Mediterranean diet is undergoing within Mediterranean societies. After the recognition of the Mediterranean diet as an intangible cultural heritage by the UNESCO in 2010, scientists have presented consensual renewed communication tools for the general public, health professionals, and stakeholders. This is a new pyramid (Figure 8.1) with which scientists hope to contribute to much better adherence to this healthy dietary pattern and its way of life in the Mediterranean area and other countries in the near future.

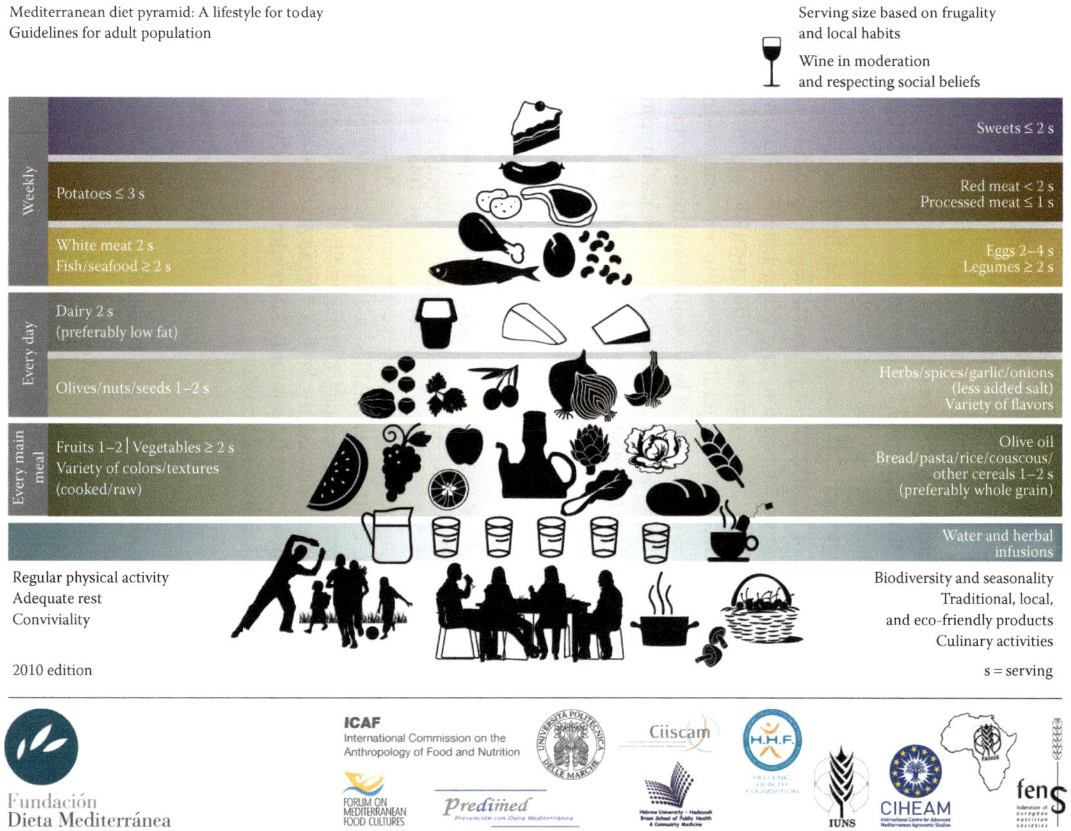

FIGURE 8.1 Mediterranean diet pyramid: A lifestyle for today. Courtesy of Fundación Dieta Mediterránea, © 2010. Use and promotion of this pyramid is recommended without any restriction.

The traditional Mediterranean diet pyramid has evolved to adopt the new way of life. As an initiative of the Mediterranean Diet Foundation and with the collaboration of numerous international entities, a wide range of experts in nutrition, anthropology, sociology, and agriculture have reached a consensus in a new richer design with the incorporation of qualitative elements. The new pyramid follows the previous pattern: at the base level, foods that should sustain the diet, and at the upper levels, foods to be eaten in moderate amounts. Moreover, social and cultural elements characteristic of the Mediterranean way of life are incorporated in the graphic design. Therefore, it is not just about prioritizing some food groups from others but also about paying attention to the way of selecting, cooking, and eating. It also reflects the composition and number of servings per meal. The health benefits of the Mediterranean diet and protective effect against chronic diseases have been well established by the scientific community. This new pyramid includes all the food groups; it is the proportions and the frequencies that determine whether a diet is healthy or unhealthy. This food consumption pattern is applicable to a healthy adult population and should be adapted to the specific needs of children, pregnant women, and other health conditions. Plant-based foods are situated at the base of the pyramid. They provide key nutrients and protective substances that contribute to the general well-being and contribute to maintaining a balanced diet, which, therefore, should be consumed in high proportions and frequency. Foods situated in the upper levels, such as those of animal origin and rich in sugars and in fats should be eaten in moderation and reserved for special occasions.

More than just food, the Mediterranean diet is also a potential model of sustainable development (Burlingame and Dernini, 2011). According to some researchers from the Mediterranean Agronomy Institute of Montpellier and Bari:

> The traditional Mediterranean Diet can be considered sustainable under different aspects. First, for the great variety of foods it includes, which guarantees its nutritional quality and biodiversity. Second, for the variety of practices and techniques used to prepare and treat food and the numerous foods whose health benefits have been demonstrated, such as olive oil, fish, fruits and vegetables, legumes, fermented milk, and spices. And finally, due to its strong cultural legacy and the traditions which are part of it; for its respect of human nature and seasonality; because of the diversity of the landscapes which contribute to wellbeing; and because it is a diet with low environmental impact, thanks to low consumption of animal products.
>
> **Burlingame *et al.*, 2012**

8.3.2 Feasible Recommendations for the Everyday Routine

The pyramid (Figure 8.1) establishes dietary daily, weekly, and occasional guidelines to be followed for a healthy and balanced diet.

- The three main meals should contain three basic elements, which can also be found throughout the day:
  - *Cereals*: One or two servings per meal in the form of bread, pasta, rice, couscous, and others; preferably whole grain, because some valuable nutrients (magnesium, phosphorus, etc.) and fiber can be lost during processing.

- *Vegetables*: One or more servings per meal—at least one of the servings should be raw; present at lunch and dinner.
- *Fruit*: One or two servings per meal; should be chosen as the most frequent dessert.

- A daily intake of 1.5–2.0 l of water should be guaranteed. Good hydration is essential to maintain the corporal water equilibrium, although needs may vary among people because of age, physical activity, personal circumstances, and weather conditions. In addition to water, nonsugar-rich herbal infusions and broths (with low fat and salt content) may complete the requirements.
- Moderate physical activity can readily be built into everyday life. It is not necessary to devote a continuous half-hour every day to moderate activity. The best choice of vigorous physical activity is that which is most enjoyable for the family or the individual—be it swimming, running, dancing, rowing, cycling, hill walking, aerobic workouts, or team games such as football and badminton. People whose work is sedentary should take special care to build moderate and vigorous physical activity into their everyday lives. It is also important to avoid long periods of sedentary behavior, such as watching television. This behavior is also often associated with consumption of energy-dense food and sugary drinks.
- Dairy products—Prefer it in the form of low-fat yogurt, cheese, and other fermented dairy products. They contribute to bone health, but can also be an important source of saturated fat.
- Olive oil is located at the center of the pyramid and should be the principal source of dietary lipids because of its high nutritional quality (especially extra virgin). Its unique composition gives it a high resistance to cooking temperatures and should be used for cooking as well as dressings (one tablespoon per person).
- Spices, herbs, garlic, and onions are a good way to introduce a variety of flavors and palatability to dishes and contribute to the reduction of salt addition. Olives, nuts, and seeds are good sources of healthy lipids, proteins, vitamins, minerals, and fiber. A reasonable consumption of olives, nuts, and seeds (such as a handful) make for a healthy snack choice.
- Respecting religious and social beliefs, a moderate consumption of wine and other fermented beverages (one glass per day for women and two glasses per day for men, as a generic reference) during meals is recommended.

A variety of plant and animal origin proteins should be consumed. Mediterranean traditional dishes do not usually have animal origin protein foods as the main ingredient but as a tasty source. This does not mean that it is necessary to give up on animal foods. There are many ways to enjoy meat and other animal foods as part of plant-based diets. Fish (two or more servings per week), white meat (two servings per week), and eggs (two to four servings per week) are good sources of animal protein. Fish and shellfish are also a good source of healthy proteins and lipids. Eating oily fish helps to provide ω-3 fatty acids, which can boost immunity and help protect against heart problems.

Consumption of red meat (less than two servings per week, preferably lean cuts) should be in smaller quantity and frequency. Substantial amounts of meat are not needed to sustain adequate consumption of protein and iron. All flesh foods are high in protein, and for people who consume varied diets without any flesh foods, more than adequate protein can be derived from a mixture of pulses (legumes) and cereals (grains). Iron is present in many plant foods, as well as in meat. It is best that processed meats are avoided. They are generally energy dense and can also contain high levels of salt. They also tend to be preserved by smoking, curing, or salting, or by addition of chemical preservatives.

In the top vertex of the pyramid are the foods rich in sugars and unhealthy fats (sweets). Sugar, candies, pastries, and beverages such as sweetened fruit juices and soft drinks should be consumed in small amounts and set aside for special occasions. These foods are energy dense and are likely to contribute to weight gain.

The following *smart tips* are adapted from WCRF and AICR report, *Food, Nutrition, Physical Activity, and the Prevention of Cancer: A Global Perspective*:

- Maintain a healthy weight.
  - Eat healthily by lowering the energy density of your diet—the following list shows how.
  - Move more—increase your physical activity levels. Aim for at least 30 minutes of moderate activity every day.
  - Keep an eye on your portion sizes—try not to finish up leftovers when you are clearing away.
  - When out in a restaurant, order a smaller portion or share with a friend and do not feel pressurized to finish everything on your plate.
  - Avoid sugary drinks—they have little nutritional value but contain a lot of calories.
- Increase your physical activity throughout the day.
  - Walk more—try getting off the bus one stop early and briskly walking to your destination. Aim to take 10,000 steps a day. Use a pedometer to keep count of your steps.
  - Take the stairs instead of the lift.
  - If you are seeing a friend for coffee, suggest a walk first.
  - Take your children or grandchildren swimming or to the park—by doing this not only are you increasing your physical activity but you are helping children to develop good habits for the future.
- Reduce the energy density of your diet.
  - Enjoy a salad or soup at the beginning of a meal. These are high in water and fiber, which help to fill you up, leaving less room for energy-dense foods. Go for noncreamy soups and lightly dressed salads.
  - Add plenty of pulses, whole grains, vegetables, and fruits to your meals. Try to add a variety of these water- and fiber-rich foods into your breakfasts, pasta sauces, omelettes, sandwiches, or side dishes.
  - Lower the fat content. If you use less oil, butter, and mayonnaise in your dishes, you will immediately reduce the energy density of your

meal without changing your usual portion size. Choose lower fat dairy products, trim visible fat from meat, and steam, grill, or bake foods instead of frying them.
    - Keep portion sizes of energy-dense foods small. Research shows that we tend to eat more when served more without even realizing it. Use smaller plates at home and skip value meals and supersized portions, especially those containing processed foods, when eating out.
- Base your meals around plant foods.
    - Remember your "*5 a day*." Opt for a colorful variety of vegetables and fruits every day.
    - Snack on carrots and other *crudités* before dinner to help curb your appetite. Make coleslaw with cabbage and grated carrot and a little low-fat dressing rather than mayonnaise. Enjoy fresh salad in a sandwich or as a side dish.
    - Store your vegetables and fruits correctly—salad vegetables and some fruits are best stored in the fridge and others in a cool place—and try to use them up before they go off. You could add a few extras when cooking a hearty soup or stew, or use overripe fruit in your blender to make a smoothie.
    - Cut your vegetables into larger chunks and avoid any unnecessary peeling or slicing.
    - Minimize the amount of water you add to your vegetables when cooking them. Try cooking them in a microwave or steaming them, rather than presoaking or boiling them in large volumes of water.
    - Try not to overcook your vegetables. Crunchy vegetables retain more nutrients than those that are overcooked.
- Think of red meat as one of the many protein sources you could choose.
- Try homemade bean burgers, falafels, or fishcakes.
- Try eating more fish—it is quick and easy to prepare.
- Swap red meat for lean poultry such as skinless chicken or turkey.
- Eggs are a good source of protein and can play an important part in a healthy diet when eaten in moderation, twice a week.
- Dairy foods can replace a portion of meat and are a good source of protein, vitamin D, and calcium for bone health.
- Tips for sensible drinking are as follows:
    - When ordering drinks, opt for the smallest serving size. Avoid double measures of spirits, which are often encouraged as "better value."
    - Do not drink alcohol when you are thirsty because you are likely to drink more. Have a glass of water or a nonalcoholic soft drink to quench your thirst before having an alcoholic drink.
    - Sip your drink slowly to pace yourself and make it last longer.
    - Avoid drinking on an empty stomach. The alcohol will be absorbed into the bloodstream more quickly.
    - Make sure you drink water before and after you drink alcohol to rehydrate yourself.

- Tips for reducing salt in your diet are as follows:
  - Take the salt off the table. If you gradually reduce the amount of salt you add to food during cooking and at the table, your taste buds should adjust within a few weeks.
  - Replace salt with spices, herbs, garlic, and lemon. Herbs and spices such as chilli powder, ginger and basil, and other flavors such as garlic, lemon, and black pepper all add flavor to food.
  - Make your own meals from scratch. Cooking your own meals with fresh ingredients gives you more opportunity to control the amount of salt in your food.
  - Look for lower-salt foods when shopping. Check food labels and select products with less salt or sodium. Choose tinned or packaged foods with no added salt (or sugar).
  - Eat fresh meat rather than processed meat. Avoid eating processed meats such as bacon, cured meats, and some sausages.
  - Limit the amount of salty snacks you eat. Replace salty snacks such as crisps and salted nuts with small portions of dried fruit or unsalted nuts.

REFERENCES

Alexandratos N. The Mediterranean diet in a world context. *Public Health Nutr.* **2006**, *9*(1A), 111–117.

Anitschkow N. Experimental atherosclerosis in animals. In Cowdry E.V. (ed.), *Arteriosclerosis: A Survey of the Problem.* Macmillan, New York, **1933**, 271–322.

Bach A., Serra-Majem L., Carrasco J.L., Roman B., Ngo J., Bertomeu I., Obrador B. The use of indexes evaluating the adherence to the Mediterranean diet in epidemiological studies: A review. *Public Health Nutr.* **2006**, *9*(1A), 132–146.

Bach-Faig A., Berry E.M., Lairon D., Reguant J., Trichopoulou A., Dernini S., Medina F.X. *et al.* Mediterranean diet pyramid today. Science and cultural updates. *Public Health Nutr.* **2011**, *14*(12A), 2274–2284.

Balanza R., Garcia-Lorda P., Perez-Rodrigo C., Aranceta J., Bonet M.B., Salas-Salvado J. Trends in food availability determined by the Food and Agriculture Organization's food balance sheets in Mediterranean Europe in comparison with other European areas. *Public Health Nutr.* **2007**, *10*(2), 168–176.

Bazzano L.A., He J., Ogden L.G., Loria C., Vupputuri S., Myers L., Whelton P.K. Legume consumption and risk of coronary heart disease in US men and women: NHANES I Epidemiologic Follow-up Study. *Arch. Intern. Med.* **2001**, *161*(21), 2573–2578.

Bazzano L.A., Thompson A.M., Tees M.T., Nguyen C.H., Winham D.M. Non-soy legume consumption lowers cholesterol levels: A meta-analysis of randomized controlled trials. *Nutr. Metab. Cardiovasc. Dis.* **2011**, *21*(2), 94–103.

Blackburn H. On the trail of heart attacks in seven countries. The Country Press, Middleborough, KY, **1995**.

Burlingame B., Dernini S. Sustainable diets: The Mediterranean diet as an example. *Public Health Nutr.* **2011**, *14*(12A), 2285–2287.

Burlingame B.A., Dernini S. Food and Agriculture Organization of the United Nations; Biodiversity International (Organization). *Sustainable Diets and Biodiversity: Directions and Solutions for Policy, Research and Action.* FAO, Rome, Italy **2012**, 307.

Church T.S., Thomas D.M., Tudor-Locke C., Katzmarzyk P.T., Earnest C.P., Rodarte R.Q., Martin C.K., Blair S.N., Bouchard C. Trends over 5 decades in U.S. occupation-related physical activity and their associations with obesity. *PLoS One* **2011**, *6*(5), e19657.

Dawber T.R., Meadors G.F., Moore F.E., Jr. Epidemiological approaches to heart disease: The Framingham Study. *Am. J. Public Health Nations Health* **1951**, *41*(3), 279–281.

Epstein F.H. Cardiovascular disease epidemiology—A journey from the past into the future. *Circulation* **1996**, *93*(9), 1755–1764.

Feart C., Samieri C., Rondeau V., Amieva H., Portet F., Dartigues J.F., Scarmeas N., Barberger-Gateau P. Adherence to a Mediterranean diet, cognitive decline, and risk of dementia. *JAMA* **2009**, *302*(6), 638–648.

Garcia-Lopez M., Toledo E., Beunza J.J., Aros F., Estruch R., Salas-Salvado J., Corella D. *et al.* Mediterranean diet and heart rate: The PREDIMED randomised trial. *Int. J. Cardiol.* **2014**, *171*(2), 299–301.

Hilakivi-Clarke L., Forsen T., Eriksson J.G., Luoto R., Tuomilehto J., Osmond C., Barker D.J. Tallness and overweight during childhood have opposing effects on breast cancer risk. *Br. J. Cancer* **2001**, *85*(11), 1680–1684.

International Obesity Task Force. *EU Platform on Diet, Physical Activity and Health-Briefing Paper*. Prepared in collaboration with the European Association for the Study of Obesity, Brussels, Germany, **2005**.

Istituto Nazionale di Ricerca per gli Alimenti e la Nutrizione (INRAN). 150 Anni di Unità d'Italia, 75 anni di ricerca su alimenti e nutrizione, Rome, Italy, **2012**.

Keys A. Mediterranean diet and public health: Personal reflections. *Am. J. Clin. Nutr.* **1995**, *61*(Suppl 6), 1321S–1323S.

Keys A., Grande F. Role of dietary fat in human nutrition. III. Diet and the epidemiology of coronary heart disease. *Am. J. Public Health Nations Health* **1957**, *47*(12), 1520–1530.

Keys A., Keys M. *Eat Well and Stay Well*. Doubleday & Co., Garden City, NY, **1959**.

Keys A., Menotti A., Karvonen M.J., Aravanis C., Blackburn H., Buzina R., Djordjevic B.S., Dontas A.S., Fidanza F., Keys M.H. The diet and 15-year death rate in the seven countries study. *Am. J. Epidemiol.* **1986**, *124*(6), 903–915.

Knoops K.T., de Groot L.C., Kromhout D., Perrin A.E., Moreiras-Varela O., Menotti A., van Staveren W.A. Mediterranean diet, lifestyle factors, and 10-year mortality in elderly European men and women: The HALE project. *JAMA* **2004**, *292*(12), 1433–1439.

Kromhout D., Keys A., Aravanis C., Buzina R., Fidanza F., Giampaoli S., Jansen A., Menotti A., Nedeljkovic S., Pekkarinen M. Food consumption patterns in the 1960s in seven countries. *Am. J. Clin. Nutr.* **1989**, *49*(5), 889–894.

Lewington S., Whitlock G., Clarke R., Sherliker P., Emberson J., Halsey J., Qizilbash N., Peto R., Collins R. Blood cholesterol and vascular mortality by age, sex, and blood pressure: A meta-analysis of individual data from 61 prospective studies with 55,000 vascular deaths. *Lancet* **2007**, *370*(9602), 1829–1839.

Lourida I., Soni M., Thompson-Coon J., Purandare N., Lang I.A., Ukoumunne O.C., Llewellyn D.J. Mediterranean diet, cognitive function, and dementia: A systematic review. *Epidemiology* **2013**, *24*(4), 479–489.

Ludwig D.S., Peterson K.E., Gortmaker S.L. Relation between consumption of sugar-sweetened drinks and childhood obesity: A prospective, observational analysis. *Lancet* **2001**, *357*(9255), 505–508.

Martinez-Lapiscina E.H., Clavero P., Toledo E., Estruch R., Salas-Salvado J., San Julian B., Sanchez-Tainta A., Ros E., Valls-Pedret C., Martinez-Gonzalez M.A. Mediterranean diet improves cognition: The PREDIMED-NAVARRA randomised trial. *J. Neurol. Neurosurg. Psychiatry* **2013**, *84*(12), 1318–1325.

Mozaffarian D., Micha R., Wallace S. Effects on coronary heart disease of increasing polyunsaturated fat in place of saturated fat: A systematic review and meta-analysis of randomized controlled trials. *PLoS Med.* **2010**, *7*(3), e1000252.

Panagiotakos D.B., Chrysohoou C., Siasos G., Zisimos K., Skoumas J., Pitsavos C., Stefanadis C. Sociodemographic and lifestyle statistics of oldest old people (>80 years) living in Ikaria island: The Ikaria study. *Cardiol. Res. Pract.* **2011**, 679187.

Psaltopoulou T., Kyrozis A., Stathopoulos P., Trichopoulos D., Vassilopoulos D., Trichopoulou A. Diet, physical activity and cognitive impairment among elders: The EPIC-Greece cohort (European Prospective Investigation into Cancer and Nutrition). *Public Health Nutr.* **2008**, *11*(10), 1054–1062.

Riboli E., Kaaks R. The EPIC Project: Rationale and study design. European Prospective Investigation into Cancer and Nutrition. *Int. J. Epidemiol.* **1997**, *26*(Suppl 1), S6–S14.

Romaguera D., Guevara M., Norat T., Langenberg C., Forouhi N.G., Sharp S., Slimani N. *et al.* Mediterranean diet and type 2 diabetes risk in the European Prospective Investigation into Cancer and Nutrition (EPIC) study: The InterAct project. *Diabetes Care* **2011**, *34*(9), 1913–1918.

Rossi M., Turati F., Lagiou P., Trichopoulos D., Augustin L.S., La Vecchia C., Trichopoulou A. Mediterranean diet and glycaemic load in relation to incidence of type 2 diabetes: Results from the Greek cohort of the population-based European Prospective Investigation into Cancer and Nutrition (EPIC). *Diabetologia* **2013**, *56*(11), 2405–2413.

Salas-Salvado J., Bullo M., Babio N., Martinez-Gonzalez M.A., Ibarrola-Jurado N., Basora J., Estruch R. *et al.* Reduction in the incidence of type 2 diabetes with the Mediterranean diet: Results of the PREDIMED-Reus nutrition intervention randomized trial. *Diabetes Care* **2011**, *34*(1), 14–19.

Scarmeas N., Stern Y., Tang M.X., Mayeux R., Luchsinger J.A. Mediterranean diet and risk for Alzheimer's disease. *Ann. Neurol.* **2006**, *59*(6), 912–921.

Schoenhofen E.A., Wyszynski D.F., Andersen S., Pennington J., Young R., Terry D.F., Perls T.T. Characteristics of 32 supercentenarians. *J. Am. Geriatr. Soc.* **2006**, *54*(8), 1237–1240.

Schulze M.B., Manson J.E., Ludwig D.S., Colditz G.A., Stampfer M.J., Willett W.C., Hu F.B. Sugar-sweetened beverages, weight gain, and incidence of type 2 diabetes in young and middle-aged women. *JAMA* **2004**, *292*(8), 927–934.

Serra-Majem L., Roman B., Estruch R. Scientific evidence of interventions using the Mediterranean diet: A systematic review. *Nutr. Rev.* **2006**, *64*(2 Pt 2), S27–S47.

Slattery M.L., Randall D.E. Trends in coronary heart disease mortality and food consumption in the United States between 1909 and 1980. *Am. J. Clin. Nutr.* **1988**, *47*(6), 1060–1067.

Sofi F., Cesari F., Abbate R., Gensini G.F., Casini A. Adherence to Mediterranean diet and health status: Meta-analysis. *BMJ* **2008**, *337*, a1344.

Sofi F., Macchi C., Abbate R., Gensini G.F., Casini A. Effectiveness of the Mediterranean diet: Can it help delay or prevent Alzheimer's disease? *J. Alzheimers Dis.* **2010**, *20*(3), 795–801.

Solfrizzi V., Colacicco A.M., D'Introno A., Capurso C., Torres F., Rizzo C., Capurso A., Panza F. Dietary intake of unsaturated fatty acids and age-related cognitive decline: A 8.5-year follow-up of the Italian Longitudinal Study on Aging. *Neurobiol. Aging* **2006**, *27*(11), 1694–1704.

Steinberg D. The pathogenesis of atherosclerosis. An interpretive history of the cholesterol controversy, part IV: The 1984 coronary primary prevention trial ends it—almost. *J. Lipid Res.* **2006**, *47*(1), 1–14.

Threapleton D.E., Greenwood D.C., Evans C.E., Cleghorn C.L., Nykjaer C., Woodhead C., Cade J.E., Gale C.P., Burley V.J. Dietary fiber intake and risk of cardiovascular disease: Systematic review and meta-analysis. *BMJ* **2013**, *347*, f6879.

Trichopoulou A., Costacou T., Bamia C., Trichopoulos D. Adherence to a Mediterranean diet and survival in a Greek population. *N. Engl. J. Med.* **2003**, *348*(26), 2599–2608.

Trichopoulou A., Lagiou P., Kuper H., Trichopoulos D. Cancer and Mediterranean dietary traditions. *Cancer Epidemiol. Biomarkers Prev.* **2000**, *9*(9), 869–873.

Trowell H. Ischemic heart disease and dietary fiber. *Am. J. Clin. Nutr.* **1972**, *25*(9), 926–932.

Valls-Pedret C., Lamuela-Raventos R.M., Medina-Remon A., Quintana M., Corella D., Pinto X., Martinez-Gonzalez M.A., Estruch R., Ros E. Polyphenol-rich foods in the Mediterranean diet are associated with better cognitive function in elderly subjects at high cardiovascular risk. *J. Alzheimers Dis.* **2012**, *29*(4), 773–782.

Willett W.C., Sacks F., Trichopoulou A., Drescher G., Ferro-Luzzi A., Helsing E., Trichopoulos D. Mediterranean diet pyramid: A cultural model for healthy eating. *Am. J. Clin. Nutr.* **1995**, *61*(6 Suppl), 1402S–1406S.

Wolin K.Y., Yan Y., Colditz G.A., Lee I.M. Physical activity and colon cancer prevention: A meta-analysis. *Br. J. Cancer* **2009**, *100*(4), 611–616.

World Cancer Research Fund (WCRF); American Insitute for Cancer Research (AICR). *Food, Nutrition, Physical Activity, and the Prevention of Cancer: A Global Perspective*, WCRF/AICR, Washington, DC, **2007**.

World Health Organization (WHO). *World Health Statistics 2013*, WHO, Geneva, Switzerland, **2013**.

World Health Organization (WHO); Food and Agriculture Organization of the United Nations (FAO). *Diet, Nutrition and the Prevention of Chronic Diseases*, WHO, Geneva, Switzerland, **2003**.

Section III

Balancing the Food and the Balanced Food

9 Healing Influences of Spices

Luisella Verotta and Maria Pia Macchi

CONTENTS

9.1 BASIL

Scientific name: *Ocimum basilicum* L.
Family: Lamiaceae
Hindi: Babuyitulsi
Sanskrit: Arjakah
Tamil: Karpuratulasi, Tirunutpatchi

9.1.1 Origin and Distribution

Basil is originally native to India and other tropical regions of Asia having been cultivated there for more than 5000 years. It arrived first in Italy from the Middle East and then in France (in the fifteenth century) and in England (in the seventeenth century). It traveled to America with the first immigrants. Today, basil plants are particularly cultivated in Italy, France, Morocco, Egypt, Asia, and Central and South America. In the United States, the largest quantities are cultivated in California.

Several other *Ocimum* species are grown in many regions of Asia. Most of the Asian species have a clove-like flavor, which is, in general, stronger than the Mediterranean species. The most notable is *Ocimum sanctum*, holy basil or *tulsi*, a revered home-grown plant in India and Nepal. Holy basil is found throughout the lowlands of India as well as in Sri Lanka, Pakistan, Bangladesh, Myanmar, southern China, Thailand, and Malaysia. In India, *tulsi* is traditionally cultivated in the kitchen herbal garden for daily use and worship.

9.1.2 Features

Ocimum basilicum possesses thin branching roots that grow out into bushy stems with many extremely aromatic leaves. The leaves, stalked and peculiarly smooth, are soft and cool to the touch. The tiny, white flowers should be nipped off to prevent the plant from going to seed and finishing its life cycle. This will also encourage thicker foliage and hence a more abundant harvest. The seeds are black and rectangular in shape.

Ocimum sanctum is slightly woody with ovate, elliptic, and oblong hairy leaves. There are at least three types of holy basil: *Rama tulsi*, *Krishna tulsi*, and *Vana tulsi*. *Rama tulsi* has green leaves and it is the most common variety. *Krishna tulsi* bears dark green to purple leaves and has a stronger taste and smell. *Vana tulsi* is a green-leaved forest variety that often grows wild.

We can find a lot of interesting connections between the ritual use of basil and holy basil in India and Europe. If basil has a role of king among spices, holy basil is the queen of medicinal plants. The ancient Egyptians and Greeks believed that basil would open the gates of heaven for a person passing on. In Europe, *O. basilicum* is placed in the hands of the deceased to ensure a safe journey, whereas in India *O. sanctum* is placed in their mouth to ensure that they will reach God. *Ocimum sanctum* has both a medicinal and a spiritual significance in Ayurveda. In traditional Indian culture, holy basil is sacred to Lord Vishnu, the Hindu god of preservation, and it is considered a *sattvic* plant, having spiritual uplifting qualities (Puri, 2003).

One of the Sanskrit names of *O. basilicum* is *Haripriya*, dear to Vishnu; another name is *Bhutagni*, killer of demons (Majupuria and Joshi, 1997). It is also commonly called *tulsi matha*, the mother able to grant wealth, prosperity, and happiness to her children. A living *tulsi* plant is kept in the south side of many Indian homes to provide divine protection for the household. Holy basil is part of the daily Hindu rituals to ensure personal health, spiritual purity, and family well-being: rosaries made from its cut stems are commonly used as meditation beads by Lord Vishnu devotees. Leaves and flowers from the holy basil plant are added to bath water and to bowls of water kept near the home entrance. The hands of guests are sprinkled with this water as they enter a house or they are invited to clean their hands in the water. Holy basil contains compounds with antiseptic activity that would help kill germs. The scent from the plant also deters insects. Thus, flies, including mosquitoes, which can spread diseases such as malaria, would be deterred from landing on people who had bathed in holy basil water.

9.1.3 Properties and Uses in Ayurvedic System of Medicine

> Oh Narada, wherever Tulsi grows there is no misery. She is the holiest of the holy. Wherever the breeze blows her fragrance there is purity. Vishnu showers blessing on those who worship and grow Tulsi. Tulsi is sacred because Brahma resides in the roots, Vishnu resides in the stems and leaves and Rudra resides in the flowering tops.
>
> ***Padma Purana*, 24.2**

There are a lot of references to *tulsi* in many Hindu texts (*Bhagavad Gita*, *Padma Purana*, *Ramayana*, *Brahma-vaivarta Purana*, and *Vishnoyamal Tantra*). Everything associated with this plant is holy, including water given to it and the soil in which it grows as well as all its parts.

Closely related to the sweet basil plant widely available in the West, holy basil has been used for perhaps 3000 years as one of India's most powerful herbs. According to Ayurveda, *tulsi* creates purity and lightness in the body: it is classified as a *rasayana*, a herb that nourishes a person's growth to perfect health and promotes long life. The daily use of this herb is believed to help maintain the balance of the *chakras* (energy centers) of the body (Tierra, 1999).

Practitioners of traditional Indian medicine commonly regard holy basil as an adaptogen and recommend it as a daily supplement for people entering periods of extreme emotional or physical stress (Aggarwal and Yost, 2011). It has a predominantly *kapha*-reducing effect on the *dosha*s but can be used to pacify *vata* and *pitta* as well. In severely overheated individuals, basil can have a mildly *pitta*-aggravating effect. Its traditional use has been in the treatment of colds and flues as its purifying actions are believed to cleanse the respiratory tract of toxins. It is also helpful in the relief of digestive gas and bloating.

Parts used: Leaves
Taste: Pungent
Quality: Light and rough
Potency: Hot
***Dosha* effect:** VK–, P+ (in excess)
Actions: Diaphoretic, antipyretic, antidepressant, antioxidant, carminative, diuretic, expectorant, galactagogue, general tonic

9.1.4 Traditional Uses and Folklore

- In Indian folk medicine, the leaves of the holy basil plant are brewed into a tea that is used as an expectorant to treat people with excessive bronchial mucus and bronchitis. The tea is also used for people with upset stomach, biliousness, and vomiting. The leaves can be mixed with ginger and given to young children with stomach aches, or mixed with pepper to treat fevers. When used to treat chest infections, it is often mixed with honey, ginger, and onion juice.
- During the rainy season, when malaria and dengue fever are widely prevalent in India, tender leaves of holy basil boiled with tea act as preventive against these diseases. In case of acute fevers, a decoction of the leaves boiled with powdered cardamom in half a liter of water and mixed with sugar and milk brings down the temperature. *Tulsi* is an important constituent of many Ayurvedic cough syrups and expectorants. It helps to mobilize mucus in bronchitis and asthma. Chewing *tulsi* leaves relieves cold and flu. When taken as a drink, it is often mixed with lemon juice or cardamom.
- Holy basil is widely used in Siddha and Unani systems of medicine as a tonic and for the treatment of fever, cough, and skin diseases.
- To reduce tooth ache, a couple of *tulsi* leaves can be crushed well and mixed with a tiny pinch of salt and a tiny pinch of black pepper powder. A local application of this paste on the aching teeth is known to reduce pain.

Precautions: Basil is generally a very safe healing herb. Studies from the 1970s suggested that holy basil might have a mild antifertility effect in animals (Aggarwal and Yost, 2011).

It is considered hot for women, so it is better to use it with other herbs that have a cooling effect (Puri, 2003).

9.1.5 In the Kitchen

Very familiar already to Theophrastus and Dioscorides, basil is a culinary herb prominently featured in Italian cuisine and also plays a major role in Southeast Asian cuisines of Indonesia, Thailand, Vietnam, Cambodia, and Laos. Whenever possible, it is better to choose fresh basil over the dried form of the herb because it is superior in flavor. The leaves of fresh basil should look vibrant and be deep green in color. They should be free from dark spots or yellowing. Because the oils in basil are highly volatile, it is best to add the herb near the end of the cooking process, so it will retain its maximum essence and flavor (Aggarwal and Yost, 2011).

Basil can be chopped with garlic and olive oil to make a dairy-free variety of pesto that can top a variety of dishes including pasta, salmon, and whole wheat *bruschetta*.

9.1.6 Chemical Constituents and Uses as Functional Food

Basil, or sweet basil, is a common name for the culinary herb *O. basilicum*. There are many varieties of *O. basilicum*, as well as several related species or species hybrids also called basil. The type used in Italian food is typically called sweet basil, as opposed to Thai basil (*O. basilicum* var. *thyrsiflora*), lemon basil (*O.* × *citriodorum*), and holy basil (*O. sanctum*), which are used in Asia. The word *basil* comes from the Greek βασιλεύς (*basileus*), meaning "king." Nicholas Culpeper notes of basil that it is "an herb of Mars and under the Scorpion, and therefore called Basilicon," relating it to *basilisk*. Basil is commonly used fresh in cooked recipes. In general, it is added at the last moment, as cooking quickly destroys the flavor. The fresh herb can be kept for a short time in plastic bags in the refrigerator, or for a longer period in the freezer, after being blanched quickly in boiling water. Basil is one of the main ingredients in pesto—a green Italian oil-and-herb sauce. Its other main ingredients are olive oil, garlic, and pine nuts.

Various basils have such different scents because the herb has a number of different essential oils that come together in different proportions for various breeds. The strong clove scent of sweet basil is derived from eugenol, the same chemical as actual cloves. The citrus scent of lemon basil and lime basil reflects their higher portion of citral and of limonene, which gives lemon peel its scent. Based on the chemical content, basils can be divided into four groups:

1. French; *O. basilicum*, contains lower amounts of phenols
2. Exotic; contains methyl chavicol (estragole) (40%–80%)
3. Methyl cinnamate—ether 90%
4. Eugenol

Ocimum basilicum L. (sweet basil) is known to occur as several chemotypes or cultivars, which differ in their essential oil composition. The major chemical components identified were monoterpenes: linalool; methyl eugenol; estragole; methyl cinnamate; and 1,8-cineole (Govindarajan *et al.*, 2013; Klimánková *et al.*, 2008; Padalia and Verma, 2011; Verma *et al.*, 2011).

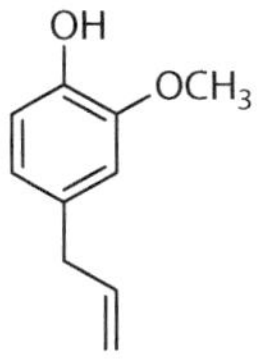

Eugenol

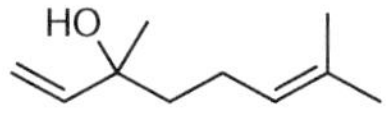

Linalool

1,8-Cineole

The essential oils exhibited broad-spectrum antibacterial (against 5 Gram-positive and 7 Gram-negative bacteria) and antifungal (against 10 fungi) activities. The bacterial species such as *Bacillus subtilis*, *Staphylococcus aureus*, *Streptococcus mutans*, and *Enterococcus faecalis*, and the fungal species such as *Epidermophyton floccosum*, *Microsporum gypseum*, and *Sporothrix schenckii* were more sensitive to the essential oils (Rao *et al.*, 2011; Soković *et al.*, 2010). The basil oil was tested against pathogenic fungi, namely, *Aspergillus niger*, *A. fumigatus*, *Penicillium italicum*, and *Rhizopus stolonifer*. Surprisingly high antifungal values were found highlighting the potential of Omani basil as a preservative in the food and medical industries (Al-Maskri *et al.*, 2011). Nine essential oils were examined for antimicrobial activity against reference and clinical strains of *Salmonella enteritidis*. Based on the size of the inhibition zone and the minimal inhibitory concentration, basil oil had the strongest antimicrobial activity against all the tested bacteria, and *S. enteritidis* SE3 was the most sensitive strain to all the tested oils. Gas chromatography/mass spectrometry analysis revealed that the major constituents of the oil were linalool (64.35%); 1,8-cineole (12.28%); eugenol (3.21%); germacrene D (2.07%); alpha-terpineol (1.64%); and rho-cymene (1.03%) (Rattanachaikunsopon and Phumkhachorn, 2010).

In vitro cytotoxicity screening against the human cervical cancer cell line (HeLa), human laryngeal epithelial carcinoma cell line (HEp-2), and NIH 3T3 mouse embryonic fibroblasts showed that basil oil has potent cytotoxicity (Kathirvel and Ravi, 2012). A study of the essential oil showed antifungal and insect-repelling properties. A similar study has confirmed that extracts from the plant are very toxic to mosquitos (Kalaivani *et al.*, 2012; Warikoo *et al.*, 2011).

Pesto, an Italian recipe (from Genoa), traditionally served with pasta

Ingredients

2 cups (500 ml) of minced fresh basil
1/4 cup (30 gm) of pine nuts
150 ml virgin olive oil
50 gm grated cheese (pecorino or Parmigiano-Reggiano)

Preparation

Mix the basil with pine nuts in a mortar and grate to make a paste, slowly adding the oil. Add the cheese. Use the sauce to make a condiment for pasta (linguine), finishing with topped grated cheese.

In a modified recipe, pasta (trofie) is added to minced boiled potatoes and green beans, and then dressed with the sauce.

REFERENCES

Aggarwal B.B., Yost D. *Healing Spices*. Sterling Publications, New York, **2011** (Italian version, Armenia, 73–81).

Al-Maskri A.Y., Hanif M.A., Al-Maskari M.Y., Abraham A.S., Al-sabahi J.N., Al-Mantheri O. Essential oil from *Ocimum basilicum* (Omani basil): A desert crop. *Nat. Prod. Commun.* **2011**, *6*(10), 1487–1490.

Govindarajan M., Sivakumar R., Rajeswary M., Yogalakshmi K. Chemical composition and larvicidal activity of essential oil from *Ocimum basilicum* (L.) against *Culex tritaeniorhynchus*, *Aedes albopictus* and *Anopheles subpictus* (Diptera: Culicidae). *Exp. Parasitol.* **2013**, *134*(1), 7–11.

Kalaivani K., Senthil-Nathan S., Murugesan A.G. Biological activity of selected Lamiaceae and Zingiberaceae plant essential oils against the dengue vector *Aedes aegypti* L. (Diptera: Culicidae). *Parasitol. Res.* **2012**, *110*(3), 1261–1268.

Kathirvel P., Ravi S. Chemical composition of the essential oil from basil (*Ocimum basilicum* Linn.) and its *in vitro* cytotoxicity against HeLa and HEp-2 human cancer cell lines and NIH 3T3 mouse embryonic fibroblasts. *Nat. Prod. Res.* **2012**, *26*(12), 1112–1118.

Klimánková E., Holadová K., Hajšlová J., Čajka T., Poustka J., Koudela M. Aroma profiles of five basil (*Ocimum basilicum* L.) cultivars grown under conventional and organic conditions. *Food Chem.* **2008**, *107*(1), 464–472.

Majupuria T.C., Joshi D.P. *Religious and Useful Plants of Nepal and India*. Gupta M. (ed.), Publishers, Lashkar, India, **1997**, 243.

Padalia R.C., Verma R.S. Comparative volatile oil composition of four *Ocimum species* from northern India. *Nat. Prod. Res.* **2011**, *25*(6), 569–575.

Puri H.S. *Rasayana. Ayurvedic Herbs for Longevity and Rejuvenation*. Taylor & Francis, New York, **2003**, 272–280.

Rao B.R., Kotharia S.K., Rajput D.K., Patel R.P., Darokar M.P. Chemical and biological diversity in fourteen selections of four *Ocimum* species. *Nat. Prod. Commun.* **2011**, *6*(11), 1705–1710.

Rattanachaikunsopon P., Phumkhachorn P. Antimicrobial activity of basil (*Ocimum basilicum*) oil against *Salmonella enteritidis in vitro* and in food. *Biosci. Biotechnol. Biochem.* **2010**, *74*(6), 1200–1204.

Soković M., Glamočlija J., Marin P.D., Brkić D., van Griensven L.J. Antibacterial effects of the essential oils of commonly consumed medicinal herbs using an *in vitro* model. *Molecules* **2010**, *15*(11), 7532–7546.

Tierra M. *Planetary Herbology*. Motilal Banarsidass Publishers, Delhi, India, **1999**, 152.

Verma R.S., Padalia R.C., Chauhan A. Variation in the volatile terpenoids of two industrially important basil (*Ocimum basilicum* L.) cultivars during plant ontogeny in two different cropping seasons from India. *J. Sci. Food Agric.* **2011**, *92*(3), 626–631.

Warikoo R., Wahab N., Kumar S. Oviposition-altering and ovicidal potentials of five essential oils against female adults of the dengue vector, *Aedes aegypti* L. *Parasitol. Res.* **2011**, *109*(4), 1125–1131.

9.2 BLACK PEPPER

Scientific name: *Piper nigrum* L.
Family: Piperaceae
Hindi: Kalimirch
Sanskrit: Maricam
Tamil: Milaku

9.2.1 Origin and Distribution

Black pepper is one of the most popular spices in the world, rightly considered the "King of Spices": the volume of international trade is the highest among all the spices known (Pruthi, 1998). The word *pepper* has its roots in the Dravidian name for long pepper: *pippali*. Ancient Greek and Latin turned *pippali* into the Latin *piper*, which was used by the Romans to refer to both black pepper and long pepper, as the Romans erroneously believed that both of these spices were derived from the same plant. In ancient Greece, pepper has held such high prestige that it was used not only as a seasoning but as a currency and a sacred offering. During the fall of ancient Rome, the invading barbarians were even honored by being given black pepper. Additionally, in the Middle Ages, the wealth of a man was often measured by his stockpile of pepper (Aggarwal and Yost, 2011).

The spice comes from the berries of a climber native to the Western Ghats of Kerala, India, where it is cultivated mostly as a mixed crop in homestead garden and it also still occurs wild in the mountains. This area is thought to be the center of origin for the crop because the diversity of cultivated forms is the largest here.

India has been a chief source of black pepper in the world markets since time immemorial. Black pepper was one of the most valuable trade goods to Europe, the Middle East, and North American countries. However, India's share in export has been declining in the recent decades. The plant spread from India to Southeast Asia as cuttings brought by Hindu colonists migrating from India to Indonesia and other countries. Today it is extensively cultivated all over the tropical regions as a commercial crop. Vietnam is actually the world's largest producer and exporter of pepper. Other major producers are India, Brazil, Indonesia, Malaysia, Sri Lanka, China, and Thailand.

9.2.2 Features

The pepper plant is a perennial climber rooting at the nodes, which grows on supporting trees. When the main stem is established, it grows lots of side shoots to create a bushy column. Flowers grow in clusters along flowering stalks known as spikes. Each spike develops around 50–60 small, round, berry-like fruits (Warrier *et al.*, 1997). Fruits are green at first but they turn red as they ripen. These fruits are picked when either green or red to produce black and white pepper.

Black pepper, green pepper, and white peppercorns are actually the same fruit (*Piper nigrum*); the difference in their color is a reflection of varying stages of development and processing methods.

Black peppercorns are made by picking the pepper berries when they are half ripe and just about to turn red. They are then left to dry, which causes them to shrivel and become dark in color. Alternatively, green peppercorns are picked when they are still unripe and green in color, whereas white peppercorns are picked when they are very ripe and subsequently soaked in brine to remove their dark outer shell leaving just the white pepper seed. Pink peppercorns are actually from a completely different plant species (*Schinus molle*) that is related to ragweed.

Black pepper is the most pungent and flavorful of all types of peppers and it is available as whole or cracked peppercorns or ground into powder.

9.2.3 Properties and Uses in Ayurvedic System of Medicine

We can find a lot of references to the medicinal use of pepper in Ayurveda. The Sanskrit name *Maricam* reminds the sun, as this spice is able to impart solar vibrations through the organism, allowing the body to resonate with light and heat (Tiwari, 1999).

As a warming spice, pepper is excellent for pacifying *kapha*, helps pacify *vata*, and increases *pitta* (Paranjipe, 2005). It is often recommended for people with *kapha dosha*, an Ayurvedic term used to describe a body type characterized by cold, wet, heavy, and slow functioning. As diabetes is believed to be a disorder of *kapha* imbalance, due to low digestive fire, black pepper is recommended for patients with diabetes.

Black pepper burns up *ama* and releases sinus congestion (Khalsa and Tierra, 2010). It is an excellent spice for problems associated with the digestive system, and frequently used to eradicate parasitic worms and as an appetizer. It is also useful to treat colds, coughs, and sore throats.

Parts used: Fruit
Taste: Pungent
Quality: Rough, light, sharp
Potency: Hot
***Dosha* effect:** KV–, P+
Actions: Appetizer, anthelmintic, carminative, digestive, aphrodisiac, diuretic

Preparations: “Trikatu” is an Ayurvedic preparation containing black pepper, long pepper, and ginger, which is prescribed routinely for a variety of diseases as part of a multidrug prescription. These herbs along with piperine (the main alkaloid of peppers) have been shown to possess diverse biological activities in mammalian systems. A review of these studies is presented, and it has been suggested that their use in the Indian system of medicine could be due to their bioavailability enhancing action on other medicaments (Johri and Zutshi, 1992).

Trikatu is used by *vata* and *pitta* types to boost digestion and to stimulate the system (Tiwari, 1999).

9.2.4 Traditional Uses and Folklore

- Stomach ailments such as dyspepsia, flatulence, constipation, and diarrhea are traditionally treated with black pepper mixed with other substances such as castor oil, cow's urine, or ghee.
- A quarter teaspoon of pepper powder and a quarter teaspoon of cumin mixed in thin buttermilk should be taken during indigestion.
- An infusion of black pepper is a common home remedy for sore throat and hoarseness. It can also be chewed to reduce throat inflammation.
- An herbal tea made of peppercorns and ginger is one of the best home remedies for cold and flu.
- Eating six black peppers with four almonds once a day with milk is a nervine tonic.
- In the morning, black pepper with honey clears *ama* from the system.
- In Unani medicine, black pepper has been described as an aphrodisiac and as a remedy to alleviate colics. A preparation called *jawa rishai thurush*, composed of pepper, ginger, salt, lemon juice, *Embelia ribes*, and mint is prescribed by Unani physicians to alleviate indigestion and stomach acidity.

Precautions: Inflammatory conditions of the digestive organs, urticaria, skin rash, and high *pitta* (Khalsa and Tierra, 2010).

9.2.5 In the Kitchen

Black pepper is extensively used in different types of cuisine, all over the world, from very ancient times. It was already a well-known and expensive spice in the Roman Empire, and it is widely used in Western recipes as it combines well with almost every other spice or herb.

In India, there are numerous references in Tamil literature and in the epic: the *Mahabharata* written in the fourth century BC describes feasts with meat sprinkled with pepper.

Both black and white pepper are widely used in South Asian cuisine, in all stages of cooking and as a table condiment to help the digestion with a warm and pungent taste. Black peppercorns and other whole herbs such as cinnamon chips, bay leaves, cloves, and cracked cardamom pods are sautéed in ghee and used to flavor basmati rice. Ground black pepper is combined with coconut milk and other spices to make sauces for vegetables. It is an integral part of the

Anglo-Indian curry powder and of *garam masala*, and can be used for almost every kind of dish, even sweets.

Black pepper is a common ingredient in *chai*, a spiced black tea beverage used throughout India that includes other warming spices such as cardamom, cinnamon, cloves, and ginger.

Pepper loses flavor and aroma through evaporation, so airtight storage is recommended. It is also better to grind whole peppercorns immediately before use as pepper's aromatics can evaporate quickly. In addition to superior flavor, buying whole peppercorns will help to ensure that you are purchasing unadulterated pepper because ground pepper is often mixed with other spices. Just like with other dried spices, it is also suggested to select only organically grown black pepper.

This exotic spice loses its sharp, woody pungent aroma and flavor when cooked for too long, so adding it toward the end of the cooking will help to preserve its flavor (Aggarwal and Yost, 2011).

9.2.6 Chemical Constituents and Uses as Functional Food

The active ingredient in black pepper is piperine (1-piperoylpiperidine), the amide of the alkaloid piperidine, and the concentration of this compound varies substantially among *Piper* species (Nakatani *et al.*, 1986). Piperine accounts for about one-half of the nonvolatile ether extract of black pepper. Safrole is a minor constituent of *P. nigrum* at an approximate concentration of 0.1%.

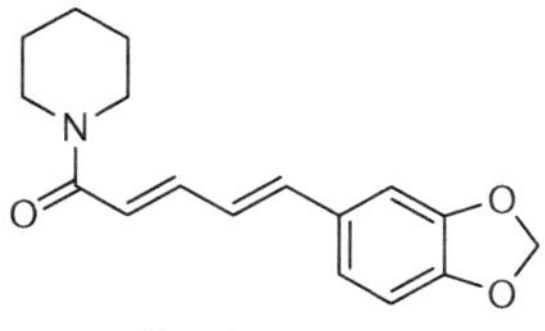

Piperine

Black pepper, either powdered or its decoction, is widely used in traditional Indian medicine and as a home remedy for relief from sore throat, throat congestion, cough, and so on. Pepper is known to cause sneezing. Piperine present in black pepper acts as a thermogenic compound. It enhances the thermogenesis of lipid, accelerates the energy metabolism in the body, and also increases the serotonin and β-endorphin production in the brain.

Black pepper (*P. nigrum*) is one of the most widely used among spices. It is valued for its distinct biting quality attributed to the alkaloid, piperine. Black pepper is used not only in human dietaries but also for a variety of other purposes such as medicine, preservatives, and perfumery. Many physiological effects of black pepper, its extracts, or its major active principle, piperine, have been reported in recent decades. Dietary piperine, by favorably stimulating the digestive enzymes of pancreas, enhances the digestive capacity and significantly reduces the gastrointestinal food transit time. Piperine has been demonstrated in *in vitro* studies to protect against oxidative damage by inhibiting or quenching free radicals and reactive oxygen species (ROS). Black pepper or piperine treatment has also been evidenced to lower lipid peroxidation *in vivo* and beneficially influence cellular thiol status, antioxidant molecules, and antioxidant enzymes in a number of experimental situations of oxidative stress. The most far-reaching attribute of piperine has been its inhibitory influence on enzymatic drug biotransforming reactions in the liver. It strongly inhibits hepatic and intestinal aryl hydrocarbon hydroxylase, and uridine 5′-diphospho-glucuronyl transferase. Piperine has been documented to enhance the bioavailability of a number of therapeutic drugs as well as phytochemicals because of this property. Piperine's bioavailability enhancing property is also partly attributed to increased absorption as a result of its effect on the ultrastructure of intestinal brush border. Although initially there were a few controversial reports regarding its safety as a food additive, such evidence has been questionable, and later studies have established the safety of black pepper or its active principle, piperine, in several animal studies. Piperine, when it is nongenotoxic, has in fact been found to possess antimutagenic and antitumor influences (reviewed in Srinivasan, 2007).

The effects of black pepper and piperine on drug-metabolizing enzymes as well as on intestinal drug absorption have been reviewed. The information provided is used to assess the clinical significance of black pepper and piperine, and optimize their effectiveness as a bioavailability enhancer (Han, 2011).

Consumption of spiced foods or herbal drinks leads to greater thermogenesis and in some cases to greater satiety. In this regard, capsaicin, black pepper, ginger, mixed spices, green tea, black tea, and caffeine are relevant examples. These functional ingredients have the potential to produce significant effects on metabolic targets such as satiety, thermogenesis, and fat oxidation. A significant clinical outcome sometimes may appear straightforwardly but also depends too strongly on full compliance of subjects. Nevertheless, thermogenic ingredients may be considered as functional agents that could help in preventing a negative energy balance and obesity (Westerterp-Plantenga *et al.*, 2006).

Pepper flavored rice, a recipe from Vaidya Kalyan Chakravarty

This recipe brings out the sharp flavor of black pepper, its main ingredient. Curry leaves and sesame seeds were used to offset the heat of black pepper.

Ingredients

3 cups of cooked white rice (each grain should be separate)
1 and 1/2 tbsp of ghee (clarified butter)
1 tsp of cumin seeds
3/4 tsp of mustard seeds
1 and 1/2 tbsp of black peppercorns
1 and 1/2 tbsp of sesame seeds
10–12 curry leaves (fresh leaves only)
Salt to taste

Preparation

Dry roast the black peppercorns, sesame seeds, and curry leaves in a pan for a few minutes on medium heat tossing them around till the flavor comes out. Once the sesame seeds change color, you can turn off the heat and make a coarse powder.

Heat ghee in a pan, and add the mustard seeds and let them pop. Add the cumin seeds and let them become brown. Add the cooked rice and mix well. Now add the ground spice powder and salt, and combine the rice with it such that the spices coat the rice well.

Note: Use only fresh peppercorns and not pre-ground black pepper powder.

REFERENCES

Aggarwal B.B., Yost D. *Healing Spices*. Sterling Publishers, New York, **2011** (Italian version, Armenia, 261–269).

Han H.K. The effects of black pepper on the intestinal absorption and hepatic metabolism of drugs. *Expert Opin. Drug Metab. Toxicol.* **2011**, *7*(6), 721–729.

Johri R.K., Zutshi U. An Ayurvedic formulation "Trikatu" and its constituents. *J. Ethnopharmacol.* **1992**, *37*(2), 85–91.

Khalsa K.P.S., Tierra M. *The Way of Ayurvedic Herbs*. Motilal Banarsidass Publishers, Delhi, India, **2010**, 105–106.

Nakatani N., Inatani R., Ohta H. *et al.* Chemical constituents of peppers (*Piper* spp.) and application to food preservation: Naturally occurring antioxidative compounds. *Environ. Health Perspect.* **1986**, 67, 135–142.

Paranjipe P. *Indian Medicinal Plants. Forgotten Healers*. Chaukhamba Sanskrit Pratishthan, Delhi, India, **2005**, 169–170.

Pruthi J.S. *Spices and Condiments*. National Book Trust, Delhi, India, **1998**, 198–209.

Srinivasan K. Black pepper and its pungent principle—Piperine: A review of diverse physiological effects. *Crit. Rev. Food. Sci. Nutr.* **2007**, *47*(8), 735–748.

Tiwari M. *Ayurveda Secrets of Healing*. Motilal Banarsidass Publishers, Delhi, India, **1999**, 434, 460.

Warrier P.K., Nambiar V.P.K., Ramankutty C. (eds.) *Indian Medicinal Plants*, vol. 4. Orient Longman, Chennai, India, **1997**, 297.

Westerterp-Plantenga M., Diepvens K., Joosen A.M., Bérubé-Parent S., Tremblay A. Metabolic effects of spices, teas, and caffeine. *Physiol. Behav.* **2006**, *89*(1), 85–91.

9.3 CARDAMOM

Scientific name: *Elettaria cardamomum* L. (Maton)
Family: Zingiberaceae
Hindi: Choti elachi
Sanskrit: Ela
Tamil: Elam

9.3.1 Origin and Distribution

The word *cardamom* is derived from the Latin *cardamomum*, a Latinization of the Greek *kardamomon*, a compound of *kardamon* "cress" and *amomon*, which was the name for an Indian spice plant. It is a tropical perennial herb growing wild in the monsoon forests of the Western Ghats in South India as well in Sikkim, Nepal, and Bhutan. It is cultivated in Sri Lanka, Laos, Vietnam, Thailand, Guatemala, Costa Rica, San Salvador, and Tanzania. The ancient Egyptians chewed cardamom seeds as a tooth cleaner; the Greeks and Romans used it as a perfume. Vikings came upon cardamom about 1000 years ago, in Constantinople, and introduced it into Scandinavia, where it remains popular to this day (Aggarwal and Yost, 2011). It was introduced in Guatemala only in 1914, but today this country is the biggest producer and exporter of cardamom in the world, followed by India. Previously, the largest world producer was Nepal.

9.3.2 Features

The plant is a perennial bushy herb with sheathed stems reaching 2–5 m in height. It has a large tuberous rhizome and long, dark green leaves growing from the plant base at the ground level. The flowers are green with a white purple-veined tip. The small, brown-black sticky seeds are contained in a pod in three double rows with about six seeds in each row (Warrier *et al.*, 1995). There are two main genera: *Elettaria*, with small and green pods, and *Amomum*, with larger and dark brown pods.

9.3.3 Properties and Uses in Ayurvedic System of Medicine

Sattvic in nature, this aromatic herb can balance the three *dosha*s, especially *vata* and *kapha*; its warm potency enkindles *agni*, the digestive fire, and removes *kapha* from the stomach and lungs. It stimulates the mind and heart, giving clarity and joy. As a *sattvic* spice, it promotes the flow of the *prana* in the body (Frawley and Lad, 1994).

Cardamom has detoxifying properties, and is effective in improving digestion and helpful in case of stomach cramps, flatulence, and gas (Paranjipe, 2005). It improves blood circulation to the lungs and can be helpful in prevention of spasms or convulsions. Hence, cardamom in small quantities is beneficial for those suffering from asthma or bronchitis. Taken after a meal, it improves digestion and refreshes the breath. It is a good tonic for general weakness.

Parts used: Seeds
Tastes: Pungent, sweet
Quality: Dry, light
Potency: Hot
***Dosha* effect:** VK–, P+ (in excess)
Actions: Warming, aromatic, digestive, diuretic, stomachic, diaphoretic, carminative, expectorant

9.3.4 Traditional Uses and Folklore

- To minimize the mucus-forming properties of milk, add a half teaspoon of cardamom powder and a half teaspoon of turmeric powder to one cup of milk.
- Cardamom herbal tea (a quarter teaspoon of powdered cardamom seeds boiled for about 3 minutes) with honey is a traditional remedy for depression (Bakhru, 2004). The herbal tea can also give relief from headache caused by indigestion.
- Infusion of cardamom, ginger, and fennel seeds in equal quantities is a home remedy for digestive disorders.

Precautions: If used in excess, cardamom can aggravate *pitta* and tends to increase the secretion of gastric acid; it is not suitable for individuals suffering from gallbladder disease and ulcer (Khalsa and Tierra, 2010).

9.3.5 In the Kitchen

Cardamom has an excellent food value; it is intensely aromatic with a strong unique taste. It is a very expensive spice, the so-called "Queen of Spices." It occupies the second or third place in the world trade, being a most-valued spice. The green seed pods of the plant are dried, and the seeds inside the pod are used in Asian cuisines, either whole or in a ground form. It is better to buy the whole pods because once the seeds are exposed or ground, they quickly lose their flavor.

Cardamom belongs to the group of sweet spices and is used predominantly to flavor sweets, baked goods, and coffee, particularly in the Arab countries (Govindarajan *et al.*, 1982). When added in coffee, it helps detoxify the caffeine in it. In the Middle East, cardamom pods are ground together with coffee beans to produce a powdered mixture of the two, which is boiled with water to make coffee. Ground cardamom is an ingredient in many Indian curries, pulses, and rice preparations, and is a primary contributor to the unique flavor of *masala chai*. Due to its high price, it is considered to be a spice for special dinners, used mostly during the festivals. It is often used in the preparations of bread and sweets in the North Europe. In Germany, it is an ingredient of the famous biscuits called *lebkuchen* (Aggarwal and Yost, 2011). The seeds can be bruised and fried before adding the main ingredients to the pan, or pounded with other spices as required. The pod itself is neutral in flavor and not generally used, imparting an unpleasant bitter flavor when left in dishes. In India, cardamom seeds are chewed as a mouth freshener after meals. Cardamom combines well with banana, milk, and nutmeg. It is excellent to give a special aroma to *banana flambé* and other desserts, with yogurt and fruits.

9.3.6 Chemical Constituents and Uses as Functional Food

Ancient Indians, more than 1000 years BC, used cardamom medicinally to treat various maladies. Cardamom, like cinnamon and ginger, is a stimulant and is said to relieve indigestion, flatulence, and headaches. The Greeks and Romans used it as a culinary spice.

Cardamom is warming, somewhat pungent, and aromatic. Though considered carminative, stimulant, and aromatic, it has been used primarily to relieve flatulence and like coriander, as an anticramping carminative herb and as a flavoring. Cardamom seeds were found to have specific antibacterial activity, especially for urinary tract infections (Ballabh *et al.*, 2008; Gilani *et al.*, 2008).

The content of essential oil in the seeds is strongly dependent on storage conditions, but may be as high as 8%. Through an ultrasound-assisted extraction, the main components of the extracted essential oil of cardamom are α-terpinyl acetate (46.0%); 1,8-cineole (27.7%); linalool (5.3%); α-terpineol (4.0%); and linalyl acetate (3.5%) (Sereshti *et al.*, 2012). Other sources report 1,8-cineol (20% to 50%); α-terpenylacetate (30%); sabinene; limonene (2%–14%); and borneol with minor amounts of geraniol, methyl eugenol, and *trans*-nerolidol.

1,8-Cineole α-Terpinyl acetate

The therapeutic properties of cardamom oil have found applications in many traditional medicines such as antiseptic, antispasmodic, carminative, digestive, diuretic, expectorant, stimulant, stomachic, and tonic. The essential oil of *Elettaria cardamomum* is a good choice for control of stored product pests (Abbasipour *et al.*, 2011).

The inhibitory activity of cardamom extract was studied on human platelets. An increase in the concentration of cardamom has decreased the malondialdehyde formation significantly. Hence, it may be said that the aqueous extract of cardamom may have component(s), which protect platelets from aggregation and lipid peroxidation (Suneetha and Krishnakantha, 2006).

Cardamom has antioxidant properties and can increase levels of glutathione, a natural antioxidant in the body. Reports are available on the antioxidant activity of the extract of cardamom from different countries (Iris *et al.*, 2006). The relative levels of antioxidant activity, total flavonoid content, total phenolic content, and reducing power of different organic and aqueous extracts obtained through sequential extractions with hexane, dichloromethane, ethyl acetate, methanol, and water of four different varieties of cardamom, namely, Mysore, Malabar, Vazhukka, and Guatemala have been studied. Ethyl acetate extract of all varieties showed greater activity. Based on the results, Malabar variety was identified as the best source of antioxidant compounds (Padmakumari Amma *et al.*, 2010).

A freshly prepared aqueous suspension of ground cardamom, which contained all the oil components of cardamom as emulsion, was found to prevent papillomagenesis on the skin and two-stage skin cancer (Das, 2012; Qiblawi *et al.*, 2012).

Cardamom extracts significantly enhance the cytotoxic activity of natural killer cells, indicating their potential anticancer effects. Cardamom exerts immunomodulatory roles and antitumor activities; hence, it manifests itself as a natural agent that can promote the maintenance of a healthy immune system (Majdalawieh and Carr, 2010).

Cardamom is a good source of minerals such as *potassium* (100 g pods contain 1119 mg of this electrolyte), calcium, and magnesium. The ratio of potassium to sodium is 100:1. Additionally, it is also an excellent source of *iron* (100 g pods contain 13.97 mg or 175% of daily-required levels of iron). Further, these aromatic pods are rich in many vital vitamins, including riboflavin, niacin, and vitamin C (http://ndb.nal.usda.gov/).

Cardamoma tea, a recipe from Vaidya Kalyan Chakravarty

Ingredients (for two cups of tea)

1 and 1/2 cups of water
2/3 cup of whole milk
2–3 pods of green cardamom seeds
2 tsp of sugar
1 heaped tsp of loose tea leaves (nonscented)

Preparation

Bring water to boil in a stainless steel. Add cardamom and tea leaves, and simmer for a minute. Add milk and sugar, and boil till desired color is obtained and a lovely aroma is released. Do not boil too much as the tea will impart a bitter flavor. Cover and let it sit for a minute. Strain into cups while piping hot.

Note: The amount of milk and water can vary according to personal preference.

Cardamom milk shake to remove tiredness, a recipe from Vaidya Kalyan Chakravarty

Ingredients

3 cups of milk (whole milk)
Sugar to taste
10–15 almonds
8 cashew nuts
1/2 tsp of cardamom seeds

Preparation

Grind cashew nuts, almond seeds, and cardamom in a blender. When they are powdered, add milk to the blender and mix thoroughly. Add sugar and serve.

Note: If this is served in summer, heat the milk and add the powdered nuts to the milk and serve warm.

REFERENCES

Abbasipour H., Mahmoudvand M., Rastegar F., Hosseinpour M.H. Fumigant toxicity and oviposition deterrency of the essential oil from cardamom, *Elettaria cardamomum*, against three stored–product insects. *J. Insect Sci.* **2011**, *11*, 165.

Aggarwal B.B., Yost D. *Healing Spices*. Sterling Publishers, New York, **2011** (Italian version, Armenia, 103–109).

Bakhru H.K. *Indian Spices & Condiments as Natural Healers*. Jaico Publishing House, Mumbai, India, **2004**, 28–32.

Ballabh B., Chaurasia O.P., Ahmed Z., Singh S.B. Traditional medicinal plants of cold desert Ladakh-used against kidney and urinary disorders. *J. Ethnopharmacol.* **2008**, *118*(2), 331–339.

Das I., Acharya A., Berry D.L., Sen S., Williams E., Permaul E., Sengupta A., Bhattacharya S., Saha T. Antioxidative effects of the spice cardamom against non-melanoma skin cancer by modulating nuclear factor erythroid-2-related factor 2 and NF-κB signaling pathways. *Br. J. Nutr.* **2012**, *108*(6), 984–997.

Frawley D., Lad V. *The Yoga of Herbs*. Motilal Banarsidass Publishers, Delhi, India, **1994**, 109.
Gilani A.H., Jabeen Q., Khan A.U., Shah A.J. Gut modulatory, blood pressure lowering, diuretic and sedative activities of cardamom. *J. Ethnopharmacol.* **2008**, *115*(3), 463–472.
Govindarajan V.S., Narasimhan S., Raghuveer K.G., Lewis Y.S. Cardamom—Production, technology, chemistry, and quality. *Crit. Rev. Food Sci. Nutr.* **1982**, *16*(3), 229–326.
Iris H., Damien Dorman H.J., Raimo H. Antioxidant activities of extracts from selected culinary herbs and spices. *Food Chem.* **2006**, *97*, 122–129.
Khalsa K.P.S., Tierra M. *The Way of Ayurvedic Herbs*. Motilal Banarsidass Publishers, Delhi, India, **2010**, 115–116.
Majdalawieh A.F., Carr R.I. *In vitro* investigation of the potential immunomodulatory and anti-cancer activities of black pepper (*Piper nigrum*) and cardamom (*Elettaria cardamomum*). *J. Med. Food.* **2010**, *13*(2), 371–381.
Padmakumari Amma K.P., Priya Rani M., Sasidharan I., Premakumari Nisha V.N. Chemical composition, flavonoid-phenolic contents and radical scavenging activity of four major varieties of cardamom. *Int. J. Biol. Med. Res.* **2010**, *1*(3), 20–24.
Paranjipe P. *Indian Medicinal Plants. Forgotten Healers*. Chaukhamba Sanskrit Pratishthan, Delhi, India, **2005**, 79–80.
Qiblawi S., Al-Hazimi A., Al-Mogbel M., Hossain A., Bagchi D. Chemopreventive effects of cardamom (*Elettaria cardamomum* L.) on chemically induced skin carcinogenesis in Swiss albino mice. *J. Med Food.* **2012**, *15*(6), 576–580.
Sereshti H., Rohanifar A., Bakhtiari S., Samadi S. Bifunctional ultrasound assisted extraction and determination of *Elettaria cardamomum* Maton essential oil. *J. Chromatogr. A*. **2012**, *1238*, 46–53.
Suneetha W.J., Krishnakantha T.P. Cardamom extract as inhibitor of human platelet aggregation. *Phytother. Res.* **2005**, *19*(5), 437–40.
Warrier P.K., Nambiar V.P.K., Ramankutty C. (eds.) *Indian Medicinal Plants*, vol. 2. Orient Longman, Chennai, India, **1995**, 360.

9.4 CINNAMON

Scientific name: *Cinnamomum zeylanicus* Blume
Family: Lauraceae
Hindi: Dalcini
Sanskrit: Tvak
Tamil: Llavangam

9.4.1 Origin and Distribution

Cinnamon is one of the oldest known spices: it is the inner bark obtained from selected shoots of a small tree *Cinnamomum verum* also called "true cinnamon," which is different from the other related species of inferior quality, such as *Cinnamomum cassia*. The botanical name *zeylanicus* is derived from Sri Lanka, the native place of the tree. The tree grows in India, Sri Lanka, Indonesia, Brazil, Vietnam, and Egypt. The quality is related, among other factors, to the region where it is growing. Cinnamon from Sri Lanka and Seychelles is considered the best (Pruthi, 1998).

Cinnamon is a plant of distinction and royalty: in the ancient world, it was a gift for monarchs and gods, more precious than gold. A number of biblical references attests to its use in ancient times, as incense and as an ingredient of holy anointing oils (Aggarwal and Yost, 2011). It was also used among several other spices in

Egyptian embalming mixtures and perfumes (Manniche, 2006). The Middle Eastern traders created elaborate myths to conceal the true origins of cinnamon and protect their monopoly. One such story included fishing cinnamon with nets at the source of the Nile, at the edge of the world. According to Herodotus, both cinnamon and cassia grew in Arabia and were guarded by winged serpents. The phoenix was reputed to build its nest from cinnamon and cassia (Brosse, 1992). Through the Middle Ages, the source of cinnamon was a mystery to the Western world. Arab traders brought the spice via overland trade routes to Alexandria in Egypt, where it was bought by Venetian traders from Italy who held a monopoly on the spice trade in Europe. The disruption of this trade by the rise of other Mediterranean powers, such as the Mamluk Sultans and the Ottoman Empire, was one of the main factors that led Europeans to search for other routes to Asia.

Portuguese traders finally landed in Ceylon (Sri Lanka) at the beginning of the sixteenth century and got the monopoly of the traditional production and management of cinnamon for over a hundred years. Dutch traders finally dislodged the Portuguese and, in 1638, took control of the Sri Lankan factories. In 1767, the East India Company established the Asia's largest estate in Kerala. The British also took control of Sri Lanka in 1796. However, the importance of the monopoly of Ceylon was already declining, as cultivation of cinnamon tree spread to other areas.

9.4.2 Features

Cinnamon is an evergreen perennial tree, which grows up to a height of 6–8 m. The leaves are large, thick, and egg shaped. The flowers are minute in large hairy clusters. Fruits are oblong or egg-shaped, dark purple, with one seed. The bark is thick, smooth, and brownish dark in color. To obtain the spice, the inner bark is carefully cured and dried. During the process, the bark shrinks and curls into a cylinder called *quill*. The quill has a delicate fragrance and a warm, sweet, and aromatic taste (Bakhru, 2004).

9.4.3 Properties and Uses in Ayurvedic System of Medicine

The Sanskrit name *tvak* literally means the "skin," being this part of the plant used in Ayurveda. It is a *sattvic* plant with a harmonizing effect, well combined with cardamon and bay leaves.

The sweet postdigestive effect alleviates *vata* and *kapha*, which tends to become imbalanced in winter. Ayurvedic physicians recommend the use of cinnamon during this season, especially for *kapha* constitution. The spice reduces excess *kapha* from the head and chest, and promotes healthy circulation in the joints and limbs. It strengthens the heart, warms the kidneys, and promotes *agni* (Frawley and Lad, 1994). It is also very helpful in the elimination of toxins as it prevents the buildup of phlegm and infections. *Pitta* constitutions can also consume cinnamon, but in relatively smaller quantities, only during the cold season.

As a spice, cinnamon is used to balance the digestion and to pacify stomach disorders. Being cardiac stimulant and *oja* increasing, it should be used in cardiac weakness.

In Ayurveda, diabetes is believed to be a disorder of *kapha* imbalance, due to low digestive fire. Cinnamon, as a warming herb, can stimulate digestive processes and regulate metabolism of fats and sugars in *kapha*, helping with type II diabetes and insulin resistance.

Parts used: Bark, leaves, buds, flowers, fruits
Tastes: Pungent, sweet
Quality: Dry, light, sharp
Potency: Hot
***Dosha* effect:** VK–, P+
Actions: *Agni* stimulant, *ama* digestive

9.4.4 Traditional Uses and Folklore

- Cinnamon bark is traditionally used as a mouth refresher and to give strength to gums (Paranjipe, 2005). A piece of bark can be chewed to stop vomiting and nausea. The hot infusion is used in *ama*-induced excessive thirst. A piece of bark can be chewed in *pitta* fever in the presence of dryness of throat.
- Combined with other warming herbs and spices such as ginger and black pepper, it can be boiled into an herbal tea to soothe discomfort associated with colds.
- The decoction is a traditional home remedy to reduce labor pain during childbirth.
- Cinnamon oil is used to pacify headaches and keep joints healthy.
- A pinch of cinnamon powder taken with honey at night helps to stop or stopping excessive urination.

Precautions: Cinnamon is a warming spice, so it is not suitable in case of high *pitta* constitution and bleeding disorders (Frawley and Lad, 1994).

9.4.5 In the Kitchen

Cinnamon is widely used in desserts and sweet dishes in the West. In the Middle East, a tea made of cinnamon with sugar is a common drink during the cold season. Another delicious drink is made with milk, powdered resin, chopped pistachios, and cinnamon (Manniche, 2006). From ancient times, it is an important spice in Persian cuisine, often mixed with rosewater or other spices to make a cinnamon-based curry powder for stews or just sprinkled on sweet treats.

In Indian cooking, it is an important ingredient for a variety of rice and vegetable preparations, especially in the south. The sticks are used to decorate rice dishes. The famous spice mixture *garam masala* and *panch puran* include cinnamon. Cinnamon is a common ingredient of *masala chai*, a spiced milk tea beverage used throughout India, which also includes other warming tonics such as cardamon, star anise, ginger, peppercorn, and cloves.

In the international market, true cinnamon is often substituted with inferior relatives such as *Cinnamomum cassia*. It is sweet and delicate, and can be powdered easily with a mortar and pestle when cassia is rough and hard. It is more suitable for sweet dishes than cassia. Sticks retain their freshness longer than powder, so it is better to grind the bark just before using it (Aggarwal and Yost, 2011).

Cinnamon combines well with many other spices, including ginger, cloves, black pepper, paprika, saffron, and nutmeg. It is a tasty and healthy dessert with peaches or apples. It will give a very special taste when added to squash, to apple cider, to preserved lemons and to pumpkin pie. Sweet potatoes cooked with dates, almonds and cinnamon are a delicious dish.

Cinnamon is also known to help balancing sugar in blood: Ayurvedic physicians advise to add some to the breakfast, with pudding, curd, or tea.

9.4.6 Chemical Constituents and Uses as Functional Food

Cinnamon, the inner bark of a tropical evergreen tree, has two main types: Ceylon cinnamon (*Cinnamomum zeylanicum* Blume) and Chinese cassia (*Cinnamomum aromaticum* Ness). When dried, it rolls into a tubular form known as a quill. It is available either in its whole quill form (cinnamon sticks) or as ground powder. Recent studies emerging from Western countries have shown many potentially beneficial health effects of cinnamon such as antiinflammatory properties, antimicrobial activity, blood glucose control, reducing cardiovascular disease, boosting cognitive function, and reducing risk of colonic cancer. Sri Lanka produces the largest quantity of *C. zeylanicum*, which is the best quality cinnamon. *Cinnamomum zeylanicum*, is also known as Ceylon cinnamon (the source of its Latin name, *zeylanicum*) or “true cinnamon.” One important difference between “true” cinnamon and the “cassia” cinnamon is the coumarin content, which is very low in Ceylon cinnamon. Coumarins are known to cause health risks if consumed regularly in high quantity; thus, their presence in food should be avoided.

The analysis of oils from different parts of *C. zeylanicum* such as leaf, stem bark, and root bark indicated the presence of more than 80 compounds. All three oils had a similar array of compounds but in varying proportions. The essential oil of cinnamon bark obtained from Sri Lanka contained cinnamaldehyde (60%–70%) and cinnamyl acetate (8%–10%) along with other minor monoterpenes. Cinnamaldehyde and camphor have been reported to be the major components of volatile oils from stem bark and root bark, respectively. *trans*-Cinnamyl acetate was found to be the major compound in fruits, flowers, and fruit stalks. The leaf oil has a major component called eugenol (reviewed in Jayaprakasha and Mohan Rao, 2011).

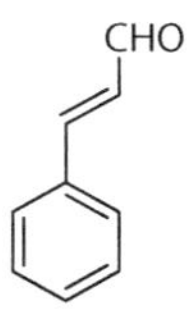

Cinnamaldehyde

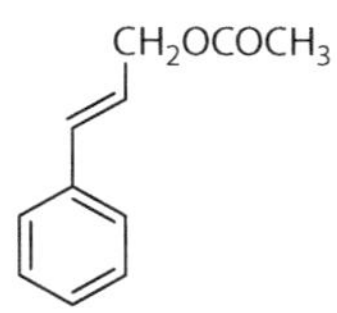

E-Cinnamyl acetate

The bark and fruits of *C. zeylanicum* were found to contain proanthocyanidins with doubly linked bisflavan-3-ol units in the molecule as nonvolatile components.

The volatile oils were found to exhibit antioxidant, antimicrobial (Bin Jantan *et al.*, 2008; Muthuswamy *et al.*, 2008; Shan *et al.*, 2007), and antidiabetic activities (reviewed in Ballali and Lanciai, 2012; Bandara *et al.*, 2012; and Deng, 2012).

Some studies have demonstrated that the herbal products have beneficial effects in patients with diabetes by improving glucose and lipid metabolism, antioxidant status, and capillary function (Bailey and Day, 1989). Cinnamon is one such a dietary component that has shown to have biologically active substances with insulin-mimetic properties. *In vitro* (Jarvill-Taylor *et al.*, 2001) and *in vivo* (Cao *et al.*, 2007; Qin *et al.*, 2003) studies have shown that cinnamon enhances glucose uptake by activating the insulin receptor kinase activity, autophosphorylation of the insulin receptor, and glycogen synthase activity. Very recent studies have demonstrated that there are insufficient data to pool the results for insulin sensitivity. No trials reported health-related quality of life, morbidity, mortality, or costs, although adverse reactions to oral cinnamon were infrequent and generally mild in nature. *Cinnamomum zeylanicum* demonstrates numerous beneficial effects both *in vitro* and *in vivo* as a potential therapeutic agent for diabetes, but there is insufficient evidence to support the use of cinnamon for type 1 or type 2 diabetes mellitus. Further trials, which address the issues of allocation concealment and blinding, are required (Leach and Kumar, 2012; Ranasinghe *et al.*, 2012).

The Ohio State University Center for Integrative Medicine has developed a specialized nutrigenomic protocol for integrative cancer care. The center uses a comprehensive nutritional and medical evaluation, including a panel of proinflammatory molecules and physiological parameters, to guide a program of individualized dietary interventions. There is growing interest in the use of integrative medicine—the combination of complementary and alternative medicine (CAM) methodologies with a conventional medical approach—for the optimization of treatment of various cancers. Cinnamon takes part in such investigation because promising due to its properties to decrease serum glucose levels (Varker *et al.*, 2012).

Cinnamon pulao, a recipe from Vaidya Kalyan Chakravarty

Ingredients

1 lb basmati rice
4 tsp of ghee
2–3 cinnamon (1-inch-long) sticks (4 g)
4 tbsp sugar
1/4 tsp of ground cumin
1/10 tsp of turmeric powder
Salt to taste
Boiling water (twice the amount of rice)

Preparation

Heat a big pan over medium heat. Add ghee. When ghee is sufficiently hot and gives off aroma, add cumin seed and cinnamon sticks, and fry for a minute. Add turmeric powder, washed rice, and salt. Stir for 2–3 minutes. Carefully, add the boiling water and bring whole mixture to boil. Cook until all water is evaporated and rice is done. Add sugar and mix well and bring out from heat. Cover rice with a lid for 5 minutes and serve with your favorite gravy.

Cinnamon tea, a recipe from Vaidya Kalyan Chakravarty

Ingredients

2 full cinnamon sticks
1/4 tsp of pepper powder
1 tsp of honey
2 cups of water

Preparation

Crush the cinnamon sticks and add them to the water. Boil the mixture for 15 minutes, or till the water becomes half (water color will change to dark brown). Pour it into a cup, add honey (only when water is 40°C), and then pepper powder. Tea is ready!

REFERENCES

Aggarwal B.B., Yost D. *Healing Spices*. Sterling Publishers, New York, **2011** (Italian version, Armenia, 92–102).

Bailey C.J., Day C. Traditional plant medicines as treatments for diabetes. *Diabetes Care*. **1989**, *12*, 553–564.

Bakhru H.K. *Indian Spices & Condiments as Natural Healers*. Jaico Publishing House, Mumbai, India, **2004** (3rd edn.), 43–48.

Ballali S., Lanciai F. Functional food and diabetes: A natural way in diabetes prevention? *Int. J. Food Sci. Nutr*. **2012**, *63*(Suppl 1), 51–61.

Bandara T., Uluwaduge I., Jansz E.R. Bioactivity of cinnamon with special emphasis on diabetes mellitus: A review. *Int. J. Food Sci. Nutr.* **2012**, *63*(3), 380–386.

Bin Jantan I., Moharam B.A.K., Santhanam, J., Jamal J.A. Correlation between chemical composition and antifungal activity of the essential oils of eight *Cinnamomum* species. *Pharm. Biol.* **2008**, *46*, 406–412.

Brosse J. *La magia delle piante*, Studio Tesi, Porddenone, Italy, **1992**, 200–204.

Cao H., Polansky M.M., Anderson R.A. Cinnamon extract and polyphenols affect the expression of tristetraprolin, insulin receptor, and glucose transporter 4 in mouse 3T3-L1 adipocytes. *Arch. Biochem. Biophys*. **2007**, *459*, 214–222.

Deng R. A review of the hypoglycemic effects of five commonly used herbal food supplements. *Recent Pat. Food Nutr. Agric*. **2012**, *4*(1), 50–60.

Frawley D., Lad V. *The Yoga of Herbs*. Motilal Banarsidass Publishers, Delhi, India, **1994**, 111–112.

Jarvill-Taylor K.J., Anderson R.A., Graves D.J. A hydroxychalcone derivative from cinnamon functions as a mimetic for insulin in 3T3-L1 adipocytes. *J. Am. Coll. Nutr.* **2001**, *20*, 327–336.

Jayaprakasha G.K., Mohan Rao L.J. Chemistry, biogenesis, and biological activities of *Cinnamomum zeylanicum*. *Crit. Rev. Food Sci. Nutr.* **2011**, *51*, 547–562.

Leach M.J., Kumar S. Cinnamon for diabetes mellitus. *Cochrane Database Syst. Rev*. **2012**, *9*, CD007170.

Manniche L. *An Ancient Egyptian Herbal*. The American University in Cairo Press, Cairo, Egypt, **2006**, 95–96.

Muthuswamy S., Rupasinghe H.P.V., Stratton G.W. Antimicrobial effect of cinnamon bark extract on *Escherichia coli* O157:H7, *Listeria innocua*, and fresh-cut apple slices. *J. Food Safety* **2008**, *28*, 534–549.

Paranjipe P. *Indian Medicinal Plants. Forgotten Healers*. Chaukhamba Sanskrit Pratishthan, Delhi, India, **2005**, 264–265.

Pruthi J.S. *Spices and Condiments*. National Book Trust, Delhi, India, **1998** (5th edn.), 97–101.

Qin B., Nagasaki M., Ren M., Bajotto G., Oshida Y., Sato Y. Cinnamon extract (traditional herb) potentiates *in vivo* insulin regulated glucose utilization via enhanced insulin signaling in rats. *Diabetes Res. Clin. Pract.* **2003**, *62*, 139–148.

Ranasinghe P., Jayawardana R., Galappaththy P., Constantine G.R., de Vas Gunawardana N., Katulanda P. Efficacy and safety of "true" cinnamon (*Cinnamomum zeylanicum*) as a pharmaceutical agent in diabetes: A systematic review and meta-analysis. *Diabet. Med.* **2012**, *29*(12), 1480–1492.

Shan B., Cai Y.Z., Brooks J.D., Corke H. Antibacterial properties and major bioactive components of cinnamon stick (*Cinnamomum burmannii*): Activity against foodborne pathogenic bacteria. *J. Agric. Food Chem.* **2007**, *55*, 5484–5490.

Varker K.A., Ansel A., Aukerman G., Carson W.E., III. Review of complementary and alternative medicine and selected nutraceuticals: background for a pilot study on nutrigenomic intervention in patients with advanced cancer. *Altern. Ther. Health Med.* **2012**, *18*(2), 26–34.

9.5 CLOVE

Scientific name: *Syzygium aromaticum* L.
Family: Myrtaceae
Hindi: Laung
Sanskrit: Lavangam
Tamil: Kirampu

9.5.1 Origin and Distribution

Clove originated in the Moluccas volcanic islands of Indonesia (historically called the Spice Islands) and grew only there until modern times. Clove seeds were found within a ceramic vessel in Syria along with evidence dating the find to 1721 BC. In the third century BC, a Chinese leader in the Han Dynasty required those who addressed him to chew cloves as to freshen their breath. Until cloves were grown outside of the Moluccas Islands, they were traded like oil, with a forced limit on exportation. As the Dutch East India Company consolidated its control of the spice trade in the seventeenth century, they sought to gain a monopoly in cloves as they had in nutmeg.

Currently, clove grows mostly in tropical belt countries. The biggest clove-producing region in the world is Zanzibar, followed by Pemba, Madagascar, and Indonesia. Other small producers are Malaysia, Sri Lanka, and Haiti. The plant was introduced in India in 1800 by the East India Company and grows in Kerala and Tamil Nadu, but not in a sufficient quantity to meet the high requirement of the country.

9.5.2 Features

Clove is one of the most important spices in the world, second only to black pepper (Pruthi, 1976). It is the air-dried, unopened flower bud obtained from an evergreen medium-sized tree. The buds are collected when they develop a crimson color and dried in the sun until they become dark brown (Bakhru, 2004).

The English name derives from the Latin *clavus*, meaning "nail" as the spice resembles a broad-headed nail.

9.5.3 Properties and Uses in Ayurvedic System of Medicine

Clove has an energizing effect, which increases *pitta* but reduces *vata* and *kapha*. It is commonly used in Ayurveda as a stimulant and a flavor enhancer for the lungs and stomach (Frawley and Lad, 1994). *Charaka Samhita* mentioned it as *agnimandya nasaka*, good for alleviating anorexia; Susruta reported the traditional use as a masticatory after food to alleviate thirst and *kapha* (Paranjipe, 2005).

Clove enhances circulation, increases *agni*, and helps counter stomach disorders such as gas, bloating, and nausea. It is a strong diaphoretic and a vasodilator with a cleaning action. It is also considered an aphrodisiac helpful in premature ejaculation.

Parts used: Leaves, dried fruits, essential oil
Tastes: Bitter, pungent
Quality: Light, unctuous, sharp
Potency: Hot
***Dosha* effect:** KV–, P+
Actions: Analgesic, aphrodisiac, digestive, aromatic, diaphoretic, antispasmodic, expectorant, carminative, antibacterial

9.5.4 Traditional Uses and Folklore

- Cloves are traditionally used with ginger and cardamom in case of nausea. They are effective in control vomiting during pregnancy.
- A decoction with honey works well for cough and colds (Paranjipe, 2005).
- In case of indigestion and flatulent colic, a decoction of six cloves in 30 ml of water is very effective.
- Clove and clove essential oil are a traditional home remedy for toothache (Aggarwal and Yost, 2011). Cloves can be chewed to disinfect and clean the teeth and mouth. Applying a cotton ball with some clove oil to the gums and teeth can reduce pain and swelling.

Precautions: Though clove is a safe spice, a small amount of individuals may be allergic to it. Those suffering from hypertension and high *pitta* should avoid clove.

9.5.5 In the Kitchen

Cloves are very aromatic, with a fine flavor and warming qualities. They are widely used in Asian, African, and Middle Eastern cooking. In Ayurveda, cloves are used both whole and ground. Ground cloves will preserve stay fresh for about 6 months, whereas whole cloves will stay fresh for about 1–2 years if stored properly. Whole cloves sautéed in ghee with other spices such as cinnamon, bay leaves, and peppercorns enhance the flavor of rice and pilafs. They are renowned for providing their uniquely warm, sweet, and aromatic taste to ginger bread and pumpkin pie, but they can also make a wonderful addition to split pea and bean soups, baked beans, and chili. Clove is an essential ingredient in curry powders and combines well with other spices such as turmeric, cinnamon, cumin, and coriander.

Clove can impart a spicy warming taste to soups, baked apples, and fruit compote. It is a typical ingredient of the south Indian rice preparation called *biryani*. It is always better to take out the clove from the preparation at the end of cooking; in the soups, we can pierce a small onion with a clove to take out easily.

Apart from food, it is used for flavoring some wines and alcoholic beverages as well. In India, it is also an ingredient of betel nut chew. In Indonesia, people often smoke cigarettes flavored with cloves.

9.5.6 Chemical Constituents and Uses as Functional Food

Cloves (*Syzygium aromaticum*) are the aromatic dried flower buds of a tree in the family Myrtaceae. It is used as a spice in cuisines all over the world, for adding flavor to meats, curries, and meat marinades. They are also used to create sweet dishes, with fruits such as apples, pears, or rhubarb, or to aromatize wine. Considered a very strong spice due to the presence of eugenol that makes up most of the clove's taste (85%), the quantity of clove used in recipes is usually small.

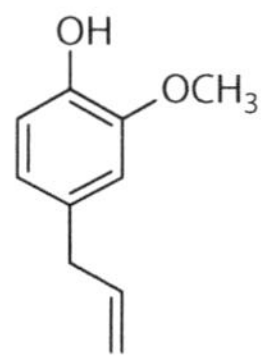

Eugenol

Eugenol is a major volatile constituent of clove essential oil obtained through hydrodistillation of mainly *Eugenia caryophyllata* (= *S. aromaticum*) buds and leaves. It is also present among the volatile constituents of basil, nutmeg, and cinnamon. It is a remarkably versatile molecule incorporated as a functional ingredient in numerous products and has found its application in the pharmaceutical, agricultural, fragrance, flavor, cosmetic, and various other industries. Its vast range of pharmacological activities has been well researched and includes antimicrobial, anti-inflammatory, analgesic, antioxidant, and anticancer activities, among others. In addition, it is widely used in agricultural applications to protect foods from microorganisms during storage, which might have an effect on human health, and as a pesticide and a fumigant. As a functional ingredient, it is included in many dental preparations, and it has also been shown to enhance skin permeation of various drugs. Eugenol is considered safe as a food additive, but due to the wide range of different applications, extensive use, and availability of clove oil, it is of mention that its general toxicity at high doses with special reference to contact dermatitis (reviewed in Kamatou *et al.*, 2012).

Eugenol is reported to possess the anticancer activity against various cancers. The molecular mechanisms of eugenol-induced apoptosis are related to increase ROS generation, depletion of reduced glutathione (GSH) and protein thiols, and a fall in the mitochondrial membrane potential in tumor cell lines (reviewed in Jaganathan and Supriyanto, 2012). In general, eugenol, as well as many other spice-derived nutraceuticals, modulates various transcription factors, growth factors, protein kinases, and inflammatory mediators. The anticancer potential through the modulation of various targets is the subject of the recent research. Although these spice-derived nutraceuticals have always been used to improve taste and color and as a preservative, they are now proposed for prevention and treatment of a wide variety of chronic inflammatory diseases, including cancer (reviewed in Sung *et al.*, 2012).

REFERENCES

Aggarwal B.B., Yost D. *Healing Spices.* Sterling Publishers, New York, **2011** (Italian version, Armenia, 110–116).

Bakhru H.K. *Indian Spices & Condiments as Natural Healers.* Jaico Publishing House, Mumbai, India, **2004**, 49–53.

Frawley D., Lad V. *The Yoga of Herbs.* Motilal Banarsidass Publishers, Delhi, India, **1994**, 112–113.

Jaganathan S.K., Supriyanto E. Antiproliferative and molecular mechanism of eugenol-induced apoptosis in cancer cells. *Molecules* **2012**, *17*, 6290–6304.

Kamatou G.P., Vermaak I., Viljoen A.M. Eugenol—From the remote Maluku Islands to the international market place: A review of a remarkable and versatile molecule. *Molecules* **2012**, *17*(6), 6953–6981.

Paranjipe P. *Indian Medicinal Plants. Forgotten Healers.* Chaukhamba Sanskrit Pratishthan, Delhi, India, **2005**, 159–160.

Pruthi J.S. *Spices and Condiments.* National Book Trust, Delhi, India, **1976**, 102–108.

Sung B., Prasad S., Yadav V.R., Aggarwal B.B. Cancer cell signaling pathways targeted by spice-derived nutraceuticals. *Nutr. Cancer* **2012**, *64*(2), 173–197.

9.6 CORIANDER

Scientific name: *Coriandrum sativum* L.
Family: Apiaceae
Hindi: Dhaniyam
Sanskrit: Dhaniakam
Tamil: Kottamalli

9.6.1 Origin and Distribution

The name *coriander* actually originated from the Greek word *Koris*, meaning "bedbug," due to the quite unpleasant smell. However, it is a very popular spice currently used and cultivated in many parts of the world, especially in Asian countries. Native of the Mediterranean, it is so extensively cultivated from Morocco to India that we cannot define exactly where the plant is wild and where it has only recently established itself.

Coriander was mentioned as a medicinal plant in Sanskrit texts dating back almost to 7000 years ago. In Egypt, it was an important ingredient in cookery and medicines, offered to the temples by the king. Some seeds were found in the tomb of Tutankhamen (Manniche, 2006). It seems to have been cultivated in Greece since at least the second millennium BC as a spice for its seeds and as an aromatic herb. The ancient Romans used both the leaves and the seeds to preserve meats. During the Middle Ages, it was an ingredient of love potions. Coriander was brought to the British colonies in North America in 1670 and was one of the first spices cultivated there.

9.6.2 Features

It is a soft, hairless aromatic plant growing to 50 cm tall. The leaves are variable in shape, broadly lobed at the base of the plant, and slender and feathery higher on the flowering stems. The flowers are small, white to pale pink, and arranged in a cluster of flowers, which looks like an umbrella. The fruit does not break open naturally when dried and has two, single-seeded units called mericarps.

9.6.3 Properties and Uses in Ayurvedic System of Medicine

Ayurvedic texts mention coriander as *agnimandya nasaka* (reduces anorexia and loss of appetite) and as *amapacaka* (digests *ama*, the toxins). It is called *hrdya*, pleasant to the mind and beneficial to the heart, and *Kustumbaru*, which alleviates various diseases (Paranjipe, 2005).

Its bitterness reduces *pitta*, whereas its pungency improves digestion and is suitable for *kapha*. Although the leaves are cooling, the seeds are slightly warming. Coriander pacifies all the three *dosha*s and does not provoke *pitta*, unlike many other spices.

Ayurveda considers coriander specific for strengthening the urinary tract (Khalsa and Tierra, 2010). Leaves when digested turn sweet and alleviate irritations in the intestines. Coriander also works on the heart *chakra*. It helps the heart by lowering bad cholesterol (low-density lipoprotein [LDL]) and raising good cholesterol (high-density lipoprotein). Furthermore, coriander works on the nervous system and can stimulate the memory.

The common use of fresh coriander leaves in Indian food can prevent heartburn, indigestion, colic, and diarrhea. It will enhance appetite, and improve digestion and absorption of nutrients.

The seeds are often combined with laxatives to prevent any griping the latter may cause. The relaxant effects of coriander (thanks to its high volatile oil content) seen in the digestive tract are also apparent in the uterus, helping to relieve menstrual problems.

Parts used: Seeds, leaves, fruits
Tastes: Bitter, pungent, astringent, sweet
Quality: Light, unctuous, rough
Potency: Cold
***Dosha* effect:** PKV=
Actions: Carminative, appetizer, cooling, alterative, diaphoretic, diuretic, tonic

9.6.4 Traditional Uses and Folklore

- From ancient times, Arabian women used to take coriander to lessen the pain of childbirth.
- In India, the juice extracted from the leaves, mixed with plant camphor, is a home remedy for menstrual problems. 10–15 ml of juice can be taken 2–3 times a day to stop excess bleeding during the menstrual cycle.
- The spice has been used as a folk medicine for the relief of anxiety and insomnia in Iran.
- Coriander is an excellent home remedy for people suffering from *pitta* aggravation due to excessive consumption of tea or coffee. A handful of coriander seeds is fried and then powdered to a nice consistency. A spoonful of this powder is then added to a glass of milk with sugar and then taken instead of tea or coffee. The leaves are an appetizer useful for anorexic patients. The juice is an ancient Ayurvedic remedy for treating nausea and morning sickness.

- Coriander decoction is recommended for patients suffering from kidney ailments. The cooling effects of coriander make it an excellent remedy for urinary disorders, particularly those with hot burning symptoms such as cystitis and urethritis. It will also help to cure mouth ulcers and swellings. Habitual intake of this decoction will also help to lessen the cholesterol levels in the bloodstream.
- In India, the roasted fennel and coriander seeds with a little salt are chewed after eating to help digestion and combat sleepiness. Through their beneficial action in the liver and digestive tract, coriander seeds can also lessen the intoxicating effect of alcohol.

Precautions: Coriander can produce an allergic reaction in some people. It is not advisable for high *vata* with nerve tissue deficiency (Frawley and Lad, 1994) and for people suffering from bronchial asthma and chronic bronchitis.

9.6.5 In the Kitchen

The leaves are widely used as a flavoring agent across many different countries, especially in India, Thailand, Greece, Mexico, and Colombia. As heat diminishes their flavor, coriander leaves are often used raw or added to the dish immediately before serving. It is better to grow the plant within easy reach of the kitchen door for a constant, fresh supply as the leaves spoil quickly when removed from the plant and lose their aroma when dried or frozen. They can be kept in the fridge for few days only, well packaged in a humid, white cloth.

Both the seeds and the leaves of coriander can be used in salads and soups. South Indian foods, especially chutneys and *rasam* (soup), are insipid without coriander.

The seeds are also used as an important ingredient in making curries and other dishes. They are used as such or ground into a powder. They can be roasted or heated on a dry pan briefly before grinding to enhance the aroma. Ground coriander seed loses flavor quickly in storage and is best ground fresh.

Regular use of coriander helps to improve the overall health and prevents many of the vitamin deficiency diseases. It is a spice specifically recommended by Ayurveda to balance *pitta*. Like cumin, it is an antidote for hot and pungent food (Frawley and Lad, 1994).

We can use it in plenty during the summer, combined with cumin, cardamom, fennel, anise, and caraway. The leaves, often called Chinese parsley, are excellent for balancing the flavor of hot spicy dishes and are used for just this purpose in Thai, Indian, Moroccan, Mexican, Chinese, Indonesian, African, and South American cooking.

Coriander kills meat-spoiling bacteria and fungi, and contains an antioxidant that prevents animal fats from turning rancid. Therefore, it is an important ingredient in meat dishes from India to Morocco. Combined with onion or garlic, it increases its ability to keep food fresh and is also thought to be an aphrodisiac. *Taklia*, a delicious Arab spice mixture, is actually a preparation of coriander and garlic mashed and fried.

9.6.6 Chemical Constituents and Uses as Functional Food

Coriander (*Coriandrum sativum* L.) is one of the oldest herbs that has been used for over 3000 years (cited in *Ebers Papyrus* of 1500 BC) for both culinary and medicinal purposes. The plant is indigenous to the Mediterranean area, presently cultivated in Central and Eastern Europe, Mediterranean regions, and Asia. India is one of the main producers of coriander seeds. The seeds of coriander are globular and aromatic with a slight bittersweet, spicy taste, and are an integral part of curry powder. The green leaves are consumed as fresh herbs, in salads, and as garnishes due to its attractive and green color and aroma. All parts of the plant are traditionally used for its medicinal properties such as antimicrobial, antioxidant, anxiolytic, sedative, neuroprotective, antidiabetic, diuretic, and gut modulatory (reviewed in Sahib *et al.*, 2012).

The seeds are a good source of lipids (25%) and are characterized by a high amount of petroselinic acid, a C18 monounsatured fatty acid, bearing the unsaturation in the rare position 6 of the alkyl chain. Other fatty acids include linoleic, oleic, and palmitic acids. The seed oils also contain tocopherols and sterols (Sriti *et al.*, 2010).

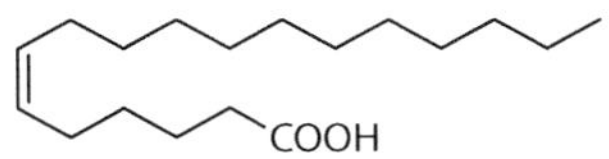

Petroselinic acid

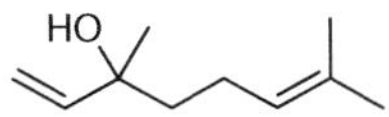

Linalool

The yield of the essential oil is around 1% with variations due to the location and harvesting conditions, whose major component is linalool (30%–80%), followed by other monoterpene alcohols and monoterpene hydrocarbons (reviewed in Sahib *et al.*, 2012). Coriander leaves are not as well studied as the seeds. Nevertheless, an essential oil, mainly constituted of aliphatic aldheydes (2*E*-decenal and decanal) and aliphatic alcohols (2*E*-decen-1-ol and *n*-decanol), flavonoids, phenolic acids (ferulic, *p*-coumaric, and vanillic acids), and polyphenols are among the most abundant constituents. The aromatic, pungent smell is due to the aliphatic aldheydes and alcohols.

The antimicrobial activity of coriander leaves and seeds, and their extracts and essential oils is one of the most reported biological properties of the plant (Matasyoh *et al.*, 2009), with marked inhibitory effect against Gram-positive and Gram-negative bacteria. The seed and leaf essential oils also show antifungal activity against *Candida* species (Begnami *et al.*, 2010).

The antioxidant activity of leaves and shoot extracts of coriander has been attributed to its high phenolic content, but the data obtained *in vitro* are not confirmed in *in vivo* models.

Preclinical evidence of anxiolytic activity (without human clinical trials) was found for *C. sativum* (reviewed in Sarris *et al.*, 2013).

Green chutney, a recipe from Vaidya Kalyan Chakravarty

Ingredients

1/2 tsp of cumin seeds

1/2 tsp of mustard seeds

1 bunch of coriander, large stems removed
1/4 cup of unsweetened grated coconut (or 1/2 cup of fresh coconut pulp)
1/2″ piece of ginger roughly chopped
2 Serrano chilies (or other spicy green chili)
2 lemons juiced
Salt to taste

Preparation

Put the cumin and mustard seeds in a small heavy bottomed pan and roast over heat until the spices are fragrant, but be careful not to burn them. The key is to keep the spices constantly moving in the pan by shaking it with a swirling motion.

Put the spices in a spice grinder and pulverize.

Add everything, including the toasted spices, into a blender and blitz until smooth. Add some water if the chutney is too thick and then adjust salt to taste.

Stored in an airtight container in the fridge, your green chutney should last at least one week.

Coriander rice, a recipe from Vaidya Kalyan Chakravarty

Ingredients

2 cups of rice, washed and soaked in water for half an hour and drained
A big bunch of coriander leaves, washed and chopped
4 green chilies
4 tbsp of grated fresh coconut
1″ ginger piece
6 garlic cloves
2 chopped tomatoes
Whole spices (6 cloves, 3 green cardamoms, 1″ pieces of cinnamon, bay leaf)
Salt to taste
2 tbsp of ghee (clarified butter)

Preparation

Heat two teaspoons of ghee in a cooking vessel, add the chopped coriander leaves, and green chilies and fry till rawness disappears (approximately 2 minutes). Remove from heat and cool.

Grind the fried coriander and green chilies along with coconut, ginger, and garlic to a paste adding little water.

Heat two tablespoons of ghee in a cooking vessel, add the whole spices, and fry for a few seconds. Add the ground coriander paste and fry for 3–4 minutes, and then add the chopped tomatoes. Fry for 3–4 minutes. Add the drained rice and mix well, and fry for 2 minutes. Add salt to taste. Add 3 1/2 cups of water and cook with lid on low-to-medium flame till rice turns soft. Garnish with fried cashew nuts (optional) and chopped coriander.

REFERENCES

Begnami A.F., Duarte M.C.T., Furletti R.V. Antimicrobial potential of *Coriandrum sativum* L. against different *Candida* species *in vitro*. *Food Chem.* **2010**, *118*, 74–77.

Frawley D., Lad V. *The Yoga of Herbs*. Motilal Banarsidass Publishers, Delhi, India, **1994**, 114.

Khalsa K.P.S., Tierra M. *The Way of Ayurvedic Herbs*. Motilal Banarsidass Publishers, Delhi, India, **2010**, 125.

Manniche L. *An Ancient Egyptian Herbal*. The American University in Cairo Press, Cairo, Egypt, **2006**, 100.

Matasyoh J.C., Maiyo Z.C., Ngure R.M., Chepkoris R. Chemical composition and antimicrobial activity of essential oil of *Coriandrum sativum*. *Food Chem.* **2009**, *113*, 526–529.

Paranjipe P. *Indian Medicinal Plants. Forgotten Healers*. Chaukhamba Sanskrit Pratishthan, Delhi, India, **2005**, 68–70.

Sahib N.G., Anwar F., Gilani A.H., Hamid A.A., Saari N., Alkharfy K.M. Coriander (*Coriandrum sativum* L.): A potential source of high-value components for functional foods and nutraceuticals—A review. *Phytother. Res.* **2013**, *27*(10), 1439–1456.

Sarris J., McIntyre E., Camfield D.A. Plant-based medicines for anxiety disorders, Part 1: A review of preclinical studies. *CNS Drugs* **2013**, *27*(3), 207–219.

Sriti J., Wannes W.A., Talou T., Mhamdi B., Handaoui G., Marzouk B. Lipid fatty acid and tocol distribution of coriander's fruits' different parts. *Indust. Crop. Prod.* **2010**, *31*, 294–300.

9.7 CUMIN

Scientific name: *Cuminum cyminum* L.
Family: Apiaceae
Hindi: Jira
Sanskrit: Jiraka
Tamil: Jirakam

9.7.1 Origin and Distribution

The English word "cumin" derives from Latin *cuminum*, which is the latinization of the Greek *kuminon*. This spice has been in use since very ancient times. Native of Upper Egypt and Syria, it was cultivated from early times in Arabia, India, China, and the Eastern Mediterranean region. The seeds excavated at the Syrian site Tell ed-Der have been dated back to the second millennium BC; they were also found in ancient Egyptian archeological sites. It is said that the ancient Egyptians, apart from using it for culinary purposes, also used it as an ingredient to mummify their dead. Originally cultivated in the Mediterranean region, cumin is mentioned in the Bible in both the Old Testament (Isa. 28:27) and the New Testament (Matt. 23:23). During the biblical times, cumin was popular as a digestive food flavor for ceremonial feasting. The ancient Greeks kept cumin at the dining table in its own container (much as pepper is frequently kept today). Cumin was also used heavily in ancient Roman cuisines. Pliny described it as the best appetizer of all seasonings. In the Middle Ages, cumin was very popular in Europe as a sign of love and devotion: people believed that it would keep poultry from wandering away and even insure fidelity between couples. Therefore, the Germans used to carry in their pockets cumin, dill, and salt during the wedding ceremony (Bakhru, 2004), and in many parts of Europe, wives sent their husbands off to war with a loaf of cumin spiced bread.

In India, cumin has been used for millennia as a traditional ingredient of innumerable *kormas*, *masalas*, curry powders, soups, and other spiced gravies. After pepper, cumin is considered to be the most important spice in the world, with India as the world's largest producer and consumer.

It is now extensively grown in India, Iran, Morocco, China, southern Russia, Indonesia, Japan, and Turkey (Bakhru, 2004). It was introduced to the Americas by Spanish and Portuguese colonists. Because cumin is often used as part of birdseed and exported to many countries, the plant can occur as a rare casual in many territories, including Britain. According to the Botanical Society of the British Isles' most recent Atlas, only one record has been confirmed since 2000. Iran is actually the major world exporter (Pruthi, 1998).

9.7.2 Features

Cumin is the dried seed of the herb *Cuminum cyminum*, a member of the parsley family. It is an annual herbaceous aromatic plant that grows up to a height of 30–50 cm. It produces stems arising from the base with many branches. The leaves are divided into long, narrow segments similar to fennel, but much smaller. The flowers are small, whitish red on a compound looking like an umbrella (Warrier *et al.*, 1994). The plant has aromatic seedlike fruits, which are oblong in shape, longitudinally ridged, and yellow-brown in color, commonly known as cumin seeds. The odor is peculiar, strong, and heavy.

9.7.3 Properties and Uses in Ayurvedic System of Medicine

Cumin is largely used in India as spice and medicine. The Sanskrit name *jiraka* means "that which helps digestion." Ayurveda considers it as one of the best digestive tonics with a lot of health benefits: it dispels gas; eliminates toxins; and strengthens the stomach, liver, and intestines. Because it is so easily digested and so effective in expelling gas from the stomach and the intestines, it is used for chronic dysentery and diarrhea. Cumin has a pungent taste, light and rough qualities, and hot potency; it kindles the gastric fire (*agni*) and improves the absorption of minerals in the intestines. According to *Sushruta Samhita*, it is an appetizer and pacifies *vata* and *pitta*. Cumin, when boiled or steeped in water, flushes toxins giving rise to glowing and radiant complexion: it is a "queen" *ama* burner. Spices such as cumin, fennel, and turmeric purify the urine daily: without at least one of them in the daily diet, *ama* buildup causes *dosha* imbalance. It is also recommended after childbirth as it increases the capacity of producing milk and also relieves pain and inflammation of the uterus (Khalsa and Tierra, 2010).

During the spring, cumin heats up and dries the body to prepare it for warmer weather ahead: its drying effect is useful for any spring dampness and excessive wetness, such as mold allergies and congested lung conditions. The seeds open the pores, revitalizing the skin after a cold winter. These dilating and purifying properties have also been traditionally applied to reduce *vata* type fevers and to reduce tightness in the chest: the *kapha* clearing and antispasmodic qualities were traditionally used during asthma attacks.

Parts used: Fruits
Tastes: Sweet, pungent
Quality: Light and rough

Potency: Cold
***Dosha* effect:** VPK=
Actions: Tonic, antiseptic, digestive, anti-inflammatory, analgesic, diuretic, mild laxative, anthelmintic, galactogogue, carminative, detoxicant

9.7.4 Traditional Uses and Folklore

In Kerala, the land of Ayurveda, *jira* water is a regular drink in many houses and even in restaurants. Instead of drinking just boiled water, a bit of *jira* is added to the water and brings good flavor and health: it balances *vata*, *pitta*, and *kapha*, helps to flush out the toxins from the body, and is helpful in strengthening our digestive and circulatory systems. According to Ayurveda, *jira* water contains good antiseptic properties that fight common cold and fevers.

For indigestion, a teaspoon of seeds could just be boiled in a glass of water, mixed with salt and a teaspoon of coriander leaf juice. This preparation should be taken twice daily after meals.

Chewing few seeds half an hour before meals helps in digestion and increases hunger, and after meals prevents dental caries, indigestion and fermentation of food in the intestine, and constipation.

9.7.5 In the Kitchen

Cumin seeds are a globally popular spice in many cuisines, particularly in South Asia, Northern Africa, and Latin America. They can be used ground or as whole seeds to add an earthy and warming feeling to food. It is better to grind the seeds just before adding to a recipe, using a coffee grinder or a mortar and pestle, to keep alive its vital life force and flavor. Cumin combines well with a wide range of other spices, including turmeric, ground fennel, ground coriander, ground dry ginger, and cinnamon. It is enriched by being sautéed in ghee or oil: just the smell starts enhancing digestion as it activates the salivary glands. It can be combined with some minced ginger, lemon juice, salt, and black pepper to make a dressing for a warm salad of boiled white beans or lightly steamed shredded carrots. It is also a flavorful addition to lentil and legume soups. It is commonly used in traditional Mexican and Brazilian cuisine, and can be found in some cheeses and traditional breads in France.

Cumin is one ingredient of Indian *lassi*, an excellent and refreshing lunchtime drink that improves the digestion. To prepare a *lassi*, just blend yogurt and water (50:50) with dry-roasted, ground cumin and salt to taste.

It is sometimes confused with *Carum carvi*, another spice from the same family. Cumin, though, is hotter to the taste, lighter in color, and larger. Many European languages do not distinguish clearly between the two.

9.7.6 Chemical Constituents and Uses as Functional Food

The flavor of seeds is warm, heavy, spicy, and curry like, dominated by the flavor component cuminaldehyde. Cumin should be stored in airtight containers, and it is best to use whole seeds and grind in a mortar just before use, as the essential oil starts vanishing immediately after grinding.

Cumin seeds contain an essential oil (3%–4%) mainly consisted of monoterpenes, the most abundant components being cuminaldehyde, cuminyl alcohol, *p*-cymene, and carvacrol, whereas sesquiterpenes are minor constituents (Sowbhagya, 2013). Cumin also contains flavonoids (e.g., apigenin and luteolin).

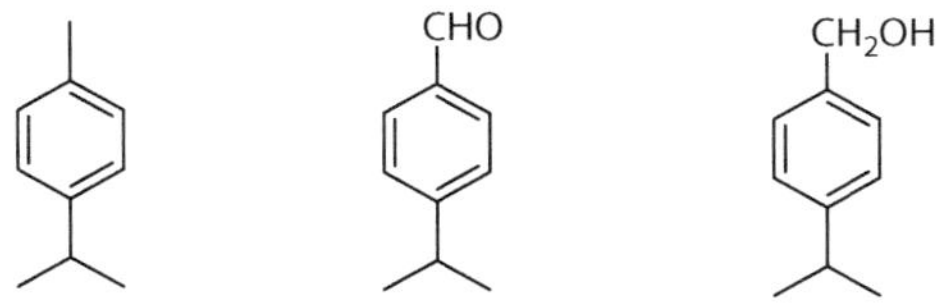

p-Cymene Cuminaldehyde Cuminyl alcohol

Traditional uses of cumin include anti-inflammatory, antibacterial, diuretic, carminative, and antispasmodic (reviewed in Johri, 2011). Cuminaldehyde, the major constituent of the essential oil, is responsible for the antimicrobial and antimutagenic properties (Burt, 2004; Vicuda-Martos *et al.*, 2008).

Cumin products (oil as well as the aqueous and solvent extracts) have shown significant antioxidant activity in several test methods, which has been attributed largely to the presence of monoterpene alcohols, flavonoids, and other polyphenolic compounds. The antiradical profile has been proposed as the underlying mechanism for their multifaceted properties.

Many studies have revealed the anticarcinogenic action of cumin, related to its potential apoptotic, antimutagenic, and antiproliferative properties, and the ability to modulate carcinogen metabolism via phase I and phase II enzymes. In all the studies, the attenuation of carcinogenicity has been attributed to their potential antioxidant action in the target tissues.

The antidiabetic and hypolipidemic effects of cumin are reported (Dhandapani *et al.*, 2002; Srinivasan, 2005). Other properties include immunomodulatory, analgesic, estrogenic, and antitussive.

Significant pharmacokinetic interaction of cumin components with antitubercular drugs (rifampicin) has been revealed. This was found to be a flavonoid glycoside from cumin endowed with permeation enhancement property (Sachin *et al.*, 2007).

Although advances in chemical and pharmacological evaluation of cumin have occurred in the past few years, a strict relationship between the composition and the total activity of the plant has not yet been accomplished. Although some identified compounds have been associated with a bioactivity, it cannot be attributed to any particular mixture of compounds. It seems reasonable to assume that the "synergy" between and within a particular class of compounds might be responsible for the bioactivity profile of the plant. In this context, there are remarkable studies, which indicate that natural products could facilitate the bioavailability of poorly bioavailable drugs.

Jira rice, a recipe from Vaidya Kalyan Chakravarty

Ingredients

1 cup of basmati rice
2 cups of water
2 tsp of cumin seeds
1 or 2 bay leaves
2 cinnamon sticks
1 black cardamom
1 1/2 tbsp of ghee
Salt to taste

Preparation

Wash rice thoroughly and soak in water for half an hour. Drain thoroughly.

In a pan, heat ghee. Add the cumin seeds, bay leaves, cardamom, and cinnamon.

When cumin seeds sizzle, add the rice. Add salt to taste.

Stir till ghee coats every grain of rice and it looks glossy. Now add water and bring it to boil. Stir once and reduce the heat to minimum and cover the pan with a lid. Leave a little gap, otherwise water will boil over. When ready, serve hot with any vegetable curry.

Jira water

Ingredients

1 tsp of cumin seeds
1 l water

Preparation

Add cumin seeds to water. Boil the water for 15–20 minutes, filter, and keep it in a thermos to drink during the day. It is good for throat irritation and digestion. It will also improve voice.

REFERENCES

Bakhru H.K. *Indian Spices & Condiments as Natural Healers*. Jaico Publishing House, Mumbai, India, **2004**, 60–64.

Burt S. Essential oils—Their antibacterial properties and potential applications in food: A review. *Int. J. Food Microbiol.* **2004**, *94*, 223–253.

Dhandapani S., Subramanian V.R., Rajagopal S., Namasivayam N. Hypolipidemic effect of *Cuminum cyminum* L. on alloxan-iduced diabetic rats. *Pharmacol. Res.* **2002**, *46*, 251–255.

Johri R.K. *Cuminum cyminum* and *Carum carvi*: An update. *Pharmacogn. Rev.* **2011**, *5*(9), 63–72.

Khalsa K.P.S., Tierra M. *The Way of Ayurvedic Herbs*. Motilal Banarsidass Publishers, Delhi, India, **2010**, 126–127.

Pruthi J.S. *Spices and Condiments*. National Book Trust, Delhi, India, **1998**, 114–120.

Sachin B.S., Sharma S.C., Sethi S., Tasduq S.A., Tikoo M.K., Tikoo A.K., Satti N.K. *et al.* Herbal modulation of drug bioavailability: Enhancement of rifampicin levels in plasma by herbal products and a flavonoid glycoside derived from *Cuminum cyminum*. *Phytother. Res.* **2007**, *21*, 157–163.

Sowbhagya H.B. Chemistry, technology, and nutraceutical functions of cumin (*Cuminum cyminum* L.): An overview. *Crit. Rev. Food Sci. Nutr.* **2013**, *53*, 1–10.

Srinivasan K. Plant foods in the management of diabetes mellitus: Spices as beneficial antidiabetic food adjuncts. *Int. J. Food Sci. Nutr.* **2005**, *56*, 399–414.

Vicuda-Martos M., Ruiz-Navajas Y., Fernandez-Lopez, J., Perez-Alvarez J.A. Antibacterial activity of different essential oils obtained from spices widely used in Mediterranean diet. *Int. J. Food Sci. Technol.* **2008**, *43*, 526–531.

Warrier P.K., Nambiar V.P.K., Ramankutty C. (eds.) *Indian Medicinal Plants*, vol. 2. Orient Longman, Chennai, India, **1994**, 241–244.

9.8 FENNEL

Scientific name: *Foeniculum vulgare* Mill.
Family: Apiaceae
Hindi: Saumph
Sanskrit: Misreya, Madhurika, Shatapushpa
Tamil: Sombu

9.8.1 Origin and Distribution

Fennel is a biennial or perennial plant that has been growing wild in the Mediterranean area and in the Asia Minor since ancient times. Today, it is cultivated in England, France, Germany, South Tyrol, Russia, Argentina, Iran, China, and India. It propagates well by the seed, and is considered an invasive species and a weed in Australia and the United States. India is the biggest producer of fennel.

There are so many mentions of the culinary and medicinal properties of fennel, a plant revered by Greeks and Romans since ancient times. The use in the treatment of eye ailments was recommended in Coptic prescriptions and in later Egyptian herbals (Manniche, 2006). In Greek mythology, Prometheus used the stalk of a fennel plant to steal the fire from the gods. During the third century BC, Hippocrates prescribed fennel to treat infant colic. Four hundred years later, Dioscorides recommended the seeds to nursing mothers to boost milk production. In the tenth century AD, the mystic Hildegard of Bingen recommended fennel seeds as a treatment for body odor.

During the Middle Ages, fennel was used together with St. John's wort and other herbs, to ward off evil spirits. It is a protective herb with many references in poetry.

> A savoury odour blown,
> Grateful to appetite, more pleased my sense
> Than smell of sweetest Fennel.

Milton

9.8.2 Features

Fennel is a stout, erect, glabrous aromatic plant with small, yellow, and fragrant flowers in compound terminal umbels. The fruits are oblong, greenish, or yellowish brown (Warrier *et al.*, 1995). The plant resembles dill, but it is easily distinguished by its sweet licorice aroma and taste.

9.8.3 Properties and Uses in Ayurvedic System of Medicine

Fennel is one of the best herbs for digestion; it is strengthens *agni* without aggravating *pitta*.

As a diuretic and urinary alkaline, it helps to remove toxic substances from the body, reducing rheumatism and swelling. It is useful in nausea and colic, and calms the nerves.

The seeds combine well with cumin and coriander: the three are cooling and can be used for all constitutions also during the summer season.

Parts used: Fruits (seeds)
Tastes: Sweet, pungent
Quality: Sharp, light
Potency: Cold
***Dosha* effect:** VKP=
Actions: Carminative, stomachic, stimulant, diuretic, antispasmodic

9.8.4 Traditional Uses and Folklore

Above the lower plants it towers,
The Fennel with its yellow flowers;
And in an earlier age than ours
Was gifted with the wondrous powers
Lost vision to restore.

- Fennel tea is a traditional eyewash and a remedy for stomach acidity, gout, cramps, colic, and spasms.
- A teaspoon of fennel seeds roasted with rock salt after meals is a good home remedy for digestive weakness in children or in the elderly (Frawley and Lad, 1994). It will also promote bowel movements.
- There is a traditional remedy that increases lactation in nursing mothers: Boil half a liter of water and add two teaspoons of fennel seeds. Allow to cool, then strain, and add a half teaspoon of raw honey to taste. Drink two to three cups daily.

9.8.5 In the Kitchen

Dried fennel seeds are an aromatic, anise-flavored spice used in cookery and sweet desserts in India, Pakistan, Afghanistan, Iran, and the Middle East.

Chewing *saumph* after meal is a common practice in India. The seeds' green color keeps the soothing qualities of the forest alive. This spice is often added to Indian recipes to aid the digestion and introduce a cooling effect in the dishes.

It is also a common spice in the Mediterranean cuisine, especially in Italian recipes. In some regions of Italy, people use not only the seeds but also the fresh tender leaves to give a special taste to soups or salads and to prepare sauces and chutneys. The leaves are delicately flavored and similar in shape to those of dill.

In some parts of India also, the leaves are used as green vegetables either alone or mixed with other vegetables. In Syria and Lebanon, they are an ingredient of a special kind of egg omelet (along with onions and flour).

9.8.6 Chemical Constituents and Uses as Functional Food

Fennel is widely cultivated, both in its native range and elsewhere, for its edible, strongly flavored leaves and fruits. Its aniseed flavor comes from anethole, an aromatic compound also found in anise and star anise, and its taste and aroma are similar to theirs, though usually not as strong. Florence fennel (*Foeniculum vulgare* Azoricum Group; syn. *F. vulgare* var. *azoricum*) is a cultivar group with inflated leaf bases, which form a bulb-like structure. It is of cultivated origin and has a mild anise-like flavor, but it is more aromatic and sweeter. It is one of the three main herbs used in the preparation of absinthe, an alcoholic mixture that originated as a medicinal elixir and became, by the late nineteenth century, a popular alcoholic drink in France and other countries. The bulb, foliage, and seeds of the fennel plant are widely used in many of the culinary traditions of the world. Dried fennel seed is an aromatic, anise-flavored spice used for cooking. The leaves are delicately flavored and similar in shape to those of dill. The bulb is a crisp vegetable that can be sautéed, stewed, braised, grilled, or eaten raw.

Foeniculum vulgare Mill. is a popular medicinal plant with various pharmacological activities mentioned in traditional medicine and modern phytotherapy such as antioxidant, cytotoxic, anti-inflammatory, antimicrobial, bronchodilatory, estrogenic, diuretic, lithontripic, galactogogue, emmenagogue, antithrombotic, hypotensive, gastroprotective, hepatoprotective, memory enhancing, and antimutagenic activities. No serious adverse events were recorded after ingestion of *F. vulgare* except some cases of allergic reactions. The estrogenic activity of *F. vulgare* brings some side effects such as decrease in protein concentration and acid and alkaline phosphatase in male genital organs, increase in weight of the mammary glands and reproductive organs in women, and premature thelarche in girls. However, no evidence of teratogenicity was recorded; it is better not to use *F. vulgare* during pregnancy due to its estrogenic activity (Rahimi and Ardekani, 2013).

The most abundant volatile component in fennel is anethole, a phenylpropanoid, found together with its isomer estragole, the main volatile component in tarragon (*Artemisia dracunculus* L.). Anethole has potent antimicrobial properties against bacteria, yeast, and fungi. Fennel is widely used as a carminative, both in humans and in veterinary medicine (e.g., dogs) to treat flatulence by encouraging the expulsion of intestinal gas. Anethole is responsible for the carminative action.

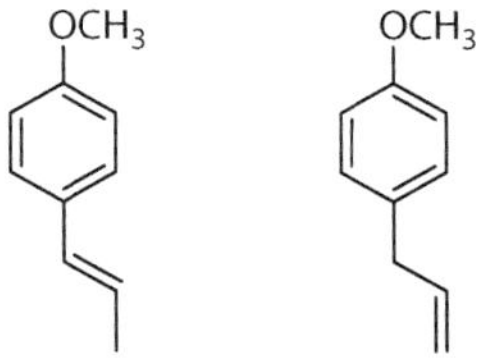

Anethole Estragole

The plants that have a substantial body of data in support of their digestion-enhancing activities mainly belong to one of three groups: bitter, aromatic, and pungent plants. Among the most important are ginger, peppermint, aniseed and fennel, citrus fruits, dandelion and artichoke, and melissa and chamomile.

The *in vitro* and *in vivo* studies in the past few decades have demonstrated that some phytochemicals derived from "natural products" such as fruits, vegetables, and certain spices, referred to as chemopreventive agents, including capsaicin, *trans*-anethole, thymoquinone, diosgenin, allicin, can not only reduce the risk of acquiring specific cancer but have also been shown to suppress cancer cell proliferation; inhibit growth factor signaling pathways; induce apoptosis; inhibit nuclear factor kappa B (NF-κB), AP-1, Akt, MAPK, Wnt, Notch, p53, AR, ER, and JAK-STAT activation pathways; inhibit angiogenesis; suppress the expression of antiapoptotic proteins; and inhibit cyclooxygenase-2 (reviewed in Rajput and Mandal, 2012).

REFERENCES

Frawley D., Lad V. *The Yoga of Herbs*. Motilal Banarsidas Publishers, Delhi, India, **1994**, 117–118.

Manniche L. *An Ancient Egyptian Herbal*. The American University in Cairo Press, Cairo, Egypt, **2006**, 111–112.

Rahimi R., Ardekani M.R. Medicinal properties of *Foeniculum vulgare* Mill. in traditional Iranian medicine and modern phytotherapy. *Chin. J. Integr. Med.* **2013**, *19*(1), 73–79.

Rajput S., Mandal M. Antitumor promoting potential of selected phytochemicals derived from spices: A review. *Eur. J. Cancer Prev.* **2012**, *21*(2), 205–215.

Warrier P.K., Nambiar V.P.K., Ramankutty C. (eds.) *Indian Medicinal Plants*, vol. 3. Orient Longman, Chennai, India, **1995**, 50–53.

9.9 FENUGREEK

Scientific name: *Trigonella foenum-graecum* L.
Family: Fabaceae
Hindi: Meti
Sanskrit: Methika
Tamil: Venthayam

9.9.1 Origin and Distribution

Indigenous to the countries on the eastern shores of the Mediterranean, it is now cultivated in many parts of the world, mainly in India, Nepal, Pakistan, Bangladesh, Turkey, China, northern Africa, France, Spain, North, Central and South America respectively. The seeds of fenugreek have been used medicinally all through the ages and were held in high repute among the Egyptians, Greeks, and Romans for medicinal and culinary purposes. The charred remains of fenugreek seeds have been found in archeological sites in Iraq, in the remains of the Bronze Age city of Lachish in Israel, and in the tomb of Tutankhamun. Fenugreek was introduced in Europe by the Benedictine monks during the ninth century: it was one of the herbs cultivated in the gardens of monasteries.

It was introduced in India by the ancient Arab traders who first landed on the Kerala coast around the second century AD (Rajaiah, 2010). Actually, India is the largest producer and exporter of fenugreek in the world. The herb is widely cultivated in the states of Rajasthan, Gujarat, Uttaranchal, Uttar Pradesh, Madhya Pradesh, Maharashtra, Haryana, and Punjab.

9.9.2 Features

The name of the genus *Trigonella* is derived from the old Greek name, meaning "three-angled," due to the form of its corolla. *Foenum-graecum* means "Greek hay," the plant being used to scent inferior hay. It is an annual erect herb, about 30–60 cm high, with tender stalks, yellow or pale whitish flowers, and yellowish seeds. The seeds (10–20) are contained in long, narrow, sickle-like pods. They emit a peculiar odor and are used as a spice. The plant is one of the so-called "Lucky Legumes" that confer wealth, health, and protection. The seeds are probably the only spice used as a defensive weapon in ancient times. During the Roman siege of Jerusalem (AD 70), the Jewish defenders added fenugreek to the boiling oil used against the enemy soldiers to make it even more slippery (Rajaiah, 2010).

9.9.3 Properties and Uses in Ayurvedic System of Medicine

Fenugreek entered into the Ayurvedic pharmacopea quite late, several centuries after the compilation of *Charaka Samhita* and *Sushruta Samhitas*. It was mentioned for the first time by Pandit Bhava Mishra, one of the physicians of the Emperor Akbar's court, in the sixteenth-century Ayurvedic text *Bhavaprakash Niganthu*. It is called *dipani* as it will increase the appetite.

Fenugreek stokes the fire, driving out cold and damp. It targets the lower parts of the body, including the kidney, lower back, pelvic area, and legs. It is used with salt to target and warm the kidneys. The seeds reduce *kapha* and *vata*, and increase *pitta*. They help in case of asthma and sinus problems by reducing excess mucus and phlegm. They can be eaten by nursing mothers to increase milk production (Aggarwal and Yost, 2011). They are considered a natural source of iron; they contain mucilage, which is known for soothing and relaxing inflamed tissues. Fenugreek improves digestion, maintains a healthy metabolism, and increases libido. The leaves are cooling and laxative, useful to reduce *pitta*.

Parts used: Seeds, leaves
Tastes: Bitter, pungent
Quality: Light, unctuous
Potency: Hot
***Dosha* effect:** VK–, P+
Actions: Anti-inflammatory, carminative, diuretic, expectorant, febrifuge, galactogogue, laxative, uterine tonic

9.9.4 Traditional Uses and Folklore

- Avicenna (Ibn Sina), the founder of the Unani system of medicine in *The Canon of Medicine*, listed fenugreek as one of the hundred of plants with great medicinal value. Arab physicians researched its medicinal properties at The Scuola Medica Salernitana, the famous ninth-century school of medicine located in Italy (Rajaiah, 2010).
- Farmers noticed that sick cattle would eat fenugreek plants even when they would not eat anything else. Therefore, it began to be used as a digestive aid and a laxative. Fenugreek seeds contain a lot of bulk and mucilage, and, when mixed with water or saliva, become gelatinous and ease sluggish bowels.
- The fiber in fenugreek fills the stomach, even when consumed in a little amount, so it is helpful in losing weight. For this purpose, it is recommended

to soak a few fenugreek seeds in water and chew them in the morning on an empty stomach.

- Chewing fenugreek seeds or sprouts helps remove bad breath Moreover, this culinary spice can be added in pickles, marinades, salads, sandwiches, pasta dishes, and soups. Including fenugreek leaves in the diet helps reduce anemia. However, regular consumption of this herb may make your sweat smell like maple syrup.
- The traditional Indian sweet *laddu* can be made with the roasted and powdered seeds of fenugreek. This preparation is an excellent tonic, beneficial for general debility and other diseases related to *vata* (Paranjipe, 2005).
- The vegetable prepared from the delicate leaves of fenugreek prevents constipation. It also purifies the blood and gives strength. It is also beneficial for people suffering from piles.
- As gruel, it will increase milk flow and promote hair growth. The sprouts are excellent for indigestion, hypofunction of the liver, and seminal debility (Frawley and Lad, 1994). The intake of a teaspoon of seeds daily is recommended to patients with diabetes. The seeds can be powdered and taken with milk (Bakhru, 2004).

Precautions: Fenugreek may cause an allergic reaction in susceptible individuals. There are reported cases of gastrointestinal discomfort and nausea; therefore, fenugreek should be used with caution. It is better to avoid fenugreek in case of hypertension or heart disease.

During pregnancy, fenugreek should be used with caution because it may cause abortion. It is not advisable for high *pitta* condition (Frawley and Lad, 1994). It may increase the risk of bleeding when taken with anticoagulants.

9.9.5 In the Kitchen

Jews customarily eat fenugreek during the meal of the first and/or second night of Rosh Hashana (The New Year).

One of the best ways to eat fenugreek is in the form of fresh sprouts. The sprout can be cooked for few minutes or added raw to the salad. Fresh tender pods and leaves have been eaten as curried vegetable since ancient times in India and Egypt. As a spice, the seeds add nutritive value and flavor to food. For this reason, fenugreek is of considerable importance in those countries where vegetarian diet is very common for cultural and religious reasons (Pruthi, 1998). The seeds are used in the pulse soups (*dal* and *sambar*), in *dosa*s, chutneys, and pickles. Fenugreek is one of the five ingredients of the famous spice combination *panch puran* and of curry powder. The seeds are often roasted to reduce bitterness and enhance taste. They combine well with fennel, cumin, coriander, turmeric, and ginger. After roasting, they are easily ground. A small amount will complement many other spices, but too much can be overpowering. If the seeds are required as part of a curry paste, they can be soaked overnight to swell and soften, to be easily mixed with the others. In India, the roasted ground seeds are infused for a coffee substitute or adulterant. A tea can be made by infusing a teaspoon of seeds with two cups of water for 5 minutes.

In Egypt, fenugreek seeds are prepared as tea, by being boiled and then sweetened. This is a popular winter drink served in coffee shops. In the United States, where maple syrup is popular, fenugreek is widely used as a substitute for maple syrup flavoring.

The seeds are an ingredient of the Middle Eastern confection *halva*. Flour mixed with ground fenugreek makes a spicy variety of bread.

The ground seeds are also utilized to an enormous extent in the manufactures of condition powders for horses and cattle.

9.9.6 Chemical Constituents and Uses as Functional Food

Fenugreek has three culinary uses: as an herb (dried or fresh leaves), as a spice (seeds), and as a vegetable (fresh leaves, sprouts). Sotolon is the chemical responsible for fenugreek's distinctive sweet smell. The distinctive cuboid-shaped, yellow-to-amber-colored fenugreek seeds are frequently encountered in the cuisines of the Indian subcontinent.

Fenugreek has a long history of medical uses in Ayurvedic and Chinese medicine, and has been used for numerous indications, including labor induction, aiding digestion, and as a general tonic to improve metabolism and health (Basch *et al.*, 2003).

The potential health benefits of legume sources that rarely appear in Western diets are often overlooked as functional foods. Fenugreek (*Trigonella foenum-graecum*) and isolated fenugreek fractions have been shown to act as hypoglycemic and hypocholesterolemic agents in both animal and human studies (Ballali and Lanciai, 2012; Roberts, 2011). The unique dietary fiber composition and high saponin content in fenugreek appear to be responsible for these therapeutic properties (Madar and Stark, 2002).

There is evidence that *T. foenum-graecum* L. (fenugreek) and its components are beneficial in the prevention and treatment of diabetes and central nervous system diseases. The pharmacological activities of trigonelline, a major alkaloid component of fenugreek, have been more thoroughly evaluated than those of fenugreek's other components, especially with regard to diabetes and central nervous system diseases. Trigonelline has hypoglycemic, hypolipidemic, neuroprotective, antimigraine, sedative, memory-improving, antibacterial, antiviral, and antitumor activities, and it has been shown to reduce diabetic auditory neuropathy and platelet aggregation. It acts by affecting β-cell regeneration, insulin secretion, activities of enzymes related to glucose metabolism, ROS, axonal extension, and neuron excitability. However, further studies are necessary to warrant the pharmacological activities of trigonelline and define the exact mechanism (Zhou *et al.*, 2012).

Numerous lactation-stimulating herbals have been identified in the literature with varying degrees of evidence, mostly anecdotal. Use of torbangun, milk thistle, and fenugreek may correlate with increased milk supply. Evidence regarding adverse effects, pharmacodynamic properties, and pharmacokinetic effects remains scarce. Despite the fact that postpartum women may turn to herbal galactagogues, scant clinical evidence exists to justify their effectiveness. Further clinical trials are needed to substantiate these findings (Mortel and Mehta, 2013; Zapantis *et al.*, 2012).

Capsicum fenugreek leaves curry, a recipe from Vaidya Kalyan Chakravarty

Ingredients

3 capsicums (110 g)
2 potatoes (125 g)
1 onion (70 g)
50 g fenugreek leaves
250 g tomatoes
2 tsp of red chili powder
1/4 tsp of turmeric powder
4–5 tsp of *ghee*
Salt to taste

Preparation

Wash, deseed, and cut the capsicums into medium-sized pieces. Take them into a steel bowl, add 10 ml water and cook them until they are 60% cooked. Wash, peel, and cut the potatoes into medium-sized pieces. Cook the potato pieces with 20 ml of water on high power until they are three-fourths cooked. The potato pieces should be cooked, but they should retain their shape and should not be mushy. Take the fenugreek leaves into a steel bowl and add 15 ml water. Cook the fenugreek leaves on high power for 1 minute. Wash and chop the tomatoes into small pieces. Cook the tomato pieces on high power for 3 minutes. Peel the onion and cut it into medium-sized pieces.

Heat a pan and add ghee to it. When the oil is hot, add the onion pieces and fry stirring well. When the onions are transparent, add the cooked capsicum and potato pieces. Add salt and mix well. Now add the turmeric powder and mix well. Cook the vegetables covered on medium heat. After 3 minutes of cooking, remove the lid and mix. Add the cooked fenugreek leaves and mix well. Fry the vegetables till dry. Now add the cooked tomato pieces and mix well. After adding tomatoes, the curry appears gravy. Fry the curry till all the moisture is absorbed. Then add the red chili powder and fry for a minute. Now the curry is ready to serve.

It is excellent with *roti*, *paratha*, or even rice.

Fenugreek lentil curry, a recipe from Vaidya Kalyan Chakravarty

Ingredients

1 cup of fenugreek leaves (washed and picked)
1 cup of lentil
1/3 tsp of turmeric
2 whole dry red chilies (deseed and tear into pieces)
5 garlic cloves slightly crushed
1 tsp of chopped ginger
10 curry leaves
1 tsp of cumin seeds
1/2 tsp of mustard seeds
1 tsp of salt

1 tsp of ghee
Chopped coriander leaves for garnish
2 cups of water

Preparation

For tempering, heat ghee in a sauce pan, add the mustard seeds, and let them splutter. Now add the cumin seeds, slightly crushed garlic, whole red chili pieces, and curry leaves, and fry for a minute. Now add the chopped ginger and green chilies, and fry for 2 minutes on medium heat. Remove from heat. Add this and turmeric to the lentils and add 2 cups of water in a pressure cooker. Now add the washed and picked fenugreek leaves on top and pressure cook.

When it is ready, add salt and mash with ladle till well combined. Garnish with coriander leaves.

Serve hot with rice and ghee to have a complete nutritious meal.

REFERENCES

Aggarwal B.B., Yosk D. *Healing Spices*. Sterling Publishers, New York, **2011** (Italian version, Armenia, 173–180).

Bakhru H.K. *Indian Spices & Condiments as Natural Healers*. Jaico Publishing House, Mumbai, India, **2004**, 80.

Ballali S., Lanciai F. Functional food and diabetes: A natural way in diabetes prevention? *Int. J. Food Sci. Nutr.* **2012**, *63*(Suppl 1), 51–61.

Basch E., Ulbricht C., Kuo G., Szapary P., Smith M. Therapeutic applications of fenugreek. *Altern. Med. Rev.* **2003**, *8*(1), 20–27.

Frawley D., Lad V. *The Yoga of Herbs*. Motilal Banarsidass Publshers, Delhi, India, **1994**, 118.

Madar Z., Stark A.H. New legume sources as therapeutic agents. *Br. J. Nutr.* **2002**, *88*(Suppl 3), S287–S292.

Mortel M., Mehta S.D. Systematic review of the efficacy of herbal galactogogues. *J. Hum. Lact.* **2013**, *29*(2), 154–162.

Paranjipe P. *Indian Medicinal Plants. Forgotten Healers*. Chaukhamba Sanskrit Pratishthan, Delhi, India, **2005**, 171–172.

Pruthi J.S. *Spices and Condiments*. National Book Trust, Delhi, India, **1998**, 131–136.

Rajaiah R. *How the Banana Goes to Heaven and Other Secrets of Health from the Indian Kitchen*. Westland, Chennai, India, **2010**, 80–85.

Roberts K.T. The potential of fenugreek (*Trigonella foenum-graecum*) as a functional food and nutraceutical and its effects on glycemia and lipidemia. *J. Med. Food.* **2011**, *14*(12), 1485–1489.

Zapantis A., Steinberg J.G., Schilit L. Use of herbals as galactagogues. *J. Pharm. Pract.* **2012**, *25*(2), 222–231.

Zhou J., Chan L., Zhou S. Trigonelline: A plant alkaloid with therapeutic potential for diabetes and central nervous system disease. *Curr. Med. Chem.* **2012**, *19*(21), 3523–3531.

9.10 GARLIC

Scientific name: *Allium sativum*
Family: Liliaceae
Hindi: Lasun
Sanskrit: Rasonah
Tamil: Vellai puntu

9.10.1 Origin and Distribution

Garlic probably originated in Central Asia, where it has been used since Neolithic times. From there, it has spread all over the world, and it is currently cultivated in temperate and tropical regions as a culinary and medicinal plant. Many different varieties have been developed to suit different climates. It is widely grown in the Latin countries bordering on the Mediterranean, as well as in India, the Philippines, China, South Korea, Ethiopia, Egypt, Kenya, Russia, the United States, Brazil, and Mexico (Bakhru, 1993). China is by far the largest producer of garlic, with about 77% of the world output. India (4.1%) and South Korea (2%) follow, with Egypt and Russia (1.6%) tied in fourth place and the United States (where garlic is grown in every state except Alaska) in sixth place (1.4%).

The name is of Anglo-Saxon origin, being derived from *gar* ("a spear") and *lac* ("a plant"), in reference to the shape of its leaves.

Garlic was known to the Chinese and Egyptians as early as in 3000 BC; it was largely consumed by the ancient Greeks and Romans. In Greece, it was placed on the piles of stones at crossroads as an offering for Hecate; according to Pliny, garlic and onion were invocated as deities by the Egyptians at the taking of oaths. Cloves of garlic have been discovered in Egyptian burials (Manniche, 2006). Herodotus reports that garlic was widely consumed by the Egyptians, especially by the pyramid builders.

Regarded as a force for both good and evil, it is thought to stimulate and warm the body, and act as an aphrodisiac. It is also *tamasic*, and for this reason, monks and priests in India usually avoid garlic (Frawley and Lad, 1994).

> In olden days, the king of the demons himself drank the elixir of immortality, which had come from the churning. Then Lord Vishnu cut off his head. When the head was severed, its windpipe remained attached to it. The drops that fell to earth were the first origin of garlic here
>
> **Wujastyk, 1998**

Garlic was born from the nectar of immortality, but it was contaminated by the contact with a demon: It is a great medicine, but it is also somehow impure for Hindu people. The Bower manuscript reports that Brahmins, who normally cannot eat garlic, may avoid this prohibition by feeding it to a cow and reaping its benefits just by consuming her milk (Wujastyk, 1998).

In an eastern Islamic myth, garlic arose from the left foot of Satan as he stepped out from the Garden of Eden. In connection with the odor associated with garlic, Islam views eating garlic and subsequently going to the mosque as inappropriate. "Whoever has eaten (garlic) should not approach our mosque," indicated Muhammad.

It was thanks to the virtues of the "yellow garlic" that Ulysses owed his escape from being changed by Circe into a pig, like his companions. In European Folklore, garlic is a symbol of protection from evil spirits, werewolves, and vampires, perhaps owing to its reputation as a potent medicine with antibacterial properties. In ancient times, it was also thought that hanging garlic bulbs on doors would check the spread of diseases such as smallpox.

9.10.2 Features

Garlic is an erect perennial herb with underground compound bulbs covered with dry membranous scales, with a pungent odor. The long, sword-shaped leaves grow from the bulb beneath the surface of the soil. The flowers rise in a cluster directly from the bulb and are small and white (Warrier *et al.*, 1994).

9.10.3 Properties and Uses in Ayurvedic System of Medicine

The Sanskrit name *rasonah* literally means "lacking one taste" as garlic contains all the tastes except sour (Paranjipe, 2005). Pungency resides in the bulb, bitterness in the leaves, salinity at the crown, astringency in the stem, and sweetness in the seeds.

It is a powerful *rasayana* mentioned in the ancient Sanskrit texts and recommended mainly for *vata* alleviation and also, to a lesser degree, for *kapha*. Generally, it should be used with ghee in *vata* diseases, with sugar in *pitta* diseases, and with honey in *kapha* diseases (Paranjipe, 2005).

It is also a powerful detoxifier useful in innumerable diseases. The *Sushruta Samhita* (Sutrasthana 46) has mentioned garlic as beneficial for the eyes, voice, and complexion of the skin.

It is very effective in the digestive disorders of *vata–kapha* origin; it augments the appetite, has a vermicide activity (Venkat *et al.*, 2012), and, if fried with ghee, stimulates the liver and has a mild laxative action. According to Ayurveda, garlic is one of the most effective antimicrobial herbs with antibacterial, antifungal, and antiviral properties. It is a carminative and effective in rheumatic pains, hypertension, and respiratory diseases, such as cough, asthma, hoarseness of voice, and even tuberculosis.

Regular use of garlic during the monsoon season protects the body from epidemics and endemics. Its use is also recommended during the winter season.

Parts used: Bulb
Tastes: Bitter, pungent, sweet, salty, astringent
Quality: Unctuous, heavy, sharp
Potency: Hot
***Dosha* effect:** VK–, P+
Actions: Aphrodisiac, rejuvenating, anthelmintic, carminative, stimulant, expectorant, antispasmodic

Preparations: Garlic has been used in hundreds of Ayurvedic preparations, and one of the best known is *gandhaka vati*. It contains ginger, black pepper, salt, cumin seeds, and garlic, and is mixed with lemon juice. Two of those pills after both principal meals relieve gas, dyspepsia, constipation, heartburn, and hyperacidity. This preparation also kills the parasites and, if taken regularly for a month or two, gives a feeling of general well-being. In a few cases, even reduction of weight has been noticed, with the disappearance of accumulation of fat on the abdominal wall.

9.10.4 Traditional Uses and Folklore

- Garlic is fully described by Pliny: Mixed with honey, it was used as an ointment for dog bites; raw garlic was prescribed as a remedy for asthma; fresh garlic with coriander was added to wine as a purgative and an aphrodisiac. With soft cheese or in gruel of peas or beans, it was used to relieve hoarseness.
- In Siddha medicine, garlic juice has been used to alleviate sinus problems.
- In Unani medicine, an extract is prepared from the dried bulb, which is inhaled to promote abortion or taken to regulate menstruations. Unani physicians also use garlic to treat paralysis, forgetfulness, tremor, colic pains, internal ulcers, and chronic fevers (Nadkarni, 2002).
- Usually, Indian home remedies are made by extracting the juice or pulping the bulb to a paste. Whooping cough in children has been treated by administering a drink made with hot water extract of the dried bulb mixed with honey or by wearing a necklace of bulbs. Hot water extracts are also taken to kill intestinal worms.
- For rheumatic disorders, it is recommended to take a garlic capsule thrice daily for 2 months (Sairam, 1998).

- Regular use of garlic is a traditional home remedy for blood infections, hypertension, and diabetes. For cold and excess of phlegm, two garlic cloves can be crushed and boiled in a cup of water along with half teaspoon of turmeric powder.
- Milk boiled with garlic bulbs gives strength, and increases vitality and vigor. This milk is called *rason siddha dugdha*.
- Garlic has also been used since ancient times in traditional Indian veterinary medicine to treat tetanus and inflammatory disorders of the lungs.

Precautions: Garlic is a very warming spice, so people with hyperacidity and high *pitta* should avoid it. When eaten at empty stomach, it may cause irritation (Sairam, 1998).

9.10.5 In the Kitchen

The pungent, spicy flavor and distinct aroma of garlic is a fundamental component of many different recipes all over the world. In the West, garlic is used in garlic bread and pasta dishes, often paired with onion, tomato, butter, and parsley.

Although highly regarded as a medicine by Ayurveda, in ancient times garlic was avoided in cookery by the Buddhists and Jains, and by the high-born Hindus. Until the seventh century AD, when the Chinese pilgrim Xuan Zang visited India, the use of garlic as food was quite unknown. This attitude changed during the period of Muslim rule when the use of garlic, ginger, and onion became a must in the South Asian cuisine. In India, it is widely used in the preparation of curries, vegetables, pickles, and chutneys.

In his *Heart of Medicine*, Vagbhata recommends the use of garlic especially during the cold season and suggests a recipe for soaking peeled garlic overnight in wine. This pickled garlic is then crushed, filtered, and added to wine, milk, or meat broths (Wujastyk, 1998).

In Korea, heads of garlic are fermented at high temperature, making a sweet, syrupy product known as black garlic.

Garlic should not be cooked for too long as it can become bitter. Often garlic heads with small cloves are stronger than larger ones.

9.10.6 Chemical Constituents and Uses as Functional Food

Garlic was in use at the beginning of recorded history and was found in Egyptian pyramids and ancient Greek temples. There are biblical references to garlic. Ancient medical texts from Egypt, Greece, Rome, China, and India each prescribed medical applications for garlic. In many cultures, garlic was administered to provide strength and increase work capacity for laborers. Hippocrates, the revered physician, prescribed garlic for a variety of conditions. Garlic was given to the original Olympic athletes in Greece, as perhaps one of the earliest "performance-enhancing" agents. It is of interest that cultures that developed without contact with one another came to similar conclusions about the efficacy of garlic (reviewed in Rivlin, 2001). In human studies, it has been confirmed that garlic produces symptomatic improvement in persons with physical fatigue, systemic fatigue due to cold, or lassitude of indefinite cause, suggesting that garlic can resolve fatigue through a variety of actions. Currently available data strongly suggest that garlic may be a promising antifatigue agent and that further studies to elucidate its application are warranted (Morihara *et al.*, 2007).

Extracts of fresh garlic that are aged over a prolonged period to produce aged garlic extract (AGE) contain antioxidant phytochemicals that prevent oxidant damage. These include unique water-soluble organosulfur compounds, lipid-soluble organosulfur components, and flavonoids, notably allicin and selenium. Long-term extraction of garlic (up to 20 months) ages the extract, creating antioxidant properties by modifying unstable molecules with antioxidant activity, such as allicin, and increasing stable and highly bioavailable water-soluble organosulfur compounds, such as *S*-allylcysteine and *S*-allylmercapto-L-cysteine. AGE exerts an antioxidant action by scavenging ROS, enhancing the cellular antioxidant enzymes such as superoxide dismutase, catalase, and glutathione peroxidase, and increasing glutathione in the cells. It inhibits lipid peroxidation, reduces ischemic/reperfusion damage, and inhibits oxidative modification of LDL, thus protecting endothelial cells from the injury by the oxidized molecules, which contributes to atherosclerosis. It inhibits the activation of the oxidant-induced transcription factor, NF-κB, which has clinical significance in human immunodeficiency virus gene expression and atherogenesis. It protects DNA against free radical-mediated damage and mutations, inhibits multistep carcinogenesis, and defends against ionizing radiation and UV-induced damage, including protection against some forms of UV-induced immunosuppression. It may have a role in protecting against loss of brain function in aging and possess other antiaging effects, as suggested by its ability to increase cognitive functions, memory, and longevity in a senescence-accelerated mouse model. It has been shown to protect against the cardiotoxic effects of doxorubicin, an antineoplastic agent used in cancer therapy and against liver toxicity caused by carbon tetrachloride (an industrial chemical) and acetaminophen, an analgesic. Substantial experimental evidence shows the ability of AGE to protect against oxidant-induced disease, acute damage from aging, radiation and chemical exposure, and long-term toxic damage. Although additional observations are warranted in humans, compelling evidence supports the beneficial health effects attributed to AGE, that is, reducing the risk of cardiovascular disease, stroke, cancer, and aging, including the oxidant-mediated brain cell damage that is implicated in Alzheimer's disease (reviewed in Banerjee *et al.*, 2003; Borek, 2001; Capasso, 2013, and Rahman, 2003).

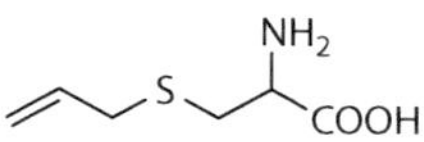

S-Allylcysteine

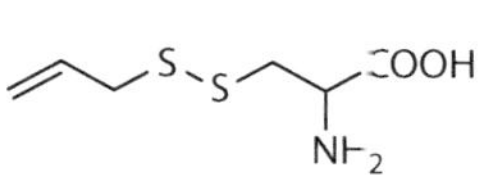

S-Allylmercaptocysteine

AGE has been shown previously to have moderate cholesterol-lowering and blood pressure-reducing effects. It exerts selective inhibition of platelet aggregation and adhesion, platelet functions that may be important for the development of cardiovascular events such as myocardial infarction and ischemic stroke (Chauhan, 2005; Steiner and Li, 2001). Human studies are also available where hypoglycemic effect of garlic is reported.

Garlic and its preparations have been widely recognized as agents for prevention and treatment of cardiovascular and other metabolic diseases, atherosclerosis, hyperlipidemia, thrombosis, hypertension, and diabetes (Padiya and Banerjee, 2013). Effectiveness of garlic in cardiovascular diseases was more encouraging in experimental studies, which prompted several clinical trials. Though many clinical trials showed a positive effect of garlic on almost all cardiovascular conditions mentioned earlier, however, a number of negative studies have recently cast doubt on the efficacy of garlic, especially its cholesterol-lowering effect of garlic. This is due to the different compositions of the preparations used (Banerjee and Maulik, 2002). Based on some double-blind, randomized, placebo-controlled clinical trials that denied the hypolipidemic effects of garlic, some meta-analyses concluded that garlic did not possess beneficial effects for hyperlipidemia. However, we should not ignore the abundant supporting data in the literature. It should be noted that the doses of garlic used in clinical trials were usually far lower than those used in animal studies, which might cover its potential effects. The type of the garlic products may be another important factor responsible for the conflicting outcomes, as different garlic products are composed of different organosulfur compounds. In addition, the biological availability of garlic products is of importance, which was omitted in many studies. Moreover, some studies indicated that different people might have a different response to garlic, and thus garlic may be more beneficial for some specific groups. Collectively, it may be inappropriate to draw a conclusion that garlic does not benefit for hyperlipidemia. Future studies with larger samples are needed to further clarify the effects of garlic used at higher but nontoxic doses on specific groups (Zeng *et al.*, 2013).

A number of studies have demonstrated the chemopreventive activity of garlic by using different garlic preparations, including fresh garlic extract, AGE, garlic oil, and a number of organosulfur compounds derived from garlic. The chemopreventive activity has been attributed to the presence of organosulfur compounds in garlic. How this is achieved is not fully understood, but several modes of action have been proposed. These include its effect on drug-metabolizing enzymes, antioxidant properties, and tumor growth inhibition. Most of these studies were carried out in the animal models. Also, recent research has focused on the antimutagenic activity of garlic. Recently, it has been observed that AGE, but not the fresh garlic extract, exhibited a radical scavenging activity. The two major compounds in AGE, *S*-allylcysteine and *S*-allylmercapto-L-cysteine, had the highest radical-scavenging activity. In addition, some organosulfur compounds derived from garlic, including *S*-allylcysteine, have been found to retard the growth of chemically induced and transplantable tumors in several animal models. Therefore, the consumption of garlic may provide some kind of protection from cancer development (Powolny and Singh, 2008; Shukla and Kalra, 2007; Thomson and Ali, 2003).

The reputation of garlic as an effective remedy for tumors extends back to the Egyptian Codex Ebers of 1550 BC. Several garlic compounds, including allicin and its corresponding sulfide, inhibit the proliferation and induce apoptosis of several human non-leukemia malignant cells, including the breast, bladder, colorectal, hepatic, prostate cancer, lymphoma, and skin tumor cell lines. Ajoene (4,5,9-trithiadodeca-1,6,11-triene-9-oxide) is a garlic-derived compound produced most efficiently from pure allicin and has the advantage of a greater chemical stability than allicin. Several clinical trials and *in vitro* studies of ajoene have demonstrated its best-known antithrombosis, antimicrobial, and cholesterol-lowering activities. Recently, topic application of ajoene has produced significant clinical response in patients with skin basal cell carcinoma. The two key antileukemia biological actions of ajoene were the inhibition of proliferation and the induction of apoptosis (Hassan, 2004). Formation of ajoene occurs when allicin is dissolved in various solvents, including edible oils. Ajoene is also found in garlic extract, and it is most stable and most abundant in macerate of garlic (chopped garlic in edible oil).

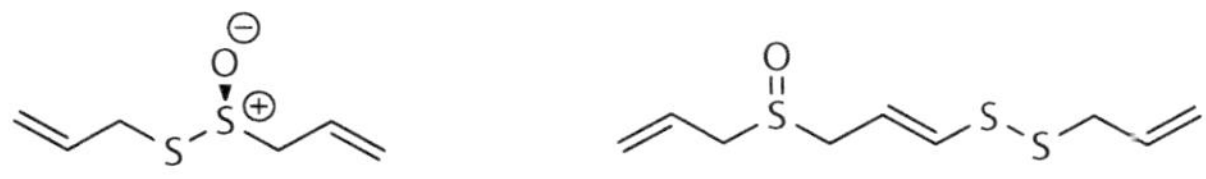

Allicin Ajoene

S-Alk(en)yl cysteine sulfoxides are odorless, nonprotein sulfur amino acids typically found in members of the family Alliaceae and are the precursors to the lachrymatory and flavor compounds found in the agronomically important genus *Allium*. Traditionally, *Allium* species, particularly the onion (*Allium cepa*) and garlic (*A. sativum*), have been used for centuries in European, Asian, and American folk medicines for the treatment of numerous human pathologies; however, recently, any significant progress has been made in determining their mechanisms of action. Indeed, our understanding of the role of *Allium* species in human health undoubtedly comes from the combination of several academic disciplines, including botany, biochemistry, and nutrition. During tissue damage, *S*-alk(en) yl cysteine sulfoxides are converted to their respective thiosulfinates or propanethial-*S*-oxide by the action of the enzyme alliinase (EC 4.4.1.4). Depending on the *Allium* species, and under differing conditions, thiosulfinates can decompose to form additional sulfur constituents, including diallyl, methyl allyl, and diethyl mono-, di-, tri-, tetra-, penta-, and hexasulfides; the vinyldithiins; and (*E*)- and (*Z*)-ajoene. Recent reports have shown that onion and garlic extracts, along with several principal sulfur constituents, can induce phase II detoxification enzymes such as glutathione-*S*-transferases (EC 2.5.1.18) and quinone reductase NAD(P) H:(quinone acceptor) oxidoreductase (EC 1.6.99.2) in mammalian tissues, as well as influence cell cycle arrest and apoptosis in numerous *in vitro* cancer cell models (Rose *et al.*, 2005).

Compounds in garlic work synergistically to produce various effects, but, because of garlic's chemical complexity, processing methods yield preparations with differing efficacy and safety. Although thiosulfinates such as allicin have been long misunderstood to be active compounds due to their characteristic odor, it is not necessary for garlic preparations to contain such odorous compounds to be effective, and they decompose and disappear during any processing. Garlic exhibits hypolipidemic, antiplatelet, and procirculatory effects. It prevents cold and flu symptoms through immune enhancement and demonstrates anticancer and chemopreventive activities. In addition, AGE possesses hepatoprotective, neuroprotective, and antioxidative activities, whereas other preparations may stimulate oxidation. Additional effects may be caused by *S*-allylcysteine, *S*-allylmercaptocysteine, saponins, *N*-α-fructosyl arginine, and other substances formed during a long-term extraction process. Although not all of active ingredients of garlic are known, and allicin-like transient components are not directly active, ample research suggests that an allicin-free garlic preparation that is standardized with a bioavailable component such as *S*-allylcysteine is active, and various effects of garlic may be attributed to it. Furthermore, various chemical constituents in garlic products, including nonsulfur compounds such as saponins, may contribute to the essential biological activities of garlic. Further studies are needed to confirm their bioavailability and associated activities (Amagase, 2006).

With the use of the US Food and Drug Administration (FDA)'s evidence-based review system for the scientific evaluation of health claims, 19 human studies were identified and reviewed to evaluate the strength of the evidence that supports a relation between garlic intake and reduced risk of different cancers with respect to food labeling. There was no credible evidence to support a relation between garlic intake and reduced risk of gastric, breast, lung, or endometrial cancer. Very limited evidence supported a relation between garlic intake and reduced risk of colon, prostate, esophageal, larynx, oral, ovary, or renal cell cancer (Kim and Kwon, 2009).

Two-year treatment with allicor (garlic powder) has a direct antiatherosclerotic effect on carotid atherosclerosis in asymptomatic men. Thus, basic findings were successfully translated into clinical practice. Because of this translation, a novel approach to antiatherosclerotic therapy was developed. The clinical trial confirmed the efficacy of both the novel approaches (Orekhov, 2013).

Bengali apple chutney, a recipe from Beeton (*Household Management*, 1866), excellent with basmati rice:

Ingredients

1 1/2 lb moist sugar
3/4 lb salt
1/4 lb garlic
1/4 lb onions
3/4 lb powdered ginger
1/4 lb dried chilies
3/4 lb dried mustard seeds
3/4 lb stoned raisins
2 bottles of best vinegar
30 large, unripe, sour apples

Preparation

Make syrup by boiling sugar and water. Pound finely the garlic, onions, and ginger in a mortar. Wash the mustard seeds in cold vinegar and dry them in the sun. Peel, core, and slice the apples, and boil them in a bottle with a half of the vinegar. When all this is done, and the apples are quite cold, put them into a large pan and gradually mix the whole of the rest of the ingredients, including the remaining half-bottle of vinegar. Stir well until the whole is thoroughly blended, and then put into bottles for use. Tie a piece of wet bladder over the mouths of the bottles, after which they are well corked.

A recipe from Vaidya Kalyan Chakravarty for heart complaints, high cholesterol, and arteriosclerosis

Ingredients

1 cup of fresh garlic juice
1 cup of fresh ginger juice
1 cup of fresh lemon juice
1 cup of green apple vinegar
3 cups of honey

Preparation

Take fresh garlic juice and add fresh ginger juice, fresh lemon juice, and green apple vinegar. Boil them on medium flame till it reduces to three cups.

Now let it cool down, add honey, and store it in a glass container.

Dose: 1 tbsp morning and evening before food.

REFERENCES

Amagase H. Clarifying the real bioactive constituents of garlic. *J. Nutr.* **2006**, *136*(3 Suppl), 716S–725S.

Bakhru H.K. *Herbs That Heal*. Orient Paperbacks, Delhi, India, **1993**, 91–96.

Banerjee S.K., Maulik S.K. Effect of garlic on cardiovascular disorders: A review. *Nutr. J.* **2002**, 1, 4. doi:10.1186/1475-2891-1-4.

Banerjee S.K., Mukherjee P.K., Maulik S.K. Garlic as an antioxidant: The good, the bad and the ugly. *Phytother. Res.* **2003**, *17*(2), 97–106.

Borek C. Antioxidant health effects of aged garlic extract. *J. Nutr.* **2001**, *131*(3), 1010S–1015S.

Capasso A. Antioxidant action and therapeutic efficacy of *Allium sativum* L. *Molecules* **2013**, *18*(1), 690–700.

Chauhan N.B. Multiplicity of garlic health effects and Alzheimer's disease. *J. Nutr. Health Aging* **2005**, *9*(6), 421–432.

Frawley D., Lad V. *The Yoga of Herbs*. Motilal Banarsidas Publishers, Delhi, India, **1994**, 119–120.

Hassan H.T. Ajoene (natural garlic compound): A new anti-leukaemia agent for AML therapy. *Leuk. Res.* **2004**, *28*(7), 667–671.

Kim J.Y., Kwon O. Garlic intake and cancer risk: An analysis using the Food and Drug Administration's evidence-based review system for the scientific evaluation of health claims. *Am. J. Clin. Nutr.* **2009**, *89*(1), 257–264.

Manniche L. *An Ancient Egyptian Herbal*. The American University in Cairo Press, Cairo, Egypt, **2006**, 76–77.

Morihara N., Nishihama T., Ushijima M., Ide N., Takeda H., Hayama M. Garlic as an anti-fatigue agent. *Mol. Nutr. Food Res.* **2007**, *51*(11), 1329–1334.

Nadkarni K.M. *Indian Materia Medica*, vol. I. Popular Prakashan, Mumbai, India, **2002**, 65–71.

Orekhov A.N. Direct anti-atherosclerotic therapy; development of natural anti-atherosclerotic drugs preventing cellular cholesterol retention. *Curr. Pharm. Des.* **2013**, *19*(33), 5909–5928.

Padiya R., Banerjee S.K. Garlic as an anti-diabetic agent: Recent progress and patent reviews. *Recent Pat. Food Nutr. Agric.* **2013**, *5*(2), 105–127.

Paranjipe P. *Indian Medicinal Plants. Forgotten Healers.* Chaukhamba Sanskrit Pratishthan, Delhi, India, **2005**, 220–222.

Powolny A.A., Singh S.V. Multitargeted prevention and therapy of cancer by diallyl trisulfide and related *Allium* vegetable-derived organosulfur compounds. *Cancer Lett.* **2008**, *269*(2), 305–314.

Rahman K. Garlic and aging: New insights into an old remedy. *Ageing Res. Rev.* **2003**, *2*(1), 39–56.

Rivlin R.S. Historical perspective on the use of garlic. *J. Nutr.* **2001**, *131*(3), 951S–954S.

Rose P., Whiteman M., Moore P.K., Zhu Y.Z. Bioactive *S*-alk(en)yl cysteine sulfoxide metabolites in the genus *Allium*: The chemistry of potential therapeutic agents. *Nat. Prod. Rep.* **2005**, *22*(3), 351–368.

Sairam T.V. *Home Remedies*, vol. I. Penguin Books, Delhi, India, **1998**, 280.

Shukla Y., Kalra N. Cancer chemoprevention with garlic and its constituents. *Cancer Lett.* **2007**, *247*(2), 167–181.

Steiner M., Li W. Aged garlic extract, a modulator of cardiovascular risk factors: A dose-finding study on the effects of AGE on platelet functions. *J. Nutr.* **2001**, *131*(3):980S–984S.

Thomson M., Ali M. Garlic [*Allium sativum*]: A review of its potential use as an anti-cancer agent. *Curr. Cancer Drug Targets* **2003**, *3*(1), 67–81.

Venkat P., Gangadharan G.G., Lakshmithathachar M.A., Alwar M.A. (eds.) *Bhojanakutuhalam.* I-AIM, Bangalore, India, **2012**, 89.

Warrier P.K., Nambiar, V.P.K., Ramankutty C. (eds.) *Indian Medicinal Plants*, vol. 1. Orient Longman, Chennai, India, **1994**, 93–98.

Wujastyk D. *The Roots of Ayurveda.* Penguin Books, Delhi, India, **1998**, 195–206.

Zeng T., Zhang C.L., Zhao X.L., Xie K.Q. The roles of garlic on the lipid parameters: A systematic review of the literature. *Crit. Rev. Food Sci. Nutr.* **2013**, *53*(3), 215–230.

9.11 GINGER

Scientific name: *Zingiber officinale* Roscoe
Family: Zingiberaceae
Hindi: Adarak
Sanskrit: Ardrakam, Shunti
Tamil: Inji

9.11.1 Origin and Distribution

Ginger is one of the five most important Indian spices and one of the earliest known in Europe. It probably originated in India as part of the ground flora of tropical lowland forests, where some of its wild relatives can still be found in the Western Ghats (Warrier *et al.*, 1996). It was introduced in China at a very early date (400 BC).

The spice dates back over 3000 years to the Sanskrit word *srngaveram*, meaning "horn root," based on its appearance. It was called *ziggiberis* in Greek and *zinziberi* in Latin. Due to its aroma and flavor, ginger was valued more for its medicinal properties than as an ingredient in cooking by the Greeks and Romans. Galen (the Greek physician) used ginger as a purification of body and to treat conditions caused by imbalances in body (Langner *et al.*, 1998). Indians and Chinese are believed to have produced ginger as a tonic root for over 5000 years to treat many ailments. Ginger was an exceedingly important article of trade and was exported over 2000 years ago from India to the Roman Empire, where it was especially valued for its medicinal properties. Arab merchants controlled the trade of ginger from the East to the West for centuries: together with black pepper, ginger was one of the most commonly traded spices during the thirteenth and fourteenth centuries. During this time in England, the value of a pound of this spice was equivalent to the cost of a sheep.

The transport and conservation of ginger is very easy being a living rhizome. Ginger requires a warm and humid climate, and thrives well from the sea level up to an altitude of 1500 m (Vijayalakshmi *et al.*, 1997). It has been successfully propagated at Kew using internodal cuttings (Keesing, 1988).

It is now extensively cultivated in India, China, Japan, Nigeria, Sierra Leone, Australia, Jamaica, and Haiti. India is actually one of the chief ginger-producing countries, with about the 50% of the total world production. The best ginger is growing in the Western Ghats of Kerala (Pruthi, 1998).

9.11.2 Features

Ginger is a slender, perennial ryzomatous herb with stem elongated and leafy. The leaves are narrow and linear in shape; the flowers are greenish yellow in color with a purplish black tip (Vijayalakshmi *et al.*, 1997). It has a distinctive thickened, branched rhizome (underground stem), which sometimes looks somewhat like a swollen hand. The rhizome has a brown corky outer layer (usually removed before use) and a pale yellow center with a spicy lemon-like scent.

9.11.3 Properties and Uses in Ayurvedic System of Medicine

The number of Sanskrit names shows the importance of this spice: ginger is the *mahaushada*, the great medicine, and *vishwabeshaja*, the universal medicine, carrying the power of the earth's fire deeply into the tissues of the body (Tiwari, 1999).

It relieves *kapha* with honey, *pitta* with rock candy, and *vata* with rock salt. Dried ginger is hotter and drier than fresh. It has a strong action in reducing *kapha* and increasing *agni*. Fresh ginger is better for deranged *vata* (Frawley and Lad, 1994).

Ginger can treat all kinds of digestive, respiratory, and inflammatory disorders. It is also an effective cardiotonic. The hot fragrance of ginger burned as incense has a warming and mobilizing effect that lets the energy flow freely to overcome blocks and improve the capacity to make decisions (Kinkele, 2005). It is thought to awaken the body's memory of the universe's original fire (Tiwari, 1999).

Parts used: Fresh and dried rhizome
Taste: Pungent
Quality: Heavy, rough, unctuous, sharp
Potency: Hot
***Dosha* effect:** VK–, P+
Actions: The raw rhizome is acrid, carminative, laxative, and digestive. The dried rhizome is emollient, appetizer, laxative, stomachic, stimulant, aphrodisiac, expectorant, anthelmintic, and carminative.

9.11.4 Traditional Uses and Folklore

- The infusion with honey thrice a day after meals is very useful in case of painful menstruation and stoppage of the flow due to cold (Bakhru, 2004). It is an excellent home remedy also for cough and asthma (Mustaquem, 1994).

Candied ginger is commonly used for cold, cough, and loss of appetite. Ginger is also traditionally used as a general tonic in postpartum debility (Paranjipe, 2005).

- Chewing a fresh piece of ginger after meals regularly improves digestion and is a traditional home remedy against nausea. Taken with salt before meals, it is an excellent carminative (Nadkarni, 2002); ginger tea is a common home remedy for cold and influenza (Dastur, 1962).
- Pastoral communities in Tamil Nadu, the southern Indian state, use ginger as a folk medicine to treat cattle suffering from gastric upset. They prepare an herbal paste to feed the cattle mixing the ginger rhizome with black pepper, asafetida (*Ferula* species), and sweet flag (*Acorus calamus*).

Precautions: Ginger is a warming spice, so it should be used in a limited quantity during the summer. Patients suffering from high fevers, ulcers, urinary burning sensation, and inflammatory diseases may avoid the use of ginger (Paranjipe, 2005). The outer skin has to be carefully discarded before use (Sairam, 1998).

9.11.5 In the Kitchen

Ginger is one of the most commonly consumed dietary condiments in the world, with a long history of use in South Asia, in both dried and fresh forms. The Hindu epic *Mahabharata* written around the fourth century BC describes a meal where meat is stewed with ginger and other spices.

Its use as a food became much more widespread in India by the thirteenth century AD with the advent of Muslim rule. It became popular to prepare meat dishes and drinks using ginger pastes. Fruit juices, tea, buttermilk, and curd products were also spiced with ginger.

Fresh ginger is actually a key ingredient in many South Asian dishes. It is a common constituent of curry powder (Bakhru, 2004). In the north of India, garlic, ginger, and onion are often pounded to make a paste, which is then fried to form the basis of vegetable and meat dishes. Preserved ginger—pieces of ginger rhizome in sugar syrup—is a popular ingredient in sweet dishes in South Asia. Fresh ginger is often pickled in vinegar in Southeast Asia.

Until recently, most ginger exports have been in the form of dried, powdered ginger. This is the form in which ginger was used in ancient Rome and medieval Europe. In Europe, ginger was used on all kinds of food in the medieval period, but, since the eighteenth century, mainly in baked goods such as gingerbread.

Ginger can be used instead of garlic and onion even in the Mediterranean diet. The outer skin needs to be peeled off. It is excellent as a pickle with lemon; as a chutney; as a jam; as a candy; with steamed apples, cheesecakes, and ice creams; in the soup; with sea fruits and fish; and with tofu, pumpkin, and sweet potatoes.

9.11.6 Chemical Constituents and Uses as Functional Food

The chemistry of ginger has been reviewed in Orasa *et al*., (2000). The steam-distilled volatile oil of ginger represents the aroma and is present generally in the range 1.0%–2.5% in the dried rhizomes. Typical ginger is characterized by a high content of sesquiterpene hydrocarbons such as α-zingiberene and monoterpenes such as geranial, neral, and camphene. The composition of the oil is highly variable depending on the location, the drying process, and the use of fresh or dried material, as for all essential oils.

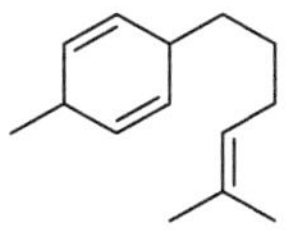

α-Zingiberene

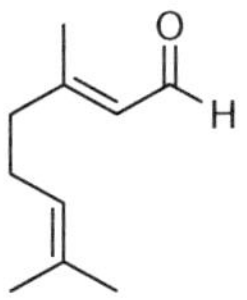

Geranial

Gingerols are the pungent principles of ginger, chemically similar to capsaicin in pepper. They are a numerous family of compounds containing the same aromatic motif, whereas the aliphatic chain of polyketide origin contains ketones, hydroxyls, and the methoxy derivatives, acetyl groups, and sulfonic acids, usually in positions 3 and 6, depending on their biogenetic origin. Shogaols are characterized by the presence of one or more double bonds in the poliketide chain.

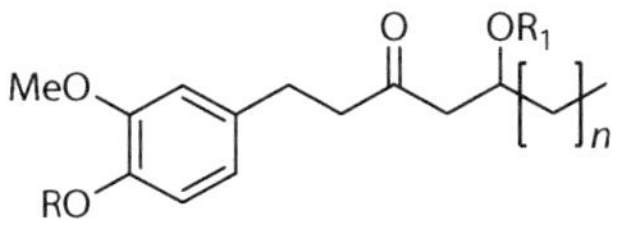

Gingerols

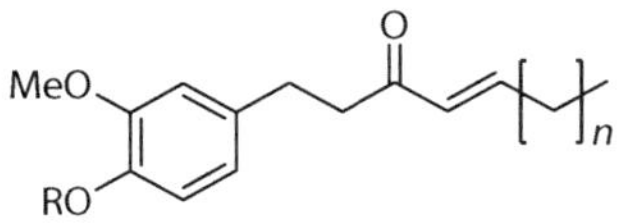

Shogaols

Diarylheptanoids (curcumin like) are also present (Jiang *et al.*, 2007; Tao *et al.*, 2008). The number of gingerols and related compounds usually averages more than 100 components (Jolad *et al.*, 2005).

Ginger and many of its chemical constituents possess many health benefits. Ginger extracts have been reported to have potential anti-inflammatory, antioxidative, and cancer chemopreventive activities (reviewed in Kubra and Rao, 2013). Phytochemicals isolated from ginger species have been documented for the treatment of chemotherapy-associated nausea, the suppression of platelet aggregation, and the inhibition of COX-2 and nitric oxide synthase. In particular, [6]-gingerol has shown antioxidant and anti-inflammatory properties to suppress cytokine formation and to promote angiogenesis. Thus, ginger and its components are affective inhibitors of the carcinogenic process.

Numerous studies in the past two decades have demonstrated that the ginger rhizome and some of its phytochemicals possess chemopreventive effects against cancers of different origins in experimental animal models. Several mechanisms are likely to account for the observed pharmacological effects—the most important being the free radical-scavenging, antioxidant, antimicrobial, and antimutagenic activities; anti-inflammatory increase in the antioxidant enzymes; modulation of phase I and II enzymes; modulation of signal transduction; transcription factors and cell cycle; and induction of selective apoptosis in neoplastic cells (reviewed in Baliga *et al.*, 2013). In spite of the number of positive results in animal results, data are still to be confirmed in human malignancies, which encounter contrasting results. This can be due to the limited number of available studies done with authenticated standardized ginger extracts. The pharmacological activity of ginger appears to be due to gingerols and shogaols, and the final ratio of these compounds is determined by many variables (location, harvesting, and processing).

Ginger shows no toxicity and is considered as "generally recognized as safe" by the US FDA. The nutraceutical properties of ginger compounds have found great interest in the food processing and pharmaceutical industries.

Due to its abundance, low cost, and safety in consumption, ginger remains a spice with tremendous potential and countless possibilities for further therapeutic applications.

Turmeric ginger elixir, a great cold and flu-preventing and immune-boosting winter drink suggested by Vaidya Kalyan Chakravarty

Ingredients

- Juice of 1 organic lemon
- 2–3 inches of fresh ginger, ginger juice, or 1 heaping tbsp of powdered ginger
- 1 tsp of turmeric
- 1 tsp of cinnamon
- 1/2 tsp of powdered clove
- 2 tsp of honey (optional)
- Basil leaves
- 1 l water

Preparation

Boil the water, add all the ingredients, and mix well. Boil again, filter, and garnish with some basil leaves. We suggest to drink it warm or to store at room temperature to keep seeping during the day.

Ginger can reduce excess *kapha* and stimulate circulation, making it the best medicine for colds and flues. Ginger, basil, cinnamon, and cloves help restore balance to *kapha* as well. Turmeric helps to keep the lungs clear of phlegm and activates the liver. Warm temperatures are important to keep both *kapha* and *vata* balanced. Honey and cinnamon strengthen the immune system and protect the body from bacterial and viral attacks.

REFERENCES

Bakhru H.K. *Indian Spices & Condiments as Natural Healers*. Jaico Publishing House, Mumbai, India, **2004**, 93–99.

Baliga M.S., Haniadka R., Pereira M.M., D'Souza J.J., Pallaty P.L., Popuri S. Update on the chemopreventive effects of ginger and its phytochemicals. *Crit. Rev. Food Sci. Nutr.* **2013**, *51*(6), 499–523.

Dastur J.F. Medicinal plants of India and Pakistan. D.B. Taraporevala sons, Mumbai, India, **1962**, 179–180.

Frawley D., Lad V. *The Yoga of Herbs*. Motilal Banarsidass Publishers, Delhi, India, **1994**, 121–122.

Jiang H., Timmermann B.N., Gang D.R. Characterization and identification of diarylheptanoids in ginger (*Zingiber officinale* Rosc.) using high-performance liquid chromatography/electrospray ionization mass spectrometry. *Rapid Commun. Mass Sp.* **2007**, *21*, 509–518.

Jolad S.D., Lantz R.C., Chen G.J., Bates R.B., Timmermann B.N. Commercially processed dry ginger (*Zingiber officinale*): Composition and effects on LPS-stimulated PGE2 production. *Phytochemistry* **2005**, *66*, 1614–1635.

Keesing J.L.S. Useful plants, in Simmonds J.B. (ed.) *Kew Gardens Book of Indoor Plants*. Philip's, Liverpool, **1988**, 186–213.

Kinkele T. *Incense and Incense Rituals*. B. Jain Publishers, Delhi, India, **2005**, 103.

Kubra I.R., Rao J.M. An impression on current developments in the technology, chemistry and biological activities of ginger (*Zingiber officinale* Roscoe). *Crit. Rev. Food Sci. Nutr.* **2013**, *52*(8), 651–688.

Langner E., Greifenberg S., Gruenwald J. Ginger: History and use. *Adv. Ther.* **1998**, *15*, 25–44.

Mustaquem M. Household remedies for Asthma. *Aryavaidyan* **1994**, *7*(3), 157–161.

Nadkarni, K.M. *Indian Materia Medica*, vol. I. Popular Prakashan, Mumbai, India, **2002**, 1308.

Orasa P., Prawat U., Pittaya T. Phytochemistry of the zingiberaceae, in Atta-ur-Rahman (ed.), *Studies in Natural Products Chemistry*. Elsevier Science Publishers, Amsterdam, the Netherlands, **2000**, 797–865.

Paranjipe P. *Indian Medicinal Plants. Forgotten Healers*. Chaukhamba Sanskrit Pratishthan, Delhi, India, **2005**, 246–247.

Pruthi J.S. *Spices and Condiments*. National Book Trust, Delhi, India, **1998** (5th edn.), 147–152.

Tao, Q.F., Xu Y., Lam R.Y.Y. *et al.* Diaryheptanoids and a monoterpenoid from the rhizomes of *Zingiber officinale*: Antioxidant and cytoprotective properties. *J. Nat. Prod.* **2008**, *71*, 12–17.

Tiwari M. *Ayurveda Secrets of Healing*. Motilal Banarsidass Publishers, Delhi, India, **1999**, 434.

Vijayalakshmi K., Subhashini B., Shivani V.K. *Plants in Pest Control. Turmeric and Ginger*. Centre for Indian Knowledge Systems, Chennai, India, **1997**, 17–19.

Warrier P.K., Nambiar V.P.K., Ramankutti C. (eds.) *Indian Medicinal* Plants, vol. 5. Orient Longman, Chennai, India, **1996**, 431–438.

9.12 PARSLEY

Scientific name: *Petroselinum crispum* Mill.
Family: Apiaceae
Hindi: Prajmoda

9.12.1 Origin and Distribution

The botanical name is a latinization of the Greek word *petroselinon* (*petra* means "rock" and *selinon* means "celery") and refers to the plant's habit of growing in rocky places.

It is native of the central Mediterranean region (southern Italy, Algeria, and Tunisia), widely cultivated in the Philippines, Malaysia, East and West Africa, Brazil, and the Caribbean. It is not commonly grown in India (Bakhru, 1993).

The ancient Greeks held parsley to be sacred, using it to not only adorn victors of athletic contests, but also decorate the tombs of their dead. The herb was dedicated to Persephone and associated with Achromorus, the Herald of Death: perhaps because of this association, they did not eat it, although they did feed their horses with it. The ancient Greek colony of Selinous in Sicily was named after wild parsley that grew abundantly there. Parsley was the symbol of the city and a leaf was depicted on Selinountian coins.

The practice of using parsley as a garnish actually has a long history that can be traced back to the civilization of the ancient Romans. The Romans used parsley as a remedy for sore eyes and as a tonic to increase the strength of their gladiators (Bakhru, 1993). Pliny complained that every sauce and salad contained it. The Romans spread parsley and soft cheese on bread, a predecessor of the modern parsley and cream cheese sandwich.

The use in the kitchen became very popular in more northern latitudes in the Middle Ages, when it was commonly grown in monasteries and Imperial gardens. Some historians credit Charlemagne with its popularization since he had it grown on his estates. The plant is now one of the most widely cultivated garden herbs, especially in California, Germany, France, and Hungary. It has been completely naturalized in various parts of England and Scotland, on old walls and rocks.

9.12.2 Features

Garden parsley is an erect, bright green, hairless, biennial, or short-lived perennial herb with a much branched stem. The leaves are large, flat, dark green, and shiny. The flowers are yellowish green or white in color. The plant normally dies after seed maturation.

There are many varieties of parsley: the most popular form is the tightly curled, green mossy variety. The Italian parsley has deeply divided flat leaves that look like the leaves of celery and has a much stronger flavor than the curly variety.

9.12.3 Properties and Uses in Ayurvedic System of Medicine

Parsley is a good herbal nutritional supplement recommended for high *kapha* and obstructed *vata* (Frawley and Lad, 1994). It is a mildly warming diuretic and helpful for draining excess *kapha* in the spring. It is also an excellent aid to digestion with a mild laxative effect: when added to the food, it relieves flatulence, reduces stomach cramps, and stimulates appetite.

Parts used: Leaves
Tastes: Bitter, pungent, sweet
Quality: Light, dry
Potency: Hot
***Dosha* effect:** VK–, P+ (in excess)
Actions: Diuretic, laxative, carminative, antispasmodic

9.12.4 Traditional Uses and Folklore

- Parsley is a protection herb, used in purification baths and to stop misfortune. In the ancient Western tradition, there is the belief that only pregnant women or witches could grow it: planting on Good Friday was the only way to ensure a good harvest.
- The tea made from the leaves of parsley is a traditional home remedy used to treat jaundice, coughs, menstrual problems, rheumatism, kidney stones, and urinary infections. It is also a breath freshener.
- The juice soothes conjunctivitis and other eye inflammations. It is also used to strengthen the kidney and the uterus (Frawley and Lad, 1994).
- Parsley is an effective food remedy for dropsy (Bakhru, 1993).
- Cramps due to menstrual irregularities are relieved by the regular use of parsley juice, especially if combined with beet, carrot, and cucumber juices (Bakhru, 1993).

Precautions: Parsley is a warming spice and should not be consumed in excess by pregnant women and in case of high *pitta*. It helps to dispel kidney and gall stones, but it should be used with care in case of kidneys inflammation or irritation. For such conditions it can be balanced with marshmallow (Frawley and Lad, 1994).

9.12.5 In the Kitchen

Today, parsley is one of the most commonly consumed herbs in Europe, the Middle East, and the United States. A fresh sprig of the curled parsley is often used as a garnish, whereas the Italian parsley is used to flavor sauces, salad, rice pilaf, "risotto," soups, and stews. It is excellent with fish, tomato dishes, and baked or boiled potatoes.

Parsley should be grown within easy reach of the kitchen door for a constant, fresh supply. The leaves are picked before flowering and keep their aroma only if fresh. Even if frozen, they are better than dried, as the flavor may be lost during drying.

By adding parsley to the dishes, people can start to realize the importance of using fresh herbs as a healthy and delicious ingredient of their daily food.

Parsley can also be taken as a beverage, simmering it gently for a few minutes and partaking of the water (Bakhru, 1993).

9.12.6 Chemical Constituents and Uses as Functional Food

Depending on how one looks at it, parsley can be considered an herb, a spice, or a vegetable. It is, however, most commonly known as an herb that can add a unique flavor or seasoning to any dish to which it is added. Indeed, it is extremely popular as a garnish, and many of the best chefs often add freshly chopped parsley atop their signature dishes before declaring them finished. The Italian pesto, a sauce originally obtained grounding in a mortar a mixture of garlic, pine nuts, parsley growing in Liguria region, and virgin olive oil, is famous all over the world.

Although parsley has been cultivated for more than 2000 years, it was used medicinally prior to being consumed as a food.

One class of clinical medicines used to lower blood pressure is known as diuretics and works by increasing the excretion of urine from the body as well as the amount of sodium in urine. *Petroselinum* species are known to have diuretic effects (Wright *et al.*, 2007).

The chemical composition of parsley is wide: it contains a number of beneficial salts with a ratio of 10:1 (potassium/sodium), vitamins (especially vitamin K, phylloquinone, and vitamin A), fiber, an essential oil, a class of polyketides named polyacetylenes, and flavonoids.

Aliphatic C(17)-polyacetylenes of the falcarinol type, which occur in common food plants of the Apiaceae family such as carrot, celeriac, parsnip, and parsley, have demonstrated interesting bioactivities, including antibacterial, antimycobacterial, and antifungal activities as well as anti-inflammatory, antiplatelet aggregatory, neuritogenic, and serotonergic effects. In addition, the cytotoxicity of falcarinol-type polyacetylenes toward human cancer cells, bioavailability, and their potential anticancer effect *in vivo* indicate that these compounds may contribute to the health effects of certain vegetables and hence could be important nutraceuticals (Christensen, 2011; Christensen and Brandt, 2006).

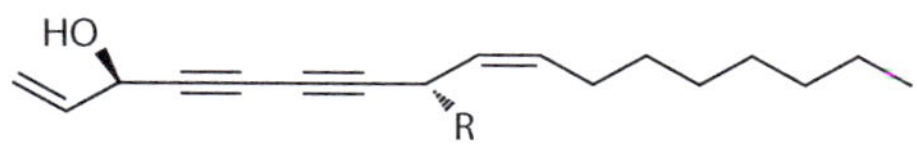

R=H, falcarinol
R=OH, falcarindiol

Galactolipids are a class of compounds widely found in the plant kingdom, including edible plants, and are an important part of the cell membranes. In plants, they consist mainly of monogalactosyldiacylglycerols and digalactosyldiacylglycerols containing one or two saturated and/or unsaturated fatty acids linked to the glycerol moiety. Several galactolipids have been shown to possess *in vitro* and/or *in vivo* antitumor-promoting and anti-inflammatory activities. Recently, it has been demonstrated that the galactolipid, 1,2-di-*O*-α-linolenoyl-3-*O*-β-D-galactopyranosyl-sn-glycerol, may be important for the anti-inflammatory activity of dog rose (*Rosa canina*), a medicinal plant with documented effect on anti-inflammatory diseases such as arthritis. This galactolipid also occurs in relative high concentrations in certain legumes (e.g., common bean, pea), leafy vegetables (e.g., kale, leek, parsley, perilla, and spinach), stem vegetables (e.g., asparagus, broccoli, brussels sprouts), and fruit vegetables (e.g., chili, bell pepper, pumpkin). Furthermore, it has been isolated from spinach and several medicinal plants as a galactolipid with possible cancer-preventive effects (Christensen, 2009).

Cancer prevention through diet may be largely achievable by increased consumption of fruits and vegetables. Considerable attention has been devoted to identifying plant-derived dietary agents, which could be developed as promising chemopreventives. One such agent is apigenin, a naturally occurring plant flavone (4′,5,7,-trihydroxyflavone) abundantly present in common fruits and vegetables, including parsley, onions, oranges, tea, chamomile, wheat sprouts, and some seasonings. Apigenin has been shown to possess remarkable anti-inflammatory, antioxidant, and anticarcinogenic properties. In the past few years, significant progress has been made in studying the biological effects of apigenin at the cellular and molecular levels (Patel *et al.*, 2007).

Parsley yields 72% juice with 64% of the total flavones extracted. Fresh parsley juice averages 281 mg flavones/100 g, present mainly as apiosylglucoside conjugates (Hostetler *et al.*, 2012).

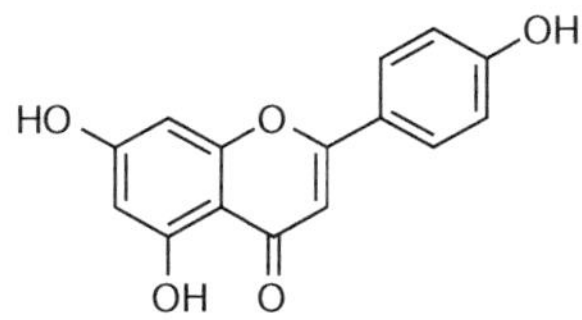

Apigenin

Tabbouleh, the national dish of Lebanon

Ingredients

1 big bunch of parsley
400 g of bulgur wheat
1 green onion (chopped)
Few mint leaves
The juice of 2 lemons
Extra virgin Olive oil to taste
Salt and pepper to taste

Preparation

Combine chopped parsley with bulgur wheat, chopped green onions (scallions), mint leaves, lemon juice, olive oil, and salt. The freshness of parsley helps to digest the heavy bulgur grains.

REFERENCES

Bakhru H.K. *Herbs That Heal*. Orient Paperbacks, Delhi, India, **1993**, 136–138.

Christensen L.P. Aliphatic C(17)-polyacetylenes of the falcarinol type as potential health promoting compounds in food plants of the Apiaceae family. *Recent Pat. Food Nutr. Agric.* **2011**, *3*(1), 64–77.

Christensen L.P. Galactolipids as potential health promoting compounds in vegetable foods. *Recent Pat. Food Nutr. Agric.* **2009**, *1*(1), 50–58.

Christensen L.P., Brandt K. Bioactive polyacetylenes in food plants of the Apiaceae family: Occurrence, bioactivity and analysis. *J. Pharm. Biomed. Anal.* **2006**, *41*(3), 683–693.

Frawley D., Lad V. *The Yoga of Herbs*. Motilal Banarsidass Publishers, Delhi, India, **1994**, 133–134.

Hostetler G.L., Riedl K.M., Schwartz S.J. Endogenous enzymes, heat, and pH affect flavone profiles in parsley (*Petroselinum crispum* var. *neapolitanum*) and celery (*Apium graveolens*) during juice processing. *J. Agric. Food Chem.* **2012**, *60*, 202–208.

Patel D., Shukla S., Gupta S. Apigenin and cancer chemoprevention: Progress, potential and promise (review). *Int. J. Oncol.* **2007**, *30*(1), 233–245.

Wright C.I., Van-Buren L., Kroner C.I., Koning M.M. Herbal medicines as diuretics: A review of the scientific evidence. *J. Ethnopharmacol.* **2007**, *114*(1), 1–31.

9.13 ROSEMARY

Scientific name: *Rosmarinus officinalis* L.
Family: Lamiaceae
Hindi: Rusmary
Sanskrit: Rusmari
Tamil: Thalisabatri

9.13.1 Origin and Distribution

Rosmarinus is a plant native of southern Europe that grows wild on the rocky hills in the Mediterranean region and is cultivated in Yugoslavia, Spain, Portugal, and other parts of Europe as well as in the United States (Pruthi, 1998). It is suitable for cultivation in India in the temperate Himalayas and Nilgiri mountains with a dry climate. The botanical name means "dew of the sea," from the Latin *ros* ("dew") and *marinus* ("of the sea").

There are many references to this plant in Greek and Christian traditions: when Aphrodite rose from the sea, born of Uranus's semen, rosemary was draped around her body. The shrub is known as the "Rose of Mary" because the Virgin Mary is said to have spread her blue cloak over a white-blossomed rosemary bush when she was resting. For this reason, the flowers turned blue (Aggarwal and Yost, 2011).

Rosemary has been used medicinally for centuries and has been referenced in various medieval texts. It is a bush that tries to be a tree and therefore embodies an energetic thrust for every process. The elemental force is the fire, able to strength consciousness and mental power (Kinkele, 2005). In ancient times, the shrub had a reputation for strengthening the memory, so it became the emblem of fidelity for lovers. It was used at weddings, as a love charm, but also at funerals, as incense in religious ceremonies and in magical spells. Students used to massage rosemary oil into the forehead and temples before examinations. In fact, rosemary stimulates the circulation, increasing the blood supply to the brain.

It was an old custom to burn rosemary in sick chambers, and in French hospitals, it is customary to burn rosemary with juniper berries to purify the air and prevent infections (Aggarwal and Yost, 2011). It was called *incensier* in France. It was freely cultivated in kitchen gardens and came to represent the dominant influence of the house mistress: "Where Rosemary flourished, the woman ruled." The Sicilians believe that young fairies, taking the form of snakes, lie among the rosemary branches. It is an auspicious plant that brings good luck to the family. It was believed that placing a sprig of rosemary under a pillow before sleep would repel nightmares, and if placed outside the home, it would protect the house. When planted during the wedding day, rosemary becomes a symbol of love and union.

9.13.2 Features

The herb has several ash-colored branches, and the bark is rather scaly and it grows up to a height of about 1 m. The evergreen leaves are linear, with an odor pungently aromatic and somewhat camphoraceous. The flowers are small and pale blue. Much of the active volatile principle resides in their calyces.

9.13.3 Properties and Uses in Ayurvedic System of Medicine

Rosemary has a warming and relaxing effect for *kapha dosha* ailments, including *kapha*-type headache, pacifies *vata*, and aggravates *pitta*. In Ayurvedic medicine, rosemary containing aromatherapy oil is used for its ability to enhance concentration and memory. Other rosemary preparations are used to treat halitosis, migraine headaches, nervous disorders, bites, stings, wounds, and upset stomach. Rosemary is also used for food poisoning prevention and to maintain healthy the reproductive system.

Parts used: Leaves, essential oil
Tastes: Bitter, pungent, astringent
Potency: Hot
***Dosha* effect:** VK–, P+
Actions: Diaphoretic, nervine tonic

9.13.4 Traditional Uses and Folklore

- Rosemary tea is helpful in case of headache caused by stomach disorders (Tierra, 1999).
- It is also an home remedy for mental fatigue and forgetfulness (Bakhru, 1992).
- The use of rosemary as aromatic herb is recommended in case of dyspepsia; it helps to digest heavy foods rich in fat, such as pork, beef, and lamb meats.

9.13.5 In the Kitchen

Fresh tender tops are used for garnishing and flavoring cold drinks, pickles, and soups. The leaves are largely employed as a condiment for meat, fish, poultry, sauces, dressings, preserves, and jams. Rosemary is a very important spice in Italy and France. It can be added to the Indian ingredients of curry giving a very special "fusion" taste.

9.13.6 Chemical Constituents and Uses as Functional Food

In many parts of the world, rosemary is a common household plant that is used for many purposes, including food flavoring, drink, and beverages, as well as cosmetic uses. The leaves, both fresh and dried, are used in traditional Mediterranean cuisines. They have a bitter, astringent taste and are highly aromatic, which complements a wide variety of foods.

Traditionally, rosemary was used for relieving renal colic, dysmenorrhea, and respiratory disorders. This was due to the antispasmodic properties of rosemary, which relaxes trachea and intestinal smooth muscles, as well as its choleretic activity (Al-Sereiti *et al.*, 1999).

The most important constituents of rosemary are caffeic acid and its derivatives such as rosmarinic acid (Petersen and Simmonds, 2003), ursolic acid, carnosic acid, and carnosol. These compounds have an antioxidant effect, and approximately 90% of the total antioxidant activity of rosemary is derived from carnosol and carnosic acid (Cheung and Tai, 2007).

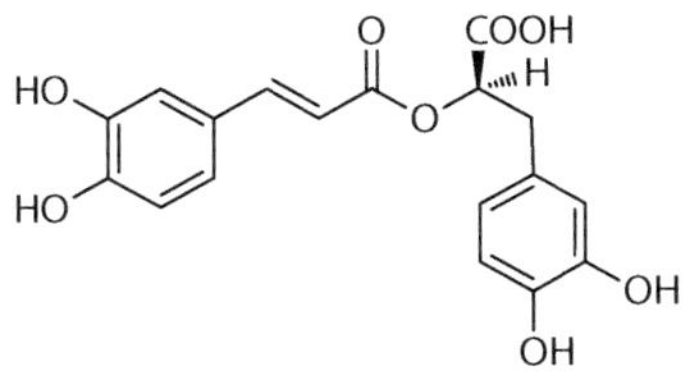

Rosmarinic acid

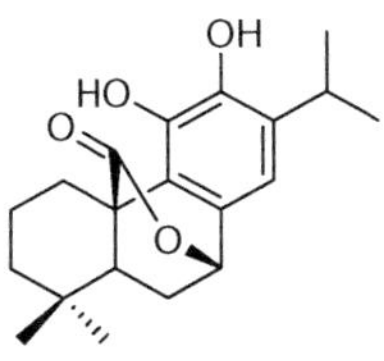

Carnosol

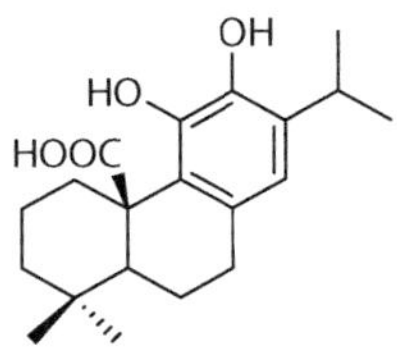

Carnosic acid

Rosemary, in particular its phenolic constituents, has also been found to exert protective effects on colonic cancer and other types of cancer in cell cultures and animal models through several mechanisms. Preclinical evidence points to the anticancer properties of rosemary extract, carnosol, carnosic acid, ursolic acid, and rosmarinic acid, and appears to arise through dose-related changes in several molecular events associated with cancers (Ngo *et al.*, 2011).

Carnosol has been evaluated for anticancer property in prostate, breast, skin, leukemia, and colon cancers with promising results. These studies have provided evidence that carnosol targets multiple deregulated pathways associated with inflammation and cancer, which include NF-κB, apoptotic-related proteins, phosphatidylinositol-3-kinase (PI3K)/Akt, androgen and estrogen receptors, as well as molecular targets. In addition, carnosol appears to be well tolerated in that it has a selective toxicity toward cancer cells versus nontumorigenic cells, and is well tolerated when administered to animals (Johnson, 2011).

Roasted potatoes with rosemary

Ingredients

1 kg small potatoes
3 tsp of olive oil
½ tsp of black pepper powder
2 garlic cloves (minced)
2 tbsp of fresh rosemary leaves (minced)
Aromatic salt to taste

Preparation

Preheat the oven to 200°C. Cut the potatoes in quarters and place in a bowl with all the dressing ingredients. Toss until the potatoes are well coated. Spread the potatoes on a baking sheet and roast in the oven for about 1 hour until they are browned and crispy.

REFERENCES

Aggarwal B.B., Yost D. *Healing Spices*. Sterling Publishers, New York, **2011** (Italian version, Armenia, 312–320).

Al-Sereiti M.R., Abu-Amer K.M., Sen P. Pharmacology of rosemary (*Rosmarinus officinalis* Linn.) and its therapeutic potentials. *Indian J. Expt Biol.* **1999**, *37*, 124–130.

Bakhru H.K. *Herbs That Heal.* Orient Paperbacks. Delhi, India, **1992**, 150.

Cheung S., Tai J. Anti-proliferative and antioxidant properties of rosemary (*Rosmarinus officinalis*). *Oncol. Rep.* **2007**, *17*, 1525–1531.

Johnson J.J. Carnosol: A promising anti-cancer and anti-inflammatory agent. *Cancer Lett.* **2011**, *305*(1), 1–7.

Kinkele T. *Incense and Incense Rituals. Health Hamony.* B. Jain Publishers, Delhi, India, **2005**, 133.

Ngo S.N.T., Williams D.B., Head R.J. Rosemary and cancer prevention: Preclinical perspectives. *Crit. Rev. Food Sci. Nutr.* **2011**, *51*(10), 946–954.

Petersen M., Simmonds M.S.J. Rosmarinic acid. *Phytochemistry* **2003**, *62*, 121–125.

Pruthi J.S. *Spices and Condiments.* National Book Trust, Delhi, India, **1998** (5th edn.), 218–220.

Tierra M. *Planetary Herbology.* Motilal Banarsidass Publishers, Delhi, India, **1999**, 112.

9.14 SAFFRON

Scientific name: *Crocus sativus* L.
Family: Iridaceae
Hindi: Kesar
Sanskrit: Kunkuma
Tamil: Kunkumapoo

9.14.1 Origin and Distribution

According to Greek mythology, handsome mortal Crocos fell in love with the beautiful nymph Smilax, but as his favors were rebuffed, he was turned into the beautiful purple crocus flower.

Saffron is the world's most expensive spice by weight, but very little is needed to flavor and color food. The English name derives from the Arab word *zafaran*, meaning "yellow."

Probably native in Greece, Asia Minor, southern Italy, and Iran, it is actually cultivated in Spain, France, Abruzzo, Sicily, Iran, and India. The La Macha belt of Spain is the largest producer of saffron in the world and contributes 80%–90% of the world production. In India, the cultivation is confined to Pampore and Kistwar areas of Jammu and Kashmir, extending to nearly 4000 acres at 1600 m altitude (Pruthi, 1976).

Saffron-based pigments have been found in 50,000-year-old paintings in northwest Iran. The Sumerians later used wild-growing saffron in their remedies and magical potions. Ancient Persians cultivated it in Derbena, Isfahan, and Khorasan by the tenth century BC. At such sites, saffron threads were woven into textiles, ritually offered to divinities, and used in dyes, perfumes, medicines, and body washes. A recent review describes the folkloric uses of saffron according to Avicenna (Hosseinzadeh and Nassiri-Asl, 2012). Alexander's troops imitated the practice of saffron bathing to cure battle wounds and introduced it in Greece. Ancient perfumers in Egypt, physicians in Gaza, town's people in Rhodes, and the Greek courtesans used saffron in their scented waters, perfumes and ointments, divine offerings, and medical treatments. Later, Cleopatra used saffron in her baths.

Dioscorides of Anazarb attributed magical medicinal properties to saffron if worn as an amulet. Pliny (first century) ascribed general panacea properties to it when taken internally. Such was the Romans' love of saffron that they took it with them when they settled in southern Gaul, where it was extensively cultivated until Rome's fall. European cultivation stopped with the end of the Roman Empire and was reintroduced in Spain and Italy by the Islamic civilization.

Some historians believe that saffron came to China with Mongol invaders from Persia. Yet saffron is mentioned in ancient Chinese medical texts dating from 200–300 BC.

Aside from its use in food and medicine, this spice has also been used by the upper class to dye clothes. Saffron is also part of Hindu rituals as a symbol of blessing and good luck. It is called *ghusrna*, as it is sacred to *Ghrsneswara*, the Lord Siva. Certain holy texts have been known to be written with saffron ink.

9.14.2 Features

Saffron is the dried yellow stigmas from a small purple crocus. Each flower provides only three stigmas, which must be carefully handpicked and then dried, with an extremely labor-intensive process.

9.14.3 Properties and Uses in Ayurvedic System of Medicine

Ayurvedic texts refer many names related to the number of properties of this expensive revitalizer: *varnya*, which improves the complexion; *sourabha*, fragrant; *caru*, flower looks like lotus; and *roudhira*, blood red in color (Paranjipe, 2005).

It is a very expensive *rasayana* traditionally used mainly by royal families. It is helpful for diseases of the head. It balances all three *dosha*s, so it can be used by anyone. It is widely used as an active ingredient in Ayurvedic preparations to increase quality and quantity of semen. It is also attributed with extraordinary properties for improving weak eyesight and highly valued as a complexion builder (Chopra *et al.*, 1956).

Saffron enjoys a great reputation as a drug that strengthens the functioning of stomach and promotes its action. It also counteracts spasmodic disorders and sustains involuntary muscle contraction. A combination of saffron and ghee is used to treat diabetes.

It is one of the best anti-*pitta* herbs and spleen–liver regulators. It is not actually a tonic itself, but even in small amounts, it can catalyze the tonic action of other herbs. Its quality is *sattva* and develops love, devotion, and compassion (Frawley and Lad, 1994). It is a strengthening agent for the heart and a cooling agent for the brain. It has also been found beneficial in the treatment of urinary problems. It is also useful in treating cold and cough.

Parts used: Flower (stigma)
Tastes: Bitter, pungent
Quality: Light, unctuous
Potency: Hot
***Dosha* effect:** VKP=
Actions: Rejuvenating, general tonic, nervine tonic, antidote to poisons, aphrodisiac, carminative, diaphoretic, diuretic, febrifuge, stimulant, sedative, styptic

9.14.4 Traditional Uses and Folklore

- In the Unani system of Medicine, saffron is used for slimming, the treatment of enlarged livers, urinary bladder and kidney infections, and menstrual disorders; for strengthening the heart; and cooling the brain. If soaked overnight in water and administered with honey, it acts as a diuretic.
- Saffron is largely used as an indigenous medicine across India. It is given along with milk to alleviate common cold. When given along with honey, it augments the hemoglobin percentage in anemia (Paranjipe, 2005).
- In sexual debility, it is considered an excellent remedy (Trivedi, 1997). Boil 500 ml of milk with four dates. When the milk is reduced to half, add five to six pieces of saffron. Add sugar to taste. Drink this milk every evening half an hour before sleeping (Puri, 2003).

Precautions: Saffron can possibly promote miscarriage, so it is better to avoid its use during pregnancy. It can also be narcotic when used in large doses (Frawley and Lad, 1994).

9.14.5 In the Kitchen

Saffron is widely used in Persian, Arab, Central Asian, European, Indian, Turkish, and Cornish cuisines. In India, it is said that one who has not eaten a dish flavored with saffron has eaten nothing in the world. Especially in the northern Mughlai dishes, this spice is very popular: serving dishes decorated with saffron is regarded as a mark of honor to the guest. The saffron-blended Indian wines are also very famous. Preparation of saffron-blended tea is an old age tradition in royal Arab kitchens.

In Europe, it is often used to flavor and color rice dishes, stews, and fish. It is also used for coloring butter, cheese, ice creams, pudding, and pastry. In foods, it performs the functions of a spice, adding its faint, delicate aroma, pleasing flavor, and magnificent yellow color to enhance palatability (Basker *et al.*, 1983).

Experts often recommend soaking the strands in very hot milk or water before using them to help release the bright, rich yellow tones and distinctive flavor. The flavor can soak into wood, so it is best to avoid wooden spoons or other utensils when making dishes with this spice. It is better to buy the whole stigma, as the powdered spice loses quickly its flavor.

Saffron is very sensitive to heat and light, so it should be stored in a cool, dark location. It is also important to keep it dry, as high humidity can cause it to smell musty and make it age faster.

Saffron has many benefits, and one of the ways to incorporate it in your diet is by taking saffron milk that can be made by dissolving one small pinch of saffron in a glass of 200 ml milk. You may add honey or sugar for taste if required.

9.14.6 Chemical Constituents and Uses as Functional Food

Saffron is widely used as a spice and as a coloring and flavoring agent in the preparation of foods and cosmetics. Due to the high costs because of a limited yield and the added value of saffron, it is one of the main food ingredients submitted to fraud and adulteration giving rise to emerging risks for population health (Moore *et al.*, 2012). In fact, the powder can be combined with other less expensive spices such as turmeric, annatto, or safflower.

The important constituents of stigmas of *C. sativus* are carotenoids (e.g., crocetin, crocins, α-carotene, lycopene, zeaxanthin), monoterpene aldehydes (e.g., picrocrocin, safranal), monoterpenoids (e.g., crocusatines), isophorones, and flavonoids (Bathaie and Mousavi, 2010; Melnyk *et al.*, 2010). Carotenoids such as crocetin and its glycosidic form digentiobioside (crocin) contribute to the color profile of this spice.

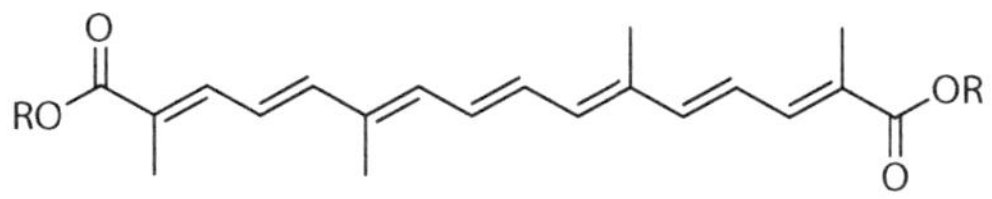

R=H, crocetin

R= 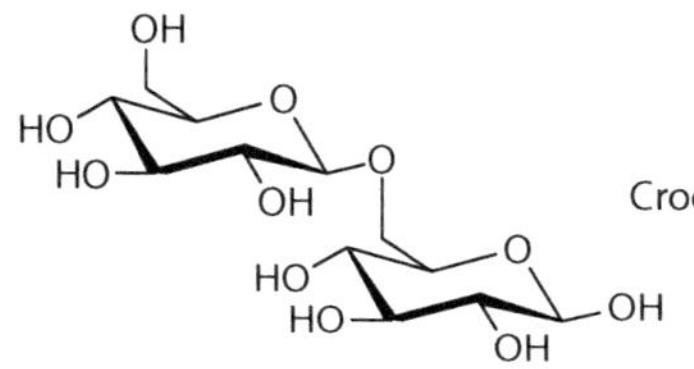

Crocin

The monoterpene aldehydes picrocrocin and its deglycosylated derivative, safranal (dehydro-β-cyclocitral), are responsible for its bitter flavor and aroma, respectively (Bhargava, 2011).

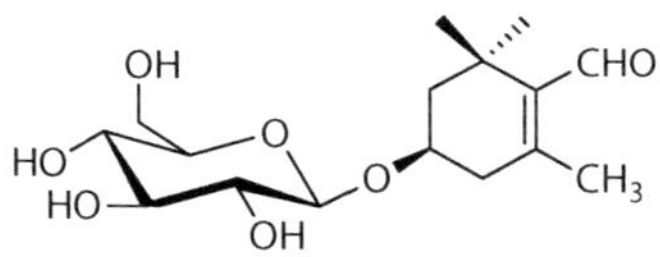

Picrocrocin

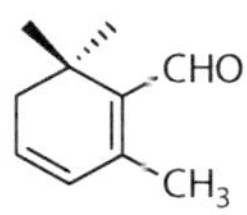

Safranal

The aroma of saffron has received much attention from scientists in recent years, not only for the compounds that make it up, but also for its glycosidic precursors. Despite it all, the volatile generation mechanisms of the spice are almost completely unknown. Only the generation of safranal, the major compound, from picrocrocin has been established. The great effort carried out to detect and identify the volatile compounds of saffron has not been enough to clarify which compounds are responsible for saffron aroma. In general, scientists have devoted little attention to the sample analyzed, taking for granted that all saffron is the same. Saffron aroma is what the consumer perceives via nasally when a container of the spice is uncovered, or it is the aroma conferred to food, normally after a thermal cooking process, and perceived retronasally (Carmona *et al.*, 2007).

Crocus sativus possesses a number of medicinally important activities such as antihypertensive, anticonvulsant, antitussive, antigenototoxic, cytotoxic, anxiolytic, aphrodisiac, antioxidant, antidepressant, antinociceptive, anti-inflammatory, and relaxant activities. It also improves the memory and learning skills, and increases the blood flow in the retina and choroid (reviewed in Srivastava *et al.*, 2010).

Preclinical evidence of anxiolytic activity is reported. Common mechanisms of action involve GABA, via direct receptor binding or ionic channel or cell membrane modulation; GABA transaminase or glutamic acid decarboxylase inhibition; a range of monoaminergic effects; and potential cannabinoid receptor modulation. Future research should focus on conducting human clinical trials on the plants reviewed with promising anxiolytic activity (reviewed in da Rocha *et al.*, 2011 and Sarris *et al.*, 2013). Saffron stigma was found to be significantly more effective than placebo and equally as efficacious as fluoxetine and imipramine. Saffron petal was significantly more effective than placebo and equally efficacious compared to fluoxetine and saffron stigma, thus showing promise in the management of mild-to-moderate depression (Dwyer *et al.*, 2011).

Saffron has shown some promising effects in clinical studies with dementia patients, with respect to cognition (Howes and Houghton, 2012; Howes and Perry, 2011), although more extensive and reliable clinical data are required.

Carthamus tinctorius (a Chinese saffron) has been one of the major medicines to treat ischemia for hundreds of years in China, Korea, and Japan. Recent experimental studies demonstrated its neuroprotective efficacy on mitigating brain infarction and global ischemia as well as preventing the neurodegeneration following ischemia (Lin, 2011).

To be mentioned are the recent studies performed by Italian researchers (Università dell'Aquila) on saffron growing in Abruzzo, the most valuable in Italy, who are studying the effect of diet doses entire spice on ocular blood and retinal function. Saffron has been found to exert neuroprotection on retinal photoreceptors subjected to oxidative stress. Patients report the amelioration of visual acuity and the ability to read (Falsini *et al.*, 2010).

Saffron has been used as an herbal remedy for various ailments, including cancer, by the ancient Arabian, Indian, and Chinese cultures. Crocetin has shown significant potential as an antitumor agent in animal models and cell culture systems. It affects the growth of cancer cells by inhibiting nucleic acid synthesis, enhancing antioxidative system, inducing apoptosis, and hindering growth factor signaling pathways (Gutheil *et al.*, 2012).

The effects of crocetin, a carotenoid (8,8′-diapo-8,8′-carotenoic acid) present in saffron and characterized by a diterpenic and symmetrical structure with seven double bonds and four methyl groups, have been examined. It has been found that this compound enhances the oxygen diffusivity through liquids, such as plasma. As a result of this property, it has been observed that crocetin increases alveolar oxygen transport and enhances pulmonary oxygenation. It improves cerebral oxygenation in hemorrhaged rats and positively acts in the atherosclerosis and arthritis treatment. It inhibits skin tumor promotion in mice (i.e., with benzo(a)pyrene); it has an inhibitory effect on intracellular nucleic acid and protein synthesis in malignant cells, as well as on protein kinase C and proto-oncogene in INNIH/3T3 cells. This is most likely due to its antioxidant activity. Furthermore, crocetin protects against oxidative damage in rat primary hepatocytes. It also suppresses aflatoxin B1-induced hepatotoxic lesions and has a modulatory effect on aflatoxin, B1 cytotoxicity, and DNA adduct formation on C3H10/T1/2 fibroblast cells. It also has a protective effect on the bladder toxicity, induced by cyclophosphamide. The toxicity reduction seems to be due to the property of crocetin to stimulate defense mechanisms from the liver cells with the increase of the glutathione-*S*-transferase. The experiments reported in the scientific literature and the interesting results obtained have been carried out *in vitro* or on laboratory animals, but not yet on man (Giaccio, 2004).

Saffron has been an important subject of interest for research teams in the past two decades due to its various biological properties. Chemical analysis has shown the presence of more than 150 components in saffron stigmas. The more powerful components of saffron are carotenoids and monoterpene aldehydes. Structure–function relationship studies show that some properties are related to deglycosylated derivatives, whereas others belong to more glycosylated ones. Saffron shows a wide range of usefulness in medicine, cosmetics, and coloring industries, so it can be used for new drug designs. However, more research about its mechanism of action is needed (Bathaie and Mousavi, 2010).

Saffron risotto (*Risotto alla Milanese*) (serves: 4)

Ingredients

1 medium onion, very finely chopped
8 tbsp of butter
1 lb (2 cups) carnaroli rice (a short-grain Italian rice)
1/2 cup of dry white wine
6 cups of hot homemade vegetable broth (the original recipe uses chicken broth); more as needed
1/4 tsp of saffron powder
1 cup of finely grated Parmigiano-Reggiano
Salt to taste

Preparation

In a heavy-based saucepan, cook the onion in two tablespoons of butter over medium heat until it is translucent and fragrant for about 5 minutes. Stir in the rice and cook it over medium heat for about 3 minutes. Add the wine and let it to evaporate, add one cup of broth and the saffron diluted in one cup of broth. Turn the heat to high until the broth comes to a simmer and then adjust the heat to maintain a steady simmer. Add salt to taste. Cook until most of the liquid has been absorbed, stirring every minute or two (there is no need to stir constantly). Add another cup of broth and keep cooking, stirring, and adding broth until the rice is *al dente*. When the rice is ready, off the heat and stir in the remaining six tablespoons of butter. Ladle onto heated plates and season with Parmigiano-Reggiano.

REFERENCES

Basker D., Negbi M., Uses of saffron. *Economic Botany* **1983**, *37*, 228–236.

Bathaie S.Z., Mousavi S.Z. New applications and mechanisms of action of saffron and its important ingredients. *Crit. Rev. Food Sci. Nutr.* **2010**, *50*(8), 761–786.

Bhargava V.K. Medicinal uses and pharmacological properties of *Crocus sativus* Linn. (saffron). *Int. J. Pharm. Pharm. Sci.* **2011**, 3, 22–26.
Carmona M., Zalacain A., Salinas M.R., Alonso G.L. A new approach to saffron aroma. *Crit. Rev. Food Sci. Nutr.* **2007**, *47*(2), 145–159.
Chopra R.N., Nayar S.L., Chopra I.C. Glossary of indian medicinal plants. Council of Scientific and Industrial Research. **1956**, 80.
da Rocha M.D., Viegas F.P., Campos H.C., Nicastro P.C., Fossaluzza P.C., Fraga C.A., Barreiro E.J., Viegas C., Jr. The role of natural products in the discovery of new drug candidates for the treatment of neurodegenerative disorders II: Alzheimer's disease. *CNS Neurol. Disord. Drug Targets* **2011**, *10*(2), 251–270.
Dwyer A.V., Whitten D.L., Hawrelak J.A. Herbal medicines, other than St. John's Wort, in the treatment of depression: A systematic review. *Altern. Med. Rev.* **2011**, *16*(1), 40–49.
Falsini B., Piccardi M., Minnella A., Savastano C., Capoluongo E., Fadda A., Balestrazzi E., Maccarone R., Bisti S. Saffron supplementation improve retinal flicker sensitivity in early age-related macular de generation. *Invest. Ophthalmol. Vis. Sci.* **2010**, *12*, 6118–6124.
Frawley D., Lad V. *The Yoga of Herbs*. Motilal Banarsidass Publishers, Delhi, India, **1994**, 142–143.
Giaccio M. Crocetin from saffron: An active component of an ancient spice. *Crit. Rev. Food Sci. Nutr.* **2004**, *44*(3), 155–172.
Gutheil W.G., Reed G., Ray A., Anant S., Dhar A. Crocetin: An agent derived from saffron for prevention and therapy for cancer. *Curr. Pharm. Biotechnol.* **2012**, *13*(1), 173–179.
Hosseinzadeh H., Nassiri-Asl M. Avicenna's (Ibn Sina) the canon of medicine and saffron (*Crocus sativus*): A review. *Phytother Res.* **2013**, *27*(4), 475–483.
Howes M.J., Houghton P.J. Ethnobotanical treatment strategies against Alzheimer's disease. *Curr. Alzheimer Res.* **2012**, *9*(1), 67–85.
Howes M.J., Perry E. The role of phytochemicals in the treatment and prevention of dementia. *Drugs Aging* **2011**, *28*(6), 439–468.
Lin B. Polyphenols and neuroprotection against ischemia and neurodegeneration. *Mini Rev. Med. Chem.* **2011**, 11(14), 1222–1238.
Melnyk J.P., Wang S.N., Marcone M.F. Chemical and biological properties of the world's most expensive spice saffron. *Food Res. Int.* **2010**, *43*, 1981–1989.
Moore J.C., Spink J., Lipp M. Development and application of a database of food ingredient fraud and economically motivated adulteration from 1980 to 2010. *Food Sci.* **2012**, *77*(4), R118–R126.
Paranjipe P. *Indian Medicinal Plants. Forgotten Healers*. Chaukhamba Sanskrit Pratishthan, Delhi, India, **2005**, 146–149.
Pruthi J.S. *Spices and Condiments*. National Book Trust, Delhi, India, **1976**, 221–226.
Puri H.S. *Rasayana. Ayurvedic Herbs for Longevity and Rejuvenation*. Taylor & Francis, London, **2003**, 164–169.
Sarris J., McIntyre E., Camfield D.A. Plant-based medicines for anxiety disorders, Part 1: A review of preclinical studies. *CNS Drugs* **2013**, *27*(3), 207–219.
Srivastava R., Ahmed H., Dixit R.K., Dharamveer, Saraf S.A. *Crocus sativus* L.: A comprehensive review. *Pharmacogn Rev.* **2010**, *4*(8), 200–208.
Trivedi M. A best aphrodisiac, saffron. *Nirogadham*, October–December **1997**.

9.15 SAGE

Scientific name: *Salvia officinalis* Linn.
Family: Lamiaceae
Hindi: Salvia
Tamil: Seemaikarpura ilai

9.15.1 Origin and Distribution

Sage grows wild along the Mediterranean coast from Spain up to the east side of the Adriatic and on the mountains and hills of Croatia and Dalmatia. It is cultivated as a spice and for medicinal purposes in Italy, France, Germany, England, Canada, and the United States. In India, it is cultivated in Jammu Kashmir and Nilgiri mountains for its essential oil. The best sage is the one growing wild in the Dalmatian region (Pruthi, 1998).

Sage was a sacred herb associated with immortality for the Egyptians, Romans, and Greeks. The Greek Theophrastus classified sage as a "coronary herb," able to flush the diseases from the body. The Hungarian gypsies believed that it can attract good and dispel devil. During the Middle Ages, the leaves, rough like the tongue, were used for oral hygiene. It is a solar plant related to Jupiter. From a mystic point of view, it is considered very effective in warding off demons and spirits, so it was cultivated in the European monasteries (Di Massimo and Di Massimo, 2005). The incense brings clarity, subdues passions, and creates a clean environment for peace of mind and meditation. The Native Americans used to burn white sage (*Salvia apiana*) for calming healing force during the sweat lodge ceremonies (Kinkele, 2005).

9.15.2 Features

It is a hardy subshrub native of southern Europe. Stems are white wholly and 15–30 cm tall; the leaves are aromatic and oblong; and the flowers are blue, purple, or white.

Sage was awarded in 2001 by the International Herb Association as "The Herb of the Year," for both its culinary and medicinal properties. There are over 500 varieties of sage, and most of them are medicinally useful.

Parts used: Whole plant, mostly leaves
Tastes: Bitter, pungent, astringent
Potency: Hot
***Dosha* effect:** KV–, P+ (in excess)
Actions: Stimulant, diaphoretic, carminative, antispasmodic

9.15.3 Properties and Uses in Ayurvedic System of Medicine

In Ayurveda, sage is classified as a mood-altering remedy that can stimulate the mental capacity (Kinkele, 2005). It has a special power to promote calmness and clarity of mind by removing emotional obstructions. It is specific for calming the heart (Frawley and Lad, 1994).

Sage has a strong action for reducing excess secretions in the body: it stops sweating and it dries up excess mucus from the nose and lungs as well as excessive salivation (Frawley and Lad, 1994).

Due to the heating potency, it is a spice to be used during the winter season.

9.15.4 Traditional Uses and Folklore

Salvia officinalis L. has been used since earliest times for flavoring, improving, and preserving food. The name *Salvia* comes from the Latin word *salvus* meaning "to heal," which sums up the folkloric belief of its "magical" therapeutic properties for many kinds of ailments and its popularity in traditional medicine. *Cur moriatur homo cui Salvia crescit in horto?* (Why should a man die whilst sage grows in his garden?) Sage has long had a reputation as an herb that mitigates mental and bodily grief, heals the nerves, counteracts fear, and protects human beings from evil influences, spiritual and physical (Buhner, 1998). Dioscorides, Pliny, and Galen all recommended sage as a diuretic, hemostatic, emmenagogue, and tonic (Kintzios, 2000).

Precautions: It should be avoided during pregnancy.

9.15.5 In the Kitchen

Sage is an important spice extensively used in Western countries, especially in Italy, England, and the United States for flavoring cheese, meats, fish, and sausage. Only during the 1500s, the European started to use the sage as a spice; previously, it was used only as a medicine. It should be used along with foods that are heavy to digest such as pork, liver, sausages, and cheese. It is excellent with potatoes, pumpkin, and apples (Aggarwal and Yost, 2011).

9.15.6 Chemical Constituents and Uses as Functional Food

Sage is an astringent herb and can be used to tighten and tone tissues. It has a special affinity for the mouth and can be used to relieve pain and heal mouth ulcers, bleeding gums, spongy gums, and cold sores. Sage is a common ingredient in many tooth powders and mouthwash recipes. The astringent and antiseptic qualities of sage make it perfect for sore throats.

The main chemical components identified in sage are essential oils (1.5%–3%), hydroxycinnamic derivatives (3%–4%), phenolic diterpenes, triterpenes, flavonoids, and other minor constituents (Cuvelier *et al.*, 1994; Lu and Yeap, 2002). Sage is highly antibacterial, antimicrobial, and antiseptic properties mainly due to its essential oil; at high concentrations, toxic effects are observed (Hammer *et al.*, 1999). It has been medicinally used for thousands of years as an antibacterial healer.

Sage's history as a cognition enhancer and treatment for cognitive decline stretches back to the ancient Greeks (Imanshahidi and Hosseinzadeh, 2006). Active components are present in the essential oil, 1,8-cineole being the most potent in terms of cholinesterase inhibition, and a number of double-blind, placebo-controlled randomized studies in healthy humans have demonstrated improved memory, attention/executive function, and mood following single doses of cholinesterase-inhibiting sage extracts or essential oils (Kennedy *et al.*, 2006; Scholey *et al.*, 2008). A number of studies have reported the antioxidant capacity of sage, due to the essential oil, the flavonoid content, the hydroxycinnamic acids (rosmarinic acid), and the diterpenes carnosol and carnosic acid (Lamien-Meda *et al.*, 2010; Lu and Yeap, 2001). Carnosic acid, carnosol, rosmanol, and caffeic acid are responsible for the free radical-scavenging activity of sage (Matsingou *et al.*, 2003).

In a small cohort of Alzheimer's disease patients, a sage tincture demonstrated improved cognitive and behavioral functioning (Akhondzadeh *et al.*, 2003). Sage and its main ingredient rosmarinic acid have demonstrated to protect neuronal cells from amyloid-β peptide-induced neurotoxicity (Iuvone *et al.*, 2005).

Salvia officinalis has been used as a traditional remedy against diabetes in Mediterranean countries, and its glucose-lowering effects have been demonstrated in animal studies. The active compounds and their possible mode of action are still unknown, although it has been suggested that diterpenes may be responsible for the antidiabetic effect, by activating peroxisome proliferator-activated receptors (PPARγ) as potential therapeutic target for many obesity-related disorders (Rau *et al.*, 2006).

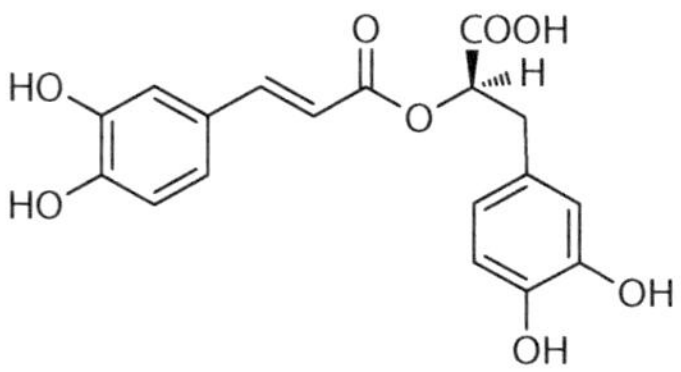

Rosmarinic acid

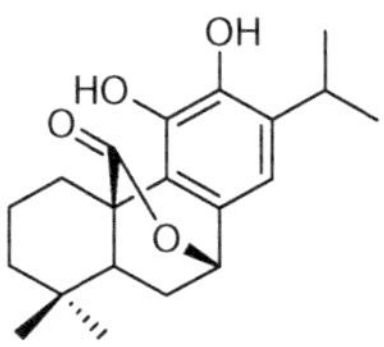

Carnosol

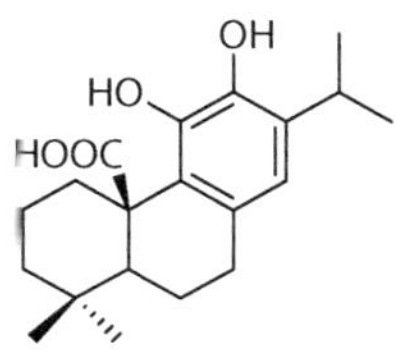

Carnosic acid

Sage pesto, a recipe from Marco Bianchi (2003)

Ingredients

2 small bunches of sage
20 g pine nuts
40 g Parmigiano-Reggiano cheese
1 garlic clove

100 ml extra virgin olive oil
Salt to taste

Preparation
Wash and dry well the sage leaves and put them with all the other ingredients in the mixer. Whip and use with boiled chickpeas.

REFERENCES

Aggarwal B.B., Yost D. *Healing Spices*. Sterling Publishers, New York, **2011** (Italian version, Armenia, 321–327).

Akhondzadeh S., Noroozian M., Mohammadi M., Ohadinia S., Jamshidi A.H., Khani M. *Salvia officinalis* extract in the treatment of patients with mild to moderate Alzheimer's disease: A double blind, randomized and placebo-controlled trial. *J. Clin. Pharm. Ther.* **2003**, *28*, 53–59.

Bianchi M. *A tavola con Marco Bianchi. Primi piatti*, vol. 1. Adriano Salari ed, Milano, Italy, **2013**, 118.

Buhner S.H. (ed.) Psychotropic and highly inebriating beers, in *Sacred and Herbal Healing Beers: The Secrets of Ancient Fermentation*. Siris Books, an imprint of Brewers Publications, Boulder, CO, **1998**.

Cuvelier M.E., Berset C., Richard H., Antioxidant constituents in sage (*Salvia officinalis*). *J. Agric. Food Chem.* **1994**, *42*, 665–669.

Di Massimo S., Di Massimo M. *Planta medica. Le erbe officinali tra Scienza e Tradizione*. Quaderni dell'Ambiente n.19, Stibu, Provincia di Pesaro e Urbino, Italy, **2005**, 281–283.

Frawley D., Lad V. *The Yoga of Herbs*. Motilal Banarsidass Publishers, Delhi, India, **1994**, 142–143.

Hammer K.A., Carson C.F., Riley T.V. Antimicrobial activity of essential oils and other plant extracts. *J. Appl. Microbiol.* **1999**, *86*, 985–990.

Imanshahidi M., Hosseinzadeh H. The pharmacological effects of *Salvia* species on the central nervous system. *Phytother. Res.* **2006**, *20*, 427–437.

Iuvone T., De Filippis D., Esposito G., D'Amico A., Izzo A.A. The spice sage and its active ingredient rosmarinic acid protect PC12 cells from amyloid-beta peptide-induced neurotoxicity. *J. Pharmacol. Exp. Ther.* **2006**, *317*, 1143–1149.

Kennedy D.O., Pace S., Haskell C., Okello E.J., Milne A., Scholey A.B., Effects of cholinesterase inhibiting sage (*Salvia officinalis*) on mood, anxiety and performance on a phychological stressor. *Neuropsychopharmacology* **2006**, *31*, 845–852.

Kinkele T. *Incense and Incense Rituals. Health Armony*. B. Jain Publishers, Delhi, India, **2005**, 134–135.

Kintzios S.E. *Sage: The Genus Salvia*. CRC Press, London, **2000**, 10–11.

Lamien-Meda A., Nell M., Lohwasser U., Börner A., Franz C., Novak J. Investigation of antioxidant and rosmarinic acid variation in the sage collection of the genebank in Gatersleben. *J. Agric. Food Chem.* **2010**, *58*, 3813–3819.

Lu Y., Yeap F.L. Antioxidant activities of polyphenols from sage (*Salvia officinalis*). *Food Chem.* **2001**, *75*, 197–202.

Lu Y., Yeap F.L. Polyphenolics of *Salvia*—A review. *Phytochemistry* **2002**, *59*, 117–140.

Matsingou T.C., Petrakis N., Kapsokefalou M., Salifoglou A. Antioxidant activity of organic extracts from aqueous infusions of sage. *J. Agric. Food Chem.* **2003**, *51*, 6696–6701.

Pruthi J.S. *Spices and Condiments*. National Book Trust, Delhi, India, **1998** (5th edn.), 226–227.

Rau O., Wurglics M., Paulke A., Zitzkowsky J., Meindl N., Bock A., Dingermann T., Abdel-Tawab M., Schubert-Zsilavecz M. Carnosic acid and carnosol, phenolic diterpene compounds of the labiatae herbs rosemary and sage, are activators of the human peroxisome proliferator-activated receptor gamma. *Planta Med.* **2006**, *7*, 881–887.

Scholey A.B., Tildesley N.T., Ballard C.G., Wesnes K.A., Tasker A., Perry E.K., Kennedy D.O. An extract of *Salvia* (sage) with anticholinesterase properties improves memory and attention in healthy older volunteers. *Phycopharmacology* (Berl.) **2008**, *198*, 127–139.

9.16 TURMERIC

Scientific name: *Curcuma longa* L.
Family: Zingiberaceae
English: Turmeric
Hindi: Haldi
Sanskrit: Haridra
Tamil: Manjal

9.16.1 Origin and Distribution

The plant is a native of southern East Asia and grows through tropical countries, especially cultivated in West Pakistan, India, Bhutan, Malaya, Sri Lanka, Bangladesh, Thailand, Taiwan, Haiti, Jamaica, Peru, and some parts of China. In India, it is an important traditional item of export (about 80% of the world production). Alleppey turmeric (Kerala) is one of the best of the world (Pruthi, 1999).

O plant, thou sprangest up at night, dusky dark colored, black in hue!
So, Rajani, recolor thou these ashy spots, this leprosy

Atharva Veda, I, 23, 1

There are a lot of references to turmeric uses in the ancient Indian literature (*Veda*, *Samhita*, and *Nighantu*) as it was used in India as a natural dye, medicine, and spice of religious significance from about 4000 years.

In 1280, Marco Polo described turmeric as a vegetable with the properties of saffron, so that it was called "the poor's saffron." Now, due to its therapeutical properties, turmeric can be called "the Indian gold" (Aggarwal and Yost, 2011). One of turmeric Sanskrit names is *kanchani*, the golden one (Rajaiah, 2010). Actually, the main use in India is as a spice (about 300–320 g head per year); only a small fraction is used as a dye, a cosmetic, and a medicine (Pruthi, 1999).

Turmeric is a very sacred plant with 108 Sanskrit names related to the number of properties: *bhadra*, "auspicious" (*Raja Nighantu*, Pip, 198/199); *krimighni*, "destroyer of germs" and *varna datri*, "natural dye" (*Raja Nighantu*, Pip, 198/199); *gauri*, "giving a beautiful color to the skin" (*Madanapaala Nighantu*, Abha, 28); *yashita priya*, "loved by women"; *yuvati*, "like a young girl"; and *Lakshmi*, "the goddess of prosperity" (*Raja Nighantu*, Pip, 198/199). It appeared in the Earth when the gods and the devils joined together to extract the nectar of immortality from the ocean of milk. First, a deadly poison came out from the ocean and Shiva drunk it to save the Earth, after that turmeric came out, named also *visaghni* (destroyer of poisons), symbolizing the power of health and longevity (*Dhanvantari Nighantu*, Gudu, 56). At last, Dhanvantari, the god of Ayurveda, appeared holding in the hands the nectar of immortality.

Turmeric is an integral part of Indian cuisine, which has been used in Hindu rituals, wedding, and festival since centuries. It is a very auspicious offering to the holy fire, and during the Hindu festivals, people use to hang up it in front of the door as it is charged with the goddess energy and grants prosperity. It is very effective in purifying the channels of the subtle body (*chakra*). As it is helpful in stretching the ligaments, it is good for the practice of *hatha yoga* (Frawley and Lad, 1994)

9.16.2 Features

It is a perennial herb with a permanent, irregularly rounded or ovate rootstock. It grows 60–90 cm in height. The stem is short with tufts of erected leaves. The leaves are very large, 30–40 cm in length and 10–15 cm in breadth, oblong, and lanceolate. The flowers are pale yellow in spikes concealed by the sheathing petioles; flowering bracts are pale green. The rhizome is cylindrical, ovoid, orange colored, and branched. The spice consists of the dried, boiled, cleaned, and polished rootstock of the plant (Warrier *et al.,* 1994).

There are 30 varieties of *Curcuma*, but *Curcuma longa* is the more important (Pruthi, 1999).

9.16.3 Properties and Uses in Ayurvedic System of Medicine

> All poison can be removed from the human body if treated with a paste of turmeric.
>
> ***Matsya Purana*, 218, 20**

Turmeric is an excellent natural antibiotic, whereas at the same time it strengthens digestion and helps improve intestinal flora. Well known for its blood-purifying quality, it wards off the toxins (Shastri, 1970, Sutra A-4) and augments the hemoglobin percentage. For this purpose, Ayurvedic physicians recommend to consume it with curd (Paranjipe, 2005).

It promotes proper metabolism and aids the digestion of protein, so it is always part of Indian cooking, especially in pulses and milk preparations. Mere sprinkling of turmeric powder in every dish provides protection against tropical germs. With herbs such as fenugreek, cumin, and black pepper, turmeric assumes synergy in combating bacteria.

Turmeric works on all the tissues of the body, and it is useful both externally and internally.

Parts used: Rhizome
Tastes: Bitter, pungent
Quality: Dry, light
Potency: Hot
***Dosha* effect:** K–, PV+
Actions: Stimulant, carminative, antibacterial, anthelmintic

9.16.4 Traditional Uses and Folklore

- In Unani system of medicine, turmeric has a long history of use as food and medicine (Bakhru, 2004). Due to its great medical benefits, it is a major ingredient of many Unani medicines as a stimulant, diuretic, carminative, and anthelmintic (Riazuddin and Siddiq, 2009).
- In the Basti and Siharthnagar districts of Uttar Pradesh, India, the rhizome is used to treat cold and cough (Khanna, 2002). Jhalda tribals (Purulia district, West Bengal) apply locally the paste of the rhizome to treat body pain (Sur and Halder, 2002). In Assam, India, the tribal women apply the paste of fresh rhizome on the skin to enhance the complexion and to protect it from infections (Dekha and Sharma, 2002).
- Turmeric is a common home remedy for anemia (with honey), asthma (with long pepper), diabetes (with laurel leaves powder and *aloe vera* gel), and for dry cough and throat infection (with cow milk).
- Daily use of turmeric powder is suggested in case of stomach pain and abdominal discomfort. Consuming about half teaspoon of turmeric powder in a glass of warm milk works as an effective antispasmodic and eases menstrual cramping.
- Consuming turmeric daily in food helps in eliminating the LDL or bad cholesterol from the body and also helps to prevent heart attacks or strokes. Because turmeric aids in the breakdown of dietary fat, adding half teaspoon of turmeric to daily cooking or to a glass of hot water can help in controlling weight.

Precaution: Turmeric is very safe—only in case of pregnancy, acute jaundice, and hepatitis. It is recommended to use only after consulting a physician. The potency is hot, so people with high *pitta* cannot consume it in excess especially during the summer (Frawley and Lad, 1994).

9.16.5 In the Kitchen

Turmeric is one of the principal ingredients of curry, an ideal addition to high-protein dishes. Its bright golden color adds an attractive tone to all the preparations. It can also be used in Western recipes, with tomato sauce, white sauce, soups, lentils, and rice dishes.

9.16.6 Chemical Constituents and Uses as Functional Food

Turmeric has been used for more than 2500 years, mostly in Asian countries to flavor and color both vegetarian and nonvegetarian food preparations, and it is one of the principal ingredients in curry powder. Turmeric is chemically diverse in composition, depending on varieties, locations, and cultivation conditions. To date, around 235 compounds, primarily phenolics and terpenoids, have been identified from this spice (Li *et al.*, 2011). The curcuminoids belonging to the group of diarylheptanoids are the major bioactive ingredients of turmeric. Commercial curcumin is a mixture of three curcuminoids, the most abundant (~70%) being curcumin followed by its demethoxy and bisdemethoxy derivatives (Pfeiffer *et al.*, 2003). The molecular bases of the pharmacological properties of turmeric against human diseases have been elucidated, and some clinical trials have unequivocally demonstrated its safety and efficacy (reviewed in Gupta *et al.*, 2012b). Two centuries ago, Vogel and Pellettier (1815) reported the isolation of "yellow coloring-matter" from the rhizomes of *C. longa* and named it curcumin. The mixture of curcuminoids in *C. longa* was later elucidated (Srinivasan, 1953), and the main component was identified as curcumin, or diferuloylmethane.

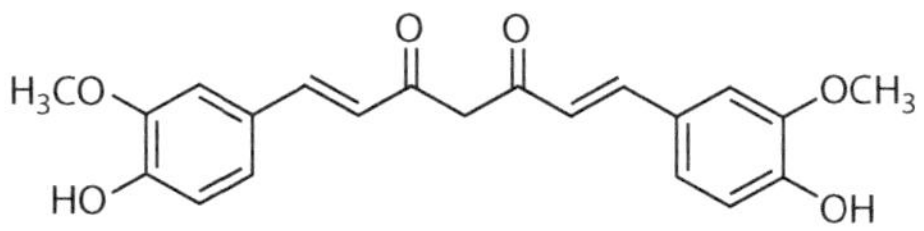

Curcumin

Although turmeric has been consumed as a dietary spice and a cure for human ailments for thousands of years in Asian countries, the biological characteristics of curcumin were not scientifically identified until the mid-twentieth century that disclosed the antibacterial properties of this compound (Schraufstatter and Bernt, 1949). In the 1970s, curcumin was subjected to detailed biological investigations that showed a number of properties such as cholesterol-lowering agent, antidiabetic, anti-inflammatory, and antioxidant. Later, curcumin was demonstrated to have anticancer activity both *in vitro* and *in vivo* (Kuttan *et al.*, 1985). We now know that curcumin can modulate multiple signaling pathways in either a direct or an indirect manner (Gupta *et al.*, 2011, 2012a).

The interest in curcumin has raised rapidly in such a way that the US FDA has approved curcumin as a "generally regarded as safe." It is considered a panacea, with great healing potential, and nowadays it is marketed as a supplement in many countries in several forms, including capsules, tablets, and even energy drinks.

Although curcumin has shown therapeutic efficacy against different human ailments, one of the major problems with its use is the poor bioavailability, essentially due to poor adsorption, rapid metabolism, and rapid systemic elimination. Promising approaches to increase curcumin bioavailability include the use of nanoparticles, liposomes, and micelles, and the synthesis of chemical derivatives (Ohori *et al.*, 2006).

At the molecular level, curcumin has been shown to modulate a wide range of signaling molecules such as indirect and direct targets (reviewed in Epstein *et al.*, 2010). One of the most important targets of curcumin are pro-inflammatory transcription factors known to regulate the expression of genes that contribute to tumorigenesis, cell survival, cell proliferation, invasion, and angiogenesis. Curcumin negatively regulates these transcription factors (Shishodia *et al.*, 2007).

Curcumin is a pleiotropic molecule because it has shown to possess *in vitro* anti-inflammatory, antioxidant, pro-apoptotic, chemopreventive, chemotherapeutic, antiproliferative, wound healing, antinociceptive, antiparasitic, and antimalarial properties (reviewed in Gupta *et al.*, 2012a). Research on animal models provides evidence for the beneficial role of curcumin against various diseases such as inflammatory conditions, diabetes, obesity, neurological disorders, cancer prevention, and treatment (reviewed in Gupta *et al.*, 2012b).

Despite its promising pharmacological profile, to date curcumin has not been approved for treatment of any human disease.

Curcumin constitutes 2%–5% of turmeric and is the most studied component. Although some of the activities of turmeric can be mimicked by curcumin, other activities are curcumin independent.

Lemon rice, a very simple and delicious recipe from South India suggested by Vaidya Kalyan Chakravarty

Ingredients

1 1/2 cups of rice
5–6 tbsp of lemon juice
1/2 cup of roasted peanuts or cashew
1 tbsp of oil
1 tsp of black gram
1 tbsp of Bengal gram
1 tbsp of chopped ginger
2 slits green chili
1/2 tsp of turmeric
10–12 curry leaves
1/2 tsp of mustard seeds
1/4 tsp of asafetida
Salt to taste

Preparation

Soak rice in water for half an hour. Cook the rice with enough salt (each grain should be separate and not sticky). Drain and spread out on a wide vessel. Add roasted peanuts or cashew to the cooked rice and mix well. Keep aside. Heat oil in a pan, add mustard seeds, and let them splutter. Add black gram, Bengal gram, chopped ginger, and green chili, and fry till the *dal* turns red (approximately 2 minutes) over medium heat. Add turmeric, curry leaves, and asafetida, and fry for 10–15 seconds. Turn off heat. Add this to the rice with lemon juice. Combine well. Adjust salt and lemon juice.

REFERENCES

Aggarwal B.B., Yost D. *Healing Spices*. Sterling Publishers, New York, **2011**, 156–172 (Italian version, Armenia).

Bakhru H.K. *Indian Spices & Condiments as Natural Healers*. Jaico Publishing House, Mumbai, India, **2004**, 149–154.

Dekha D., Sharma G.C. Traditional use of plants for beauty care by the tribal people of Goalpara District of Assam, in Trivedi P.C. (ed.), *Ethnobotany*. Aavishkar Publishers, Jaipur, India, **2002**, 366.

Epstein J., Sanderson I.R., Macdonald T.T. Curcumin as a therapeutic agent: The evidence from *in vitro*, animal and human studies. *Br. J. Nutr.* **2010**, *103*, 1545–1547.

Frawley D., Lad V. *The Yoga of Herbs*. Motilal Banarsidass Publishers, Delhi, India, **1994**, 149–150.

Gupta S.C., Patchva S., Koh W., Aggarwal B.B. Discovery of curcumin, a component of the golden spice, and its miraculous biological activities. *Clin. Exp. Pharmacol. Physiol.* **2012a**, *39*(3), 283–299.

Gupta S.C., Prasad S., Kim J.H., Patchva S., Webb L.J., Priyadarsini I.K., Aggarwal B.B., Multitargeting by curcumin as revealed by molecular interaction studies. *Nat. Prod. Rep.* **2011**, 28, 1937–1955.

Gupta S.C., Sung B., Kim J.H., Prasad S., Li S., Aggarwal B.B. Multitargeting by turmeric, the golden spice: From kitchen to clinic. *Mol. Nutr. Food. Res.* **2012b**, 1–19.

Khanna K.K. Ethnobotany of five districts of Teri region of Uttar Pradesh, in Trivedi P.C. (ed.), *Ethnobotany.* Aavishkar Publishers, Jaipur, India, **2002**, 137.

Kuttan B., Bhanumathy P., Nirmala K., George M.C. Potential anticancer activity of turmeric (*Curcuma longa*). *Cancer Lett.* **1985**, *29*, 197–202.

Li S., Yuan W., Deng G.R., Wang P., Yang P., Aggarwal B.B. Chemical composition and product quality control of turmeric (*Curcuma longa* L.). *Pharm. Crops.* **2011**, *2*, 28–54.

Ohori H., Yamakoshi H., Tomizawa M., Shibuya, M., Kakudo Y., Takahashi, A., Takahashi, S. Synthesis and biological analysis of new curcumin analogues bearing an enhanced potential for the medicinal treatment of cancer. *Mol. Cancer Ther.* **2006**, *5*, 2563–2571.

Paranjipe P. *Indian Medicinal Plants. Forgotten Healers*. Chaukhamba Sanskrit Pratishthan, Delhi, India, **2005**, 94–96.

Pfeiffer E., Hhle S., Solyom A., Metzler M. Studies on the stability of turmeric constituents. *J. Food Engineer.* **2003**, *56*, 257–259.

Pruthi J.S. *Spices and Condiments*, National Book Trust, Delhi, India, **1998** (5th edn.), 244–250.

Rajaiah R. *How the Banana Goes to Heaven. And Other Secrets of Health from the Indian Kitchen*. Westland, Chennai, India, **2010**, 130.

Riazuddin A.S., Siddiq M. *Vedic Plants. Medicinal and Other Uses*. Chaukhamha Orientalia, Varanasi, India, **2009**, 68–70.

Schraufstatter E., Bernt H. Antibacterial action of curcumin and related compounds. *Nature* **1949**, *164*, 456.

Shastri K.N. (ed.) *Charaka Samhita*. Chaukhambha Vidyabhavan, Varanasi, India, **1970**.

Shishodia S., Singh T., Chaturvedi M.M. Modulation of transcription factors by curcumin. *Adv. Exp. Med. Biol.* **2007**, *595*, 127–148.

Srinivasan K.R. A chromatographic study of the curcuminoids in *Curcuma longa*, L. *J. Pharm. Pharmacol.* **1953**, *5*, 448–457.

Sur P.R., Halder A.C. Observations on the ethnobotany of Purulia district West Bengal, in Trivedi P.C. (ed.), *Ethnobotany*. Aavishkar Publishers, Jaipur, India, **2002**, 154.

Vogel, A., Pellettier. Examen chimique de la racine de Curcuma. *J. Pharm.* **1815**, *1*, 289, 1–300.

Warrier P.K., Nambiar V.P.K., Ramankutty C. (eds.) *Indian Medicinal Plants*. Orient Longman, Chennai, India, **1994**, 259–261.

10 Healing Properties of Food

Luisella Verotta and Maria Pia Macchi

CONTENTS

10.1 FRUIT

10.1.1 APPLE

Scientific name: *Malus domestica* Borkh.
Family: Rosaceae
Hindi: Sev
Sanskrit: Sevaphalam
Tamil: Applepazham

10.1.1.1 Origin and Distribution

The apple is the pomaceous fruit of the apple tree, which belongs to the species *Malus domestica* of the rose family (Rosaceae). It is one of the most widely cultivated tree fruits and the most widely known of the many members of genus *Malus* that are used by humans. Apples have been present in the mythology and religions of many cultures, including Greek and Christian traditions. In Greece, the apple was sacred to Era and Aphrodite, and associated with love and marriage (Brosse, 1985). In the north of Europe, apples are called "the Food of Gods" and are said to provide eternal youthfulness (Bakhru, 1994).

There are more than 7500 known cultivars of apples, resulting in a range of desired characteristics. Different cultivars are available for temperate and subtropical climates, and are bred for various tastes and uses, including cooking, fresh eating, and cider production. The main producer is China with around 33,000 tons per year (2010), followed by the United States, Turkey, and Italy.

The center of diversity of the genus *Malus* is in eastern Turkey. The apple tree was perhaps the earliest tree to be cultivated, and its fruits have been improved through selection over thousands of years. Alexander the Great is credited with finding dwarfed apples in Kazakhstan in 328 BC. Winter apples, picked in late autumn and stored just above freezing, have been an important food in Asia and Europe for millennia. In India, the tree is cultivated mainly in Himachal Pradesh and in Mustang and Dolpo regions of Nepal.

10.1.1.2 Features

The tree can reach 15 m and has grayish trunk and pubescent young branches, ovate-elliptic and dentate leaves, and white or pink flowers. The fruit is fleshy, with tough skin and variable color, according to the variety.

10.1.1.3 Properties and Uses in Ayurvedic System of Medicine

Ayurvedic texts recommend eating cooked apples as they are warm, soft, and easier to digest. Raw apples can be too cold and dry for *vata* constitution, especially during winter. While cooking, it is suggested to add cinnamon and clove to enhance the apple's properties.

Apple is very nourishing, causes the formation of phlegm, increases virility and appetite, and purifies blood. It is beneficial in many diseases such as diarrhea, piles, constipation, weakness, headache, and *pitta* and *kapha* types of arthritis (Nibodhi, 2010).

Parts used: Fruit
Tastes: Sweet, astringent, sour
Quality: Light, rough
Potency: Cold
***Dosha* effect:** PK–, V+
Actions: Astringent, refrigerant, nutritive

10.1.1.4 Traditional Uses and Folklore

- Since ancient times, apple taken with honey has been considered to be a very effective cure for heart disorders.
- If an apple is taken daily after dinner, it cures constipation and other diseases of the stomach.
- Baked apples with cinnamon and clove strengthen the digestive system.

Precautions: It is very important to purchase only organic apple as the apple is subjected to a lot of treatments with pesticides.

10.1.1.5 In the Kitchen

Ayurveda suggests eating a stewed apple for breakfast: it will stimulate regular bowel movements, increase vitality and alertness, and provide a light but satisfying start to the day.

10.1.1.6 Chemical Constituents and Uses as Functional Food

Apples are often eaten raw. The whole fruit including the skin is suitable for human consumption, except for the seeds, which may affect some consumers due to the content in amygdaline, a cyanogenic glucoside. Apples can be canned or juiced. They are milled to produce apple cider (nonalcoholic, sweet cider) and filtered for apple juice. The juice can be fermented to make vinegar. Through distillation, various alcoholic beverages can be produced, such as Calvados. From seeds, oil can be obtained. Apples are an important ingredient in many desserts, such as apple pie, apple crumble, apple crisp, and apple cake. They are often eaten baked or stewed, and they can also be dried and eaten or reconstituted (soaked in water, alcohol, or some other liquid) for later use.

A typical apple serving weighs 242 g and contains 126 calories with significant dietary fiber and vitamin C content (USDA Nutrient Database).

The proverb "An apple a day keeps the doctor away," addressing the health effects of the fruit, dates from nineteenth-century Wales, and it is supported by recent studies on the benefits of this fruit (reviewed in Hyson, 2011, and Boyer and Liu, 2004).

Besides a number of *in vitro* data using cancer cell lines, a number of animal studies support the protective effect of apple products (including a clear juice preparation) on colon cancer. Apple products reduced important markers, including DNA damage and hyperproliferation. The findings of all the investigators suggest that the whole is greater than the sum of the parts in terms of a protective effect of apple products (Barth *et al.*, 2007; Koch *et al.*, 2009).

Animal feeding has shown the potential of apple products in breast cancer reduction, with a significant dose-dependent reduction in the number and onset of mammary tumors compared to control rats (daily administration of whole apples for 24 weeks). Dose-dependent inhibition of markers of cell proliferation was found (Liu *et al.*, 2009).

Dietary intake of apple products (fruit or juice) was quantified in a hospital-based human trial to look for a statistical link between disease risk and defined amounts of apple products. The consumption of one medium-sized apple (170 g) per day was associated with a reduction in risk of cancer at several sites (esophagus, colorectal, larynx, breast, ovary) compared to lower or null consumption (Gallus *et al.*, 2005). In another trial, fruit and vegetable consumption was correlated with lung cancer risk (Linseisen *et al.*, 2007). The study revealed a significant inverse association between combined apple and pear intake and lung cancer in all participants and current smokers.

Apple may reduce the risk of chronic disease by various mechanisms, including antioxidant, antiproliferative, and cell signaling effects (Gerhauser, 2008). Exposure to apples and apple products has been associated with beneficial effects on risk, markers, and etiology of cardiovascular disease (CVD), asthma, and Alzheimer's disease. Recent work suggests that these products may also be associated with improved outcomes related to cognitive decline of normal aging, diabetes, weight management, bone health, pulmonary function, and gastrointestinal protection (reviewed in Hyson, 2011).

Extracts isolated from apple show an inhibitory effect on pancreatic lipase activity, which is one of the enzymes involved in fat digestion, suggesting a potential in the management of the obesity epidemic.

Apple pomace is a by-product of the cider-processing industry, representing about 35% of the original fruit. The disposal of pomace has an added cost to the beverage industry, especially in regions with an ancient and vast tradition in cider consumption associated with high production rates, which is the case in Asturias where about 20,000 tons of pomace is generated every year (Suárez *et al.*, 2010). This product is a poor animal food due to its low-protein content, but in recent years, the nutritional value of apple pomaces has been demonstrated, and these by-products have been regarded as a food additive due to their high content in antioxidant polyphenols and dietary fiber. The antiherpes simplex virus type 1 and type 2 effects of apple pomace have also been recently demonstrated (Álvarez *et al.*, 2012).

Apples contain a wide range of polyphenols (hydroxycinnamic acids, oligomeric proanthocyanidins, flavonols) quantified in 5,300–27,000 mg/kg dry weight, mostly concentrated in the peel (Wojdylo *et al.*, 2008).

Epicuticular waxes of apples were studied as early as 1920, and ursolic acid was one of the first reported components [reviewed in Szakiel *et al.*, 2012]. There is a growing list of other triterpenoids that are also present in apple fruit cuticular wax with ursane and oleanane skeletons. The antitumor activity of triterpenoids isolated from apple peel has been investigated with a variety of tests. Ursolic acid itself is ubiquitous in medicinal and edible plants (e.g., herbs such as basil, peppermint, and rosemary), and it has attracted attention due to its diverse pharmacological properties, including antitumorigenic effects exerted by inhibition of the STAT3 pathway. This compound was shown to inhibit nuclear factor (NF)-κB activation induced by carcinogenic agents through suppression of IκBα kinase and p65 phosphorylation, and it also suppressed tumor necrosis factor (TNF)-induced expression of cyclin D_1, cyclooxygenase 2 (COX-2), and matrix metalloproteinase 9, which are involved in the initiation, promotion, and metastasis of tumors (Shishodia *et al.*, 2003).

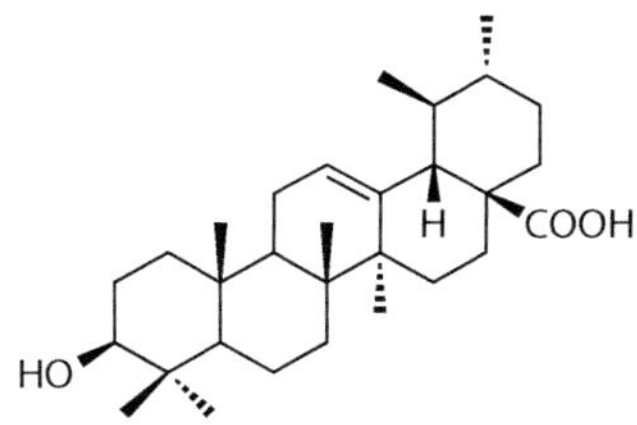

Ursolic acid

Cooked apples with cloves, a recipe from Maharishi Ayurveda

Ingredients

1 whole fresh, sweet apple, cored and peeled
5 whole cloves
1/4 cup of purified water

Preparation

Cut the apple into small pieces. Add cloves, apple pieces, and water in a covered pot. Cook the apple pieces in 1/4 cup of water until they are soft. Discard the cloves, cool slightly, and enjoy.

REFERENCES

Álvarez A.L., Melón S., Dalton K.P., Nicieza I., Roque A., Suárez B., Parra F. Apple pomace, a by-product from the Asturian cider industry, inhibits Herpes Simplex Virus Types 1 and 2 in vitro replication: Study of its mechanisms of action. *J. Med. Food* **2012**, *15*(6), 581–587.

Bakhru H.K. *Foods That Heal*. Orient Paperbacks, Delhi, India, **1994**, 35–38.

Barth S.W., Faehndrich C., Bub A., Watzl B., Will F., Dietrich H., Rechkemmer G., Briviba K. Cloudy apple juice is more effective than apple polyphenols and an apple juice derived cloud fraction in a rat model of colon carcinogenesis. *J. Agric. Food Chem.* **2007**, *55*, 1181–1187.

Boyer J., Liu R. Apple phytochemicals and their health benefits. *Nutr. J.* **2004**, *3*, 5.

Brosse J. *Les arbres de France*. Librarie Plon, Paris, France, **1985** (Italian version Ed. Studio Tesi, 1991, 127–135).

De la Garza A.L., Milagro F.I., Boque N., Campión J., Martínez J.A. Natural inhibitors of pancreatic lipase as new players in obesity treatment. *Planta Med.* **2011**, *77*(8), 773–785.

Gallus S., Talamini R., Giacosa A., Montella M., Ramazzotti V., Franceschi S., Negri E., La Vecchia C. Does an apple a day keep the oncologist away? *Ann. Oncol.* **2005**, *16*, 1841–1844.

Gerhauser C. Cancer chemopreventive potential of apples, apple juice, and apple components. *Planta Med.* **2008**, *74*, 1608–1624.

Hyson D.A. A comprehensive review of apples and apple components and their relationship to human health. *Adv. Nutr.* **2011**, *2*(5), 408–420.

Koch T.C., Briviba K., Watzl B., Fähndrich C., Bub A., Rechkemmer G., Barth S.W. Prevention of colon carcinogenesis by apple juice in vivo: Impact of juice constituents and obesity. *Mol. Nutr. Food Res.* **2009**, *53*(10), 1289–1302.

Linseisen J., Rohrmann S., Miller A., Bas Bueno-de-Mesquita H., Büchner F., Vineis P., Agudo A. *et al.* Fruit and vegetable consumption and lung cancer risk: Updated information from the European Prospective Investigation into Cancer and Nutrition (EPIC). *Int. J. Cancer.* **2007**, *121*, 1103–1114.

Liu J.-R., Dong H.-W., Chen B.-Q., Zhao P., Liu R. Fresh apples suppress mammary carcinogenesis and proliferative activity and induce apoptosis in mammary tumors of the Sprague-Dawley rat. *J. Agric. Food Chem.* **2009**, *57*, 297–304.

Nibodhi. *Annapurna's Prashad.* Mata Amritanandamayi Mission Trust, Amritapuri, India, **2010** (2nd edn.), 72–73.

Shishodia S., Majumdar S., Banerjee S., Aggarwal B.B. Ursolic acid inhibits nuclear factor-κB activation induced by carcinogenic agents through suppression of IκBα kinase and p65 phosphorylation: Correlation with down-regulation of cyclooxygenase 2, matrix metalloproteinase 9, and cyclin D1. *Cancer Res.* **2003**, *63*, 4375–4383.

Suárez B., Álvarez A.L., Diñeiro Y., del Barrio G., Picinelli A., Parra F. Phenolic profiles, antioxidant activity and in vitro antiviral properties of apple pomace. *Food Chem.* **2010**, *120*, 339–342.

Szakiel A., Pączkowski C., Pensec F., Bertsch C. Fruit cuticular waxes as a source of biologically active triterpenoids. *Phytochem. Rev.* **2012**, *11*, 263–284.

USDA Nutrient Database. http://ndb.nal.usda.gov/.

Wojdylo A., Oszmianski J., Laskowski P. Polyphenolic compounds and antioxidant activity of new and old apple varieties. *J. Agric. Food Chem.* **2008**, *56*, 6520–6530.

10.1.2 Banana

Scientific name: *Musa paradisiaca* L.
Family: Musaceae
Hindi: Kela
Sanskrit: Kadali
Tamil: Vazhaipazham

10.1.2.1 Origin and Distribution

The plant originated in the Indo-Malaysian region reaching to northern Australia. It was well known in the Mediterranean area in the third century BC and was first carried to Europe probably in the tenth century AD. Early in the sixteenth century, Portuguese mariners transported the plant from the West African coast to South America. At present, the plant grows in 135 countries across the tropics and subtropics.

There are two different versions of the name *Musa*: it can be in memory of Antonius Musa, the physician of Augustus Caesar, who is said to have first promoted the fruit in Europe, or it is simply derived from the Sanskrit name *mocha*, meaning "juicy" and also "ascetic." *Paradisiaca* is a Linnaeus reference to the old belief that the forbidden fruit of the Paradise was not the apple, but the banana (Rajaiah, 2010). It is said that banana leaves were the first garment of Adam and Eve.

According to a South Asian legend, the plant is regarded as an incarnation of the Goddess Parvati, as it fertilizes itself without cross-pollination (Isomaki and Gandhi, 2004). There is also an ancient belief that Lakshmi, the goddess of wealth, resides on a banana leaf, so the fruit is associated with prosperity and fertility. A banana plant is often installed in the corner of a rice field as a protective charm (Gupta, 2001). The leaves are used as decoration for temples and houses during marriages, and the fruits are an offering during various rituals (Majupuria and Joshi, 1997). Buddhist sculptures show banana leaves, and a drink called *mochapana* is mentioned in the Buddhist book of monastic rules.

10.1.2.2 Features

A tall, herbaceous plant with aerial pseudostem dying after flowering, with oblong leaves, flowers unisexual in spike, and fruits berries in several clusters, generally golden yellow color on ripening (Warrier *et al.*, 1996).

10.1.2.3 Properties and Uses in Ayurvedic System of Medicine

Sri Raghunatha Suri's *Bhojanakutuhalam* refers to different properties of the various varieties of banana, which can have different effects on the *doshas*, but all are very nutritious and stimulate the digestive fire. They are heavy and difficult to digest (Venkat *et al.*, 2012).

The banana fruit strengthens the muscle tissue (*mamsa dhatu*) (Nibodhi, 2010). Ripe banana may raise mood and alleviate insomnia, treat anemia and hypertension, and protect against stroke.

The flowers are good for dysentery, diabetes, ascites, and dropsy. The stem is very specific for renal and vesicular calculi (Warrier *et al.*, 1996).

Parts used: Fruits, flowers, stem
Tastes: Sweet, sour (ripe); astringent (unripe)
Quality: Heavy, oily
Potency: Cold
***Dosha* effect:** Ripe banana V–, PK+; unripe banana PK–, V+
Actions: Astringent, refrigerant, laxative, nutritive, tonic, strengthening, aphrodisiac

10.1.2.4 Traditional Uses and Folklore

- In ancient Persia, the banana was regarded as a nature's secret of perpetual youth (Bakhru, 1994).
- Regular intake of ripe bananas is beneficial for constipation, body weakness, heart weakness, and intestinal ulcers.
- A common home remedy for indigestion is a ripe banana along with a cup of milk before sleep.
- For piles, it is suggested to boil a mashed ripe banana in one teacup of milk and take it twice or thrice a day.
- A diet of bananas only for 3 or 4 days is advised in case of arthritis and gout.
- Unripe fruit is a valuable article of diet, used as a vegetable especially for those suffering from hemoptysis and diabetes. Flour made of green plantains dried in the sun is used as *chappatis* in cases of dyspepsia with flatulence and acidity. A slight gruel made of banana flour mixed with milk is beneficial in cases of gastritis. Cooked flowers are also used in diabetes. A soup made of flowers is given in convalescence after diarrhea (Nadkarni, 2002).

Precautions: Ripe banana fruits in excess can cause dysentery. Unripe banana is very difficult to digest, so it is usually cooked.

10.1.2.5 In the Kitchen

Banana leaves are widely used in South India as plates, for lining cooking pits and for wrapping food for cooking or storage. Traditionally, there is a belief that by eating food on a banana leaf platter, one wards off infections and food poisoning (Sairam, 1998). The practice started in ancient times for hygienic reasons. The tradition of cooking in banana leaves is actually common through

Asia, Latin America, and Central Africa. The food steamed in the banana leaf is healthy and delicious.

10.1.2.6 Chemical Constituents and Uses as Functional Food

Banana is very low in saturated fat, cholesterol, and sodium. It is also a good source of dietary fiber, vitamin C, potassium, and manganese, and a very good source of vitamin B_6 (USDA Standard Reference-21). The cultivars do not influence the chemical composition, except for insoluble fiber content. Variations of the chemical composition were observed in the bananas from different areas according to cultivation method (greenhouse and outdoors) and farming style (conventional and organic). The concentrations of Na, K, Ca, Mg, Fe, Cu, Zn, and Mn also vary depending on the location of cultivation (Forster *et al.*, 2002a, 2002b).

Banana cultivars rich in provitamin A carotenoids may offer a potential food source for alleviating vitamin A deficiency, particularly in developing countries. Many factors are associated with the presently known food sources of vitamin A that limit their effectiveness in improving vitamin A status. Acceptable carotenoid-rich banana cultivars have been identified in Micronesia, and some carotenoid-rich bananas have been identified elsewhere (Davey *et al.*, 2007; Englberger *et al.*, 2003, 2006).

Banana inflorescences (from *Musa acuminata*) are popularly known as navels, and they are used in Brazil as nutritional complements. The dehydrated inflorescences were found to contain a significant nutritive complement based on their high content of potassium (5008.26 mg/100 g) and fiber 49.83% (lignin, cellulose, and hemicelluloses), revealing important functional and nutritional properties (Fingolo *et al.*, 2012).

Six different cultivars of banana flowers (*Musa paradicicus*) (Kathali, Bichi, Shingapuri, Kacha, Champa, and Kalabou) were analyzed for the content of polyphenol expressed as gallic acid equivalent and flavonoid expressed as quercetein equivalent, and the *in vitro* total antioxidative activities of the flower extracts were compared with standard and expressed as trolox equivalent. The Kacha variety contains the maximum amount of polyphenol (11.94 ± 0.03 mg of gallic acid equivalent/g of dry weight) and flavonoid (0.174 ± 0.001 g of quercetin equivalent/g of polyphenol). It also has the highest total antioxidant capacity (Vijayakumar *et al.*, 2008). The study suggests that the flowers of *M. paradicicus* that are found in India and consumed as vegetable can provide valuable functional ingredients that help in the prevention of oxidative stress (China *et al.*, 2011). Banana flower and pseudostem are by-products of banana cultivation and are rich sources of dietary fiber associated with polyphenols, which could promote health beneficial effects (Bhaskar *et al.*, 2012). Recent findings also suggest that banana flower and pseudostem have antidiabetic and anti-advanced glycation end product properties, and are beneficial as food supplements for diabetics (Bhaskar *et al.*, 2011b). The mechanism of action appears due to banana flower and pseudostem extracts' ability to promote glucose uptake into the cells, presumably through glucose transporters. Glucose uptake is likely promoted by phenolic acids besides other bioactives. It can be hypothesized that consumption of nutraceutical-rich extract of banana flower and pseudostem could replace some amount of insulin being taken for diabetes (Bhaskar *et al.*, 2011a).

The unripe banana flour (UBF) presents a high amount of total dietary fiber (56.24 g/100 g), which consisted of resistant starch (48.99 g/100 g) and fructans (0.05 g/100 g). The contents of available starch and soluble sugars were low. Appropriate procedure used to obtain the UBF resulted in the recovery of undamaged starch granules and in a low-energy product (597 kJ/100 g) (Menezes *et al.*, 2011). In general, flour prepared from unpeeled green banana (*Musa acuminata*) was found to show enhanced nutrition values with higher contents of mineral, dietary fiber, and total phenolics (Haslinda *et al.*, 2009).

Organic acids have been found to be slowly released from the food matrix during simulated digestion of both banana and sweet potato, either raw or after freeze-drying. In the Caco-2 cell assay, malic and oxalic acids were absorbed more than citric acid. Oxalic and citric acids, but not malic acid, were transported across the cell monolayer. Organic acids from plant food have been shown to play an important role in the prevention of chronic diseases (osteoporosis, obesity), inherent to Western diets. Partial uptake is consistent with a dual nutritional role for organic acids as alkalinizing agents (fraction that is taken up) and as modulators of large intestinal function (fraction that is not taken up in the small intestine) (Sabboh-Jourdan *et al.*, 2011).

REFERENCES

Bakhru H.K. *Foods That Heal.* Orient Paperbacks, Delhi, India, **1994**, 31–35.

Bhaskar J.J., Salimath P.V., Nandini C.D. Stimulation of glucose uptake by *Musa* sp. (cv. elakki bale) flower and pseudostem extracts in Ehrlich ascites tumor cells. *J. Sci. Food Agric.* **2011a**, *91*(8), 1482–1487.

Bhaskar J.J., Shobha M.S., Chilkunda N.D., Salimath P.V. Banana (*Musa* sp. var. elakki bale) flower and pseudostem: Dietary fiber and associated antioxidant capacity. *J. Agric. Food Chem.* **2012**, *60*(1), 427–432.

Bhaskar J.J., Shobha M.S., Sambaiah K., Salimath P.V. Beneficial effects of banana (*Musa* sp. var. elakki bale) flower and pseudostem on hyperglycemia and advanced glycation end-products (AGEs) in streptozotocin-induced diabetic rats. *J. Physiol. Biochem.* **2011b**, *67*(3), 415–425.

China R., Dutta S., Sen S., Chakrabarti R., Bhowmik D., Ghosh S., Dhar P. In vitro antioxidant activity of different cultivars of banana flower (*Musa paradicicus* L.) extracts available in India. *J. Food Sci.* **2011**, *76*(9), C1292–C1299.

Davey M.W., Stals E., Ngoh-Newilah G., Tomekpe K., Lusty C., Markham R., Swennen R., Keulemans J. Sampling strategies and variability in fruit pulp micronutrient contents of West and Central African bananas and plantains (*Musa* species). *J. Agric. Food Chem.* **2007**, *55*(7), 2633–2644.

Englberger L., Darnton-Hill I., Coyne T., Fitzgerald M.H., Marks G.C. Carotenoid-rich bananas: A potential food source for alleviating vitamin A deficiency. *Food Nutr. Bull.* **2003**, *24*(4), 303–318.

Englberger L., Schierle J., Aalbersberg W., Hofmann P., Humphries J., Huang A., Lorens A. *et al.* Carotenoid and vitamin content of Karat and other Micronesian banana cultivars. *Int. J. Food Sci. Nutr.* **2006**, *57*(5/6), 399–418.

Fingolo C.E., Braga J.M., Vieira A.C., Moura M.R., Kaplan M.A. The natural impact of banana inflorescences (*Musa acuminata*) on human nutrition. *An. Acad. Bras. Cienc.* **2012**, *84*(4), 891–898.

Forster M.P., Rodríguez Rodríguez E., Díaz Romero C. Differential characteristics in the chemical composition of bananas from Tenerife (Canary Islands) and Ecuador. *J. Agric. Food Chem.* **2002a**, *50*(26), 7586–7592.

Forster M.P., Rodríguez Rodríguez E., Martín J.D., Díaz Romero C. Statistical differentiation of bananas according to their mineral composition. *J. Agric. Food Chem.* **2002b**, *50*(21), 6130–6135.

Gupta S.M. *Plant Myths and Traditions in India.* Munshiram Manoharlal Publishers, Delhi, India, **2001**, 44–45.

Haslinda W.H., Cheng L.H., Chong L.C., Noor Aziah A.A. Chemical composition and physicochemical properties of green banana (*Musa acuminata* × *balbisiana* Colla cv. Awak) flour. *Int. J. Food Sci. Nutr.* **2009**, *60*(Suppl 4), 232–239.

Isomaki R., Gandhi M. *The Book of Trees.* Other India Press, Goa, India, **2004**, 202–205.

Majupuria T.C., Joshi D.P. *Religious and Useful Plants of Nepal and India.* Gupta M.(ed.), Publishers, Lashkar, India, **1997**, 243.

Menezes E.W., Tadini C.C., Tribess T.B., Zuleta A., Binaghi J., Pak N., Vera G., Dan M.C., Bertolini A.C., Cordenunsi B.R., Lajolo F.M. Chemical composition and nutritional value of unripe banana flour (*Musa acuminata*, var. Nanicão). *Plant Foods Hum. Nutr.* **2011**, *66*(3), 231–237.

Nadkarni K.M. *Indian Materia Medica*, vol. 1. Popular Prakashan, Mumbai, India, **2002**, 822–827.

Nibodhi. *Annapurna's Prashad.* Mata Amritanandamayi Mission Trust, Amritapuri, India, **2010** (2nd edn.), 75–76.

Rajaiah R. *How the Banana Goes to Heaven. And Other Secrets of Health from the Indian Kitchen.* Westland, Chennai, India, **2010**, 167–176.

Sabboh-Jourdan H., Valla F., Epriliati I., Gidley M.J. Organic acid bioavailability from banana and sweet potato using an in vitro digestion and Caco-2 cell model. *Eur. J. Nutr.* **2011**, *50*(1), 31–40.

Sairam T.V. *Home Remedies*, vol. I. Penguin Books, Delhi, India, **1998**, 174–182.

USDA Standard Reference-21. http://ndb.nal.usda.gov/.

Venkat P., Gangadharan G.G., Lakshmithathachar M.A., Alwar M.A. (eds.) *Bhojanakutuhalam.* I-AIM, Bangalore, India, **2012**, 138–140.

Vijayakumar S., Presannakumar G., Vijayalakshmi N.R. Antioxidant activity of banana flavonoids. *Fitoterapia.* **2008**, *79*(4), 279–282.

Warrier P.K., Nambiar V.P.K., Ramankutty C. (eds.) *Indian Medicinal Plants*, vol. 4. Orient Longman, Chennai, India, **1996**, 78–85.

10.1.3 Berries

10.1.3.1 Properties and Uses in Ayurvedic System of Medicine

Berries are generally considered a liver tonic. Unripe berries counter excessive urination, nocturnal emissions, sexual debility, and diarrhea (Nibodhi, 2010). Tierra (1999) includes a number of berries in a group of highly nurturing herbs, used by Ayurveda and Chinese medicine to treat anemia, blood deficiency, and liver problems. As the blood is intimately related to *pitta*, berries are specific also for *pitta* disorders. These fruits tend to nourish a feeling of tolerance and love; they foster a maternal kind of patience and devotion.

- *Blueberries* are *tridoshic* in moderation.
- *Blackberries* are used against diarrhea and hemorrhoids.
- *Raspberries* are relatively *tridoshic*. The berries are nutritive and refrigerant. The fruit and juice are taken for anemia (Tierra, 1999). The leaves are astringent and are used to prepare an herbal tea that increases the muscle tone of the uterine wall and eases the pain of childbirth. When ripe, they are

a great blood and liver cleanser. They should not be eaten in quantity and never with dairy products (Nibodhi, 2010).
Tastes: Sweet, astringent (unripe), sour
Potency: Cold
***Dosha* effect:** PK–, V+ (slightly, when dry)
Actions: Refrigerant, astringent

- *Cranberries* help to build the immune defenses and alleviate *pitta*-related disorders such as skin rashes, burning urination, and edema.
Tastes: Sour, astringent
Potency: Hot
***Dosha* effect:** P–, VK+
Action: Diuretic

- *Strawberries* are a traditional diuretic with a healing effect on the kidneys. They alleviate *vata*-related disorders, such as rheumatism, gout, hypertension, and constipation. They also relieve many *pitta* disorders such as herpes simplex and skin problems. It has been suggested that strawberries need to be consumed alone, that is, nothing else should be consumed along with them (Nibodhi, 2010).
Tastes: Sour, sweet
Potency: Cold (heat when sour)
***Dosha* effect:** VP–, K+ (in excess)
Actions: Refrigerant, mild astringent, relieves thirst

- *Mulberry* (*Morus alba*) belongs to the Moraceae family. It is a tonic used for anemia and insomnia. It is very good for constipation due to excess of *vata*. It is sweet and has a cold potency (Tierra, 1999).
- *Black currant* (*Ribes rubrum*) belongs to the family of Saxifragaceae. It has a sweet and sour taste and a cooling potency. It has a demulcent and nutritive action. It is traditionally used to treat anemia, inflammatory diseases, excessive thirst, and sore throat with dry cough (Tierra, 1999).

10.1.3.2 Chemical Constituents and Uses as Functional Food

Berries are an important source of provitamin A, minerals, vitamin C, and B complex vitamins. Berry fruits contain about 15% soluble solids (mainly sugars) and a high level of fructose. They are very low in fat and sodium. Berries contain high levels of a diverse range of phytochemicals, most of which are phenolic molecules (reviewed in Nile and Park, 2014). These polyphenols comprise a wide variety of compounds, divided into several classes such as hydroxybenzoic acids, hydroxycinnamic acids, anthocyanins, proanthocyanidins, flavonols, flavones, flavanols, flavanones, isoflavones, stilbenes, and lignans. Among the colorful fruits, berries such as blackberry (*Rubus* species), black raspberry (*Rubus occidentalis*), blueberry, bilberry, cranberry (*Vaccinium* species), red raspberry (*Rubus idaeus*), and strawberry (*Fragaria ananassa*) are the richest in anthocyanidins.

Many laboratory and animal studies have shown that berries have anticancer, antioxidant, and antiproliferative properties. The bioactive components of berries impart anticancer effects through various complementary and overlapping mechanisms of action, including the induction of metabolizing enzymes, the modulation of gene expression, and their effects on cell proliferation, apoptosis, and subcellular signaling pathways. Some berries, such as strawberries and black raspberries, have been identified as sources of phenolic compounds such as gallic and ellagic acids, which have potential cancer chemopreventive activity (Xue *et al.*, 2001).

From research laboratories, worldwide there is growing evidence that berries are an important part of a healthy diet. Berry phenolics are best known for their ability to act as antioxidants, but the biological activities exerted by berry phytochemicals *in vivo* extend beyond antioxidation. In fact, a large and growing body of evidence shows that berry phytochemicals regulate the activities of metabolizing enzymes; modulate nuclear receptors, gene expression, and subcellular signaling pathways; and repair DNA oxidative damage. However, berry phenolics are extensively metabolized and also further converted by the colonic microflora into related molecules. These compounds may persist *in vivo*, accumulate in target tissues, and contribute significantly to the biological effects that have been observed for berry fruits.

Among commonly consumed berries, blueberries and cranberries contain predominantly anthocyanins (Chu *et al.*, 2011; Neto and Vinson, 2011), whereas blackberries, black raspberries, red raspberries, and strawberries contain predominantly ellagitannins (Giampieri *et al.*, 2012). Therefore, the class (and specific chemical structures) of tannins present in a particular berry type may contribute significantly to the unique biological properties. For example, the bacterial antiadhesive properties observed for the cranberry appear to be unique among berry fruits. Similarly, the distinct biological effects observed for blueberries (a proanthocyanidin-rich fruit) and strawberries (an ellagitannin-rich fruit) on neuronal function and behavior in aging animals may be due to the effects of the individual classes of tannins in different regions of the brain.

Ellagitannin-enriched extracts, prepared from *Rubus fruticosus* L. (blackberry) and *R. idaeus* L. (raspberry) have recently demonstrated a preventing effect in gastric inflammation and support for their use in dietary regimens against peptic ulcer (Sangiovanni *et al.*, 2013).

REFERENCES

Chu W.-K., Cheung S.C.M., Lau R.A.W., Benzie I.F.F. Bilberry (*Vaccinium myrtillus* L.), in Benzie I.F.F., Wachtel-Galor S. (eds.), *Herbal Medicine: Biomolecular and Clinical Aspects*. CRC Press, Boca Raton, FL, **2011** (2nd edn.), chapter 4.

Giampieri F., Tulipani S., Alvarez-Suarez J.M., Quiles J.L., Mezzetti B., Battino M. The strawberry: Composition, nutritional quality, and impact on human health. *Nutrition* **2012**, *28*, 9–19.

Neto C.C., Vinson J.A. Cranberry, in Benzie I.F.F., Wachtel-Galor S. (eds.), *Herbal Medicine: Biomolecular and Clinical Aspects*. CRC Press, Boca Raton, FL, **2011** (2nd edn.), chapter 6.

Nibodhi. *Annapurna's Prasad*. Mata Amritanandamayi Mission Trust, Amritapuri, India, **2010**, 76–79.
Nile S.H., Park S.W. Edible berries: Bioactive components and their effect on human health. *Nutrition* **2014**, *30*, 134–144.
Sangiovanni E., Vrhovsek U., Rossoni G., Colombo E., Brunelli C., Brembati L., Trivulzio S. *et al.* Ellagitannins from *Rubus* berries for the control of gastric inflammation: In vitro and in vivo studies. *PLoS One* **2013**, *8*(8), e71762.
Tierra M. *Planetary Herbology*. Motilal Banarsidass Publishers, Delhi, India, **1999**, 311–318.
Xue H., Aziz R.M., Sun N., Cassady J.M., Kamendulis L.M., Xu Y., Stoner G.D., Klaunig J.E. Inhibition of cellular transformation by berry extracts. *Carcinogenesis* **2001**, *22*(2), 351–356; erratum in: Carcinogenesis 2001, 22(5), 831–833.

10.1.4 Citrus Fruits

10.1.4.1 Grapefruit

Scientific name: *Citrus paradisi* Macfad
Family: Rutaceae

10.1.4.1.1 Origin and Distribution

The grapefruit has been discovered quite recently, in the eighteenth century, in the Barbados and described for the first time by the Rev. Griffith Hughes, with the name of "forbidden fruit" (Hughes, 1750). The botanical name is probably a reference to the heavenly taste. The English name alludes to the appearance of the fruits in clusters similar to grapes.

The fruit is probably the result of a natural cross-breeding occurred between the sweet orange (*Citrus sinensis*) and the pomelo (*Citrus maxima*), brought from Indonesia to the Barbados in the seventeenth century. Grapefruit trees were planted in Florida in the early nineteenth century, and Florida is still a major producer of grapefruits, such as California, Arizona, and Texas. The United States are now the top producer of grapefruits, followed by China, South Africa, Mexico, Syria, Israel, Turkey, India, Argentina, and Cuba.

Approximately 60% of the grapefruit crop is used for the manufacture of juice and canned grapefruit, whereas the rest is sold fresh.

10.1.4.1.2 Features

Citrus paradisi is a subtropical evergreen tree. It is usually 5–6 m tall, although it can reach even 13–15 m. The leaves are dark green, long, and thin. The flowers are white and four petaled. The fruit is yellow-orange skinned, with a diameter of about 10–15 cm. The flesh is segmented, with a bitter and sour taste with a typical underlying sweetness. There are different varieties, with the pulp of different color: white, pink, and ruby. Generally, the redder varieties are sweeter. Some varieties are seedless.

10.1.4.1.3 Properties and Uses in Ayurvedic System of Medicine

Grapefruit reduces phlegm in the morning time, aids the digestion of fats and sugars, helps to lose weight, and stimulates liver and pancreatic secretions (Nibodhi, 2010).

Grapefruit is good for calming *vata* (Morningstar and Desai, 1999).

Parts used: Fruit
Taste: Sour
Quality: Liquid
Potency: Hot
***Dosha* effect:** V–, P+, K=
Actions: Stimulant, expectorant, astringent

10.1.4.1.4 Traditional Uses and Folklore

- Eating a grapefruit 1–2 hours before sleep counteracts insomnia.
- The pulp or the juice of grapefruit helps to recover easily from fever. It reduces the burning sensation that occurs during high temperature.
- A glass full of fresh squeezed grapefruit in the morning is the best remedy to control the constipation.

10.1.4.1.5 In the Kitchen

Grapefruits are available throughout the year, but they are in season and at their best from winter through early spring. They can be eaten raw or made into jam, marmalade, or jelly. The fruit peel is also made into sweet candies.

An excellent salad can be made with fennel cuts mixed with olive oil and salt, covered with fresh grated ginger and pink grapefruit cuts.

REFERENCES

Hughes G. *The Natural History of Barbados*. Selbstverl, London, **1750**, Book V, 127.

Morningstar A., Desai U. *The Ayurvedic Cookbook*. Motilal Banarsidass Publishers, Delhi, India, **1999**, 178.

Nibodhi. *Annapurna's Prashad*. Mata Amritanandamayi Mission Trust, Amritapuri, India, **2010** (2nd edn.), 82–83.

10.1.4.2 Orange

Scientific name: *Citrus sinensis* L.
Family: Rutaceae
Hindi: Narengi
Sanskrit: Swadu-naringa, Nagranga
Tamil: Narangam

10.1.4.2.1 Origin and Distribution

The name of orange derives from the Sanskrit word *nagranga*, probably derived from the Dravidian term *narangam*. The botanical name is due to the origin of the fruit.

Orange is native to China where it was already cultivated in 2500 BC. It has probably been a hybrid between pomelo (*Citrus maxima*) and mandarin (*Citrus reticulata*) cultivated since very ancient times. It was introduced in the Mediterranean area through Moorish Spain and later in Europe by Italian and Portuguese merchants. The fruit was very appreciated in Europe as a luxury item, and wealthy people grew oranges in private conservatories called *orangeries*. Spaniards spread the cultivation of oranges in the Americas in the mid-1500s. Portuguese, Spanish, and Dutch sailors planted citrus trees along trade routes to prevent scurvy. Oranges grow well in tropical and subtropical climates, and have been the most cultivated fruit trees in the world since 1987.

Brazil is the world's leading orange producer, with an output almost as high as that of the next three countries combined (the United States, India, and China). The United States is the second largest producer, with many groves located in Florida, California, Arizona, and Texas.

Other top producers are Mexico, Spain, Egypt, Italy, Indonesia, Turkey, Pakistan, and Iran.

10.1.4.2.2 Features

Orange is a tropical to semitropical, evergreen tree high about 9–10 m, with oval leaves alternately arranged and small flowers. The fruits are green when unripe and orange when fully ripe. Oranges can be sweet or bitter. The bitter variety has a greater medicinal value; the sweet one is the most commonly consumed.

10.1.4.2.3 Properties and Uses in Ayurvedic System of Medicine

Orange is the safest of all the acid fruits. Juice is valuable in bilious affections and stops bilious diarrhea (Nadkarni, 2002). It is an appetizer full of nourishment, gives strength, pacifies inflammation, and helps in food digestion.

It is excellent for treating over acidity, constipation, or a sluggish intestinal tract; it harmonizes the stomach, purifies the blood, and counteracts cough, diabetes, bronchitis, liver and heart diseases, and vomiting (Nibodhi, 2010). It is quite relieving for *vata* (Morningstar and Desai, 1999).

Parts used: Fruit
Tastes: Sweet, sour
Quality: Heavy
Potency: Hot
***Dosha* effect:** V–, PK+ (the sour varieties or if taken in excess)
Actions: Stimulant, expectorant, appetizer

10.1.4.2.4 Traditional Uses and Folklore

- For children suffering from anemia and nervous debility, fresh grape juice mixed with orange juice is an excellent tonic (Nadkarni, 2002).
- Taking one orange in the evening before sleep and one in the morning can cure chronic constipation.

10.1.4.3 Lemon

Scientific name: *Citrus limon* (L.) Burm. f.
Family: Rutaceae
Hindi: Jamiri nimbu
Sanskrit: Jambirah
Tamil: Elumiccai

10.1.4.3.1 Origin and Distribution

The origin of the word *limon* may be from the Arabic *laymun* or *limun* and from the Persian *limun*, a generic term for citrus fruit.

Lemon tree has been cultivated in Southeast Asia since ancient times: it is probably a hybrid between sour orange and citron, which first grew in Assam, northern Myanmar, and China. It was introduced in the south of Italy in the first century AD, during the time of Ancient Rome and in Persia, Iraq, and Egypt around AD 700. The first record is a reference to lemon as an ornamental plant of the Islamic gardens in a tenth-century Arabic treatise on farming. The tree became common in the Arab and Mediterranean regions between 1000 and 1150. The first cultivation in Europe was in Genoa in the middle of the fifteenth century. The plant was introduced to the Americas by Columbus, and the first cultivations started in Florida and California in the nineteenth century. In India, lemon tree is mainly cultivated in plains and hills in areas up to 1200 m elevation. India is the biggest producer in the world, followed by Mexico, Argentina, and Brazil.

10.1.4.3.2 Features

Citrus limon is a much branched thorny shrub with spreading branches and flowers white in short racemes; fruits are large, globose berries with thick or thin rind and pale and very acid pulp (Warrier *et al.*, 1994).

10.1.4.3.3 Properties and Uses in Ayurvedic System of Medicine

Lemon is one of the most alkaline foods available: it protects against stomach and intestinal fermentation and creates germ invulnerability. It is excellent for fever and flues, and speeds up a cleansing process of the body (Nibodhi, 2010). It helps in removing excessive accumulation of *kapha* in the stomach. It is very useful in vitiated conditions of *vata* and *kapha*. In excess, it can aggravate *pitta*.

Lemon ups *agni*, so it is just right for a lethargic appetite (Khalsa and Tierra, 2010). A small amount of lemon added to a dish after cooking can stimulate *agni*, cleanse *ama*, and clear the palate (Morningstar and Desai, 1999).

Parts used: Fruit
Tastes: Sour, astringent
Quality: Light, sharp
Potency: Hot
***Dosha* effect:** VK–, P+
Actions: Expectorant, laxative, astringent, relieves thirst, digestive stimulant, anthelmintic, antiseptic

10.1.4.3.4 Traditional Uses and Folklore

- For internal cleansing, it is suggested to mix the juice of a freshly squeezed half lemon to a large glass of warm water and drink it first thing each morning (Khalsa and Tierra, 2010). This is also a famous home remedy for weight loss, as it will decrease *kapha* accumulation.
- Fresh lemon juice is recommended to be taken in the evening for the relief of dyspepsia with vomiting and bilious headaches. Preserved with sugar or honey, lemons are a traditional home remedy for sore throat and act as detergent (Nadkarni, 2002).
- With salt, lemon counteracts sunstroke, so the *nimbu pani* is a very common drink all over India during the hot summer.
- For common cold, cough, and asthma, a traditional home remedy is the lemon juice mixed with some hot water and honey (Khare, 2008).
- Small pieces of lemon mixed with powdered black pepper and powdered dry ginger and kept in the sunlight are a common home remedy for indigestion.

10.1.4.3.5 In the Kitchen

In almost all countries, lemon or lime juice is part of the daily diet: it can be added to herbal tea, green tea, beans, lentil, spinach, or salad.

In India, lemon is much used in a wide variety of chutneys and pickles as a popular and effective remedy for indigestion caused by overeating. Therefore, it is always present in the traditional meal (*thali*). An excellent pickle can be prepared by adding lemon and ginger together.

Lemon juice is an ingredient of soft drinks and cocktails; it is used in marinade for fish and meat, and in the United Kingdom, it is frequently added to pancakes.

Lemon juice is also used as a short-term preservative on certain foods that tend to oxidize and turn brown after being sliced, such as apples, bananas, and avocados, where its acid denatures the enzymes that cause browning and degradation.

Lemon juice and rind are also used to make marmalade. The grated outer rind of the fruit is used to add flavor to baked goods, puddings, rice, and other dishes.

10.1.4.4 Chemical Constituents and Uses as Functional Food

Citrus genus is the most important fruit tree crop in the world and lemon is the third most important *Citrus* species. Lemons and limes have particularly high concentrations of citric acid. Oranges are good sources of ascorbic acid (vitamin C). Blood oranges are a natural mutation of *Citrus sinensis*, although today the majority of them are hybrids. High concentrations of anthocyanins give the rind, flesh, and juice of the fruit their characteristic dark red color. Blood oranges were first discovered and cultivated in Sicily in the fifteenth century, and are grown under the name *sanguinella*. Grapefruit is also a good source of vitamin C, and the pink and red hues contain the antioxidant lycopene. It contains a number of polyphenolic compounds, including the flavanone naringin, along with the two furanocoumarins (bergamottin and dihydroxybergamottin). Furanocoumarins irreversibly inhibit a cytochrome P450 enzyme called CYP3A4, a metabolizing enzyme for almost 50% of drugs (He *et al.*, 1998).

The sweet orange (*C. sinensis* L. Osbeck) fruit contains a variety of phytochemicals that contribute to the overall flavor and properties of the fruit. These substances, delivered by the juice, include sugars such as sucrose, fructose, and glucose; organic acids (primarily citric, malic, and isocitric acids); carotenoids such as xanthophylls and carotenes; vitamins such as vitamin C, A, B_1, B_6, and B_3; flavor compounds, including various esters, alcohols, ketones, lactones, and volatile hydrocarbons; and polyphenols such as flavonoids and hydroxycinnamic acids. It is noteworthy that the content of these substances differs significantly among cultivars, and it is also affected by fruit maturation and environmental factors such as climate, soil, and agricultural procedures.

The positive effects of citrus fruits as chemopreventive agent are related to the presence of flavonoids, in particular flavanones and their glycosides.

Citrus flavonoids, including naringenin, hesperidin, nobiletin, and tangeretin, have emerged as promising therapeutic agents for the treatment of metabolic dysregulation, through inhibition of hepatic fatty acid synthesis and increased fatty acid oxidation demonstrated in animal models (Assini *et al.*, 2013). Citrus flavonoids are known to improve dyslipidemia, normalize glucose homeostasis, prevent oxidative stress, and attenuate inflammation, which collectively have the ability to enhance metabolic health and improve cardiovascular risk (reviewed in Mulvihill and Huff, 2012).

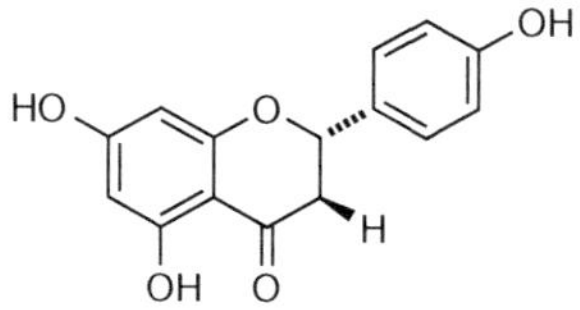

Naringenin (flavanone)

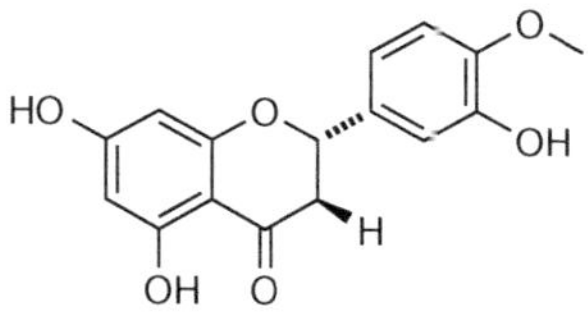

Hesperitin

Epidemiological studies reported an inverse relationship between the intake of citrus and the risk of CVDs. Clinical and experimental data further showed the antihypertensive, lipid-lowering, insulin-sensitizing, antioxidative, and anti-inflammatory properties of citrus flavanones, which could explain their antiatherogenic action in animal models. Although flavanones may be promising compounds that are particularly active in CVD prevention, clinical data are still scarce and most *in vitro* data have been obtained under nonphysiologically relevant conditions. Moreover, the mechanisms responsible for flavanone action are not fully elucidated (reviewed in Chanet *et al.*, 2012).

Pooled results from observational studies showed an inverse association between the intake of citrus fruits and the risk of breast, pancreatic, and stomach cancers (reviewed in Song and Bae, 2013). Experimental studies have provided growing evidence supporting the beneficial action of citrus flavonoids on multiple cancer-related biological pathways (carcinogen bioactivation, cell signaling, cell cycle regulation, angiogenesis, and inflammation) (reviewed in Benavente-García *et al.*, 2007). Several examples of citrus flavonoids that are potential as chemotherapeutic agents are tangeretin, nobiletin, hesperetin, hesperidin, naringenin, and naringin. These flavonoids have been shown to possess inhibition activity on certain cancer cells' growth through various mechanisms. Moreover, citrus flavonoids also show promising effect in combination with several chemotherapeutic agents against the growth of cancer cells. Some mechanisms involved in those activities are through cell cycle modulation, anti-angiogenic effect, and apoptosis induction. Previous studies showed that tangeretin suppressed the growth of T47D breast cancer cells by inhibiting ERK phosphorylation. When combined with tamoxifen, doxorubicin, and 5-fluorouracil, respectively, it was proven to be synergist on several cancer cells. Hesperidin and naringenin increased cytotoxicity of doxorubicin on MCF-7 cells and HeLa cells (reviewed in Meiyanto *et al.*, 2012).

Citrus flavonoids have also shown estrogenic effect *in vivo*. One example is hesperidin that has the ability to decrease the concentration of serum and hepatic lipid, and reduce osteoporosis of ovariectomized rats. Those studies showed the great potential of citrus fruits as natural products to be developed as the source of not only co-chemotherapeutic agents but also phytoestrogens.

Recent studies have shown an association between dietary polyphenols and the prevention of osteoarthritis-related musculoskeletal inflammation. Polyphenols from citrus fruits exert a positive effect on osteoarthritis with an emphasis on molecular antiosteoarthritic mechanisms (reviewed in Shen *et al.*, 2012a). Animal studies strongly suggest that commonly consumed antioxidant-rich fruits have a pronounced effect on the bone, as shown by higher bone mass, trabecular bone volume, number, and thickness, and lower trabecular separation through enhancing bone formation and suppressing bone resorption, resulting in greater bone strength. Such osteoprotective effects seem to be mediated via antioxidant or anti-inflammatory pathways and their downstream signaling mechanisms, leading to osteoblast mineralization and osteoclast inactivation (reviewed in Shen *et al.*, 2012b).

Recent attention has been given to the influence of dietary factors on health and mental well-being. The main citrus flavonoids can traverse the blood–brain barrier; hence, they are promising candidates for intervention in neurodegeneration and are constituents in brain foods.

Multiple neuroprotection mechanisms and antioxidant and signaling properties of citrus flavonoids are reported, and receptor-mediated neuroprotective actions and parallel signaling pathways have been explored. Hesperitin induces cellular defense proteins against oxidative stress and neurotoxicity. It is suggested that citrus fruits, which are rich in abundant sources of hesperitin and other flavonoids, are promising for the development of general food-based neuroprotection and brain foods (reviewed in Hwang *et al.*, 2012).

Flavonoids, such as those found in cocoa, tea, berries, and citrus, have been shown to be highly effective in preventing age-related cognitive decline and neurodegeneration in both animals and humans. Evidence suggests that flavonoids may express such ability through a multitude of physiological functions, including the ability to modulate the brain's immune system. Their regulation of these immune events appears to be mediated by their actions on intracellular signaling pathways, including the NF-κB cascade (reviewed in Spencer *et al.*, 2012).

There is evidence for the actions of flavonoids from apples, berries, and citrus on cognitive behavior and the underlying cellular architecture. Although the precise mechanisms by which these flavonoids act within the brain remain unresolved, they are able to protect vulnerable neurons and enhance the function of existing neuronal structures, two processes known to be influenced by flavonoids and also known to underpin neurocognitive function (reviewed in Spencer, 2010).

Anthocyanidins are contained in blood oranges *moro*, a variety of sweet orange (*C. sinensis*) with an intense red pigmentation. A recent study has demonstrated that blood orange consumption inhibits fat accumulation in mice. Furthermore, the administration of protocatechuic acid, the major *in vivo* metabolite of anthocyanidins, reduces the activity of lipogenic enzymes in the liver, thus leading to decreased hepatic lipid accumulation. *Moro* juice counteracted liver steatogenesis in mice with diet-induced obesity, through modulation of enzymes involved in lipogenesis and lipid oxidation. Thus, *moro* juice consumption may represent a promising dietary option for prevention of fatty liver (Salamone *et al.*, 2012).

Dietary supplementation of *moro* juice, but not Navelina juice, significantly reduced body weight gain and fat accumulation regardless of the increased energy intake because of sugar content. The anti-obesity effect of *moro* juice on fat accumulation cannot be explained only by its anthocyanin content, suggesting that multiple components present in the *moro* orange juice might act synergistically to inhibit fat accumulation (Titta *et al.*, 2010).

Citrus fruits and its closely related genera contain about 36 limonoid aglycones and 17 limonoid glucosides. They are mostly bitter in taste and account for the scent of fresh peels of citrus fruits (Roy and Saraf, 2006). Evaluation of the biological activity of citrus limonoids has indicated the potential of these compounds to improve human health as anticancer, cholesterol-lowering, and antiviral agents (reviewed in Manners, 2007).

Ginger lemon tea

It is a very common drink in India with a digestive action as it restores the proper flow of *apana vata* and enhances the digestive fire. It pacifies *vata* and *kapha dosha*s.

Preparation: Boil 300 mL of water with a teaspoon of fresh grated ginger. After boiling, add six teaspoons of lemon juice, filter, and add honey or a pinch of salt and raw sugar as for your taste.

REFERENCES

Assini J.M., Mulvihill E.E., Huff M.W. Citrus flavonoids and lipid metabolism. *Curr. Opin. Lipidol.* **2013**, *24*(1), 34–40.

Benavente-García O., Castillo J., Alcaraz M., Vicente V., Del Río J.A., Ortuño, A. Beneficial action of Citrus flavonoids on multiple cancer-related biological pathways. *Curr. Cancer Drug Targets* **2007**, *7*(8), 795–809.

Chanet A., Milenkovic D., Manach C., Mazur A., Morand C. Citrus flavanones: What is their role in cardiovascular protection? *J. Agric. Food Chem.* **2012**, *60*(36), 8809–8822.

He K., Iyer K.R., Hayes R.N., Sinz M.W., Woolf T.F., Hollenberg P.F. Inactivation of cytochrome P450 3A4 by bergamottin, a component of grapefruit juice. *Chem. Res. Toxicol.* **1998**, *11*(4), 252–259.

Hwang S.L., Shih P.H., Yen G.C. Neuroprotective effects of citrus flavonoids. *J. Agric. Food Chem.* **2012**, *60*(4), 877–885.

Khalsa K.P.S., Tierra, M. *The Way of Ayurvedic Herbs.* Motilal Banarsidass Publishers, Delhi, India, **2010**, 54–55.

Khare A. Traditional Remedies from your kitchen. *Heritage Amruth.* **2008**, 4(6), 35.

Manners G.D. Citrus limonoids: Analysis, bioactivity, and biomedical prospects. *J. Agric. Food Chem.* **2007**, *55*(21), 8285–8294.

Meiyanto E., Hermawan A., Anindyajati. Natural products for cancer-targeted therapy: Citrus flavonoids as potent chemopreventive agents. *Asian Pac. J. Cancer Prev.* **2012**, *13*(2), 427–436.

Morningstar A., Desai U. *The Ayurvedic Cookbook.* Motilal Banarsidass publishers, Delhi, India, **1999**, 178.

Mulvihill E.E., Huff M.W. Citrus flavonoids and the prevention of atherosclerosis. *Cardiovasc. Hematol. Disord. Drug Targets* **2012**, *12*(2), 84–91.

Nadkarni K.M. *Indian Materia Medica*, vol. 1. Popular Prakashan, Mumbai, India, **2002**, 339–342, 346–348.

Nibodhi. *Annapurna's Prashad.* Mata Amritanandamayi Mission Trust, Amritapuri, India, **2010** (2nd edn.), 84–85, 87–88.

Roy A., Saraf S. Limonoids: Overview of significant bioactive triterpenes distributed in plants kingdom. *Biol. Pharm. Bull.* **2006**, *29*(2), 191–201.

Salamone F., Li Volti G., Titta L., Puzzo L., Barbagallo I., La Delia F., Zelber-Sagi S., Malaguarnera M., Pelicci P.G., Giorgio M., Galvano F. *Moro* orange juice prevents fatty liver in mice. *World J. Gastroenterol.* **2012**, *18*(29), 3862–3868.

Shen C.L., Smith B.J., Lo D.F., Chyu M.C., Dunn D.M., Chen C.H., Kwun I.S. Dietary polyphenols and mechanisms of osteoarthritis. *J. Nutr. Biochem.* **2012a**, *23*(11), 1367–1377.

Shen C.L., von Bergen V., Chyu M.C., Jenkins M.R., Mo H., Chen C.H., Kwun I.S. Fruits and dietary phytochemicals in bone protection. *Nutr. Res.* **2012b**, *32*(12), 897–910.

Song J.K., Bae J.M. Citrus fruit intake and breast cancer risk: A quantitative systematic review. *J. Breast Cancer.* **2013**, *16*(1), 72–76.

Spencer J.P. The impact of fruit flavonoids on memory and cognition. *Br. J. Nutr.* **2010**, *104*(Suppl 3), S40–S47.

Spencer J.P., Vafeiadou K., Williams R.J., Vauzour D. Neuroinflammation: Modulation by flavonoids and mechanisms of action. *Mol. Aspects Med.* **2012**, *33*(1), 83–97.

Titta L., Trinei M., Stendardo M., Berniakovich I., Petroni K., Tonelli C., Riso P. *et al.* Blood orange juice inhibits fat accumulation in mice. *Int. J. Obesity* **2010**, *34*, 578–588.

Warrier P.K., Nambiar V.P.K., Ramankutty C. (eds.) *Indian Medicinal Plants*, vol. 2. Orient Longman, Chennai, India, **1994**, 97–100.

10.1.5 Date

Scientific name: *Phoenix dactylifera* L.
Family: Palmae
Hindi: Khajur
Sanskrit: Kharjurah
Tamil: Pareecham pazham

10.1.5.1 Origin and Distribution

The species name *dactylifera*, meaning "date-bearing," comes from the ancient Greek *dáktulos*, meaning "date" (also "finger") and *ferō*, meaning "I bear." The date palm is one of the earliest fruit trees cultivated by man: it is a very nourishing fruit, called the *bread of Sahara*. We can find carved figures of date palms in the ancient monuments of Egypt, where the fruit was used to prepare a wine. The Bible contains numerous references to the palm, the fruit, and its virtues. In ancient Greece, the palm tree is related to the birth of Apollo. Prophet Mohammed mentions the origin of the first palm date made from the tempered dust, which remained after the formation of Adam (Bakhru, 1994). The fruit is sacred in both Hindu and Islamic religions (Majupuria, 1977). The powder of dried fruits has been used in Hindu rituals since very early days (Sharma, 2009).

Although its place of origin is unknown because of long cultivation, it probably originated from lands around Iraq. It is probable that the date palm was originally domesticated in the Fertile Crescent region of Mesopotamia, approximately 7000 years ago. At present, it mainly grows in Iraq and other Arabian countries. The domestic date palm was produced from a wild version of the same name, some of which have been found today in North Africa, the Middle East, and the Near East. The Middle East nations are the major producers and exporters of date fruits.

10.1.5.2 Features

The tree is a tall palm (25–35 m high). The trunk is about 1 m in diameter, covered by petiole bases, the base surrounded by a mass of root suckers. The leaves are 3.5 m long, grayish green, pinnate. The fruits are oblong berries that are brownish yellow in color, with a single seed (Paranjipe, 2005). Dates reach their full fruit-bearing potential at about 10 years of age, and, on average, continue to bear fruits for 40–50 years. Some trees can continue producing fruits for 100 years, but as such trees grow so tall and the fruit becomes more and more difficult to reach, they are often cut down.

Dates are high-energy fruits and can be picked fresh and eaten, as well as easily sun dried for long-term storage.

In deserts, date palms grow in oases, where the trees help create and sustain a microclimate suitable for agriculture and livestock raising. The stems of the trees can be used as construction wood, the leaves' midribs are suitable for fencing material, and the leaves themselves can be woven into baskets.

10.1.5.3 Properties and Uses in Ayurvedic System of Medicine

Date is considered by Ayurveda a *rasayana* as it increases vital fluids and strength. It is given wherever there is general weakness with a loss of vital body energy (*ojas*) (Nibodhi, 2010). It is a very suitable food for the old and debilitated: it provides strength to the heart, liver, and kidneys.

It prevents premature graying of hair and gives a healthy look. For patients affected by anemia and weakness, Ayurveda suggests eating daily few dates followed by milk in the morning before breakfast and in the evening before sleep (Puri, 2003).

According to Bhavaprakasha, date is cooling, sweet, unctuous, an appetizer, nourishing, a refresher, a cardiac tonic, and aphrodisiac (Khare, 2004); it is beneficial to the heart and is used in diseases such as tuberculosis, cardiac and sexual debility, and burning sensation. In convalescence and general debility, it boosts up the energy level.

Parts used: Fruit
Tastes: Sweet, astringent
Quality: Heavy, oily
Potency: Cold
***Dosha* effect:** VP–, K+
Actions: Demulcent, expectorant, nutrient, emetic, laxative, aphrodisiac, brain sedative, nerve tonic, good for heart (Warrier, 1996)

10.1.5.4 Traditional Uses and Folklore

- *Kharjura mantha* is one of the simplest preparations with dates: the fruits are crushed and churned with water and are very beneficial to athletes for quick energy regain (Paranjipe, 2005).
- To strengthen the heart, dates can be soaked overnight in water and crushed in the same water in the morning. They have also a laxative effect.
- Date is also a very important ingredient of Unani tonics.

10.1.5.5 In the Kitchen

Date is a very auspicious fruit, essential part of many religious functions. With this fruit, Muslims break the Ramadan fast. During the marriage ceremony, it is given with milk in the nuptial night. Mahatma Gandhi used to eat date soaked in water for breakfast (Puri, 2003).

The date has a sticky surface that can attract dust and impurities. It is advisable to buy only the best quality in good packing conditions and wash the fruits before use.

Ayurveda suggests drinking buttermilk after eating dates to have good digestion. Rich Arabs eat dates filled with butter.

Dates can be used to prepare sweets, vegetables, pickles, and sour chutneys. They are delicious when filled with almonds, walnuts, or candied orange, and covered with chocolate. Date nut bread, a type of cake, is very popular in the United States, especially around holidays.

The sparkling date juice is a nonalcoholic version of Champagne in some Islamic countries.

10.1.5.6 Chemical Constituents and Uses as Functional Food

The fruit of date palm (edible flesh of the fruit) is especially delicious and highly nutritious. Date flesh is found to be low in fat and protein but rich in sugars, mainly fructose and glucose. It is a high source of energy, as 100 g of flesh can provide an average of 314 kcal. Ten minerals were reported, the major being selenium, copper, potassium, and magnesium, with average values 6–10 times higher than apple, orange, and wild berries (USDA National Nutrient Database for Standard Reference). The consumption of 100 g of dates can provide over 15% of the recommended daily allowance from these minerals. Vitamins B complex and C are the major vitamins in dates. High in dietary fiber (8.0 g/100 g), insoluble dietary fiber was the major fraction of dietary fiber in dates.

Dates are a good source of antioxidants, mainly carotenoids and phenolics (reviewed in Al-Farsi and Lee, 2008).

The major class of phytochemicals found in the lipid fraction of date fruits is carotenoid, the most abundant being β-carotene (3%–10%) and lutein (89%–94%) (Boudries *et al.*, 2007).

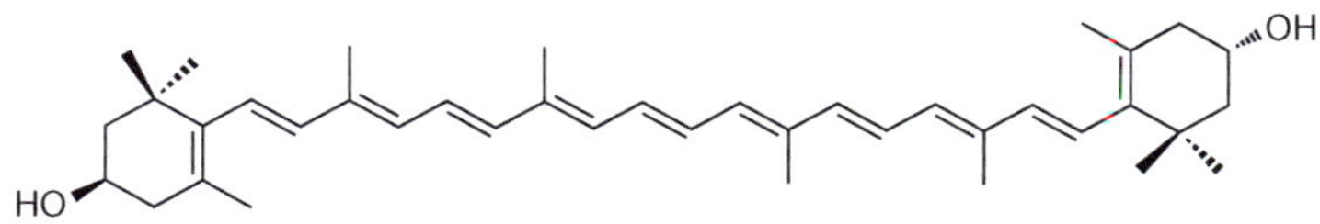

Lutein

Date fruits are also rich in simple phenolic acids and proanthocyanidins, with a total amount of 3.0 g/100 g of polyphenols (Duke and Beckstrom-Sternberg, 2007).

Date fruit may be considered as one of the highly nutritious fruits available on the earth for human consumption, having the best nutrient score among fresh apricots, cranberries, figs, grapes, and plums (Vinson, 2005). The nutritional composition of dates cultivated in different parts of the world is updated in the USDA National Nutrient Database for Standard Reference.

Fully ripened date fruit contains more than twice the amount of fructose compared to any other consumed fruits. Thus, the low glycemic index (GI) value of date fruit may be attributed partly to the high concentration in fructose, and it has been reported that there exists an inverse correlation between fructose and the GI values of dates (Ali *et al.*, 2009). North African varieties appear to have the lowest GI, compared to Australian dates (Denyer and Dickinson, 2005).

In vitro and animal studies on the antioxidant and anti-hypocholesterolemic effect of date fruits have been substantiated in healthy subjects, showing that consumption of 100 g/day of North African dates for 4 weeks does not alter the body mass index, glucose, or total cholesterol, but reduces serum triacylglycerols and very low-density lipoprotein (LDL) (Rock *et al.*, 2009).

Based on the available evidence from the literature (reviewed in Vayalil, 2012), date fruit is a fruit of high nutritional value. Though it is a sugar-packed fruit (fresh or dried), consumption of certain varieties does not induce any metabolic and inflammatory marker associated with metabolic syndrome and related diseases. Date fruit consumption may serve as a single food source providing sugars, minerals, and vitamins that meet the daily nutritional requirements. Therefore, date fruit consumption could be a good alternative for infants' malnutrition furnishing a basic nutritional diet required to fight against deficiency-related diseases and infections.

In the folklore, date fruits have been ascribed to have many medicinal properties when consumed either alone or in combination with other herbs. Although fruit of the date palm served as the staple food for millions of people around the world for several centuries, studies on the health benefits are still poor, hardly recognizing it as a healthy food.

REFERENCES

Al-Farsi M.A., Lee C.Y. Nutritional and functional properties of dates: A review. *Crit. Rev. Food Sci. Nutr.* **2008**, *48*(10), 877–887.

Ali A., Al-Kindi Y.S.M., Al-Said F. Chemical composition and glycemic index of three varieties of Omani dates. *Int. J. Food Sci. Nutr.* **2009**, 60, 51–62.

Bakhru H.K. *Foods That Heal*. Orient Paperbacks, Delhi, India, **1994**, 35–38.

Boudries H., Kefalas P., Hornero-Mendez D. Carotenoid composition of Algerian date varieties (*Phoenix dactylifera*) at different maturation stages. *Food Chem.* **2007**, *101*(4), 1372–1377.

Denyer G., Dickinson S. The Glycemic Index and GI Database. University of Sydney, Sydney, Australia, **2005** (www.glycemicindex.com).

Duke J.A., Beckstrom-Sternberg S. Dr. Duke's Ethnobotanical Databases. **2007** (Accessed from http://www.ars-grin.gov/duke/plants.html, on 2007).

Khare C.P. (ed.) *Indian Herbal Remedies*. Springer-Verlag, Berlin, Germany, **2004**, 356–357.

Majupuria T.C., Joshi D.P. *Religious and Useful Plants of Nepal and India*. Gupta M.(ed.), Publishers, Lashkar, India, **1997**, 243.

Nibodhi. *Annapurna's Prashad*. Mata Amritanandamayi Mission Trust, Amritapuri, India, **2010** (2nd edn.), 80–81.

Paranjipe P. *Indian Medicinal Plants. Forgotten Healers*. Chaukhamba Sanskrit Pratishthan, Delhi, India, **2005**, 135–136.

Puri H.S. *Rasayana. Ayurvedic Herbs for Longevity and Rejuvenation*. Taylor & Francis, London, **2003**, 102–104.

Rock W., Rosenblat M., Borochov-Neori H., Volkova N., Judeinstein S., Elias M., Aviram M. Effects of date (*Phoenix dactylifera* L., Medjool or Hallawi Variety) consumption by healthy subjects on serum glucose and lipid levels and on serum oxidative status: A pilot study. *J. Agric. Food Chem.* **2009**, *57*(17), 8010–8017.

Sharma P.V. *Fruits and Vegetables in Ancient India*. Chaukhamba Orientalia, Varanasi, India, **2009**, 40.

USDA National Nutrient Database for Standard Reference. http://ndb.nal.usda.gov/.

Vayalil P.K. Date fruits (*Phoenix dactylifera* Linn): An emerging medicinal food. *Crit. Rev. Food Sci. Nutr.* **2012**, *52*(3), 249–271.

Vinson J.A., Zubik L., Bose P., Samman N., Proch J. Dried fruits: Excellent in vitro and in vivo antioxidants. *J. Am. Coll. Nutr.* **2005**, *24*(1), 44–50.

Warrier P.K., Nambiar V.P.K., Ramankutty C. (eds.) *Indian Medicinal Plants*, vol. 4. Orient Longman, Chennai, India, **1996**, 240–243.

10.1.6 Fig

Scientific name: *Ficus carica* L.
Family: Moraceae
Hindi: Anjir
Sanskrit: Anjira
Tamil: Shimeatti, Anjura

10.1.6.1 Origin and Distribution

The fig belongs to a family of sacred trees: Buddha achieved enlightenment under the *Ficus religiosa*; in ancient Egypt, the *Ficus sycomorus* was the cosmic tree (Brosse, 1985). The holiness of figs is not circumstantial as they played a central role in the agricultural revolution. Figs were the first plant cultivated by humans as early as 9400 BC, at least a thousand years before wheat, barley, and legume cultivation.

Native to western Asia and eastern Mediterranean, the plant grows today in subtropical, Mediterranean, or warm temperate areas. In the United States, the commercial cultivation is concentrated in California and Texas. In India, it is cultivated in Maharashtra, Karnataka, Uttar Pradesh, and Gujarat (Bakhru, 1994).

In the Genesis, Adam and Eve hid their nakedness with fig leaves. Prophet Mohammed greatly appreciated the fig tree: "If I had to mention a fruit that descended from paradise, I would say this is it because the paradisiacal fruits do not have pits [...] eat from these fruits for they prevent hemorrhoids and help gout." In ancient Greece, the fig was sacred to Demeter and Dionysus. When a man hospitably received the goddess Demeter in his house, she rewarded him with the creation of the cultivated fig tree. The fruit was considered a very nutritious food source by Greeks and Romans. Plato suggests that philosophers eat figs "as they improve intelligence" (Brosse, 1985). Cato the Elder, in his *De Agri Cultura*, lists several strains of figs grown at the time. The fruits were used by Romans, among other things, to fatten geese for the production of a precursor of *fois gras*.

10.1.6.2 Features

The fig is a small tree with alternate, long petioled leaves. The stem and leaves contain an acrid milky juice (Sairam, 1998). Fruits are fleshy and pear shaped, with a large number of small seeds sticking to the wall of the cavity (Bakhru, 1994). Figs are not actually a fruit but an inflorescence called a "multiple fruit" because the flowers and seeds grow together in a single mass. They ripen twice a year: once in the spring and once in the late summer or fall. The late summer and fall crop is usually the largest and best tasting. Being highly perishable, the fruit is sold mostly in its dry form.

> Figs are restorative and the best fruit that can be taken by those who are brought low by long sickness.
>
> **Pliny**

10.1.6.3 Properties and Uses in Ayurvedic System of Medicine

The fig is very nutritious and helps to maintain intestinal regularity due to laxative and diuretic properties. It strengthens the gums, teeth, and tongue. It is an excellent tonic for the liver, stimulates the immune system, and provides a large amount of energy to the body (Nibodhi, 2010). In Sanskrit, it is called *sramahara* (alleviating fatigue) (Sharma, 2009).

Parts used: Fruit
Tastes: Sweet, astringent
Quality: Heavy (fresh figs)
Potency: Cold
***Dosha* effect:** VP–, K+ (fresh figs); PK–, V+ (dried figs)
Actions: Nutritive, demulcent, expectorant, laxative

10.1.6.4 Traditional Uses and Folklore

- Dried stewed figs are specific for constipation (Morningstar and Desai, 1999).
- The juice of fresh figs is a traditional home remedy for kidney stones. Intake of two to three figs after each meal is recommended in case of constipation due to diabetes.
- For eye problems, due to *pitta* aggravation, it is recommended to take every morning during breakfast two to three figs boiled with one tablespoon of raisins in a cup of milk. This remedy is also a brain tonic.
- For physical weakness, *pitta* aggravation, liver problems, dry cough, and constipation, two to three dried figs can be soaked overnight in one teacup of water and eat along with one tablespoon of honey the next morning (Sairam, 1998). The same remedy is excellent for piles also (Bakhru, 1994).
- Together with ghee and other *ojas* building fruits such as dates and almonds, figs are a traditional home remedy for sexual debility.

10.1.6.5 In the Kitchen

It is always better to soak the dried figs in water as they will be easier to digest.

10.1.6.6 Chemical Constituents and Uses as Functional Food

Fig fruit has been a typical component in the health-promoting Mediterranean diet for millennia. It is highly caloric (352 kcal/100 g) but also a rich source of minerals (potassium) and vitamins (USDA Nutrient Database).

To study the potential health-promoting constituents of fig fruits, six commercial fig varieties differing in color (black, red, yellow, and green) were analyzed for total polyphenols, total flavonoids, the antioxidant capacity, and the amount and profile of anthocyanins. Hydrolysis revealed cyanidin as the major aglycon. Cyanidin-3-*O*-rhamnoglucoside (cyanidin-3-*O*-rutinoside) is the main anthocyanin in all fruits. The color appearance of fig extract correlated well with total polyphenols, flavonoids, anthocyanins, and antioxidant capacity. Extracts of darker varieties showed higher contents of phytochemicals compared to lighter colored varieties. The skin of the fruit contributed most of the above phytochemicals and antioxidant activity compared to the pulp of the fruit (Solomon *et al.*, 2006).

Metabolite profiling on the leaves, pulps, and peels of two Portuguese white varieties of *Ficus carica* (*Pingo de Mel* and *Branca Tradicional*) showed that all samples presented a similar phenolic profile composed by 3-*O*- and 5-*O*-caffeoylquinic acids, ferulic acid, quercetin-3-*O*-glucoside, quercetin-3-*O*-rutinoside, psoralen, and bergapten. The profile of leaves' organic acids presented oxalic, citric, malic, quinic, shikimic, and fumaric acids, whereas in pulps and peels, quinic acid was absent. The antioxidant potential of the different plant parts was checked. All materials exhibited activity against 2,2-diphenylpicrylhydrazyl and nitric oxide radicals in a concentration-dependent way. However, only the leaves presented capacity to scavenge superoxide radical (Oliveira *et al.*, 2009).

Fresh figs are subjected to two different drying processes: sun drying and oven drying. In figs, monomer sugars predominate, which is an important nutritional information, and the content of sugars as well as organic acids in fresh figs was lower than in dried fruits. However, the best sugar/organic acid ratio was measured after the sun-drying process. Analysis of individual phenolic compounds revealed a higher content of all phenolic groups determined after the oven-drying process, with the exception of cyanidin-3-*O*-rutinoside. Similarly, higher total phenolic content and antioxidant activity were detected after the drying process. With these results, it can be concluded that the differences in analyzed compounds in fresh and dried figs are significant. The results indicate that properly dried figs can be used as a good source of phenolic compounds (Slatnar *et al.*, 2011).

A hexane extract of the Tunisian common Jrani caprifig latex was assayed *in vitro* for antibacterial activity against several Gram-positive and Gram-negative bacteria. A strong bactericidal effect was demonstrated, the most sensitive bacteria being *Staphylococcus saprophyticus* clinical isolate, and *Staphylococcus aureus* ATCC 25923, with a minimum inhibitory concentration of 19 μg/mL. The chemical composition of the extract was also investigated. Thirty-six compounds of the extract were identified; 90.56% of the total area of peaks were coumarins (Lazreg-Aref *et al.*, 2012).

The hexane and hexane–ethyl acetate (v/v) extracts also inhibited multiplication of viruses by tested techniques at concentrations of 78 μg/mL. These two extracts are possible candidates as herbal medicines for herpes virus, echovirus, and adenovirus infectious diseases. All extracts had no cytotoxic effect on Vero cells at all tested concentrations (Lazreg-Aref *et al.*, 2011).

Natural rubber was extracted from the fig tree (*F. carica*) cultivated in Korea as part of a survey of rubber-producing plants (Kang *et al.*, 2003).

REFERENCES

Bakhru H.K. *Foods That Heal*. Orient Paperbacks, Delhi, India, **1994**, 38–40.

Brosse J. *Les arbres de France*. Librarie Plon, Paris, France, **1985** (Italian version Ed. Studio Tesi, 1991, 87–92).

Kang H., Kang M.Y., Han K.H. Identification of natural rubber and characterization of rubber biosynthetic activity in fig tree. *Plant Physiol*. **2000**, *123*(3), 1133–1142.

Lazreg-Aref H., Gaaliche B., Fekih A., Mars M., Aouni M., Pierre Chaumon J., Said K. In vitro cytotoxic and antiviral activities of Ficus carica latex extracts. *Nat. Prod. Res.* **2011**, *25*(3), 310–319.

Lazreg-Aref H., Mars M., Fekih A., Aouni M., Said K. Chemical composition and antibacterial activity of a hexane extract of Tunisian caprifig latex from the unripe fruit of *Ficus carica*. *Pharm. Biol.* **2012**, *50*(4), 407–412.

Morningstar A., Desai U. *The Ayurvedic Cookbook*. Motilal Banarsidass Publishers, Delhi, India, **1999**, 179.

Nibodhi. *Annapurna's Prashad*. Mata Amritanandamayi Mission Trust, Amritapuri, India, **2010** (2nd edn.), 81–82.

Oliveira A.P., Valentão P., Pereira J.A., Silva B.M., Tavares F., Andrade P.B. *Ficus carica* L.: Metabolic and biological screening. *Food Chem. Toxicol.* **2009**, *47*(11), 2841–2846.

Sairam T.V. *Home Remedies*, vol. I. Penguin Books, Delhi, India, **1998**, 121–126.

Sharma P.V. *Fruits and Vegetables in Ancient India*. Chaukhamba Orientalia, Varanasi, India, **2009**, 61.

Slatnar A., Klancar U., Stampar F., Veberic R. Effect of drying of figs (*Ficus carica* L.) on the contents of sugars, organic acids, and phenolic compounds. *J. Agric. Food Chem.* **2011**, *59*(21), 11696–11702.

Solomon A., Golubowicz S., Yablowicz Z., Grossman S., Bergman M., Gottlieb H.E., Altman A., Kerem Z., Flaishman M.A. Antioxidant activities and anthocyanin content of fresh fruits of common fig (*Ficus carica* L.). *J. Agric. Food Chem.* **2006**, *54*(20), 7717–7723.

10.1.7 Grape

Scientific name: *Vitis vinifera* L.
Family: Vitaceae
Hindi: Drakh
Sanskrit: Draksha
Tamil: Drakshai

10.1.7.1 Origin and Distribution

Grape is one of the earliest fruit vine cultivated in the world. It is probably indigenous to Caucasia and its surrounding area; from there, it gradually spread to western Asia, southern Europe, Algeria, and Morocco (Bakhru, 1994). According to the Food and Agriculture Organization (FAO), the grape's cultivation covers 75,866 km^2 of the world. Approximately 71% of the world grape production is used for wine, 27% as fresh fruit, and 2% as dried fruit. The biggest producers are now China, Italy, the United States, Spain, France, Turkey, Chile, Argentina, India, and Iran. In India, it is cultivated mainly in Tamil Nadu, Maharashtra, Andhra Pradesh, Punjab, and Haryana. In the south, it grows through the year.

The grape's medicinal and nutritional properties have been greatly appreciated since antiquity. The earliest archeological evidence dates from 8000 years ago in Georgia. Ancient Egyptian hieroglyphics record the cultivation of purple grapes. Greeks, Phoenicians, and Romans grew purple grapes for both eating and wine production. In Greece, the plant was sacred to the god of agriculture, Dionysus, who was said to have created the first grapevines and handed down to mankind the art of winemaking. Grapes are especially significant in the Christian religion as wine represents the blood of Jesus.

10.1.7.2 Features

Vitis vinifera is a climber with broad leaves and spherical or ovoid fruits, which grows in cluster and can be of innumerable varieties differing in shape, size, color, aroma, and taste (Bakhru, 1994).

10.1.7.3 Properties and Uses in Ayurvedic System of Medicine

Grape is mentioned in the *Charaka Samhita* and *Vagbhata Samgraha* as the best among the fruits (*palottama*) and has been used as food and medicine for centuries

(Sharma, 2009). There are many varieties of grapes, but in Ayurveda, dry, big, black grapes with seeds are mainly used.

Draksha is considered a *rasayana* with nutritive and energizing properties; it is an important ingredient of most fermented Ayurvedic preparations (*asav* and *aristha*) (Puri, 2003). It is a powerful tonic with a mineralizing, cleaning, and rejuvenating effect on all the *dhatus*. *Charaka Samhita* (Shastri, 1970) categorized grape as *sramahara*, a relief to fatigue, and *kanthya*, beneficial for the throat. *Sushruta Samhita* (Acharya, 1980) mentioned it as *hrdya*, beneficial for the heart. Grape strengthens the lungs and relieves the mucus. The raisins help to regain the body weight (Paranjipe, 2005).

Sri Raghunatha Suri in the *Bhojanakutuhalam* reports that grape is aphrodisiac and highly nourishing: "it will instill desire for love-sport even in those that have completely lost their virility." Grape is used for vitiation of *pitta* treats burning sensation and urinary obstruction, and imparts taste (Venkat *et al.*, 2012).

Parts used: Fruits
Tastes: Sweet, sour
Quality: Soft, oily
Potency: Cold
***Dosha* effect:** VP–, K+ (in excess)
Actions: Rejuvenating, digestive, diuretic, refrigerant, laxative, aphrodisiac, expectorant, tonic (Warrier *et al.*, 2011)

Preparation: A very common tonic preparation is *draksharistha*, prescribed for insomnia, mental and physical fatigue.

10.1.7.4 Traditional Uses and Folklore

- Grapes increase metabolism and provide immediate nourishment soon after consumption. As a blood tonic, they are useful for women during menstruation, pregnancy, postpartum, or breast-feeding.
- The black raisins (dried fruits) are recommended to relieve headache, due to vitiation of *pitta dosha* because of constipation. The decoction of raisins is laxative and aphrodisiac, and improves the acuity of vision (Paranjipe, 2005).
- Strained grape juice in teaspoon doses night and morning is given to children for constipation during teething and also to prevent convulsions due to constipation. It is also a good diet consumed in the morning and evening during jaundice (Nadkarni, 2002).
- In case of frequent constipation, it is suggested to chew a handful of black raisins at bedtime, and then drink a glass of warm water. Raisins eliminate the excess *pitta* from the digestive system and reduce flatulence by lubricating the internal mucosal lining of the intestines.
- In case of dry cough, grapes pacify *vata*, lubricate the constricted bronchi, and reduce chest pain and uneasiness caused by coughing.

10.1.7.5 In the Kitchen

Sweet purple grapes can be eaten in moderation by all *doshas*; they are light enough to be eaten by *kapha* also. Dried grapes (raisins) when well soaked in water can balance each *dosha* (Morningstar and Desai, 1999).

10.1.7.6 Chemical Constituents and Uses as Functional Food

Due to the economical interest of the plant, the chemical composition of grape has been accurately determined. The twigs contain prevalently tannins (2%–3%), carbohydrates (1%), tartaric acid (1%) and other organic acids, and minerals (1%) and nitrogen compounds. The skin of the fruit is covered by a waxy substance and contains tartaric acid (0.2%–0.3%) and other organic acids (1%), mineral salts (1.5%–2%) with a particular abundance of iron, and in addition flavonoids, terpenic alcohols, and other volatile compounds. Grape seeds contain high amounts of lipids (10%–12%) and tannins (5%). The pulp of the fruit is particularly rich in carbohydrates (more than 20%), mainly glucose and fructose, and organic acids, principally malic, tartaric, and citric acids. It also contains fats (0.6%), proteins (0.7%), minerals (K, P, Ca, Mg, Na, Fe, and Zn), vitamins (B_6, B_1, B_2, C, A, and E), folate, and niacin (USDA Nutrient Database).

The main bioactive compounds, besides tannins, are flavonoids such as anthocyanidins (e.g., peonidin, delphinidin, petunidin, and malvidin) and their glycosides anthocyanins (up to 0.3% of dry weight), which are responsible for the red color of the leaves in autumn and of the fruits of the red varieties (Flamini, 2003). White grapes are derived from red grapes through gene mutations that turn off the production of anthocyanidins. Other flavonoids include flavonol glycosides and catechin oligomers known as proanthocyanidins (Baltenweck-Guyot *et al.*, 2000). Flavonoids are particularly abundant in seeds and skin of the fruit. Anthocyanidins and other polyphenol pigments are also responsible for the varying shades of purple in red wines.

The stem, root, and fruit peels contain the stilbene derivative resveratrol (*trans*-3,4′,5-trihydroxystilbene), stilbene dimers (viniferins), and glycosides (astringin and piceid) (Bavaresco *et al.*, 1999; Pezet *et al.*, 2003; Vitrac *et al.*, 2005; reviewed in Rivière *et al.*, 2012). These compounds are a response from the plant to a hostile environment (phytoalexins) and are produced under stress conditions; thus, their content is usually low in comparison with other metabolites. The seeds also contain polyphenols (5%–8%) such as flavonoids, gallic acid, flavan-3-ols (catechin, epicatechin, gallocatechin, epigallocatechin, epicatechin-3-*O*-gallate), and proanthocyanidin oligomers (Peng *et al.*, 2001).

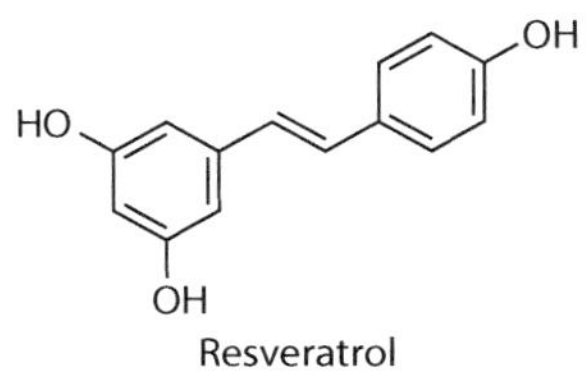

Resveratrol

During wine maturation, polymeric pigments are formed from the reaction of anthocyanins with other flavonoids and low-molecular-weight compounds, such as pyruvic acid. These anthocyanin derivatives include pyranoanthocyanins, vitisins, and anthocyanin vinylflavanols (Flamini, 2005; Rentzsch *et al.*, 2007). The strong presence of polyphenols confers the astringent effects (Kiesewetter *et al.*, 2000).

The leftovers of wine pressing are used to extract grapeseed oil. The oil has a fine texture and is almost odorless. It contains antioxidants such as vitamin E, vitamin C, and β-carotene, and fair amounts of the essential fatty acid, linoleic acid. Other fatty acids are palmitic, palmitoleic, stearic, oleic, α-linolenic, icosanoic, icosenoic, and docosanoic acids.

A great number of therapeutic properties of grape have been recognized (reviewed in Yang and Xiao, 2013; Vislocky and Fernandez, 2010, and Williamson and Carughi, 2010).

Resveratrol is known for its purported antioxidant activity in the vascular and nervous systems. In contrast to its direct antioxidant role within the central nervous system, recent research supports a protective mechanism through increasing endogenous cellular antioxidant defenses, which triggers a cascade of parallel neuroprotective pathways (Mullin, 2011). A growing body of *in vitro* and *in vivo* evidence indicates that resveratrol acts through multiple pathways and reduces ischemic damage in vital organs, such as the heart and the brain, in various rodent models. Rigorous clinical trials evaluating the effects of resveratrol in stroke are absent, though the general population consumption appears to be relatively safe. Resveratrol has shown potential for treating stroke in laboratory animals and *in vitro* human cell studies, yet there is still a need for human research in preclinical settings (reviewed in Singh *et al.*, 2013, and Wu and Hsieh, 2011).

A number of epidemiological studies provide strong evidence highlighting the cardioprotective benefits of the Mediterranean diet enriched in green vegetables, fruits, fish, and grape wine. Regular consumption of these products leads to an accumulation of nitrate/nitrite/NO, polyunsaturated fatty acids (PUFAs), and polyphenolic compounds in the human body. Studies have confirmed that these constituents are bioactive exogenous mediators, which induce strong protection against myocardial infarction (MI). The cardioprotective pathways mediated by nitrite/NO, PUFAs, and phenolic compounds of grape wines discovered in the recent years, including cross talk between different mechanisms and compounds, have been thoroughly studied (reviewed in Nadtochiy and Redman, 2011).

Fruit juices and extracts are under investigation as treatments for cognitive impairment. Blueberry, strawberry, blackberry, grape, and plum juices or extracts have been successfully tested in cognitively impaired rodents. Published trials of the benefits of grape and blueberry juices in the treatment of small numbers of cognitively impaired persons have recently appeared. The benefits of fruit products are thought to be a result of its polyphenol content. The grape polyphenol found in grapes, resveratrol, now being studied in humans, and the one in grapes and blueberries, pterostilbene, have been found to improve cognition in rodents. Strong research is still needed to understand the appropriate form (e.g., juice, powder, or individual polyphenol) on which to base the treatment to initiate the experimental therapy long before the onset of symptoms, due to the poor bioavailability of these products (reviewed in Cherniack, 2012).

As key regulators involved in numerous cellular signaling pathways, sirtuins are emerging as potential targets in various neurodegenerative diseases such as Alzheimer's disease and Parkinson's disease, thus suggesting that the modulation of sirtuin activity could provide an interesting and novel therapeutic option. In particular, much attention has been raised by neuroprotective effects attributed to SIRT1 activation due to genetically induced sirtuin overexpression or administration of resveratrol, a natural compound found in the skin of red grapes. Similarly, also sirtuin inhibitors display benefits in various neuropathologic disease models (reviewed in Huber and Superti-Furga, 2011).

REFERENCES

Acharya J.T. (ed.) *Sushruta Samhita*. Chaukhambha Orientalia, Varanasi, India, **1980**, Sutrasthana A-38.

Bakhru H.K. *Foods That Heal*. Orient Paperbacks, Delhi, India, **1994**, 41–44.

Baltenweck-Guyot R., Trendel J.M., Albrecht P., Schaeffer A. Glycosides and phenylpropanoid glycerol in *Vitis vinifera* cv. Gewurztraminer wine. *J. Agric. Food Chem.* **2000**, *48*, 6178–6182.

Bavaresco L., Fregoni C., Cantu E., Trevisan M. Stilbene compounds: From the grapevine to wine. *Drugs Exp. Clin. Res.* **1999**, *25*, 57–63.

Cherniack E.P. A berry thought-provoking idea: The potential role of plant polyphenols in the treatment of age-related cognitive disorders. *Br. J. Nutr.* **2012**, *108*(5), 794–800.

Flamini R. Mass spectrometry in grape and wine chemistry. Part I: Polyphenols. *Mass Spectrom. Rev.* **2003**, *22*, 218–250.

Flamini R. Some advances in the knowledge of grape, wine and distillates chemistry as achieved by mass spectrometry. *J. Mass Spectrom.* **2005**, *40*, 705–713.

Huber K., Superti-Furga G. After the grape rush: Sirtuins as epigenetic drug targets in neurodegenerative disorders. *Bioorg. Med. Chem.* **2011**, *19*(12), 3616–3624.

Kiesewetter H., Koscielny J., Kalus U. *et al.* Efficacy of orally administered extract of red vine leaf AS 195 (folia *Vitis viniferae*) in chronic venous insufficiency (stages I–II). A randomized, double-blind, placebo-controlled trial. *Arzneimittelforschung* **2000**, *50*, 109–117.

Morningstar A., Desai U. *The Ayurvedic Cookbook*. Motilal Banarsidass Publishers, Delhi, India, **1999**, 177.

Mullin G.E. Red wine, grapes, and better health—Resveratrol. *Nutr. Clin. Pract.* **2011**, *26*(6), 722–723.

Nadkarni K.M. *Indian Materia Medica*, vol. 1. Popular Prakashan, Mumbai, India, **2002**, 1285–1289.

Nadtochiy S.M., Redman E.K. Mediterranean diet and cardioprotection: The role of nitrite, polyunsaturated fatty acids, and polyphenols. *Nutrition* **2011**, *27*(7/8), 733–744.

Paranjipe P. *Indian Medicinal Plants. Forgotten Healers*. Chaukhamba Sanskrit Pratishthan, Delhi, India, **2005**, 77–78.

Peng Z., Hayasaka Y., Iland P.G., Sefton M., Hoj P., Waters E.J. Quantitative analysis of polymeric procyanidins (tannins) from grape (*Vitis vinifera*) seeds by reverse phase high-performance liquid chromatography. *J. Agric. Food Chem.* **2001**, *49*, 26–31.

Pezet R., Perret C., Jean-Denis J.B., Tabacchi R., Gindro K., Viret O. δ-Viniferin, a resveratrol dehydrodimer: One of the major stilbenes synthesized by stressed grapevine leaves. *J. Agric. Food Chem.* **2003**, *51*, 5488–5492.

Puri H.S. *Rasayana. Ayurvedic Herbs for Longevity and Rejuvination*. Taylor & Francis, London, **2003**, 105–106.

Rentzsch W., Schwarz M., Winterhalter P. Pyranoanthocyanins—An overview on structures, occurrence, and pathways of formation. *Trends Food Sci. Technol.* **2007**, *18*, 526–534.

Rivière C., Pawlus A.D., Mérillon J.M. Natural stilbenoids: Distribution in the plant kingdom and chemotaxonomic interest in Vitaceae. *Nat. Prod. Rep.* **2012**, *29*(11), 1317–1333.

Sharma P.V. *Fruits and Vegetables in Ancient India.* Chaukhamba Orientalia, Varanasi, India, **2009**, 31–32.

Shastri K.N. (ed.) *Charaka Samhita.* Chaukhambha Vidyabhavan, Varanasi, India, **1970**, Sutrasthana A-4.

Singh N., Agrawal M., Doré S. Neuroprotective properties and mechanisms of resveratrol in in vitro and in vivo experimental cerebral stroke models. *ACS Chem. Neurosci.* **2013**, *4*(8), 1151–1162.

USDA Nutrient Database. http://ndb.nal.usda.gov/.

Venkat P., Gangadharan G.G., Lakshmithathachar M.A., Alwar M.A. (eds.) *Bhojanakutuhalam.* I-AIM, Bangalore, India, **2012**, 145–146.

Vislocky L.M., Fernandez M.L. Biomedical effects of grape products. *Nutr. Rev.* **2010**, *68*(11), 656–670.

Vitrac X., Bornet A., Vanderlinde R., Valls J., Richard T., Delaunay J.C , Mérillon J.M., Teissédre P.L. Determination of stilbenes (δ-viniferin, *trans*-astringin, *trans*-piceid, *cis*- and *trans*-resveratrol, ε-viniferin) in Brazilian wines. *J. Agric. Food Chem.* **2005**, *53*, 5664–5669.

Warrier P.K., Nambiar V.P.K., Ramankutty C. (eds.) *Indian Medicinal Plants*, vol. 5. Orient Longman, Chennai, India, **1996**, 396–403.

Williamson G., Carughi A. Polyphenol content and health benefits of raisins. *Nutr. Res.* **2010**, *30*(8), 511–519.

Wu J.M., Hsieh T.C. Resveratrol: A cardioprotective substance. *Ann. N. Y. Acad. Sci.* **2011**, *1215*, 16–21.

Yang J., Xiao Y.Y. Grape phytochemicals and associated health benefits. *Crit. Rev. Food Sci. Nutr.* **2013**, *53*(11), 1202–1225.

10.1.8 Indian Gooseberry

Scientific name: *Phyllanthus emblica* L.
Family: Euphorbiaceae
Hindi: Amalak
Sanskrit: Amalaki
Tamil: Nellikai

One can survive by consuming the fruit juice of amalaki only

***Vamana Purana*, 91.51**

10.1.8.1 Origin and Distribution

The tree is indigenous to India and grows wild in South India and at the foot of Himalayas in the north. It is cultivated in plains throughout the country up to 200 m (Warrier *et al.*, 1995). It is also cultivated in China, Sri Lanka, and Malaysia.

It is a long-living plant with *sattvic* quality, considered very auspicious and sacred to Hindus as it is believed to increase the *prana* (the life energy). As the fruits are very nutritious, the tree is called *dhatri*, the nursing mother. An ancient myth associates the plant with the goddesses Lakshmi and Parvati: it was created by their tears of devotion to the gods Shiva and Vishnu and the leaves are still now used in the worship of the two gods (Gupta, 2001).

10.1.8.2 Features

Indian Gooseberry is a medium-sized deciduous tree with smooth greenish-gray exfoliating bark. The leaves resemble those of tamarind. The flowers are greenish yellow in color. The fruits are pale yellow, globular, small, round, six-lobed, thick, and hard in consistency (Paranjipe, 2005). There are two varieties: the wild one, with very small fruits, and the cultivated one, with bigger fruits. Both have medicinal properties (Sairam, 1998).

10.1.8.3 Properties and Uses in Ayurvedic System of Medicine

There are various Sanskrit names for this excellent fruit mentioned in all the ancient scriptures of Ayurveda: it is called *shreephala*, "the fruit of prosperity"; *Shiva*, "the auspicious"; *divyadhara*, "with divine qualities"; *sayastha*, "able to arrest old age"; and *amrta*, "ambrosia" (Nectar of immortality).

Amalaki is one of the best *rasayana*s: it is called *acharasayana* as it clears all the three *dosha*s and overcomes the degenerative effects of old age. A regular use can improve memory, health, youthfulness, and diseases resistance (Puri, 2003). The *Sushruta Samhita* mentions Indian gooseberry as the best of all acid fruits. It improves appetite, cleanses the intestines, and regulates blood sugar (Upadhyaya, 1970).

Parts used: Fruit
Tastes: Sweet, bitter, pungent, astringent, sour (The sour taste is dominant.)
Quality: Light, dry
Potency: Cold
***Dosha* effect:** PV–, K, *ama*+ (in excess)
Actions: The fresh fruit is refrigerant, tonic, diuretic, and laxative. The dried fruit is an excellent astringent, refrigerant, stomachic, and blood purifier (Dastur, 1962).

Preparations: *Amalaki* is the base of *chyavanprasha*, one of the best Ayurvedic tonics. This tonic takes the name from an ancient sage who was rejuvenated by the Asvinikumaras by a preparation, chiefly containing *amalaki* fruits (Sharma, 2009).

Another very important Ayurvedic medicine is *triphala*, a combination of the powder of Indian gooseberry with the powder of *Terminalia chebula* and *Terminalia bellirica.*

10.1.8.4 Traditional Uses and Folklore

- Hakims of Unani system of Medicine use *amla* as a good medicine for the heart and also for external applications due to the cooling and astringent properties.
- Five grams of the powder, mixed in one cup of warm water taken twice a day are an excellent general tonic and rejuvenative (Frawley and Lad, 1993).

Precautions: Fresh *amalaki* has a mild laxative effect, so it is suggested to avoid the use in case of diarrhea and acute dysentery, or use only if properly directed by a physician (Frawley and Lad, 1994).

10.1.8.5 In the Kitchen

Indian people eat the berry raw with a little salt, but it can also be cooked as a vegetable, in the form of pickle or marmalade.

In Ayurveda, a healthy diet should contain a combination of all six tastes. For this reason, it is suggested to eat *amalaki* with salt because she has all the five other tastes excluding the salty one. With salt it is a perfect and complete food. In the southern parts of India, people cook it in the traditional soup called *sambar.*

10.1.8.6 Chemical Constituents and Uses as Functional Food

It has been shown that *amla* does not contain any significant amount of vitamin C, but other organic acids (such as malic and mucic acids) and various common small tannin-like molecules and some unique tannins, one or more of which were mistaken for vitamin C in the initial tests conducted more than 50 years ago. Even so, at the levels described (at about 0.5%–0.7% of the fruit pulp), the amount of vitamin C consumed would be small by modern standards. The dominant active constituent of *amla* is a group of tannins derived from gallic and ellagic acids, which make up a large portion of the extractable nonnutritive constituents (Ghosal *et al.*, 1996; Majeed *et al.*, 2009). A recent investigation of *Emblica* leafflower fruits from China has identified more than 100 tannin-like compounds. Mucic acid gallate, mucic acid lactone gallate, monogalloylglucose, gallic acid, digalloylglucose, putranjivain A, galloyl-hexahydroxydiphenoyl-glucose, elaeocarpusin, and chebulagic acid were suggested to be the most abundant compounds in the crude methanol extracts of the fruits (Yang *et al.*, 2012).

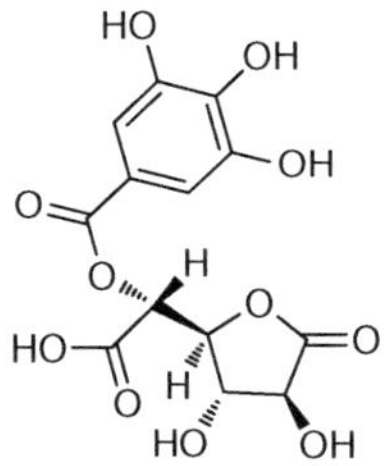

Mucic acid 1,4-lactone 5-*O*-gallate

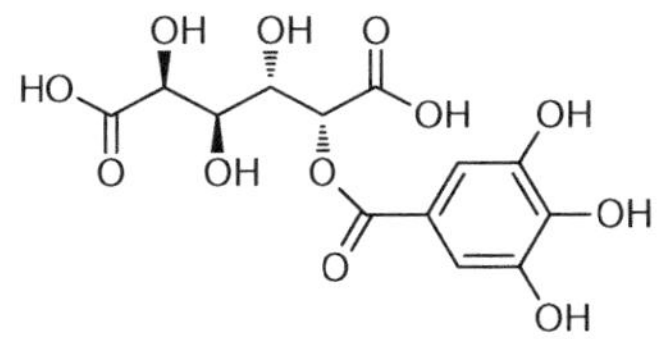

Mucic acid 2-*O*-gallate

The contents of the dry fruit include roughly carbohydrates 70%–75% (fiber, about 17% and sugars/starches/gums, about 25%), polyphenols 28%, and minerals 4%–6% (calcium, magnesium, potassium, sodium, zinc, iron, etc.).

Preclinical studies have shown that *amla* possesses antipyretic, analgesic, antitussive, antiatherogenic, adaptogenic, cardioprotective, gastroprotective, antianemia, antihypercholesterolemia, wound healing, antidiarrheal, antiatherosclerotic, hepatoprotective, nephroprotective, and neuroprotective properties. In addition, experimental studies have shown that *amla* and some of its phytochemicals also possess antineoplastic effects (reviewed in Baliga and Dsouza, 2011, and Krishnaveni and Mirunalini, 2010).

Crude extracts of *Emblica officinalis* have been found to induce specifically programmed cell death of mature osteoclasts without altering the process of osteoclastogenesis, specifically increasing the expression levels of Fas, a critical member of the apoptotic pathway. *Emblica officinalis* extracts specifically compete with the binding of a transcription factor involved in osteoclastogenesis (NF-κB) to its specific target DNA sequences (Piva *et al.*, 2009).

Emblica officinalis was found to offer protection against biological, physical, and chemical stressors, as judged by using markers of stress responses and objective parameters for stress manifestations. Using a model of cisplatin-induced alterations in gastrointestinal motility, the ability of *E. officinalis* to exert a normalizing effect, irrespective of direction of pathological change, was measured. *Emblica officinalis* strengthened the defense mechanisms against free radical damage induced during stress (Rege *et al.*, 1999; Scartezzini and Speroni, 2000).

REFERENCES

Baliga M.S., Dsouza J.J. Amla (*Emblica officinalis* Gaertn), a wonder berry in the treatment and prevention of cancer. *Eur. J. Cancer Prev.* **2011**, *20*(3), 225–239.

Dastur J.F. *Medicinal Plants of India and Pakistan*. Taraporevala Sons and Co., Mumbai, India, **1962**, 80.

Frawley D., Lad, V. *The Yoga of Herbs*. Motilal Banarsidass Publishers, Delhi, India, **1993**, 157–158.

Ghosal S., Tripathi V.K., Chauhan S. Active constituent of *Emblica officinalis*: Part 1. The chemistry and antioxidant effects of two new hydrolysable tannins, emblicanin A and B. *Indian J. Chem.* **1996**, *35B*, 941–948.

Gupta S.M. *Plant Myths and Traditions in India*. Munshiram Manoharlal Publishers, Delhi, India, **2001**, 26.

Krishnaveni M., Mirunalini S. Therapeutic potential of *Phyllanthus emblica* (amla): The ayurvedic wonder. *J. Basic Clin. Physiol. Pharmacol.* **2010**, *21*(1), 93–105.

Upadhyaya Y. (ed.) *Ashtanga Hridayam*. Chaukhambha Orientalia, Varanasi, India, **1970**, Uttara Sthana, A-40, 46–58.

Majeed M., Bhat B., Jadhav A., Srivastava J., Nagabhushanam K. Ascorbic acid and tannins from *Emblica officinalis* Gaertn. fruits—A revisit. *J. Agric. Food Chem.* **2009**, *57*(1), 220–225.

Paranjipe P. *Indian Medicinal Plants. Forgotten Healers*. Chaukhamba Sanskrit Pratishthan, Delhi, India, **2005**, 7–9.

Piva R., Penolazzi L., Borgatti M., Lampronti I., Lambertini E., Torreggiani E., Gambari R. Apoptosis of human primary osteoclasts treated with molecules targeting nuclear factor-κB. *Ann. N. Y. Acad. Sci.* **2009**, *1171*, 448–456.

Puri H.S. *Rasayana. Ayurvedic Herbs for Longevity and Rejuvination*. Taylor & Francis, London, **2003**, 22–42.

Rege N.N., Thatte U.M., Dahanukar S.A. Adaptogenic properties of six rasayana herbs used in Ayurvedic medicine. *Phytother. Res.* **1999**, *13*(4), 275–291.

Sairam T.V. *Home Remedies*, vol. I. Penguin Books, Delhi, India, **1998**, 210–217.

Scartezzini P., Speroni E. Review on some plants of Indian traditional medicine with antioxidant activity. *J. Ethnopharmacol.* **2000**, *71*(1/2), 23–43.

Sharma P.V. *Fruits and Vegetables in Ancient India*. Chaukhamba Orientalia, Varanasi, India, **2009**, 10–11.

Warrier P.K., Nambiar V.P.K., Ramankutty C. (eds.) *Indian Medicinal Plants*, vol. 4. Orient Longman, Chennai, India, **1995**, 256–263.

Yang B., Kortesniemi M., Liu P., Karonen M., Salminen J.-P. Analysis of hydrolyzable tannins and other phenolic compounds in emblic leafflower (*Phyllanthus emblica* L.) fruits by high performance liquid chromatography–electrospray ionization mass spectrometry. *J. Agric. Food Chem.* **2012**, *60*(35), 8672–8683.

10.1.9 Mango

Scientific name: *Mangifera indica* L.
Family: Anacardiaceae
Hindi: Aam
Sanskrit: Amrah
Tamil: Mankai, Mamaram

The year in which I don't eat enough mangoes should be deleted from my life-span.

Rabindranath Tagore

10.1.9.1 Origin and Distribution

Mango is the national fruit of India, a fruit beloved by everybody. The tree is one of the most sacred of the country: the Sanskrit name *amra* means "of the people." *Indica* means "originating from India" and *Mangifera* is derived from the Tamil name.

The center of origin is the northeast India, Myanmar, and Bangladesh. Fossil records were found there dating back 25–30 million years ago. The tree has been cultivated in the Indian subcontinent for over 4000 years. Buddhist monks took mango plants on voyages to Malaya and eastern Asia in the fourth and fifth centuries BC. By the tenth century AD, Persian traders had taken the fruit to the Middle East and East Africa. With the arrival of the Portuguese in India in the fifteenth century, it was later spread to South America, to the Philippines, and to West Africa. Besides India, the plant is now cultivated commercially in China, Pakistan, Bangladesh, the Philippines, Haiti, Mexico, and Brazil. In India alone, there are at least 500 commercial varieties of mango (Bakhru, 1994).

Mango is considered very auspicious and sacred by both Hindus and Buddhists: it is often referred to as the king of fruits and forms an integral part of many traditional functions. The fruit is supposed to confer prosperity and wealth. During all the religious occasions, Hindus use the twigs as tooth brushes (except on Sundays and Tuesdays) and the leaves as spoons for pouring libations in the sacrificial fire. The leaves are also used to decorate new houses, temples, and marriage halls as a symbol of fertility and good luck (Majupuria and Joshi, 1997). One of the names of Shiva is Ekambareswarar (The Lord of the Mango tree) as Hindus believe that the Lord himself took the tree to the earth to please his wife Parvati, who was craving for it. It is still believed that benevolent spirits reside in the mango groves.

Lord Buddha was once presented by the courtesan Amrapali with a mango grove where he used to rest and meditate. According to a Burmese legend, Buddha also caused a mango tree to sprout instantly from a seed after eating the fruit. Therefore, the Buddhists consider the tree holy (Gupta, 2001).

Since ancient times, mango orchards have been a sign of prosperity. The Mughal emperor Akbar had 100,000 mango trees in his famous Lakhi Bagh in Bihar (Rajaiah, 2010).

10.1.9.2 Features

Mango is a large, evergreen tree growing to a height of 10–45 m. The leaves are spirally arranged on branches, linear-oblong or elliptic-lanceolate, reddish, and thinly flaccid when first formed. They release an aromatic odor when crushed. The inflorescence occurs in panicles consisting of about 3000 tiny whitish red or yellowish green flowers. The fruit is a well-known large drupe but shows a great variation in shape and size. It contains a thick yellow pulp, single seed, and thick yellowish red skin when ripe. The seed is solitary, ovoid, or oblong, encased in a hard, compressed fibrous endocarp (Warrier *et al.*, 1995).

10.1.9.3 Properties and Uses in Ayurvedic System of Medicine

Mangifera indica has been an important plant in the Ayurvedic and indigenous medical systems for over 4000 years. According to Ayurveda, a wide number of medicinal properties are attributed to the different parts of mango tree. Mango possesses antidiabetic, antioxidant, antiviral, cardiotonic, hypotensive, and anti-inflammatory properties (Shah *et al.*, 2010).

Ayurveda considers the ripe fruit to be hot, calming, nourishing, and full of *prana*. It is better to eat it with milk to balance the hot potency. Sri Raghunatha Suri in the *Bhojanakutuhalam* reports that mango mixed with milk alleviates *vata* and *pitta*, imparts taste, promotes growth, confers strength, and is aphrodisiac.

> It is best to eat mango by sucking it since doing so mitigates its sour taste and augments its sweet taste: further, mango eaten in this manner imparts taste, strength and vitality … it is cold in potency, undergoes metabolism quickly and greatly alleviates *vata* and *pitta*. The distilled juice of mango is strengthening, heavy, alleviates *vata* and is laxative
>
> **Venkat *et al.*, 2012**

The raw mango is cooling and very effective in treating gastrointestinal disorders and acidity due to *pitta*. It will also protect the body from heat stroke and dehydration. For this reason, *aamchar*, the dry raw mango powder is a very popular condiment in North India (Rajaiah, 2010).

Parts used: Fruits
Tastes: Sweet (ripe mango), sour and astringent (unripe mango)
Quality: Unctuous, heavy
Potency: Hot (ripe mango), cold (unripe mango)
***Dosha* effect:** VP–, K+ (ripe mango), VPK= (unripe mango)
Actions: Nutritive, demulcent, diuretic, astringent, refrigerant, aphrodisiac

10.1.9.4 Traditional Uses and Folklore

- All parts of the mango plant are used in traditional South Asian medicine, but the fruits are most important.
- Unani physicians recommend the use of mango for strengthening the nervous and blood systems, ridding the body of toxins and treating anemia.
- For spleen enlargement, a teacup of ripe mango pulp with one teaspoon of honey thrice a day is a traditional home remedy. For digestive and liver trouble, it is recommended to suck a ripe mango followed by a glass of milk. For body weakness, mango has to be eaten with one teaspoon of honey and a pinch of saffron, cardamom, and rose water (Sairam, 1998).

Precautions: Over indulgence in mango can lead to weakening of digestive fire, intermittent fever, disorders of blood, and ophthalmic disorders. One must therefore not eat mangoes excessively (Venkat *et al.*, 2012).

10.1.9.5 In the Kitchen

A confection made with the juice of the ripe fruit, sugar, and aromatics is a nice restorative tonic. When nearly ripe, the fruit is made into a curry having a sweet and sour taste. Unripe fruits are much used in India in pickles and chutneys (Nadkarni, 2002). They are also peeled, sliced, dried, and ground to make the spice called *amchur*. This spice is a concentrate of mango properties, much used in chutneys and curries (Aggarwal and Yost, 2011).

The contents of mango seeds have been eaten in times of food scarcity in India. The seeds can be roasted or boiled, or ground into flour.

10.1.9.6 Chemical Constituents and Uses as Functional Food

Mango is the national fruit of India and the Philippines. It is the largest fruit tree in the world, found in the wild (Belem, in Brazilian Amazon, is called the city of mangoes) and cultivated in the tropical regions.

It is native to tropical Asia and has been cultivated in Indian subcontinent for over 4000 years. Various parts of the plant are used medicinally in traditional medicines in India and China (Shah *et al.*, 2010). A number of phytochemicals are described from stem bark, leaf, and pulp, which include characteristic *C*-glycosidic xanthones (mangiferin and congeners) and phytosterols (mangsterol, mangoleanone, mangiferonic acid). The edible sweet pulp is rich in vitamin C, vitamin A, and β-carotene.

A number of biological activities of mangiferin (mainly present in young leaves) have been suggested, including antidiabetic and anti-inflammatory abilities. These might be explained by its antioxidant ability as well as its ability to modulate several key inflammatory pathways. Mangiferin has also been shown to be an effective inhibitor of NF-κB signaling pathway (Vyas *et al.*, 2012).

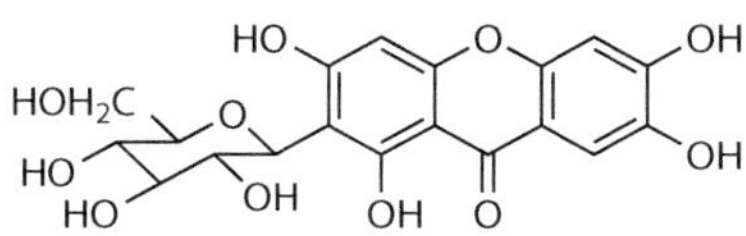

Mangiferin

Vimang is an aqueous extract of *M. indica* traditionally used in Cuba for its analgesic, anti-inflammatory, antioxidant, and immunomodulatory properties. Several formulations are available for Vimang, and also for mangiferin, its major component. Preclinical studies demonstrated that these products prevented the binding of NF-κB to DNA, which induces the transcription of genes implicated in the expression of some mediators and enzymes involved in inflammation, pain, oxidative stress, and synaptic plasticity. Thus, Vimang is proposed of potential utility in the neuropathic pain treatment (Garrido-Suárez *et al.*, 2010).

REFERENCES

Aggarwal B.B., Yost D. *Healing Spices.* Sterling Publishers, New York, **2011** (Italian version, Armenia, 52–57).

Bakhru H.K. *Foods That Heal.* Orient Paperbacks, Delhi, India, **1994**, 63–68.

Garrido-Suárez B.B., Garrido G., Delgado R., Bosch F., del C Rabí M. A *Mangifera indica* L. extract could be used to treat neuropathic pain and implication of mangiferin. *Molecules* **2010**, *15*(12), 9035–9045.

Gupta S.M. *Plant Myths and Traditions in India.* Munshiram Manoharlal Publishers, Delhi, India, **2001**, 41–42.

Majupuria T.C., Joshi D.P. *Religious and Useful Plants of Nepal and India.* Gupta M.(ed.), Publishers, Lashkar, India, **1997**, 243.

Nadkarni K.M. *Indian Materia Medica*, vol. 1. Popular Prakashan, Mumbai, India, **2002**, 764–769.

Rajaiah R. *How the Banana Goes to Heaven. And Other Secrets of Health from the Indian Kitchen.* Westland, Chennai, India, **2010**, 140–149.

Sairam T.V. *Home Remedies*, vol. I. Penguin Books, Delhi, India, **1998**, 147–153.

Shah K.A., Patel M.B., Patel R.J., Parmar P.K. *Mangifera indica* (mango). *Pharmacogn. Rev.* **2010**, *4*(7), 42–48.

Venkat P., Gangadharan G.G., Lakshmithathachar M.A., Alwar M.A. (eds.) *Bhojanakutuhalam.* I-AIM, Bangalore, India, **2012**, 163–168.

Vyas A., Syeda K., Ahmad A., Padhye S., Sarkar F.H. Perspectives on medicinal properties of mangiferin. *Mini Rev. Med. Chem.* **2012**, *12*(5), 412–425.

Warrier P.K., Nambiar V.P.K., Ramankutty C. (eds.) *Indian Medicinal Plants*, vol. 3. Orient Longman, Chennai, India, **1995**, 380–390.

10.1.10 Papaya

Scientific name: *Carica papaya* L.
Family: Caricaceae
Hindi: Pappaya
Sanskrit: Brahmairandah
Tamil: Pappali

10.1.10.1 Origin and Distribution

Papaya is one of the most valuable tropical fruits, called the "fruit of the angels" by Christopher Columbus. It is probably native to Mexico and Costa Rica. It was taken by the Spaniards to Manila in the mid-sixteenth century, and from there it gradually spread to many other tropical and subtropical countries. It is now cultivated in India, China, Sri Lanka, Malaya, the Philippines, Australia, Mexico, Brazil, Peru, Venezuela, Central Africa, and South Africa (Bakhru, 1994). India is the leading producer, with 42% of the world's crop.

10.1.10.2 Features

It is a small, soft-wooded, fast-growing, short-lived tree with a tuft of palm-like leaves at the top. Flowers are unisexual, males in long dropping panicles, and females in short clusters. The fruit is a large, oblong or nearly spherical fleshy berry with yellow-orange rind like a gourd, with many seeds (Warrier *et al.*, 1994).

10.1.10.3 Properties and Uses in Ayurvedic System of Medicine

Ayurveda recommends the use of papaya in vitiated conditions of *vata*. The fruit cleans the digestive system, has a tonic effect on the stomach, and helps to maintain the acid/alkaline balance in the body. It regulates sugar metabolism and helps in digestive difficulties. It is also good for cough and lung disorders (Nibodhi, 2010).

Parts used: Fruit
Tastes: Sweet, sour
Quality: Light, unctuous, sharp
Potency: Hot

***Dosha* effect:** V–, PK+

Actions: Demulcent, laxative, digestive aid, tonic, aphrodisiac, appetizer, anthelmintic, anti-inflammatory

10.1.10.4 Traditional Uses and Folklore

- The ripe fruit if eaten regularly corrects habitual constipation. It is useful in bleeding piles and dyspepsia. The dried and salted fruit reduces enlarged spleen and liver (Nadkarni, 2002).
- For liver and spleen inflammation, it is suggested to eat daily a big slice of papaya with honey (Sairam, 1998).
- The seeds are one of the best ways to eliminate roundworms in children. (Morningstar and Desai, 1999).

Precautions: Avoid consuming papaya during pregnancy. Consumption of green papaya is a very common remedy for contraception and abortion in India, Bangladesh, Pakistan, and Sri Lanka.

10.1.10.5 In the Kitchen

Fresh ripe papaya can be used in fruit salad, with lime juice or to prepare a milk shake. It is better to purchase already ripe papayas. Green and hard papayas are good only if cooked. They are an excellent ingredient of a jam (mixed with ginger, lemon, and apple).

The fruit can be boiled and mixed with lime juice and sugar to prepare a sauce. Unripe and green fruits are made into curry and eaten by women to stimulate secretion of milk. The fruit can also be pickled (Nadkarni, 2002). In some parts of Asia, the young leaves of the papaya are steamed and eaten like spinach.

The black dried seeds are a common and cheap substitute for black pepper as they have a pungent taste.

10.1.10.6 Chemical Constituents and Uses as Functional Food

Carica papaya L. (Caricaceae) is one of the major fruit crops cultivated in tropical and subtropical zones. Papaya fruit is a source of nutrients such as provitamin A, carotenoids, vitamin C, folate, and dietary fiber. One half of a small papaya can provide 150% of the recommended dietary intake of vitamin C. It is low in calories, fat free, cholesterol free, and a good source of potassium, folate, and fiber. Papaya skin, pulp, and seeds also contain a variety of phytochemicals, including lycopene and polyphenols. Papaya is a good source of nutrients and some phytochemicals such as β-cryptoxanthin and benzyl isothiocyanates (ITCs) (Nguyen *et al.*, 2013).

In *in vitro* studies, ITCs induced apoptosis in various cancer cell lines and experimental rodents via modulation of multiple signal transduction pathways and apoptosis intermediates. Recently, Japanese researchers have shown the benzyl ITC extracted from papaya could also induce toxicity more preferentially in the proliferating human colon epithelial cells than in the quiescent cells (Nakamura and Miyoshi, 2006). Several epidemiological studies also indicated that the dietary consumption of ITCs or ITC-containing foods inversely correlates with the risk of developing lung, breast, and colon cancers, providing evidence that they have a potential to prevent cancer in humans. Various ITCs are effective chemoprotective agents against chemical carcinogenesis in experimental animals. ITCs are capable of inhibiting both the formation and the development of a cancer cell through multiple pathways, that is, inhibition of carcinogen-activating cytochrome P450 mono-oxygenases, induction of carcinogen-detoxifying phase 2 enzymes, induction of apoptosis, and inhibition of cell cycle progression (Nakamura, 2009).

No clinical or animal cancer studies were identified, and only seven *in vitro* cell culture-based studies were reported; these indicate that *C. papaya* extracts may alter the growth of several types of cancer cell lines. However, many studies focused on specific compounds in papaya and reported bioactivity, including anticancer effects (Nguyen *et al.*, 2013).

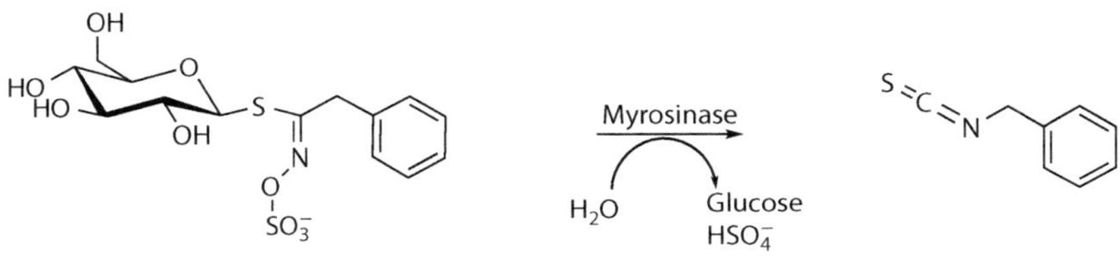

Benzyl glucosinolate

Benzyl isothiocyanate

In traditional medicine, different parts of *C. papaya*, including its leaves, barks, roots, latex, fruit, flowers, and seeds, have a wide range of reputed medicinal application. Many of these traditional uses have been validated by scientific studies. Experiments have shown that *C. papaya* possesses anthelmintic, antiprotozoan, antibacterial, antifungal, antiviral, anti-inflammatory, antihypertensive, hypoglycemic and hypolipidemic, wound healing, antitumor, free-radical scavenging, antisickling, neuroprotective, diuretic, abortifacient, and antifertility activities [reviewed in Krishna *et al.*, 2009, and Singh *et al.*, 2010].

REFERENCES

Bakhru H.K. *Foods That Heal*. Orient Paperbacks, Delhi, India, **1994**, 72–75.

Krishna K.L., Paridhavi M., Patel J.A. Review on nutritional, medicinal and pharmacological properties of papaya (*Carica papaya* Linn.). *Nat. Prod. Rad.* **2008**, *7*, 364–373.

Morningstar A., Desai U. *The Ayurvedic Cookbook*. Motilal Banarsidass Publishers, Delhi, India, **1999**.

Nadkarni K.M. *Indian Materia Medica*, vol. 1. Popular Prakashan, Mumbai, India, **2002**, 273–277.

Nakamura Y. Chemoprevention by isothiocyanates: Molecular basis of apoptosis induction. *Forum Nutr.* **2009**, *61*, 170–181.

Nakamura Y., Miyoshi N. Cell death induction by isothiocyanates and their underlying molecular mechanisms. *Biofactors* **2006**, *26*(2), 123–134.

Nguyen T.T., Shaw P.N., Parat M.O., Hewavitharana A.K. Anticancer activity of *Carica papaya*: a review. *Mol. Nutr. Food Res.* **2013**, *57*(1), 153–164.

Nibodhi. *Annapurna's Prashad*. Mata Amritanandamayi Mission Trust, Amritapuri, India, **2010** (2nd edn.), 88–89.

Sairam T.V. *Home Remedies*, vol. I. Penguin Books, Delhi, India, **1998**, 49–54.

Singh D., Jaiswal P., Kumar P., Singh V. *Carica papaya* Linn: A potential source for various health problems. *J. Pharm. Res.* **2010**, *3*, 998–1003.

Warrier P.K., Nambiar V.P.K., Ramankutty C. (eds.) *Indian Medicinal Plants*, vol. 1. Orient Longman, Chennai, India, **1994**, 383–385.

10.1.11 Pomegranate

Scientific name: *Punica granatum* L.
Family: Punicaceae
Hindi: Anar
Sanskrit: Dadimah
Tamil: Madalai

10.1.11.1 Origin and Distribution

Pomegranate is considered to have originated in the Himalayas in northern India, but it has been cultivated since ancient times throughout the Middle East and the entire Mediterranean region. Carbonized exocarp of the fruit has been identified in Early and Late Bronze Age in the Middle East. A large, dry pomegranate was found in the tomb of Djehuty, the butler of Queen Hatshepsut in Egypt; Mesopotamian cuneiform records mention pomegranates from the mid-third millennium BC onward. Today the plant is found wild in Persia, Arabia, Afghanistan, and Baluchistan, and is widely cultivated throughout the Mediterranean region of southern Europe, the Middle East and Caucasian regions, the northern and tropical Africa, the Latin America, the Indian subcontinent, and the drier parts of Southeast Asia. It is a very common garden plant in China, as it is believed to ward off the evil eye (Sairam, 1998). Since very ancient times, it has been considered an emblem of fertility and numerous progeny.

The name "pomegranate" is derived from the Latin words *pomum* and *granatus*, which means "a seeded apple." The genus name *Punica* refers to the Phoenicians, who were active in broadening its cultivation, partly for religious reasons. Pliny called it *malum punicum*, the apple of Carthage.

We can find a number of references to the fruit in Babylonian texts, the Book of Exodus, the Homeric Hymns, and the Quran. In ancient Egypt, it was regarded as a symbol of prosperity and ambition. It was the fruit brought to Moses to demonstrate the fertility of the Promised Land. In the ancient Greek mythology, it is known as the "fruit of the dead," tasted by Persephone in the Underworld. It was sacred to Hera and Aphrodite, and even today the pomegranate has strong symbolic meanings for the Greeks. It is a symbol of abundance, fertility, and good luck, used during weddings and funerals. According to the Quran, pomegranates grow in the gardens of paradise. Prophet Mohammed advised his followers to eat the fruits as a way to purge the spirit from envy (Isomaki and Gandhi, 2004). The pomegranate is the symbol of Armenia, where it represents fertility, abundance, and marriage. In Iran and India, it also symbolizes prosperity and fertility. In Hindu religion, it is associated with Bhomidevi (the earth goddess), Ganesha, and Lakshmi and Vishnu (Majupuria and Joshi, 1997). The Parsis use its twigs to make their sacred broom and squeeze the juice of the fruit into the mouth of dying people (Isomaki and Gandhi, 2004).

10.1.11.2 Features

It is a glabrous shrub or small tree, with narrowly elliptic or lanceolate leaves and bright red flowers. The fruits are large globose berry, yellowish red when ripe with a woody rind. The seeds are very numerous, entirely filling the fruit (Warrier *et al.*, 1995).

10.1.11.3 Properties and Uses in Ayurvedic System of Medicine

Pomegranate is mentioned in all the ancient texts of Ayurveda; *Sushruta Samhita* classified it as *hrdya*, heart tonic, appetite stimulant, antiemetic, and antidiarrheal (Paranjipe, 2005).

Sri Raghunatha Suri in the *Bhojanakutuhalam* reports that the sweet fruit treats cough; alleviates vitiated *vata*, *kapha*, and *pitta*; stimulates the digestive fire; and removes fatigue (Venkat *et al.*, 2012).

The fruit has a calming effect on the mind and is very effective in heart diseases and anemia. Sour pomegranate increases the appetite. Sweet pomegranate increases strength and intelligence.

Parts used: Fruit
Tastes: Sweet, sour
Quality: Light, oily
Potency: Cold
***Dosha* effect:** The sweet variety alleviates all the three *dosha*s but may increase *ama*; the sour variety may aggravate *pitta* (Frawley and Lad, 1994).
Actions: Astringent, tonic, alterative, refrigerant, stomachic

10.1.11.4 Traditional Uses and Folklore

- In the Unani system of medicine, the pomegranate is used as a remedy for diabetes.
- During the summer season, a glass of pomegranate juice is very beneficial in neutralizing the effect of sunstroke and also checks the bleeding of gum and cures indigestion. Mixed with rock salt and honey, the juice improves the appetite.
- A cup of pomegranate juice with a quarter teaspoon of cinnamon and two teaspoons of honey is a traditional home remedy for anemia. The fresh juice prevents *pitta* aggravation and diabetes (Sairam, 1998).
- The fruit effectively controls hyperacidity and thirst, when given with rock candy (Paranjipe, 2005). It is recommended as a diet during the convalescence after diarrhea (Nadkarni, 2002).
- Many other home remedies for internal and external applications are also made using the rind, flowers, and seeds. The use of seeds and rind as a contraceptive and abortifacient is recorded in the ancient Indian literature, medieval sources, and modern folk medicine.

Precautions: Pomegranate has anti-diarrheal properties. In excess can cause constipation (Sairam, 1998).

10.1.11.5 In the Kitchen

Wild pomegranate seeds are dried and used as a spice in Indian and Pakistani cuisines. The spice is called *anardana* and is added to curries and chutneys to give a typical sweet and sour taste (Aggarwal and Yost, 2011).

The seeds of pomegranate are often added to rice preparation as a tasteful and healthy garnish.

10.1.11.6 Chemical Constituents and Uses as Functional Food

After the pomegranate is opened by scoring it with a knife and breaking it open, the arils (seed casings) are separated from the peel and internal white pulp membranes. A very sweet and sour juice is prepared from squeezing the arils, with sour notes from the acidic tannins. Pomegranate juice has long been a popular drink in Armenian, Persian, and Indian cuisines, and is now present in the market all over the world for its nutritious value. Dried pomegranate arils, found in some natural specialty food markets, still contain the seed and residual aril water, maintaining a natural sweet and tart flavor. Dried arils can be used in several culinary applications: as a topping for salad, yogurt, or ice cream, and in bakery.

Pomegranate arils provide 12% of the daily value (DV) for vitamin C and 16% of the DV for vitamin K per 100 g serving with a number of healthy minerals such as manganese, phosphorus, and zinc.

Pomegranates have been shown to contain more than 120 different phytochemicals, and some of them act in concert to exert antioxidant and anti-inflammatory effects. The main components in fruits and juices are ellagic acid and ellagitannins (punicalagin and punicalin) (Heber, 2011), together with flavonoids, anthocyanins, and phenolic acids.

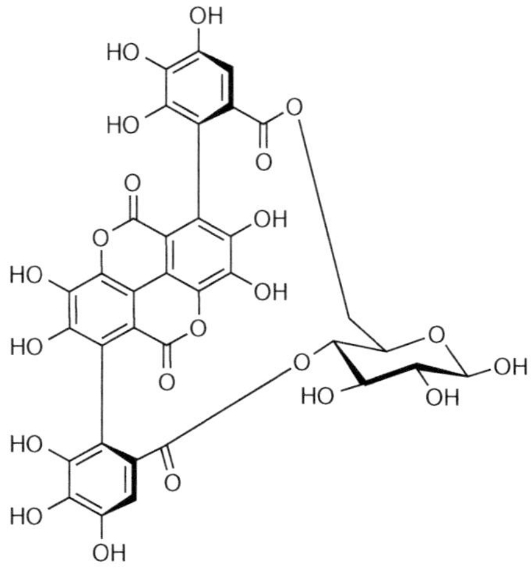

Punicalin

Seeds are a rich source of lipids and their oil, which constitutes 12%–20% of the total seed weight and contains a number of unique fatty acids (punicic acid represents 70%–76% of the seed oil) (Vroegrijk *et al.*, 2011). Sugars, fruit organic acids, and vitamin C are common constituents of seed, peel, and juice (reviewed in Viladomiu *et al.*, 2013).

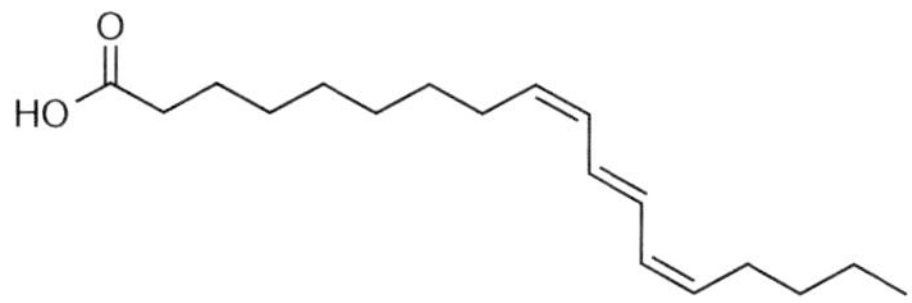

Punicic acid

Ellagitannins are a family of bioactive polyphenols present in fruits (most of the known berries) and nuts. They are hydrolizable tannins releasing ellagic acid on acid hydrolysis and form urolithins following metabolism by the gut flora, and owe their bioactivity to the potent antioxidant action. The hydrolysis process allows a prolonged release of ellagic acid in the bloodstream after ingestion of pomegranate. By pressing the whole fruit during the squeezing process, ellagitannins are extracted into pomegranate juice in significant quantities, reaching levels of >2 g/L (Gil *et al.*, 2000). Ellagitannins are not absorbed, but ellagic acid is detected in the bloodstream 1 hour after ingestion of pomegranate juice (Mertens-Talcott *et al.*, 2006). Plasma is cleared 4 hours after consumption (Seeram *et al.*, 2004) and urolithin metabolites persist in the urine up until 48 hours after ingestion of pomegranate juice (Seeram *et al.*, 2006). Urolithins present antioxidant and anti-inflammatory properties *in vitro* (Bialonska *et al.*, 2009; González-Sarrías *et al.*, 2010). Thus, it is suggested that ellagitannins are hydrolyzed in the stomach, some ellagic is absorbed into circulation and further metabolized to urolithins and other metabolites through colonic microbial metabolism.

Ellagitannins in pomegranate interact with the gut microflora (Bialonska *et al.*, 2010); in particular, they increase the growth of some bifidobacteria as well as the production of short fatty acids, which have been shown to elicit beneficial effects through the activation of peroxisome proliferator-activated receptors (PPARs) that are ligands for lipidic molecules and molecular targets for drugs against type 2 diabetes, and represent promising new targets for the treatment and prevention of inflammatory disorders. Functionally, PPARs regulate inflammation, immunity, and metabolism (Daynes and Jones, 2002). Multiple targeted pathways occur with the assumption of therapeutically active doses of pomegranate.

Emerging science has demonstrated anticancer effects of pomegranate, with the most impressive data so far in prostate cancer. Ellagitannins and urolithins can accumulate in the intestine and prostate (Seeram *et al.*, 2007). Pomegranate juice given daily for 2 years to 40 prostate cancer patients with increasing prostate-specific antigen levels provides evidence for the possible utilization of a nontoxic option for prevention or delay of prostate carcinogenesis (Heber, 2011). The synergistic action of the pomegranate constituents appears to be superior compared to individual constituents.

In vivo and *in vitro* studies have demonstrated that pomegranate exerts hypoglycemic effects, including increased insulin sensitivity, inhibition of α-glucosidase, and impact on glucose transporter type 4 function, but is also responsible for the reduction of total cholesterol and the improvement of blood lipid profiles, as well as anti-inflammatory effects through the modulation of PPAR pathways. These effects may also explain how pomegranate-derived compounds function in the amelioration of adverse health effects caused by metabolic syndrome (reviewed in Medjakovic and Jungbauer, 2013). Although the fruit is consumed in many countries, epidemiological and clinical studies are unavailable. Additional research is necessary to corroborate the promise of current *in vivo* and *in vitro* findings.

Promising results have so far been obtained for the treatment of certain diseases, including obesity, insulin resistance, intestinal inflammation, and cancer. Although moderate consumption of pomegranate does not result in adverse effects, future studies are needed to assess safety and potential interactions with drugs, which may alter the bioavailability of bioactive constituents of pomegranate as well as drugs.

REFERENCES

Aggarwal B.B., Yost D. *Healing Spices*. Sterling Publishers, New York, **2011** (Italian version, Armenia, 230–239).

Bialonska D., Kasimsetty S.G., Khan S.I., Ferreira D. Urolithins, intestinal microbial metabolites of pomegranate ellagitannins, exhibit potent antioxidant activity in a cell-based assay. *J. Agric. Food Chem.* **2009**, *57*(21), 10181–10186.

Bialonska D., Ramnani P., Kasimsetty S.G., Muntha K.R., Gibson G.R., Ferreira D. The influence of pomegranate by-product and punicalagins on selected groups of human intestinal microbiota. *Int. J. Food Microbiol.* **2010**, *140*(2/3), 175–182.

Daynes R.A., Jones D.C. Emerging role of PPARs in inflammation and immunity. *Nature Rev. Immunol.* **2002**, *2*(10), 748–759.

Frawley D., Lad V. *The Yoga of Herbs*. Motilal Banarsidas Publishers, Delhi, India, **1994**, 135–136.

Gil M.I., Tomás-Barberán F.A., Hess-Pierce B., Holcroft D.M., Kader A.A. Antioxidant activity of pomegranate juice and its relationship with phenolic composition and processing. *J. Agric. Food Chem.* **2000**, *48*(10), 4581–4589.

González-Sarrías A., Larrosa M., Tomás-Barberán F.A., Dolara P., Espín J.C. NF-κB-dependent anti-inflammatory activity of urolithins, gut microbiota ellagic acid-derived metabolites, in human colonic fibroblasts. *Br. J. Nutr.* **2010**, *104*(4), 503–512.

Heber D. Pomegranate ellagitannins, in, Benzie I.F.F., Wachtel-Galor S. (eds.), *Herbal Medicine: Biomolecular and Clinical Aspects*. CRC Press, Boca Raton, FL, **2011** (2nd edn.), chapter 10.

Isomaki R., Gandhi M. *The Book of Trees*. Other India Press, Goa, India, **2004**, 210–211.

Majupuria T.C., Joshi D.P. *Religious and Useful Plants of Nepal and India*. Gupta M. (ed.), Publishers, Lashkar, India, **1997**, 243.

Medjakovic S., Jungbauer A. Pomegranate: A fruit that ameliorates metabolic syndrome. *Food Funct.* **2013**, *4*(1), 19–39.

Mertens-Talcott S.U., Jilma-Stohlawetz P., Rios J., Hingorani L., Derendorf H. Absorption, metabolism, and antioxidant effects of pomegranate (*Punica granatum* L.) polyphenols after ingestion of a standardized extract in healthy human volunteers. *J. Agric. Food Chem.* **2006**, *54*(23), 8956–8961.

Nadkarni K.M. *Indian Materia Medica*, vol. 1. Popular Prakashan, Mumbai, India, **2002**, 1031–1035.

Paranjipe P. *Indian Medicinal Plants. Forgotten Healers*. Chaukhamba Sanskrit Pratishthan, Delhi, India, **2005**, 64–65.

Sairam T.V. *Home Remedies*, vol. I. Penguin Books, Delhi, India, **1998**, 235.

Seeram N.P., Aronson W.J., Zhang Y., Henning S.M., Moro A., Lee R.P., Sartippour M. *et al.* Pomegranate ellagitannin-derived metabolites inhibit prostate cancer growth and localize to the mouse prostate gland. *J. Agric. Food Chem.* **2007**, *55*(19), 7732–7737.

Seeram N.P., Henning S.M., Zhang Y., Suchard M., Li Z., Heber D. Pomegranate juice ellagitannin metabolites are present in human plasma and some persist in urine for up to 48 hours. *J. Nutr.* **2006**, *136*(10), 2481–2485.

Seeram N.P., Lee R., Heber D. Bioavailability of ellagic acid in human plasma after consumption of ellagitannins from pomegranate (*Punica granatum* L.) juice. *Clin. Chim. Acta* **2004**, *348*(1/2), 63–68.

Venkat P., Gangadharan G.G., Lakshmithathachar M.A., Alwar M.A. (eds.) *Bhojanakutuhalam*. I-AIM, Bangalore, India, **2012**, 145.

Viladomiu M., Hontecillas R., Lu P., Bassaganya-Riera J. Preventive and prophylactic mechanisms of action of pomegranate bioactive constituents. *Evid. Based Complement. Alternat. Med.* **2013**, ID 789764.

Vroegrijk I.O., van Diepen J.A., van den Berg S., Westbroek I., Keizer H., Gambelli L., Hontecillas R. *et al.* Pomegranate seed oil, a rich source of punicic acid, prevents diet-induced obesity and insulin resistance in mice. *Food Chem. Toxicol.* **2011**, *49*(6), 1426–1430.

Warrier P.K., Nambiar V.P.K., Ramankutty C. (eds.) *Indian Medicinal Plants*, vol. 4. Orient Longman, Chennai, India, **1995**, 396–402.

10.1.12 *Prunus* Species

Prunus is a genus of trees and shrubs, including plums, cherries, peaches, apricots, and almonds. *Prunus armeniaca* is the common apricot, *Prunus domestica* is the plum, *Prunus avium* is the sweet cherry, *Prunus amygdalus* is the almond, and *Prunus persica* is the peach.

They are linked by the chemical composition of the kernels that are characterized by the presence of cyanogenic glycosides (amygdalin, the compound responsible for their bitterness). Nearly 570 chemical compounds have so far been isolated from several *Prunus* species.

10.1.12.1 Apricot

Scientific name: *Prunus armeniaca* L.
Family: Rosaceae
Hindi: Jardalu
Sanskrit: Urumana
Tamil: Sarkkarai badami

10.1.12.1.1 Origin and Distribution

The Chinese associate the apricot with education and medicine, as apricot kernels are an important ingredient of their traditional medicine. Physicians are still poetically referred to as "Experts of the Apricot Grove" in relation to the story of Dong Feng, a physician living during the Three Kingdoms period, who required no payment from his patients, except that they plant apricot trees in his orchard on recovering from their illnesses, resulting in a large grove of apricot trees and a steady supply of medicinal ingredients.

According to Vavilov, the tree originated in the Chinese region, where it was domesticated. Other sources say that the apricot was first cultivated in India around 3000 BC. In India, the plant is almost naturalized in the northwestern Himalayas, Deccan, and Mysore. Its introduction in Greece is attributed to Alexander the Great, and from Greece the Arabs took the plant to North Africa. The name apricot derives from the Arabic *albarqúq*. The Spanish colonists took it to the United States, where the tree was perfectly adapted to the Californian climate. In Armenia, the plant has been cultivated since ancient times, and we can find now about 50 varieties of apricots there. The plant was imported from Armenia to the southern parts of Europe by the Romans (106–57 BC), hence the botanical name *Prunus armeniaca*.

The first producer of apricots in the world is Turkey, which hosts more than 20% of the world production. It is followed by Iran, Italy, Spain, Morocco, and the United States. Other apricot-producing countries are Russia, Greece, France, Pakistan, Syria, China, South Australia, and Tasmania.

10.1.12.1.2 Features

The apricot is a small tree, about 10 m tall, with a reddish bark and a dense, spreading canopy. The leaves are ovate and 5–9 cm long, with a rounded base, a pointed tip, and a finely serrated margin. The flowers are pinkish white and appear much in advance of the foliage. The fruit is a drupe similar to a small peach, with a yellow skin overlaid with red on the side most exposed to the sun. The flesh is firm and sweet with a yellowish orange color (Sharma, 2009).

10.1.12.1.3 Properties and Uses in Ayurvedic System of Medicine

Apricot prevents and treats constipation, alleviates fever and cough, and counters muscle and nerve disorders (Nibodhi, 2010).

Parts used: Fruit
Tastes: Sweet, sour
Quality: Heavy
Potency: Warm
***Dosha* effect:** VK–, P+ (in excess)
Actions: Laxative, tonic, nutritive (Nadkarni, 2002)

10.1.12.1.4 Traditional Uses and Folklore

The antioxidant properties of apricots were outlined in a study conducted by Dr. Paul D. White and Dr. Edward G. Toomey in the community of Hunza, located in a remote Himalayan region. The diet of this community is mainly vegetarian and focuses especially on apricots. This indigenous community is very healthy and has a life expectancy of about 90–110 years. Probably, their quality of life is due to many other factors and not only to the high consumption of apricot: They have a very simple life, totally in tune with nature, and they strictly avoid eating anything they cannot produce or find in their environment. The abundance of apricots must anyway influence their "buoyant health." It is interesting that the Hunza measure the economic level of a person by the number of apricot trees they own. They eat the fresh fruits in abundance at the time of production and dry the excess in the sun to keep them throughout the year. They even prepare ice cream out of apricots and snow. The seeds are crushed and pressed to extract oil used for food, for their lamps, or as a cosmetic (Bircher, 1980).

10.1.12.1.5 In the Kitchen

Apricots can be dried, cooked into pastry, or used to prepare jam. The fruits are also distilled into brandy and liqueur. The dried apricots are usually treated with sulfur dioxide during processing and also with sulfites to extend their shelf life and also give apricots a bright orange color. The brown color unsulfured dried apricots are a much healthier choice. Sulfite reactions can be particularly acute in people who suffer from asthma. Organic dried apricots are not treated with any preservative.

10.1.12.1.6 Chemical Constituents and Uses as Functional Food

Apricots possess the highest levels and widest variety of carotenoids. Catechin, chlorogenic acid, and neochlorogenic acid were the predominant phenolic compounds, and β-carotene was the predominant carotenoid compound. The carotenoid content increased with ripening and postharvest storage, whereas changes in phenolic content and antioxidant capacity were variety dependent. The carotenoid content is higher in cultivars grown in warmer climates (Campbell *et al.*, 2013). Seven carotenoids were identified: antheraxanthin, lutein, zeaxanthin, β-cryptoxanthin, 13-*cis*-β-carotene, all-*trans*-β-carotene, and 9-*cis*-β-carotene; among these, all-*trans*-β-carotene was found to be about 50% of total carotenoids. No significant loss of carotenoid content was found following the drying processes even at gentle microwave heating (60°C) (Fratianni *et al.*, 2013).

Due to their high content in fibers, dried apricots are used to relieve constipation.

REFERENCES

Bircher R. *Gli Hunza. Un popolo che ignora la malattia*. Quaderni d'Ontignano, Fiesole, **1980**.

Campbell O.E., Merwin I.A., Padilla-Zakour O.I. Characterization and the effect of maturity at harvest on the phenolic and carotenoid content of northeast USA apricot (*Prunus armeniaca*) varieties. *J. Agric. Food Chem.* **2013**, *61*(51), 12700–12710.

Fratianni A., Albanese D., Mignogna R., Cinquanta L., Panfili G., Di Matteo M. Degradation of carotenoids in apricot (*Prunus armeniaca* L.) during drying process. *Plant Foods Hum. Nutr.* **2013**, *68*(3), 241–246.

Nadkarni K.M. *Indian Materia Medica*, vol. 1. Popular Prakashan, Mumbai, India, **2002**, 1013–1014.

Nibodhi. *Annapurna's Prashad.* Mata Amritanandamayi Mission Trust, Amritapuri, India, **2010** (2nd edn.), 73–74.

Sharma P.V. *Fruits and Vegetables in Ancient India.* Chaukhamba Orientalia, Varanasi, India, **2009**, 76.

10.1.12.2 Plum

Scientific name: *Prunus domestica* L.
Family: Rosaceae
Hindi: Alubukhara
Sanskrit: Arukam
Tamil: Aluppukkarappalam

10.1.12.2.1 Origin and Distribution

Plums are one of the first fruits domesticated by humans, such as figs, grapes, and olive. *Prunus domestica* is native of the Caucasus and trans-Caucasus regions (Sharma, 2009).

Actually, the plant is cultivated around the world, and China is the top producer. Serbia, Romania, Chile, Iran, the United States, Turkey, Spain, India, and Italy are other important producers. In India, it is cultivated at high altitudes in Nilgiri mountains and Himachal Pradesh (Warrier *et al.*, 1995).

10.1.12.2.2 Features

It is a small tree growing up to 12 m in height with glabrous branches, leaves simple, alternate, and flowers white or cream colored, usually in clusters. The fruits are drupes varying in colors from green and golden yellow to red and dark purple, with a wide variety of different shapes (Warrier *et al.*, 1995).

10.1.12.2.3 Properties and Uses in Ayurvedic System of Medicine

Plums are described by *Charaka* (Shastri, 1970). They are beneficial against immune deficiency, fever, and dry cough. They help to clean the liver and the digestive system (Nibodhi, 2010).

Parts used: Fruit
Taste: Sweet
Quality: Light
Potency: Cold
***Dosha* effect:** VP–, K+
Actions: Laxative, demulcent, digestive, tonic

10.1.12.2.4 Traditional Uses and Folklore

Plums may be taken in the morning for breakfast in case of acid dyspepsia (Nadkarni, 2002).

10.1.12.2.5 In the Kitchen

Stewed plums are recommended by Ayurveda for breakfast as an excellent cleanser. They can be used daily by *pitta* and *kapha* and for a short period (not more than 10 days) by *vata* (Morningstar and Desai, 1999).

10.1.12.2.6 Chemical Constituents and Uses as Functional Food

Most dried prunes are produced from cultivar d'Agen, especially in California and France, where the cultivar originated. Fresh prune-making plums are a rich source of phenolic compounds, many of them concentrated in the skin of the fruit, which contains about 5 times more phenolics per unit of weight than the pulp. The skin contains mainly anthocyanins (6.3 mg/g dry weight): cyanidin 3-rutinoside (44%) and peonidin 3-rutinoside (42%). Anthocyanins in the skin give the fresh prune plums their purple color at pH 3.6, which is typical for plums (Raynal and Moutounet, 2002). The content of neochlorogenic acid (3′-*O*-caffeoylquinic acid) is equal to that of anthocyanins. Exocarp contains significant quantities of rutin (2.2 mg/g dry weight) and catechins (0.74 mg/g dry weight). Chlorogenic acid (5′-*O*-caffeoylquinic acid) is also prominent (1.37 mg/g dry weight), but there are only traces of free caffeic acid (Macheix *et al.*, 1998). The pulp does not contain anthocyanins or rutin, and very little catechins. Hydroxycinnamates (caffeoylquinic acids) constituted 86% of the total phenolics (Donovan *et al.*, 1998).

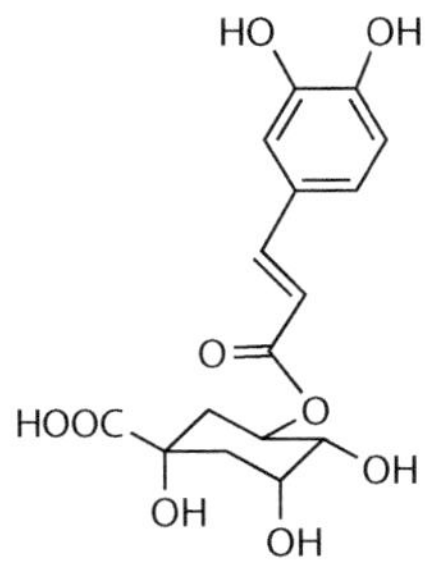

3-*O*-Caffeoylquinic acid

Dried prunes contain higher amounts of phenolic compounds (184 mg/100 g of fruit) than prune-making plums, because dehydration concentrates the constituents despite partial degradation. Neochlorogenic acid represents 71% of the total phenolics and chlorogenic acid represents 24%, raising the content of hydroxycinnamates to 95% of all phenolic compounds.

Compared with fresh prune plums, the concentration of sugar increases in dried prunes, because of the dehydration, but there are also qualitative changes in the proportion of individual sugars. The most striking change is the nearly total disappearance of sucrose, which is hydrolyzed to glucose and fructose during processing. Galacturonan, which was partially substituted with methyl and *O*-acetyl ester groups, and rhamnogalacturonan were the main constituents of the linear regions of the sugar chains of a pectic polysaccharide from fresh plums. The ramified region contained mainly 1,4-linked β-D-galactopyranose residues and, to a lesser extent, 1,5-linked α-L-arabinofuranose residues (Popov *et al.*, 2014). Prune's polysaccharides act as gastroprotective agents in rats (Cantu-Jungles *et al.*, 2014). Prunes are a good source of energy in the form of simple sugars, but do not mediate a rapid rise in blood sugar concentration, possibly because of high fiber, fructose, and sorbitol content. The combination of soluble and insoluble fiber in dried prunes probably acts in a gentle way in the lower intestine, softening the stool, increasing its bulk, and promoting intestinal motility. It was noticed that the laxative principle in dried prunes is water soluble and therefore present in prune juice, which is also very effective in promoting bowel function, although it contains very little dietary fiber.

According to the USDA food composition table, fresh plums contain 0.6 g, dried prunes 2.0 g, and prune juice only 0.01 g of fiber per 100 g. The provisional USDA table of fiber content estimates that dried prunes contain 6–7 g of the total dietary fiber per 100 g.

The USDA Carotenoid Database, which is currently used in most epidemiological studies, accepted only a Finnish study for the estimate of carotenoids in fresh plums and dried prunes, which found 240 μg of lutein and 430 μg of β-carotene per 100 g in fresh plums but only 120 μg of lutein, 140 μg of β-carotene, and 31 μg of α-carotene in dried prunes. It seems that considerable degradation of carotenoids occurs during processing of prune plums to dried prunes.

Prunes contain a high amount of potassium (745 mg/100 g). Dried prunes are an important source of boron, which is postulated to play a role in the prevention of osteoporosis. A serving of prunes (100 g) fulfills the daily requirement for boron (2–3 mg). The amount of boron in a 100-g serving of dried prunes is equal to an average daily intake for adult males. No losses of minerals during the dehydration process occur, and the amount in dried prunes should reflect, in concentrated form, the proportions present in prune-making plums. A 100-g serving of dried prunes will deliver 20% of the daily reference value for potassium, 20% of the reference daily intake (RDI) for copper, 14% of the RDI for iron, about 10% of the RDI for magnesium and zinc, and 10% of safe and adequate intake for manganese. Because of their sweet flavor and well-known mild laxative effect, prunes are considered to be an epitome of functional foods. The laxative action of both prune and prune juice could be explained by their high sorbitol content (14.7 and 6.1 g/100 g, respectively).

A recent study compared the effects of dried plum on the bone to other dried fruits and further explored the potential mechanisms of action through which dried plum may exert its osteoprotective effects. Dried plum was unique in its ability to downregulate osteoclast differentiation coincident with upregulating osteoblast and glutathione activity. These alterations in bone metabolism and antioxidant status compared to other dried fruits provide insight into dried plum's unique effects on the bone (Rendina *et al.*, 2013). In another study, it was demonstrated that pregnant mice treated with plum extract had fetuses and newborn mice with higher osteogenesis index than those of the controls (Monsefi *et al.*, 2013).

Prunes have long been considered a functional food merely related to bowel function. Now there is justification for a network of physiologically important functions for most constituents in prunes (dietary fiber, sorbitol, potassium, copper, boron, and phenolic compounds), with each of them playing a role in many aspects of human health (reviewed in Stacewicz-Sapuntzakis *et al.*, 2010).

REFERENCES

Cantu-Jungles T.M., Maria-Ferreira D., da Silva L.M., Baggio C.H., Werner M.F., Iacomini M., Cipriani T.R., Cordeiro L.M. Polysaccharides from prunes: Gastroprotective activity and structural elucidation of bioactive pectins. *Food Chem.* **2014**, *146*, 492–499.

Donovan J.L., Meyer A.S., Waterhouse A.L. Phenolic composition and antioxidant activity of prunes and prune juice (*Prunus domestica*). *J. Agric. Food Chem.* **1998**, *46*, 1247–1252.

Macheix J.J., Fleuriet A. Phenolic acids in fruits, in Rice-Evans, C.A., Packer, L. (eds.), *Flavonoids in Health and Disease*. Marcel Dekker, New York, **1998**, 35–59.

Monsefi M., Parvin F., Farzaneh M. Effects of plum extract on skeletal system of fetal and newborn mice. *Med. Princ. Pract.* **2013**, *22*(4), 351–356.

Morningstar A., Desai U. *The Ayurvedic Cookbook*. Motilal Banarsidass Publishers, Delhi, India, **1999**, 183.

Nadkarni K.M. *Indian Materia Medica*, vol. 1. Popular Prakashan, Mumbai, India, **2002**, 1015.

Nibodhi. *Annapurna's Prashad*. Mata Amritanandamayi Mission Trust, Amritapuri, India, **2010** (2nd edn.), 92.

Popov S.V., Ovodova R.G., Golovchenko V.V., Khramova D.S., Markov P.A., Smirnov V.V., Shashkov A.S., Ovodov Y.S. Pectic polysaccharides of the fresh plum *Prunus domestica* L. isolated with a simulated gastric fluid and their anti-inflammatory and antioxidant activities. *Food Chem.* **2014**, *143*, 106–113.

Raynal J., Moutounet M. Intervention of phenolic compounds in plum technology: II. Mechanisms of anthocyanin degradation. *J. Agric. Food Chem.* **1989**, 37, 1051–1053.

Rendina E., Hembree K.D., Davis M.R., Marlow D., Clarke S.L., Halloran B.P., Lucas E.A., Smith B.J. Dried plum's unique capacity to reverse bone loss and alter bone metabolism in postmenopausal osteoporosis model. *PLoS One* **2013**, *8*(3), e60569.

Sharma P.V. *Fruits and Vegetables in Ancient India*. Chaukhamba Orientalia, Varanasi, India, **2009**, 18–19.

Shastri K.N. (ed.) *Charaka Samhita*. Chaukhambha Vidyabhavan, Varanasi, India, **1970**, Sutrasthana 27.138.

Stacewicz-Sapuntzakis M., Bowen P.E., Hussain E.A., Damayanti-Wood B.I., Farnsworth N.R. Chemical composition and potential health effects of runes: A functional food? *Crit. Rev. Food Sci. Nutr.* **2010**, *41*(4), 251–286.

Warrier P.K., Nambiar, V.P.K., Ramankutty, C. (eds.) *Indian Medicinal Plants*, vol. 4. Orient Longman, Chennai, India, **1995**, 356–360.

10.2 VEGETABLES

10.2.1 Artichoke

Scientific name: *Cynara cardunculus* L. subsp. *scolymus* (L.) Hayek
Family: Asteraceae
Hindi: Kunjor, Hatichuk
Tamil : Koonaippoo

10.2.1.1 Origin and Distribution

The wild variety of the species is a perennial plant native to the Mediterranean region, called a cardoon. The globe artichoke (*Cynara cardunculus* var. *scolymus*) is a perennial thistle widely cultivated in the countries bordering the Mediterranean Sea. Artichokes have been cultivated in Sicily since the time of the ancient Greeks. The botanical name partly derives from the tradition of fertilizing the plant with ashes (Latin: *cinis*, *cineris*) and partly from the Greek *skolymos*, meaning "thistle" from the spines found on the bracts that enclose the flower heads, forming the edible part of the plant. This plant was appreciated by the ancient Egyptians, Greeks, and Romans, who used it both as a food and as a medicine (for the beneficial effects against hepatobiliary diseases and as a digestive aid). According to Roy Ladurie, in his book *Les Paysans de Languedoc*, "the blossoms of the thistle were improved by the Arabs, passed from Naples to Florence in mid-1400. In 1500 it was noted in France where was named *carchofas* (Italian *carciofo*)." The Dutch introduced artichokes in England, and they were brought to the United States in the nineteenth century.

According to the FAO, the top 10 artichoke-producing countries are Italy, Egypt, Spain, Peru, Argentina, China, Morocco, France, the United States, and Chile.

10.2.1.2 Features

Globe artichoke can grow up to a height of 1 m, with long leaves and branched stems that are spiny. The flowers are violet or white in color: they develop in a large head from an edible bud with numerous triangular scales.

10.2.1.3 Properties and Uses in Ayurvedic System of Medicine

Artichokes are a food that can be eaten by all the *dosha*s without loss of balance. They need to be well cooked with warm aromatic herbs (such as bay leaves) to be easy assimilable by *vata* (Morningstar and Desai, 1999). Beneficial to immunity, they are excellent for cleansing the liver and gall bladder (Nibodhi, 2010).

Parts used: Flower head
Tastes: Sweet, astringent
Quality: Dry, light
Potency: Hot
***Dosha* effect:** PK–, P+ (in excess)
Actions: Diuretic, digestive

10.2.1.4 In the Kitchen

There are many different recipes around the world, using different spices and aromatic herbs. A recipe well known in Rome is Jewish-style artichokes, which are deep-fried whole.

When cut, artichokes can turn brown due to the enzymatic browning and chlorophyll oxidation. Placing them in water slightly acidified with lemon juice can prevent the discoloration.

10.2.1.5 Chemical Constituents and Uses as Functional Food

Artichoke is an important vegetable in Mediterranean cultures; the main European producer is Italy, with the spiny species growing in Sardinia, whereas the big globoid called "romanesco" is typical of the central regions.

Nutritional and pharmaceutical properties of both artichoke heads and leaves are linked to their chemical composition, which includes high levels of polyphenolic compounds and inulin. Inulin represents a major carbohydrate reserve in plants; it is a fructose-based polysaccharide that is not digested in the small intestine, because humans lack the enzymes required for the hydrolysis of fructans. Inulin also shows a positive influence on gut microflora, beneficially affecting mineral absorption, blood lipid composition, and prevention of colon cancer (Costabile *et al.*, 2010; Lattanzio *et al.*, 2009).

Several studies have demonstrated the efficiency and safety of artichoke extracts in the treatment of hepatobiliary dysfunctions and digestive complaints in animals and human subjects. Several clinical investigations showed the ability of artichoke extract to prevent the oxidative modification of blood lipoproteins and to reduce blood cholesterol levels (Bundy *et al.*, 2008; Qiang *et al.*, 2012). Moreover, leaf extracts have been reported to show antioxidant and protective properties against hyroperoxide-induced oxidative stress in cultured rat hepatocytes (Azzini *et al.*, 2007).

The beneficial effects of artichoke are related to its chemical constituents, mainly polyphenols belonging to the class of caffeoylquinic acids and flavonoids such as luteolin and apigenin glycosides. Artichoke accumulates various caffeic acid esters of quinic acid, the most well known, although not the most abundant, being cynarin (1,3-*O*-dicaffeoylquinic acid) (Negro *et al.*, 2012; Panizzi and Scarpati, 1965; Schütz *et al.*, 2006a). The flavones luteolin and apigenin are also present, with minor amounts of anthocyanidins present in the capitula. These compounds are considered minor constituents (about 10% or less) of artichoke tissues. Nevertheless, luteolin is a strong antioxidant that protects low-density lipoproteins from oxidation (Brown and Rice-Evans, 1998), whereas anthocyanins, besides their health promoting properties, play an important role in pigmentation of some cultivars (*Violetto di Chioggia*-Italy; *Violet de Provence*-France) (Negro *et al.*, 2012; Schütz *et al.*, 2006b).

The dried leaves of artichoke have long been used in folk medicine for their choleretic and hepatoprotective activities (Valussi, 2012), which are often related to the cynarin content. In various pharmacological test systems, artichoke leaf extracts have shown hepatoprotective, anticarcinogenic, antioxidative, antibacterial, anti-HIV, bile-expelling, and urinative activities, as well as the ability to inhibit cholesterol biosynthesis and LDL oxidation (reviewed in Lattanzio *et al.*, 2009). These broad therapeutic indications are ascribed to several active compounds, acting sinergically, and include caffeoylquinic acids, flavonoids, and inulin.

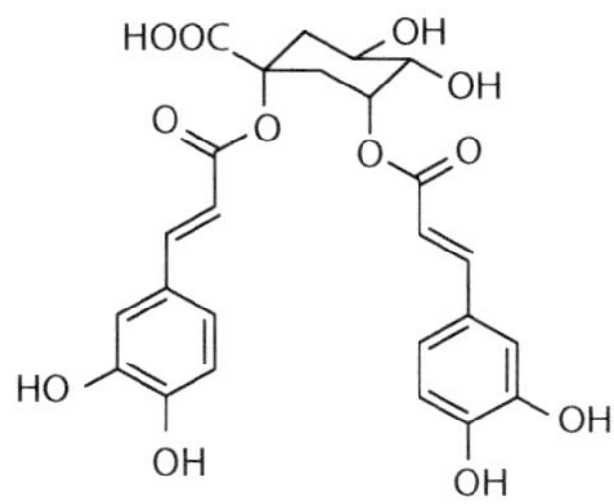

Cynarin

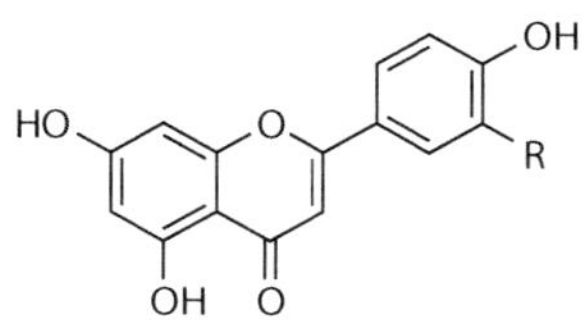

R=H apigenin
R=OH luteolin

In a recent study, the absorption and metabolism of bioactive molecules after oral consumption of cooked edible portion of artichoke heads in human subjects were evaluated. It was demonstrated that some artichoke compounds are able to cross the gastric barrier and reach biologically effective concentrations in the bloodstream (Azzini *et al.*, 2007). An increase in caffeic acid and other hydroxycinnamic acids was observed due to the hydrolysis of the original polyphenolic esters by gut microflora with hydrogenated and methylated metabolites. No circulating levels of luteolin and apigenin were measured (Azzini *et al.*, 2007).

The efficacy of a dietary supplementation with an extract from *Cynara scolymus* (Cs) on the glucose pattern in a group of patients with naïve impaired fasting glycemia (IFG) has been recently evaluated. Compared with the placebo, the supplemented group showed a significant difference in fasting blood glucose, homeostatic metabolic assessment, and lipidic pattern. These data demonstrate the efficacy of Cs extract on the reduction of glycometabolic parameters in overweight subjects with IFG (Rondanelli *et al.*, 2014).

Lasagne with artichokes, green beans, and mushrooms, a recipe from Marco Bianchi (2013)

Ingredients

500 g lasagne without eggs
200 g mushrooms
300 g green beans
500 g artichokes (cleaned)
2 garlic cloves
2 tsp of Extra virgin olive oil
1 tsp of corn starch or potato flour
Salt and pepper to taste

Preparation

Blanch the lasagne sheets in lightly salted boiling water, making sure that they do not stick together.

Meanwhile cut the green beans and the artichokes and boil them. Then sáute them for about 10 minutes with little olive oil, a couple of garlic cloves, and the mushrooms. Next take out a handful of green beans, one of mushrooms and one of artichokes, which will be used to make the paste. The rest will be used to stuff the lasagne.

To make the paste, whip the ingredients, add olive oil and corn starch or potato flour and whip again.

Then prepare the lasagne, alternating the layers of pasta with the layers of paste and pan-fried vegetables. Spray with pepper and then oven-bake for 20–30 minutes at 180°C–200°C. Serve lukewarm.

REFERENCES

Azzini E., Bugianesi R., Romano F., Di Venere D., Miccadei S., Durazzo A., Foddai M.S., Catasta G., Linsalata V., Maiani G. Absorption and metabolism of bioactive molecules after oral consumption of cooked edible heads of *Cynara scolymus* L. (cultivar Violetto di Provenza) in human subjects: A pilot study. *Br. J. Nutr.* **2007**, *97*, 963–969.

Bianchi M. *A tavola con Marco Bianchi. Primi piatti*, vol. 1. Adriano Salari ed., Milano, Italy, **2013**, 32.

Brown J.E., Rice-Evans C.A. Luteolin-rich artichoke extracts protect low density lipoprotein from oxidation *in vitro*. *Free Rad. Res.* **1998**, *29*, 247–255.

Bundy R., Walker A.F., Middleton R.W., Wallis C., Simpson H.C. Artichoke leaf extract (*Cynara scolymus*) reduces plasma cholesterol in otherwise healthy hypercholesterolemic adults: A randomized, double blind placebo controlled trial. *Phytomedicine* **2008**, *15*(9), 668–675.

Costabile A., Kolida S., Klinder A., Gietl E., Bäuerlein M., Frohberg C., Landschütze V., Gibson G.R. A double-blind, placebo-controlled, cross-over study to establish the bifidogenic effect of a very-long-chain inulin extracted from globe artichoke (*Cynara scolymus*) in healthy human subjects. *Br. J. Nutr.* **2010**, *104*(7), 1007–1017.

Lattanzio V., Kroon P.A., Linsalata V., Cardinali A. Globe artichoke: A functional food and source of nutraceutical ingredients. *J. Funct. Foods* **2009**, *1*, 131–144.

Morningstar A., Desai U. *The Ayurvedic Cookbook*. Motilal Banarsidass Publishers, Delhi, India, **1999**, 150.

Negro D., Montesano V., Grieco S., Crupi P., Sarli G., De Lisi A., Sonnante G. Polyphenol compounds in artichoke plant tissues and varieties. *J. Food Sci.* **2012**, *77*, C244–C252.

Nibodhi. *Annapurna's Prashad*. Mata Amritanandamayo Mission Trust, Amritapuri, India, **2010** (2nd edn.), 94–95.

Panizzi L., Scarpati M.L. Sugli acidi 1,4- e 1,5-dicaffeilchinici. *Gazz. Chim. Ital.* **1965**, *95*, 71–82.

Qiang Z., Lee S.O., Ye Z., Wu X., Hendrich S. Artichoke extract lowered plasma cholesterol and increased fecal bile acids in Golden Syrian hamsters. *Phytother Res.* **2012**, *26*(7), 1048–1052.

Rondanelli M., Opizzi A., Faliva M. Sala P., Perna S., Riva A., Morazzoni P., Bombardelli E., Giacosa A. Metabolic management in overweight subjects with naive impaired fasting glycaemia by means of a highly standardized extract from *Cynara scolymus*: A double-blind, placebo-controlled, randomized clinical trial. *Phytother Res.* **2014**, *28*(1), 33–41.

Schütz K., Muks E., Carle R., Schieber A. Quantitative determination of phenolic compounds in artichoke-based dietary supplements and pharmaceuticals by high-performance liquid chromatography. *J. Agric. Food Chem.* **2006a**, *54*, 8812–8817.

Schütz K., Persike M., Carle R., Schieber A. Characterization and quantification of anthocyanins in selected artichoke (*Cynara scolymus* L.) cultivars by HPLC-DAD-ESI-MS^n. *Anal. Bioanal. Chem.* **2006b**, *384*, 1511–1517.

Valussi M. Functional foods with digestion-enhancing properties. *Int. J. Food Sci. Nutr.* **2012**, *63*(Suppl 1), 82–89.

10.2.2 Beetroot

Scientific name: *Beta vulgaris* L.
Family: Amaranthaceae
Hindi: Cukkandar
Sanskrit: Palakya
Tamil: Cencirai

10.2.2.1 Origin and Distribution

Beetroot, well known as sugar beet, is native of the sea coasts of the Mediterranean and now widely cultivated in Europe, West Asia, and the United States. In India, it is cultivated in gardens for using leaves and roots as food (Nadkarni, 2002).

The first factory specialized in sugar production from beetroot was established in Poland in 1801, as a result of British blockade on sugar cane during the Napoleonic Wars.

10.2.2.2 Features

Beetroot is an annual or biennial tuberous herb with condensed stem, leaves simple in rosettes, ovate, and cauline. Flowers usually appear only during the second year: they are two to three in a cluster or sometimes solitary on a long slender spike. The tuberous taproot when fully grown is fleshy and purplish red in color (Warrier *et al.*, 1994). There are several varieties of different shapes and colors (Bakhru, 1994).

10.2.2.3 Properties and Uses in Ayurvedic System of Medicine

The swollen roots are sweet, warm, expectorant, and tonic, recommended during the autumn to calm *vata*. They are traditionally utilized for uterine disorders, constipation, and hemorrhoids. The juice has a therapeutic effect for liver problems related to *pitta*, but in excess it can aggravate *agni* (Morningstar and Desai, 1999).

The leaves are sweet, acrid, cooling, diuretic, anti-inflammatory, and tonic. They are useful in liver and spleen disorders, constipation, vitiated conditions of *pitta*, and general debility (Warrier *et al.*, 1994).

Parts used: Taproot, leaves
Taste: Sweet
Potency: Taproots are warm; leaves are cold
***Dosha* effect:** VK–, P+ (in excess)
Actions: Expectorant, tonic

10.2.2.4 Traditional Uses and Folklore

- The juice of the root is good for the liver. A decoction of the root at bedtime or in the early morning an hour before breakfast is a traditional home remedy for habitual constipation and hemorrhoids (Nadkarni, 2002). With carrot and cucumber juice, it is highly beneficial for kidney and gall bladder disorders (Bakhru, 1994).
- The ancient Romans used beetroots to treat fevers and constipation: in the *Apicius De re coquinaria*, we can find three soup recipes with beetroot recommended as a laxative (Flower and Rosenbaum, 1958).
- The Bedouins of the Middle East use the raw root and leaves to treat urinary tract infections.

10.2.2.5 In the Kitchen

The chopped beetroots are cooked with ghee and spices as a common side dish in India. In the south, they are often cooked with grated coconut. In Europe, they are boiled, grilled, roasted, or added raw to the salad as a tasty touch of color. In South America, the juice is very popular.

The leaves are also edible. They are boiled or steamed and served with oil, salt, and lemon juice.

If we want to keep the beneficial properties of beetroot, we have to be careful not to overcook it, as the group betalains are unstable in heat. Therefore, the root should not be cooked without the skin for longer than 15 minutes or roasted for longer than an hour.

10.2.2.6 Chemical Constituents and Uses as Functional Food

Beta vulgaris L. has a long history of cultivation along the Mediterranean area stretching back to the second millennium BC. It was later spread to Babylonia by the eighth century BC and as far east as China by AD 850.

In the Central and Eastern Europe, red beetroot (*B. vulgaris* var. rubra) is a popular vegetable, known for a long time for its beneficial health effects: stimulation of hematopoietic and immune systems, kidney and liver protection, and as a special diet during cancer treatment, due to its antioxidant, anti-inflammatory, hepatoprotective, and antitumor properties (Escribano *et al.*, 1998; Kapadia *et al.*, 2003; Georgiev *et al.*, 2010; Winkler *et al.*, 2005). Besides diverse polyphenols, red beetroot contains betalains, a family of nonphenolic and water-soluble antioxidants that comprise red betacyanins and yellow betaxanthins (Azeredo, 2009; Kanner *et al.*, 2001). Betalains have several applications in foods, such as desserts, confectioneries, dry mixes, and dairy and meat products. The food colorant known as "beetroot red," extracted from beetroots, is commercialized in the European Union and the United States as food colorant.

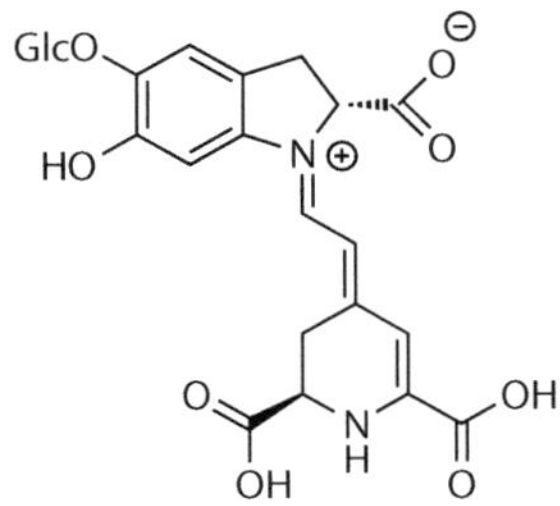

Betanin

Several works have demonstrated the potent antiradical-scavenging activity of betalains *in vitro*. Some reports based on *in vitro* tests, rank beet among the 10 most potent vegetables with respect to their antioxidant activity, betalains being responsive at least in part for these beneficial properties (reviewed in Azeredo, 2009).

Recent studies concerning the effect of beetroot juice on *N*-nitroso compounds, well known as hepatic carcinogens and cause of liver necrosis, indicate that beetroot juice has a protective effect against liver damage in rats, by reducing DNA damage as well as the biomarkers of liver injury (Krajka-Kuźniak *et al.*, 2012; Szaefer *et al.*, 2014). The starting experiments on biological and health-promoted properties of betalains appear to be directly related to their antioxidant properties, because the chemical and *in vitro* antiradical properties have been positively corroborated by very scarce clinical trials in humans (Moreno *et al.*, 2008).

Significant hypotensive effects of a low-dose (100 g) beetroot, which was unaffected by processing or the presence of betacyanins, were demonstrated, and this is attributed to the presence of high amounts of nitrates. These data strengthen the evidence for cardioprotective blood pressure-lowering effects of dietary nitrate-rich vegetables (Hobbs *et al.*, 2012).

Spaghetti with beetroot sauce, a recipe from Marco Bianchi (2013)

Ingredients

400 g whole spaghetti
100 mL extra virgin olive oil
100 g garlic
400 g precooked beetroot
Chili and salt to taste

Preparation

Pour the olive oil in a pan and add the peeled garlic; heat gently covering the pan, until the garlic takes on a golden brown color. Then after a few minutes, when the smell invades the kitchen, switch off and remove the garlic. Whip the beetroot, add water to obtain a creamy consistency, and then add salt and chili. Mix the paste with the warm oil and cook for 3–4 minutes. Meanwhile cook the pasta in salted water and, once drained, let it sauté in the beetroot paste. The color will definitely be peculiar and intense and the taste unique.

REFERENCES

Azeredo H.M.C. Betalains: Properties, sources, applications, and stability—A review. *Int. J. Food Sci. Tech.* **2009**, *44*, 2365–2376.

Bakhru H.K. *Foods That Heal.* Orient Paperbacks, Delhi, India, **1994**, 91–94.

Bianchi M. *A tavola con Marco Bianchi. Primi piatti*, vol. 1. Adriano Salari ed., Milano, Italy, **2013**, 58.

Escribano J., Pedreno M.A., Garcia-Carmona F., Munoz R., Characterization of the antiradical activity of betalains from *Beta vulgaris* L. roots. *Phytochem. Anal.* **1998**, *9*, 124–127.

Flower B., Rosenbaum E. (trans.) *The Roman Cookery Book: A Critical Translation of the Art of Cooking by Apicius for Use in the Study and the Kitchen.* Harrap, London, **1958**, chapter II, 69.

Georgiev G.V., Weber J., Kneschke E.M., Denev P.N., Bley T., Pavlov, A.I. Antioxidant activity and phenolic content of betalain extracts from intact plants and hairy root cultures of red beetroot *Beta vulgaris* cv. Detroit dark red. *Plant Foods Hum. Nutr.* **2010**, *65*, 105–111.

Hobbs D.A., Kaffa N., George T.W., Methven L., Lovegrove J.A. Blood pressure-lowering effects of beetroot juice and novel beetroot-enriched bread products in normotensive male subjects. *Br. J. Nutr.* **2012**, *108*(11), 2066–2074.

Kanner J., Harel S., Granit R. Betalains: A new class of dietary cationized antioxidants. *J. Agric. Food Chem.* **2001**, *49*, 5178–5185.

Kapadia G.J., Azuine M.A., Sridhar R., Okuda Y., Tsuruta A., Ichiishi E., Mukainake T. *et al.* Chemoprevention of DMBA-induced UV-B promoted, NOR-1-induced TPA promoted skin carcinogenesis, and DEN-induced phenobarbital promoted liver tumors in mice by extract of beetroot. *Pharmacol. Res.* **2003**, *47*, 141–148.

Krajka-Kuźniak V., Szaefer H., Ignatowicz E., Adamska T., Baer-Dubowska W. Beetroot juice protects against N-nitrosodiethylamine-induced liver injury in rats. *Food Chem. Toxicol.* **2012**, *50*, 2027–2033.

Moreno D.A., García-Viguera C., Gil J.I., Gil-Izquierdo A. Betalains in the era of global agri-food science, technology and nutritional health. *Phytochem. Rev.* **2008**, *7*, 261–280.

Morningstar A., Desai U. *The Ayurvedic Cookbook.* Motilal Banarsidass Publishers, Delhi, India, **1999**, 165.

Nadkarni K.M. *Indian Materia Medica*, vol. 1. Popular Prakashan, Mumbai, India, **2002**, 197–198.

Szaefer H., Krajka-Kuźniak V., Ignatowicz E., Adamska T., Baer-Dubowska W. Evaluation of the effect of beetroot juice on DMBA-induced damage in liver and mammary gland of female Sprague-Dawley rats. *Phytother Res.* **2014**, *28*(1), 55–61.

Warrier P.K., Nambiar V.P.K., Ramankutty C. (eds.) *Indian Medicinal Plants*, vol. 1. Orient Longman, Chennai, India, **1994**, 265–267.

Winkler C., Wirleitner B., Schroecksnadel K., Schennach H., Fuchs D. In vitro effects of beetroot juice on stimulated and unstimulated peripheral blood mononuclear cells. *Am. J. Biochem. Biotechnol.* **2005**, *1*, 180–185.

10.2.3 Bitter Melon

Scientific name: *Momordica charantia* L.
Family: Cucurbitaceae
Hindi: Karela
Sanskrit: Karavellam
Tamil: Pavakkay

10.2.3.1 Origin and Distribution

Native of the Indian subcontinent, bitter melon is cultivated in India up to an altitude of 1500 m. It is widely cultivated in Indonesia, Sri Lanka, Malaysia, the Philippines, China, Central and South America, Africa, and Australia; it grows well in Italy also.

10.2.3.2 Features

It is an annual climber with many branches, yellow flowers, pendulous, fusiform, and fleshy fruits, and wrinkled skin, resembling a dehydrated cucumber. There are many varieties with some difference in the shape, size, and bitterness of the fruit.

10.2.3.3 Properties and Uses in Ayurvedic System of Medicine

Bitter melon has excellent immune boosting properties. It is one of the best fruits for diabetes and all kinds of kidney disorders; it is also reputed excellent for cleansing the liver, bile blood, and lymph (Nibodhi, 2010). It increases appetite and stimulates *agni*. In high *pitta* conditions, it can be used with a bit of ghee.

Parts used: Whole plant
Taste: Bitter
Quality: Light, dry
Potency: Cold
***Dosha* effect:** PK–, V+ (in excess)
Actions: Tonic, alterative, antipyretic, stomachic, anthelmintic, depurative, purgative, appetizing, antidiabetic, carminative, digestive (Khalsa and Tierra, 2010; Warrier *et al.*, 1995)

10.2.3.4 Traditional Uses and Folklore

- Traditional physicians use to prescribe bitter melon for patients with diabetes. Smaller varieties are the best. The patient has to take 1 tablespoon each of Indian gooseberry juice and bitter melon fruit juice every morning for 2 months (Sairam, 1998).
- The juice of the leaves has many traditional uses as a home remedy for piles, blood disorders, and alcohol intoxication. The regular use is prescribed by Unani physicians as a preventive medicine against leprosy (Bakhru, 1994).

Precautions: Bitter melon is very bitter, so people with an excess of vata are advised to avoid it or to eat only occasionally in small quantity. Excess intake can cause diarrhea and vomiting (Sairam, 1998).

10.2.3.5 In the Kitchen

The fruit is very bitter and has to be steeped in saltwater before using it. Bitterness is concentrated in the seeds and tubercles (Nadkarni, 2002).

In India, people cut it into small round pieces, add salt, and dry them in the sun. The dried fruits are then fried in coconut oil. Usually, the fruit is cooked with many spices, some sugar, and a lot of onions to mitigate the bitter taste. It is excellent with potatoes and pumpkin.

The tender shoots and leaves can be cooked as greens (Bakhru, 1994).

10.2.3.6 Chemical Constituents and Uses as Functional Food

Momordica charantia Descourt (bitter melon) is a tropical and subtropical vine widely grown in Asia, Africa, and the Caribbean for its edible fruit, which is among the most bitter of all fruits. Bitter melon originated in the Indian subcontinent and was carried to China in the fourteenth century.

The chemical composition of *M. charantia* is complex; more than 200 compounds have been isolated and identified from its fruit, leaves, vines, seeds, and roots. Cucurbitane-type triterpenes and triterpene glycosides are the major chemical constituents of *M. charantia* and are considered its marker compounds (reviewed in Pawar *et al.*, 2013). Fatty acids (α-eleostearic acid, linolenic acid) and a number of minor components such as phenols and proteins are also present (Kumar *et al.*, 2011).

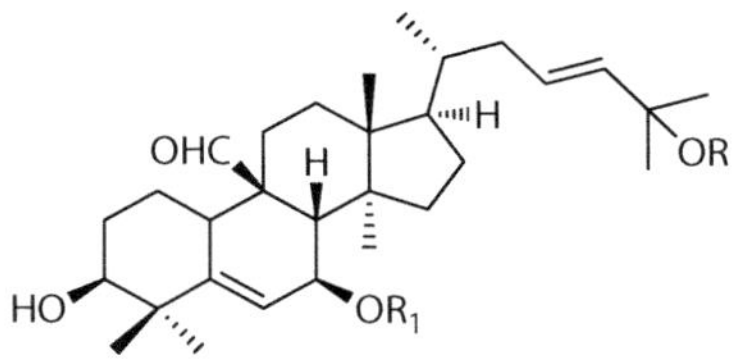

Momordicoside K, R=CH_3, R_1=Glc
3,7,25-Trihydroxycucurbita-5,23(*E*)-dien-19-al, R=R_1=H

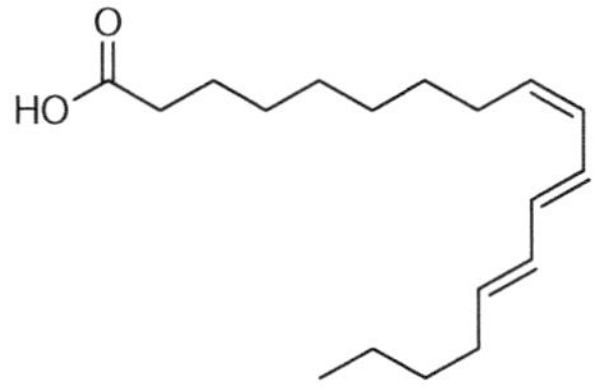

α-Eleostearic acid

The antitumor activity of bitter melon has recently begun to emerge. Several groups of investigators have reported that the treatment of bitter melon-related products in a number of cancer cell lines induces cell cycle arrest and apoptosis without affecting normal cell growth (reviewed in Nerurkar and Ray, 2010). α-Eleostearic acid has been found to induce programmed cell death of fat cells and HL60 leukemia cells *in vitro* (Kobori *et al*., 2008). Diets containing 0.01% bitter melon seed oil (0.006% as α-eleostearic acid) were found to prevent azoxymethane-induced colon carcinogenesis in rats.

For centuries, Ayurveda has recommended the use of bitter melon as a functional food to prevent and treat diabetes and associated complications.

Over the years, there has been a worldwide interest in bitter melon as a dietary supplement due to its various beneficial effects, including lowering diabetes and lipidemia (Nerurkar *et al*., 2006, 2008). All parts of the plant (fruit, seed, and leaves) were shown to possess hypoglycemic properties in normal and diabetic animal models. Bitter melon extract is well tolerated and has been termed as relatively safe in acute, subchronic, and chronic doses in animal studies. Moreover, clinical studies have used bitter melon fruit or bitter melon tea to evaluate its effects in patients with diabetes. More than 100 cucurbitane-type triterpenes and triterpene glycosides have been isolated and identified from different structures of this plant. Some of them have shown an antidiabetic activity in *in vitro* and *in vivo* assays (reviewed in Singh *et al*., 2011).

Nutritional value of bitter melon is attributed to its high mineral and vitamin contents, whereas bitterness is attributed to momordicosides K and L, and has no health effects (Donya *et al*., 2007). 3β,7β,25-Trihydroxycucurbita-5,23(*E*)-dien-19-al has been reported to show blood hypoglycemic effects in diabetes-induced mice (Pawar *et al*., 2013).

A recent Cochrane study reports that there is insufficient evidence on the effects of *M. charantia* for type 2 diabetes mellitus. Further studies are required to address the issues of standardization and the quality control of preparations. For medical nutritional therapy, further observational trials evaluating the effects of *M. charantia* are needed before randomized controlled trials are established to guide any recommendations in clinical practice (Ooi *et al.*, 2012).

Bitter melon curry, a recipe by Vaidya Kalyan Chakravarty

Ingredients

250 g bitter melon
20 g tamarind paste
40 g grated jaggery
Few curry leaves
3 fenugreek seeds
1 1/2 tsp of Bengal gram
1 tsp of black gram
3/4 tsp of mustard seeds
3/4 tsp of cumin seeds
1/4 tsp of turmeric powder
2 tsp of red chili powder
4–6 tsp of ghee
Salt to taste

Preparation

Wash, clean, and cut the bitter melons into medium-sized pieces. Remove the seeds. Take these pieces in a steel bowl and add some water. Place this bowl on a stove and cover it partially. Cook the bitter melons on high power for 10 minutes. Take tamarind in a steel bowl and add some water. Place this bowl on a stove and cook on high power until it reaches a boil. Cool the cooked tamarind and grind along with grated jaggery into a paste. Wash and chop curry leaves. Heat a pan and add ghee to it. When it is hot, add fenugreek seeds, Bengal gram, and black gram. Fry till the seeds start changing color. Then add mustard seeds and cumin seeds. Fry till the seeds crackle. Then add the chopped curry leaves and fry well. Now add the cooked bitter melon pieces and mix well. Add salt, turmeric powder, and red chili powder. Mix well and add the ground tamarind paste. Mix well and add some (10–15 mL) water and cook covered on low flame for 3 minutes. After 3 minutes, remove the lid and cook for a minute. Transfer the curry to a serving bowl. Serve hot with plain rice.

REFERENCES

Bakhru H.K. *Foods That Heal*. Orient Paperbacks, Delhi, India, **1994**, 94–96.

Donya A., Hettiarachchy N., Liyanage R., Lay J., Jr., Chen P., Jalaluddin M. Effects of processing methods on the proximate composition and momordicosides K and L content of bitter melon vegetable. *J. Agric. Food Chem.* **2007**, *55*, 5827–5833.

Khalsa K.P.S., Tierra M. *The Way of Ayurvedic Herbs*. Motilal Banarsidass Publishers, Delhi, India, **2010**, 59–60.

Kobori M., Ohnishi-Kameyama M., Akimoto Y., Yukizaki C., Mitsuru Yoshida M. α-Eleostearic acid and its dihydroxy derivative are major apoptosis-inducing components of bitter gourd. *J. Agric. Food Chem.* **2008**, *56*(22), 10515–10520.

Kumar S.R., Ashish J., Satish N. *Momordica charantia* Linn.: A mini review. *Int. J. Biomed. Res.* **2011**, *2*(11), 579–587.

Nadkarni K.M. *Indian Materia Medica*, vol. 1. Popular Prakashan, Mumbai, India, **2002**, 805–807.

Nerurkar P.V., Lee Y.K., Linden E.H., Lim S., Pearson L., Frank J. Lipid lowering effects of *Momordica charantia* (bitter melon) in HIV-1-protease inhibitor-treated human hepatoma cells, HepG2. *Br. J. Pharmacol.* **2006**, *148*, 1156–1164.

Nerurkar P.V., Lee Y.K., Motosue M., Adeli K., Nerurkar V.R. *Momordica charantia* (bitter melon) reduces plasma apolipoprotein B-100 and increases hepatic insulin receptor substrate and phosphoinositide-3 kinase interactions. *Br. J. Nutr.* **2008**, *100*, 751–759.

Nerurkar P.V., Ray R.B. Bitter melon: Antagonist to cancer. *Pharm. Res.* **2010**, *27*, 1049–1053.

Nibodhi. *Annapurna's Prashad*. Mata Amritanandamayi Mission Trust, Amritapuri, India, **2010** (2nd edn.), 96–97.

Ooi C.P., Yassin Z., Hamid T.A. *Momordica charantia* for type 2 diabetes mellitus. *Cochrane Database Syst. Rev.* **2012**, *8*, CD007845. doi:10.1002/14651858.

Pawar R.S., Tamta H., Ma J., Krynitsky A.J., Grundel E., Wamer W.G., Rader J.I. Updates on chemical and biological research on botanical ingredients in dietary supplements. *Anal. Bioanal. Chem.* **2013**, *405*(13), 4373–4384.

Sairam T.V. *Home Remedies*, vol. I. Penguin Books, Delhi, India, **1998**, 164.

Singh J., Cumming E., Manoharan G., Kalasz H., Adeghate E. Medicinal chemistry of the anti-diabetic effects of *Momordica charantia*: Active constituents and modes of actions. *Open Med. Chem. J.* **2011**, *5*(Suppl 2-M2), 70–77.

Warrier P.K., Nambiar V.P.K., Ramankutty C. (eds.) *Indian Medicinal Plants*, vol. 4. Orient Longman, Chennai, India, **1995**, 48–51.

10.2.4 Cabbage

Scientific name: *Brassica oleracea* L.
Family: Brassicaceae or Cruciferae
Hindi: Bantgopi
Sanskrit: Kembukah
Tamil: Muttaikkos

10.2.4.1 Origin and Distribution

The species *Brassica oleracea* includes cabbage, broccoli, cauliflower, kale, Brussels sprouts, and Chinese kale. In its uncultivated form, it is known as wild cabbage; it is native to the southern Europe and the Mediterranean region. It was a very popular vegetable in ancient Greece and Rome. It was introduced by the Romans in the lands which they conquered. It is now widely cultivated in the hilltops of India, Indonesia, Malaysia, the Philippines, the Central, East, and West Africa, and the Central and South America (Bakhru, 1994).

10.2.4.2 Features

Cabbage is a suffruticose biennal with stems quite glabrous, branching in the upper part, leaves fleshy and glabrous, flowers large and pale yellow in racemes, fruits irregularly spreading or ascending linear pods, and mostly one seeded (Warrier *et al.*, 1994). In the Canary Islands, some cabbage cultivars can grow up to 3 m tall and yield fresh leaves through the year.

10.2.4.3 Properties and Uses in Ayurvedic System of Medicine

Cabbage is excellent for bleeding stomach ulcers and regulates the spleen, pancreas, and stomach; it treats abdominal spasms, pain, and ulcers; helps alleviate constipation; and removes toxins from the colon; it counters depression and irritability (Nibodhi, 2010). The cabbage family is cold and heavy: it will need a strong *agni* to be digested, so it is not recommended for *vata*. Broccoli can be eaten sometimes by *vata* as they are the lightest and easiest to digest (Morningstar and Desai, 1999). They need anyway to be well cooked with spices such as asafetida.

Parts used: Leaves
Tastes: Sweet, astringent

Quality: Heavy
Potency: Cold
***Dosha* effect:** PK–, V+
Actions: Cooling, stomachic, emollient, diuretic, anthelmintic, antibacterial

10.2.4.4 Traditional Uses and Folklore

The juice of red cabbage made into syrup is recommended for chronic coughs, bronchitis, and asthma. Raw cabbage is used as a traditional remedy for intestinal worms (Nadkarni, 2002). It is also recommended for constipation, with black pepper, salt, and lemon juice (Bakhru, 1994).

10.2.4.5 Chemical Constituents and Uses as Functional Food

The Cruciferae (also known as the Brassicaceae) are the family of plants that include the various members of the species *B. oleracea* (e.g., broccoli, cabbage, cauliflower, kale, collard greens, brussels sprouts, kohlrabi, and sprouting broccoli) widely consumed in various parts of the world. Forage and root brassicas are widely used as winter feeds for cattle and sheep.

It is difficult to trace the exact history of Brassicaceae, but they were most likely domesticated somewhere in Europe before 1000 BC. By the Middle Ages, they were a prominent part of European cuisine. At least four different cabbage cultivars are present in Italian cuisine as different subspecies of *B. oleracea* L.

A striking and characteristic chemical property of cruciferous plants is their high content of glucosinolates (GSLs; more than 120 types), which often approaches 1% or more of their dry weight (reviewed in Moreno *et al.*, 2006). The interest devoted to this group of natural products is caused by the appreciable biological effects of both the intact GSLs and especially the complex group of GSL transformation products produced in nonenzymatic and enzymatic reactions. Depending on the concentration and structural types of these compounds, their biological effects can be toxic, antinutritional, or beneficial to health. Most serious economic problems in livestock seem to result from rapeseed meal arising from GSLs or their breakdown products. By contrast, GSLs and their hydrolysis products are reportedly well-known protectors against carcinogenesis. GSLs play further protective and evolutionarily important roles in plants. These include allelopathy (suppression of growth of neighboring plants), specific positive and negative feeding cues for some insects, and broad antibiotic properties, including nematocidal, antimicrobial, antifungal, antiprotozoal, and insecticidal activities (reviewed in Assayed and Abd El-Aty, 2009).

Consumption of broccoli has long been considered to play a role in a healthy diet. Broccoli accumulates significant amounts of the phytonutrient glucoraphanin (4-methylsulfinylbutyl GSL), which is metabolized *in vivo* to the biologically active sulforaphane. The preponderance of evidence available from *in vitro*, animal, and human studies supports the association of sulforaphane with the induction of phase II detoxifying enzymes, and thus, it is an indirect, long-acting antioxidant. Sulforaphane also modulates phase 1 cytochrome P450 enzymes, thereby inhibiting the activation of procarcinogens and preventing the generation of DNA adducts during the initiation stage of cancer. The overall net effect on phase

1 and 2 enzymes is an increase in metabolism and detoxification of chemical carcinogens (a monograph on sulforaphane GSL is reported in AMR, 2010).

Glucoraphanin is enzymatically hydrolyzed by myrosinase, an enzyme compartmentally separated from glucoraphanin in plant cells, which is released when the plant is chewed or processed. Heat partially denatures myrosinase, and the intact glucoraphanin ingested from cooked or raw vegetables is hydrolyzed by myrosinase-producing bacteria present in the gut. Sulforaphane is absorbed in the small intestine (jejunum), metabolized in a pathway that involves the liver, small intestine, and kidneys, the liver being the primary site of activity, where sulforaphane is conjugated with glutathione and is further converted to *N*-acetylcysteine conjugates (Conaway *et al.*, 2000). All these conjugates act as indirect antioxidants and carcinogen-detoxifying agents.

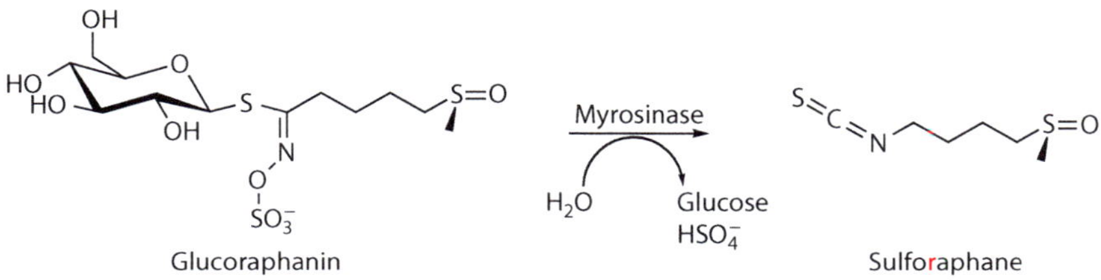

This has provided the impetus for developing varieties of broccoli, both sprouts and whole heads, which are rich in glucoraphanin. The cancer-preventive properties of cruciferous vegetables, especially broccoli, have been studied for decades. However, the evidence of broccoli directly affecting cancer incidence or progression is ambiguous, in part due to the presence of substantial polymorphisms in enzymes that metabolize sulforaphane.

The concept that vegetables and fruits are relevant sources of cancer-preventive substances is strongly supported by population studies. Among others, cruciferous vegetables such as broccoli, cabbage, cauliflower, and Brussels sprouts are thought to affect the development of various types of cancers, and especially prostate tumors. Yet, the identification of the molecular mechanisms by which the "active" compounds are contained in these vegetables mediate their anticancer activity has historically lagged behind. Accordingly, direct laboratory evidence of how individual nutrients affect cancer genes and the pathways they control remains the major obstacle to progress in this research field (Appendino and Bardelli, 2010).

Because broccoli sulforaphane is one of the most potent inducers of phase II enzymes, exploration of the impact of broccoli on other areas of human health, such as cardiovascular health and upper airway immunity, has been suggested (James *et al.*, 2012).

Phenolic compounds (flavonoids and hydroxycinnamic acids) are also present in all Brassicaceae, which also contribute to the antioxidant properties of Brassica species (reviewed in Cartea *et al.*, 2011).

Steaming and drying result in an apparent increment of sulforaphane content as well as antioxidant activity, most likely due to an increase of the extractability of antioxidants and sulforaphane. Freezing and boiling diminish polyphenol concentration, mainly due to volatilization and leaching into the cooking water (Mahn and Reyes, 2012).

Cabbage coconut curry, a recipe by Vaidya Kalyan Chakravarty

Ingredients

300 g cabbage
1/2 coconut, small
2–3 green chilies
1/4 tsp of turmeric powder
2 tsp of Bengal gram
1 1/2 tsp of black gram
3/4 tsp of mustard seeds
3/4 tsp of cumin seeds
3–5 tsp of ghee
Small piece of ginger
Few curry leaves
Salt to taste

Preparation

Slice the cabbage finely and put it in a steel bowl with a lid and add little water. Cook it on high power for 5–10 minutes. After cooking the cabbage, spread it in a plate to cool. Grind the coconut pieces in a dry mixer to get grated coconut. Wash and clean green chilies and ginger, and grind them to a coarse paste. Wash curry leaves and chop them finely. Take the cooked cabbage and remove excess water by pressing hard between your palms.

Heat a pan and add ghee to it. When ghee is hot, add Bengal gram and black gram, and fry well. Fry till the seeds change color. Then add mustard seeds and cumin seeds, and fry till the seeds splutter. Add the chopped curry leaves and coarsely ground chili–ginger pastel and fry well. Add turmeric powder and mix once. Add the water-removed cabbage to the pan and mix. Add salt, mix, and cook for few seconds. Then add grated coconut and mix well. Transfer this curry to a serving bowl. Serve with hot rice.

REFERENCES

AMR. Sulforaphane glucosinolate monograph. *Altern. Med. Review* **2010**, *15*(4), 352–360.

Appendino G., Bardelli A. Broccoli, PTEN deletion and prostate cancer: Where is the link? *Mol Cancer.* **2010**, *9*, 308.

Assayed M.E., Abd El-Aty A.M. Cruciferous plants: Phytochemical toxicity versus cancer chemoprotection. *Mini Rev. Med. Chem.* **2009**, *9*(13), 1470–1478.

Bakhru H.K. *Foods That Heal*. Orient Paperbacks, Delhi, India, **1994**, 98–101.

Cartea M.E., Francisco M., Soengas P., Velasco P. Phenolic compounds in Brassica vegetables. *Molecules* **2011**, *16*, 251–280.

Conaway C.C., Getahun S.M., Liebes L.L., Pusateri D.J., Topham D.K., Botero-Omary M., Chung F.L. Disposition of glucosinolates and sulforaphane in humans after ingestion of steamed and fresh broccoli. *Nutr. Cancer* **2000**, *38*, 168–178.

James D., Devaraj S., Bellur P., Lakkanna S., Vicini J., Boddupalli S. Novel concepts of broccoli sulforaphanes and disease: Induction of phase II antioxidant and detoxification enzymes by enhanced-glucoraphanin broccoli. *Nutr. Rev.* **2012**, *70*(11), 654–665.

Mahn A., Reyes A. An overview of health-promoting compounds of broccoli (*Brassica oleracea* var. *italica*) and the effect of processing. *Food Sci. Technol. Int.* **2012**, *18*(6), 503–514.

Moreno D.A., Carvajal M., López-Berenguer C., García-Viguera C. Chemical and biological characterisation of nutraceutical compounds of broccoli. *J. Pharm. Biomed. Anal.* **2006**, *41*, 1508–1522.

Morningstar A., Desai U. *The Ayurvedic Cookbook*. Motilal Banarsidass Publishers, Delhi, India, **1999**, 161.

Nadkarni K.M. *Indian Materia Medica*, vol. 1. Popular Prakashan, Mumbai, India, **2002**, 217–218.

Nibodhi. *Annapurna's Prashad*. Mata Amritanandamayi Mission Trust, Amritapuri, India, **2010** (2nd edn.), 98.

Warrier P.K., Nambiar V.P.K., Ramankutty C. (eds.) *Indian Medicinal Plants*, vol. 1. Orient Longman, Chennai, India, **1994**, 306–308.

10.2.5 Carrot

Scientific name: *Daucus carota* L.
Family: Apiaceae
Hindi: Gajar
Sanskrit: Garjarah
Tamil: Gajjarakkilamgu

10.2.5.1 Origin and Distribution

Carrot is a popular root vegetable and an important fodder crop with the world production of over 20 million megatons. More than 80 collections of *Daucus carota* seeds are present in the Kew's Millenium Seed Bank based at Wakehurst in West Sussex.

Wild carrot has probably originated on the Iranian Plateau (an area that now includes Afghanistan, Pakistan, and Iran), where it was grown for its aromatic leaves and seeds. It was in fact included in a list of aromatic herbs (rather than that of vegetables) grown in the royal garden of Babylon in the eighteenth century BC. Wild carrot was taken to Asia Minor in the tenth or eleventh century, to Spain in the twelfth century, to northwestern Europe in the fifteenth century, to China in the thirteenth or fourteenth century, and to Japan in the seventeenth century. Since then, selective breeding has increased sweetness, reduced bitterness, and minimized the woody core. The color also changed from dark purple to orange. The carrot is now widely cultivated across much of western Asia and Europe, with an immense variety of root shapes, sizes, and colors. Dark red and purple carrots are still grown in Afghanistan.

10.2.5.2 Features

Daucus carota is a biennial herb growing up to 150 cm tall with a grooved, hairless, or bristly stem. The leaves are divided with a feathery appearance and a typical odor. The flowers are produced after the second year of growth. The spiny fruits attach to the fur of passing animals, aiding seed dispersal.

10.2.5.3 Properties and Uses in Ayurvedic System of Medicine

Carrot imparts taste, alleviates vitiated *pitta* and *kapha*, and treats bloating of abdomen, worm infestation, colic, burning sensation, and thirst (Venkat *et al.*, 2012).

It is wonderful for all the conditions related to the liver, increases the blood flow, and helps to build new blood (Nibodhi, 2010). It is an appetizer recommended for *kapha* and *vata*. It is a kidney tonic, very helpful in phlegmatic diseases. Carrot increases *agni* and in large quantity can aggravate *pitta*. Juice is fine for *pitta* when balanced with generous amounts of cooling celery, cucumber, or lettuce (Morningstar and Desai, 1999).

Parts used: Roots, leaves
Tastes: Sweet, pungent, astringent
Quality: Light, sharp
Potency: The root is heating; the juice is cooling
***Dosha* effect:** VK–, P+ (in excess)
Actions: Appetizer, carminative, constipating, nutritive, digestive, diuretic, aphrodisiac, expectorant (Warrier *et al.*, 1994)

10.2.5.4 Traditional Uses and Folklore

- The soup is an effective home remedy for diarrhea (Bakhru, 1994).
- The regular use of carrot prevents digestive disorders and purifies the blood. It is recommended in case of skin diseases such as eczema.
- Carrot juice with spinach juice and a small amount of lemon juice is a home remedy for constipation (Bakhru, 1994).

10.2.5.5 In the Kitchen

Carrots are eaten raw in salads or cooked in soups, shredded in carrot cake, juiced, or puréed in baby food. The salad is very popular throughout the Middle East. In India, a sweet, excellent dessert called carrot *halwa* is prepared by cooking carrots with ghee and milk. The greens and seeds of the carrot are also edible. Carrot juice is popular as a healthy drink, both alone or mixed with other vegetable and fruit juices. A tasty combination is with apple and ginger juice.

10.2.5.6 Chemical Constituents and Uses as Functional Food

Carrot is the most important source of dietary β-carotene that is further metabolized in the small intestine into retinol (vitamin A). β-Carotene is released from cell fragments through both mastication and gentle cooking. A strategy to reduce the incidence of vitamin A deficiency is to improve precursor bioavailability from meals. Because vitamin A precursors are fat soluble, it was observed that carotenoids are more easily absorbed from food if prepared in such a way that the food matrix containing provitamin A (β-carotene) is sufficiently fat rich. Increased 75% bioavailability was found when the carrots were stir fried, compared to raw material (Ghavami *et al.*, 2012).

To predict the carotenoid bioaccessibility in carrot-derived purées, a model for digestion was set up. Adding olive oil (2%) during digestion, especially as an oil-in-water emulsion, resulted in a substantial increase in carotenoid uptake in the micellar phase (Moelants *et al.*, 2012).

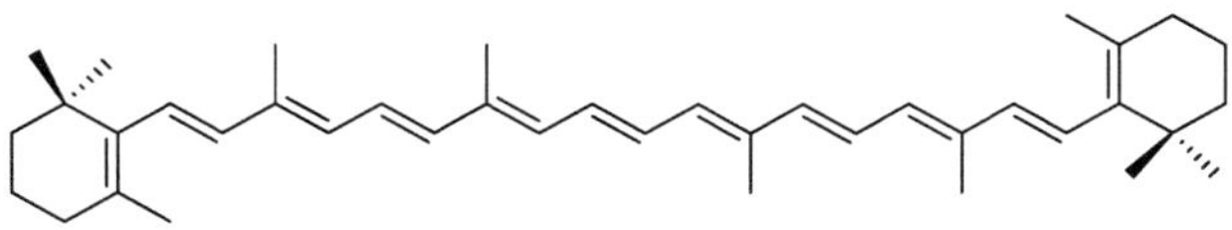

β-Carotene

β-Carotene *in vitro* bioaccessibility also increases with increasing processing temperature and time until steady-state conditions were reached after prolonged heating (Lemmens *et al.*, 2011). The effect of thermal processing (85°C–130°C) on the stability and isomerization of β-carotene in both an olive oil/carrot emulsion and an olive oil phase enriched with carrot β-carotene was studied. During processing, the total β-carotene concentration decreased quickly, after which a plateau value was reached, which was dependent on the applied temperature (Knockaert *et al.*, 2012; Desobry *et al.*, 1998).

The presence of food acidulants and antioxidant spices (turmeric/onion) proved to increase β-carotene bioacessibility (Veda *et al.*, 2008).

It was hypothesized that regular carrot juice intake would be associated with reductions in oxidative stress (8-*iso*-prostaglandin $F_{2\alpha}$) and inflammation (thromboxane B_2, prostaglandin E_2 metabolites, and high-sensitivity C-reactive protein). Evidence suggests that higher plasma carotenoid concentrations are protective in relation to breast cancer recurrence (Butalla *et al.*, 2012).

Carotenoids are involved in the regulation of gene expression or effect cell functions. These biological effects are independent of the provitamin A activity and have been attributed to the function of β-carotene as a free radical-trapping agent and single oxygen quencher. It shows antimutagenic, chemopreventive, photoprotective, and immunoenhancing properties (reviewed in Sharma *et al.*, 2012).

Carrots also contain aliphatic C(17)-polyacetylenes (falcarinol and falcarindiol) (Christensen, 2011). A hormesis effect was seen for all isolated polyacetylenes when added to cancer cells (CaCo-2 origin). Hormesis is a dose–response phenomenon characterized by a low-dose stimulation and high-dose inhibition. Extracts of carrots, containing different amounts of falcarinol, falcarindiol, and falcarindiol 3-acetate, had significant inhibitory effects on both normal and cancer cell proliferation. In cancer cells, the extracts containing the highest concentration of falcarinol tended to have the highest growth inhibitory effect, in accordance with a higher potency of falcarinol than falcarindiol (Purup *et al.*, 2009).

Carrots are a good source of carbohydrates and minerals, with a limited energy value (126 kJ/100 g).

Carrot soup

Ingredients

2 carrots
1 onion
2 cloves
2 garlic cloves
1/2 cup of milk
A pinch of pepper powder

2 tbsp of chopped parsley leaves
1/2 tsp of ghee
Himalayan salt to taste

Preparation

Heat the ghee and add the cloves, garlic, and onions. Sauté for few minutes, add carrots and one cup of water and let cook up for about 10 minutes. Discard the cloves and churn well. Add some more water and salt to taste. Let boil again for few minutes, add the milk, bring a boil again, and finally add the pepper and coriander leaves.

REFERENCES

Bakhru H.K. *Foods That Heal*. Orient Paperbacks, Delhi, India, **1994**, 101–104.

Butalla A.C., Crane T.E., Patil B., Wertheim B.C., Thompson P., Thomson C.A. Effects of a carrot juice intervention on plasma carotenoids, oxidative stress, and inflammation in overweight breast cancer survivors. *Nutr. Cancer.* **2012**, *64*(2), 331–341.

Christensen L.P. Aliphatic C17-polyacetylenes of the falcarinol type as potential health promoting compounds in food plants of the Apiaceae family. *Recent Pat. Food Nutr. Agric.* **2011**, *3*, 64–77.

Desobry S.A., Netto F.M., Labuza T.P. Preservation of β-carotene from carrots. *Crit. Rev. Food Sci. Nutr.* **1998**, *38*(5), 381–396.

Ghavami A., Coward W.A., Bluck L.J. The effect of food preparation on the bioavailability of carotenoids from carrots using intrinsic labelling. *Br. J. Nutr.* **2012**, *107*(9), 1350–1366.

Knockaert G., Pulissery S.K., Lemmens L., Van Buggenhout S., Hendrickx M., Van Loey A. Carrot β-carotene degradation and isomerization kinetics during thermal processing in the presence of oil. *J. Agric. Food Chem.* **2012**, *60*(41), 10312–10319.

Lemmens L., Colle I.J., Van Buggenhout S., Van Loey A.M., Hendrickx M.E. Quantifying the influence of thermal process parameters on in vitro β-carotene bioaccessibility: A case study on carrots. *J. Agric. Food Chem.* **2011**, *59*(7), 3162–3167.

Moelants K.R., Lemmens L., Vandebroeck M., Van Buggenhout S., Van Loey A.M., Hendrickx M.E. Relation between particle size and carotenoid bioaccessibility in carrot- and tomato-derived suspensions. *J. Agric. Food Chem.* **2012**, *60*(48), 11995–12003.

Morningstar A., Desai U. *The Ayurvedic Cookbook*. Motilal Banarsidass Publishers, Delhi, India, **1999**, 168.

Nibodhi. *Annapurna's Prashad*. Mata Amritanandamayi Mission Trust, Amritapuri, India, **2010** (2nd edn.), 98–99.

Purup S., Larsen E., Christensen L.P. Differential effects of falcarinol and related aliphatic C(17)-polyacetylenes on intestinal cell proliferation. *J. Agric. Food Chem.* **2009**, *57*(18), 8290–8296.

Sharma K.D., Karki S., Thakur N.S., Attri S. Chemical composition, functional properties and processing of carrot—A review. *J. Food Sci. Technol.* **2012**, *49*(1), 22–32.

Veda S., Platel K., Srinivasan K. Influence of food acidulants and antioxidant spices on the bioaccessibility of β-carotene from selected vegetables. *J. Agric. Food Chem.* **2008**, *56*(18), 8714–8719.

Venkat P., Gangadharan G.G., Lakshmithathachar M.A., Alwar M.A. (eds.) *Bhojanakutuhalam*. I-AIM, Bangalore, India, **2012**, 88.

Warrier P.K., Nambiar V.P.K., Ramankutty C. (eds.) *Indian Medicinal Plants*, vol. 2. Orient Longman, Chennai, India, **1994**, 311–312.

10.2.6 Celery

Scientific name: *Apium graveolens* L.
Family: Apiaceae
Hindi: Ajmod
Sanskrit: Ajmoda
Tamil: Asamtavomam

10.2.6.1 Origin and Distribution

Celery has been cultivated since more than 2000 years for its medicinal properties. We find reference to its use in the Iliad and the Odyssey. It was sacred to Zeus and Poseidon. The ancient Greeks believed that the plant was sprouted from the blood of Kadmilos, the father of the Cabeiri divinities, so the plant was used during funerary rituals. The winners of the Nemean and Isthmian games were crowned with wild celery.

It was also used during funerary rituals; celery leaves and inflorescences were part of the garlands found in the tomb of pharaoh Tutankhamun. Only after seventeenth century, celery has started to be used also in the kitchen, to improve taste of soups and salads. The ancient Romans used to wear celery garlands as a "home remedy" after a very rich and heavy dinner (Aggarwal and Yost, 2011).

The herb is native to Europe and Asia, and now grows in many parts of the world in temperate regions and tropical mountains. In India, it is widely cultivated in Punjab and Uttar Pradesh for the seed production.

10.2.6.2 Features

Celery is an annual plant with leaves compounding with a long, big, and succulent stalk. The fruits are small, dark brown in color, with a peculiar flavor when cut open: the seeds are brown in color (Bakhru, 1993).

10.2.6.3 Properties and Uses in Ayurvedic System of Medicine

The seeds are commonly used in Ayurvedic and Unani systems of medicine for debility related to cold, flatulence, gout, urinary problems, and constipation. They are tonic and aphrodisiac.

The stalk is light and easy to digest; it helps in burning up *ama* while building the digestive fire. Its roughage scrapes away excess mucus, improves circulation, reduces lymphatic stagnation, and acts as diaphoretic and diuretic.

Parts used: Root, seeds
Tastes: Bitter, pungent
Quality: Dry, light
Potency: Hot
***Dosha* effect:** K–, VP+
Actions: Carminative, diuretic, nervine, vasodilator

10.2.6.4 Traditional Uses and Folklore

- The seeds are used to treat chronic kidney and bladder problems of a cold nature (Tierra, 1999).
- To treat arthritis, the fresh juice of the leaves and the stem is recommended. For rheumatism and gout, the fluid extract of the seeds is more effective than the raw vegetable. An abundant use of celery and carrot juice is beneficial for nervous afflictions and anemia. The regular use prevents stone formation in the kidneys and gall bladder. Celery juice with honey is a traditional remedy for insomnia (Bakhru, 1993).

Precautions: It is contraindicated for pregnant women, because it is strongly abortive.

According to the Environmental Working Group's 2013 report "Shopper's Guide to Pesticides," conventionally grown celery is among the top 12 fruits and vegetables on which pesticide residues have been most frequently found. Therefore, it is strongly suggested to use it only when it is grown organically.

10.2.6.5 In the Kitchen

Celery is an excellent basic food: it can be eaten raw in salad or juice as well as in soups and sauces. It combines well with tomato, onions, carrots, and bell peppers. The seeds are bitter and become sweet after cooking. They are excellent with omelets or fried eggs. For the salad dressing, they can be slightly roasted. In the northern India, the wild celery is used in the preparation of curries, pickles, and chutneys (Aggarwal and Yost, 2011).

10.2.6.6 Chemical Constituents and Uses as Functional Food

Apium graveolens L. belonging to the family of Apiaceae is an annual or biennial herb widely used as a spice and seasoning in food. The direct ancestors of celery were cultivated in Europe and the Mediterranean as early as 1000 BC, and we have evidence of celery being used as a medicinal plant in ancient Egypt. Celery is often served as a major plate vegetable rather than an additive to salads or soups. The root (celeriac) is also widely appreciated as food.

A broad range of biological activities has been attributed to *A. graveolens*. These include antimicrobial activity, larvicidal activity, hepatoprotective activity, nematicidal and mosquito repellent potential, and antihyperlipidemic activity. The stalks of celery have shown effects on bioassays related to epilepsy, hyperlipemia (Iyer and Patil, 2011), hypertension, and memory enhancement, and the seeds are used for treating bronchitis, asthma, and liver and spleen diseases (Zhou *et al.*, 2009). The green parts of the plant are known to lower cholesterol (Tsi and Tan, 2000) and to have anti-inflammatory (Mencherini *et al.*, 2007; Powanda and Rainsford, 2011), antimicrobial, and anticancer activities (Ovodova *et al.*, 2009; Sultana *et al.*, 2005). An ethanol extract of *A. graveolens* significantly protects the gastric mucosa and suppresses the basal gastric secretion in rats, possibly through its antioxidant potential (Al-Howiriny *et al.*, 2010).

The essential oil of *A. graveolens* was strongly inhibitory against *Escherichia coli* and moderately inhibitory against *Pseudomonas aeruginosa* and *Staphylococcus aureus*. The major components identified were β-pinene, camphene, cumene, limonene, α-thuyene, α-pinene, β-phellendrene, *p*-cymene, γ-terpinene, sabinene, and terpinolene (Baananou *et al.*, 2013).

The essential oil shows the presence of sedanenolide (Oguro and Watanabe, 2011), neocnidilide, and neophytadiene as main components and may be useful in the clinical treatment of fungal diseases (Marongiu *et al.*, 2013). These phthalides may also act as diuretics, also helping to lower blood pressure (Wright *et al.*, 2007). They are a major contributor responsible for the unique flavor of this vegetable.

L-3-*n*-Butylphthalide extracted from seeds of *A. graveolens* L. (Chinese celery), has been demonstrated to have neuroprotective effects on ischemic, vascular dementia, and amyloid-β-infused animal models, showing promising preclinical potential as a multitarget drug for the prevention and/or treatment of Alzheimer's disease (Peng *et al.*, 2010).

The chemical composition of celery is wide; it contains a number of beneficial salts with a ratio of 3:1 (potassium/sodium), vitamins (especially vitamin A), fiber, an essential oil constituted of terpenes and phthalides, a class of polyketides named polyacetylenes (Christensen, 2011; Christensen and Brandt, 2006), and flavonoids.

Considerable attention has been devoted to identifying plant-derived dietary agents, which could be developed as promising chemopreventives.

Celery yields 79% juice with 62% of the total flavones extracted. Fresh celery juice averaged 28.5 mg flavones per 100 g, present mainly as malonyl apiosylglucoside conjugates (Hostetler *et al.*, 2012b). The most abundant flavone is apigenin (4′,5,7-trihydroxyflavone) abundantly present in common fruits and vegetables, including parsley, onions, oranges, tea, chamomile, wheat sprouts, and some seasonings (López-Lázaro, 2009). Apigenin has been shown to possess remarkable anti-inflammatory, antioxidant, and anticarcinogenic properties. In the past few years, significant progress has been made in studying the biological effects of apigenin at cellular and molecular levels. Recent progress has been made in testing the efficacy of apigenin in preclinical models of cancer (Patel *et al.*, 2007). Although the bioavailability of apigenin is limited, the slow pharmacokinetics may lead to possible accumulation of this flavonoid in the tissues to effectively impart its chemopreventive effects.

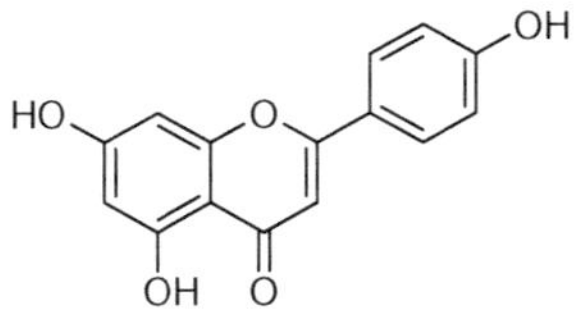

Apigenin

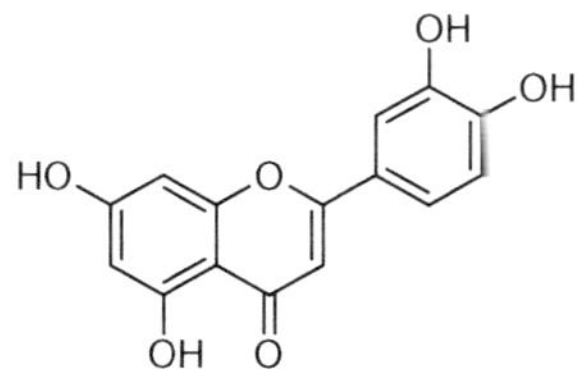

Luteolin

Deglycosylation increases absorption of dietary flavones *in vivo* and modulates inflammation by reducing TNF-α and NF-κB, suggesting the potential use of functional foods rich in flavones for the treatment and prevention of inflammatory diseases (Hostetler *et al.*, 2012a).

The major phenolic acids identified in celery extracts are caffeic acid *p*-coumaric acid, and ferulic acid, whereas the flavonoids are apigenin, luteolin, and kaempferol. The antioxidant activity is strictly correlated with the amount of these phenolics (Yao *et al.*, 2010).

Celery increases its antioxidant capacity in all the cooking methods, except boiling where it loses 14% (Jiménez-Monreal *et al.*, 2009).

Celery pesto, a recipe from Marco Bianchi (2013)

Ingredients

80 g leaves and a small piece of celery stem
20 g almonds
20 g pine nuts
1 or 2 garlic cloves
A big spoonful of low-fat soft ricotta
5 tsp extra virgin olive oil
A pinch of coarse-grained salt

Preparation

Wash the leaves and stem of the celery and dry them. Dry roast the almonds and pine nuts in a small pan. Put the celery leaves, garlic, almonds, pine nuts, salt, and four spoonfuls of water in the mixer. Whip the ingredients, then add the olive oil and the low-fat ricotta, and whip a little further. Add salt to taste and add a little water to make the compound creamy. Pour over pasta.

REFERENCES

Aggarwal B.B., Yost D. *Healing Spices*. Sterling Publishers, New York, **2011** (Italian version, Armenia, 328–333).

Al-Howiriny T., Alsheikh A., Alqasoumi S., Al-Yahya M., ElTahir K., Rafatullah S. Gastric antiulcer, antisecretory and cytoprotective properties of celery (*Apium graveolens*) in rats. *Pharm. Biol.* **2010**, *48*(7), 786–793.

Baananou S., Bouftira I., Mahmoud A., Boukef K., Marongiu B., Boughattas N.A. Antiulcerogenic and antibacterial activities of *Apium graveolens* essential oil and extract. *Nat. Prod. Res.* **2013**, *27*(12), 1075–1083.

Bakhru H.K. *Herbs That Heal.* Orient Paperbacks, Delhi, India, **1993**, 59–62.

Bianchi M. *A tavola con Marco Bianchi. Primi piatti*, vol. 1. Adriano Salari ed., Milano, Italy, **2013**, 118.

Christensen L.P. Aliphatic C(17)-polyacetylenes of the falcarinol type as potential health promoting compounds in food plants of the Apiaceae family. *Recent Pat. Food Nutr. Agric.* **2011**, *3*(1), 64–77.

Christensen L.P., Brandt K. Bioactive polyacetylenes in food plants of the Apiaceae family: Occurrence, bioactivity and analysis. *J. Pharm. Biomed. Anal.* **2006**, *41*(3), 683–693.

Hostetler G.L., Riedl K.M., Cardenas H., Diosa-Toro M., Arango D., Schwartz S., Doseff A.I. Flavone deglycosylation increases their anti-inflammatory activity and absorption. *Mol. Nutr. Food Res.* **2012a**, 56(4), 558–569.

Hostetler G.L., Riedl K.M., Schwartz S.J. Endogenous enzymes, heat, and pH affect flavone profiles in parsley (*Petroselinum crispum* var. *neapolitanum*) and celery (*Apium graveolens*) during juice processing. *J. Agric. Food Chem.* **2012b**, *60*, 202–208.

Iyer D., Patil U.K. Effect of chloroform and aqueous basic fraction of ethanolic extract from *Apium graveolens* L. in experimentally-induced hyperlipidemia in rats. *J. Complement. Integr. Med.* **2011**, *8*. pii:/j/jcim.2011.8.issue-1/1553-3840.1529/1553-3840.1529.xml.

Jiménez-Monreal A.M., García-Diz L., Martínez-Tomé M., Mariscal M., Murcia M.A. Influence of cooking methods on antioxidant activity of vegetables. *J. Food Sci.* **2009**, *74*(3), H97–H103.

López-Lázaro M. Distribution and biological activities of the flavonoid luteolin. *Mini Rev. Med. Chem.* **2009**, *9*(1), 31–59.

Marongiu B., Piras A., Porcedda S., Falconieri D., Maxia A., Frau M.A., Gonçalves M.J., Cavaleiro C., Salgueiro L. Isolation of the volatile fraction from *Apium graveolens* L. (Apiaceae) by supercritical carbon dioxide extraction and hydrodistillation: Chemical composition and antifungal activity. *Nat. Prod. Res.* **2013**, *27*(17), 1521–1527.

Mencherini T., Cau A., Bianco G., Della Loggia R., Aquino R.P., Autore G. An extract of *Apium graveolens* var. dulce leaves: Structure of the major constituent, apiin, and its anti-inflammatory properties. *J. Pharm. Pharmacol.* **2007**, *59* (6), 891–897.

Oguro, D., Watanabe H. Asymmetric synthesis and sensory evaluation of sedanenolide. *Biosci. Biotechnol. Biochem.* **2011**, *75*(8), 1502–1505.

Ovodova R.G., Golovchenko V.V., Popov S.V., Popova G.Y., Paderin N.M., Shashkov A.S., Ovodov Y.S. Chemical composition and anti-inflammatory activity of pectic polysaccharide isolated from celery stalks. *Food Chem.* **2009**, *114*(2), 610–615.

Patel D., Shukla S., Gupta S. Apigenin and cancer chemoprevention: Progress, potential and promise (review). *Int. J. Oncol.* **2007**, *30*(1), 233–245.

Peng Y., Sun J., Hon S., Nylander A.N., Xia W., Feng Y., Wang X., Lemere C.A. L-3-n-Butylphthalide improves cognitive impairment and reduces amyloid-beta in a transgenic model of Alzheimer's disease. *J Neurosci.* **2010**, *30*(24), 8180–8189.

Powanda M.C., Rainsford K.D. A toxicological investigation of a celery seed extract having anti-inflammatory activity. *Inflammopharmacology* **2011**, *19*(4), 227–233.

Sultana S., Ahmed S., Jahangir T., Sharma S. Inhibitory effect of celery seeds extract on chemically induced hepatocarcinogenesis: Modulation of cell proliferation, metabolism and altered hepatic foci development. *Cancer Lett.* **2005**, *221*(11), 11–20.

Tierra M. *Planetary Herbology.* Motilal Banarsidass Publishers, Delhi, India, **1999**, 304.

Tsi D., Tan B.K. The mechanism underlying the hypocholesterolaemic activity of aqueous celery extract, its butanol and aqueous fractions in genetically hypocholesterolaemic RICO rats. *Life Sci.* **2000**, *66*(8), 755–767.

Wright C.I., Van-Buren L., Kroner C.I., Koning M.M. Herbal medicines as diuretics: A review of the scientific evidence. *J. Ethnopharmacol.* **2007**, *114*(1), 1–31.

Yao Y., Sang W., Zhou M., Ren G. Phenolic composition and antioxidant activities of 11 celery cultivars. *J. Food Sci.* **2010**, *75*(1), C9–C13.

Zhou K., Zhao F., Liu Z., Zhuang Y., Chen L., Qiu F. Triterpenoids and flavonoids from celery (*Apium graveolens*). *J. Nat. Prod.* **2009**, *72*, 1563–1567.

10.2.7 Dandelion

Scientific name: *Taraxacum officinale* Weber
Family: Compositae
Hindi: Dudhal
Sanskrit: Dugddhapheni
Tamil: Seemai kattu mullangi

10.2.7.1 Origin and Distribution

The herb probably originated in Greece and Central Asia. It is very common in Europe and in the hills of northwest Himalayas in India, Tibet, Bhutan, and Nepal, from 300 to 5400 m.

The botanical name comes from the Greek words *taraxo* ("disorder") and *takos* ("pain or remedy"), in relation to the plant's great healing properties. Some other authors suggest a relation with the Greek word *trogemon*, meaning "edible." Dandelion has been considered an important food and medicine since very ancient times: it was the food offered by Hecate to Theseus in Greek mythology. In European folklore, it is a symbol of strength and fidelity, as it is said to contain the spirit of the fairies (Di Massimo and Di Massimo, 2005). Like the fairies, the herb likes the sun and has flowers of the same color of fairies' dresses. There is also the belief that it is able to fulfill the lovers' desires.

10.2.7.2 Features

It is an erect perennial herb with thick taproot and long, opposite, lanceolate, and deeply serrated leaves, with feathered dentate shape. Flowers are yellow, clustered in axillary panicles (Dastur, 1962).

10.2.7.3 Properties and Uses in Ayurvedic System of Medicine

Dandelion is a *rasayana* with an excellent detoxifying action for *pitta* constitution, working on elimination of accumulated *ama*. It enkindles *agni*, cleanses the liver, and dispels the stagnated *pitta* and bile (Frawley, 1994). It is useful in constipation, colic, flatulence, gout, insomnia, jaundice, biliary calculi, hepatic diseases, and general debility (Warrier *et al.*, 1996).

Parts used: Tender leaves, root, flowers
Tastes: Pungent, bitter, sweet, astringent
Quality: Light, dry, sharp
Potency: Cold
***Dosha* effect:** PK–, V+
Actions: Diuretic, laxative, anthelmintic, tonic

10.2.7.4 Traditional Uses and Folklore

- The decoction is a home remedy for kidney diseases, constipation, general debility, and jaundice.
- One or two tablespoons of juice extracted from tender leaves is an effective remedy in case of bleeding from the nose, lungs, and uterus.
- In Nepal, the infusion of flowers is used for heart troubles (Joshi and Joshi, 2001). Tibetan medicine uses the herb against chronic fever, gastroenteritis, and *pitta* aggravation (Gyal, 2006).

Precautions: It is not indicated for high *vata*. Diuretic properties of the leaf can interact with orthodox diuretics and hypotensives. The root may reduce serum levels of some drugs.

10.2.7.5 In the Kitchen

The cooked leaves, often mixed with spinach or with other greens, are used as side dish, with a tonic action. The tender leaves can be added to salad: they combine well with chicory. They can also be added to soups or mixed with other ingredients (ginger, black olives, and pistachio) to prepare a very special chutney. During the spring, dandelion helps to eliminate the toxins accumulated in the body because of a diet too rich in fat and fried foods (Frawley, 1994). The roots can be dried, roasted, grounded, and added to coffee (Nadkarni, 2002). Like chicory, dandelion can be a healthy alternative to tea and coffee.

10.2.7.6 Chemical Constituents and Uses as Functional Food

Dandelion's bitterness is due to its sesquiterpene lactones, mostly of the eudesmanolide and germacranolide types (taraxacolides, dihydrolactucin, ixerin, taraxinic acids, and ainslioside), which are unique to this plant. The plant also contains terpenoids (phytosterols, taraxasterol), polysaccharides and inulin, several phenylpropanoids, and flavonoids, and, among them, luteolin and chicoric acid are the most abundant phenolic compounds (Schutz *et al.*, 2005, 2006; Williams *et al.*, 1996). Recently, 4-hydroxyphenylacetic acid derivatives of inositol have been isolated from the roots (Kenny *et al.*, 2014).

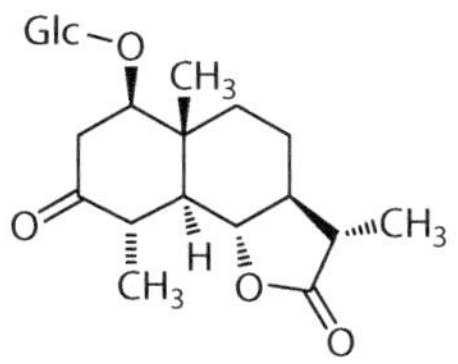

Taraxacolid-β-D-glucoside

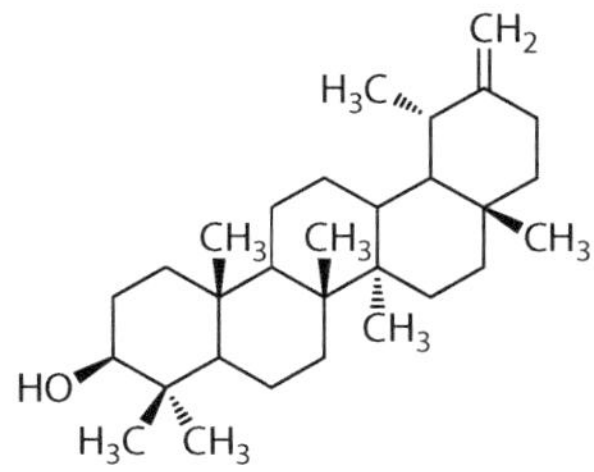

Taraxasterol

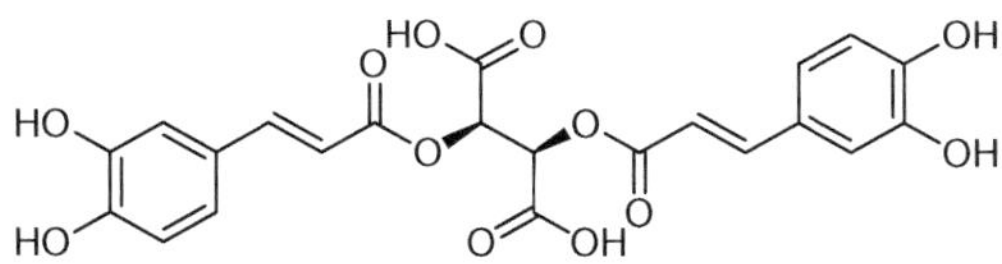

Chicoric acid

Dandelion consists of 1.5% lipids (total weight) and has higher proportions of unsaturated fatty acids (oleic, palmitoleic, linoleic, and linolenic acids) than either lettuce or spinach. It is also one of the richest green vegetable sources of β-carotene (Souci *et al.*, 2008).

Dandelion is a rich source of vitamins (A, C, D, E, and B), choline, inositol, lecithin, minerals, and oligoelements (calcium, sodium, magnesium, iron, silicon, copper, phosphorus, zinc, and manganese). Moreover, it contains a high level of potassium (~83 mg/kg plant tissue; or 59 mg per 236-mL cup).

Dandelion is a nontoxic herb that can be potentially exploited for its choleretic, diuretic, antirheumatic, and anti-inflammatory properties. Dandelion and its formulations might have potential to prevent or ameliorate the outcome of several degenerative diseases such as atherosclerosis and coronary artery disease, obesity, diabetes mellitus, and cancer (Yarnell and Abascal, 2009).

The mechanisms of action through which dandelion's phytochemicals exert their activities involve multiple cell signaling pathways and molecules such as NF-κB, Akt, mitogen-activated protein kinase kinase/extracellular signal-regulated kinase, soluble vascular cell adhesion molecule 1, mitogen-activated protein kinase, matrix metalloproteinase, TNF, and interleukin. An extensive review on the different biological activities of dandelion is reported in González-Castejón *et al.* (2012).

Dandelion is an entirely edible plant, and its leaves, roots, and flowers are incorporated into different food products. A survey of the literature shows that the most widely used daily dose of crude dried dandelion roots or leaves is in the range of 4–10 g, whereas that of fresh roots or leaves habitually consumed as plant foods is approximately 50 g/day or more, varying with different culinary habits.

An analysis of the nutrient content of dandelion reveals high amounts of minerals, proteins, fiber, and vitamins, and a balanced combination of trace elements making dandelion an interesting source of micronutrients (Escudero *et al.*, 2003).

REFERENCES

Dastur J.F. *Medicinal Plants of India and Pakistan*. Taraporevala Sons, Mumbai, India, **1962**, 172–173.

Di Massimo S., Di Massimo M. *Planta medica. Le erbe officinali tra scienza e tradizione*. Quaderni dell'Ambiente, Provincia di Pesaro e Urbino, **2005**, vol 19, 355–357.

Escudero N.L., De Arellano M.L., Fernandez S., Albarracín G., Mucciarelli S. *Taraxacum officinale* as a food source. *Plant Foods Hum. Nutr.* **2003**, *58*, 1–10.

Frawley D., Lad V. *The Yoga of Herbs*. Motilal Banarsidas Publishers, New Delhi, India, **1994**, 114–115.

González-Castejón M., Visioli F., Rodriguez-Casado A. Diverse biological activities of dandelion. *Nutr. Rev.* **2012**, *70*(9), 534–547.

Gyal Y. (Namdul Tenzin trans. & ed.) *Tibetan Medical Dietary Book*, vol. 1. Men-Tse-Khang, Dharamsala, India, **2006**, 90.

Joshi K.K., Joshi S.D. *Genetic Heritage of Medicinal Plants of Nepal Himalayas*. Buddha Academic Publishers & Distributors, Kathmandu, Nepal, **2001**, 163.

Kenny O., Smyth T.J., Hewage C.M., Brunton N.P., McLoughlin P. 4-Hydroxyphenylacetic acid derivatives of inositol from dandelion (*Taraxacum officinale*) root characterised using LC-SPE-NMR and LC-MS techniques. *Phytochemistry* **2014**, *98*, 197–203.

Nadkarni K.M. *Indian Materia Medica*, vol. 1. Popular Prakashan, Mumbai, India, **2002**, 1195–1196.

Schutz K., Carle R., Schieber A. *Taraxacum*—A review on its phytochemical and pharmacological profile. *J. Ethnopharmacol.* **2006**, *107*, 313–323.

Schutz K., Kammerer D.R., Carle R., Schieber A. Characterization of phenolic acids and flavonoids in dandelion (*Taraxacum officinale* WEB. ex WIGG.) root and herb by high-performance liquid chromatography/electrospray ionization mass spectrometry. *Rapid Commun. Mass Spectrom.* **2005**, *19*, 179–186.

Souci S.W., Fachmann W., Kraut H. *Food Composition and Nutrition Tables*. MedPharm Scientific Publishers, Stuttgart, Germany. **2008** (7th edn.).

Warrier P.K., Nambiar V.P.K., Ramankutty C. (eds.) *Indian Medicinal Plants*, vol. 5. Orient Longman, Chennai, India, **1996**, 243–244.

Williams C.A., Goldstone F., Greenham J. Flavonoids, cinnamic acids and coumarins from the different tissues and medicinal preparations of *Taraxacum officinale*. *Phytochemistry* **1996**, *42*, 121–127.

Yarnell E., Abascal K. Dandelion *Taraxacum officinale* and *T. mongolicum*. *Integr. Med.* **2009**, *8*, 35–38.

10.2.8 Drumstick

Scientific name: *Moringa oleifera* Lam.
Family: Moringaceae
Hindi: Sahijan, Mungana
Sanskrit: Sigruh
Tamil: Murunkai

10.2.8.1 Origin and Distribution

Moringa is a common tree native to northern India and cultivated throughout subtropical areas from West Africa to Fiji. It is probably the fastest growing tree known: it can reach 3 m high in just 10 months after it is planted (Khalsa and Tierra, 2010). The name derives from the Tamil word *murunkai*. The English name is due to the fruit shape.

10.2.8.2 Features

Drumstick is an unarmed medium-sized graceful tree, with branches easily breakable; leaves usually tripinnate; flowers white in bunches; fruits pods up to 45 cm long, pendulous, greenish, triangular, and nine-ribbed; and seeds trigonous, winged at angles (Warrier *et al.*, 1995).

10.2.8.3 Properties and Uses in Ayurvedic System of Medicine

The Sanskrit name *sigruh* literally means "that which possesses sharp attribute." In the ancient Ayurvedic texts, drumstick is also called *mocaka*, meaning "that which makes free from diseases": in fact, it is reported that *Moringa* leaves can prevent 300 diseases.

The fruit pods are cooked as vegetable and recommended for loss of appetite, cardiac debility, and vitiation of *vata* and *kapha* (Paranjipe, 2005).

In the *Bhojanakutuhalam*, Sri Raghunatha Suri says that the leaves are used as vegetable: they are wholesome, impart taste, stimulate the digestive fire, kill worms, and favor digestion (Venkat *et al.*, 2012).

Parts used: Leaves, fruits, seeds, flowers
Tastes: Sweet, bitter, pungent
Quality: Light, dry, sharp
Potency: Hot
***Dosha* effect:** KV–, P+
Actions: Nutritive, laxative, diuretic, blood purifier, brain tonic, anti-inflammatory, anthelmintic

10.2.8.4 Traditional Uses and Folklore

- For anemia, chest congestion, senility, and sexual debility, this recipe is a good traditional remedy: fry half teaspoon of black pepper powder in one teaspoon of ghee; add two cups of fresh drumstick leaves and stir fry for 3 minutes. It is suggested to eat with rice or *chappati*s for 40 days. For body heat, excessive menstruation, tuberculosis, and watery sperm, fry one cup of very tender fruits with ghee and eat for 1 month (Sairam, 1998).
- The leaf juice along with carrot juice is a home remedy for high blood pressure and urine retention.
- Leaves ground into a paste with a few pods of garlic, a bit turmeric, salt, and pepper are a remedy for catarrhal affections.
- In the Unani system of medicine, the drumstick is referred to as a laxative and antispasmodic: the leaves are used as external application for laryngitis. The pods (fruits) made into a soup are prescribed by Hakims as a diet in subacute cases of enlarged liver and spleen, articular pains, debility of nerves, and paralysis. Flowers are sometimes boiled with milk as aphrodisiac. Mahomedan writers consider the flowers useful in cold humors and swelling (Nadkarni, 2002).
- The decoction of leaves is diaphoretic and useful in fever (Paranjipe, 2005).

Precautions: The fruit has a warming potency, so people with high *pitta* can eat drumstick only in limited quantity, especially during the summer.

10.2.8.5 In the Kitchen

Moringa oleifera has an impressive range of dietary benefits. The fruit is a delicious ingredient of the south Indian traditional soup called *sambar*. It is highly nutritious and widely used in different recipes. As a vegetable, the young, unripe pod acts as a preventive against intestinal worms (Nadkarni, 2002).

The leaves are very rich in vitamins and iron, but unfortunately rarely used as vegetables. The main use is as a fodder for cattle or as a green manure (Sairam, 1998).

Leaf powder can be added to sauces and soups; the seeds can be cooked as peas or roasted and eaten like peanuts. Charitable organizations in India are using the leaves as a food integrator to prevent vitamin A deficiency and childhood blindness in developing areas (Khalsa and Tierra, 2010).

10.2.8.6 Chemical Constituents and Uses as Functional Food

Moringa oleifera Lam., an edible tree found worldwide in the dry tropics, is increasingly being used for nutritional supplementation. Its nutrient-dense leaves are high in protein quality, leading to its widespread use by doctors, healers, nutritionists, and community leaders to treat undernutrition and a variety of illnesses. Despite the fact that no rigorous clinical trial has tested its efficacy for treating undernutrition, the adoption of *M. oleifera* continues to increase (Thurber and Fahey, 2009).

The crude protein content of *Moringa* leaves is 24.42%–25.29%, suggesting that leaves are a good source of protein. On a wet basis, it is higher (5.4%) than the protein content of alfalfa sprout (3.7%), sweet potato leaves (3.3%), and mung bean sprout (3.1%). *Moringa* leaf is also a good plant source of fat (1.19%–2.77% wet basis) (Shih *et al.*, 2011).

Moringa oleifera has an impressive range of medicinal uses with high nutritional value. Different parts of this plant contain a profile of important minerals and are a good source of protein, vitamins, β-carotene, amino acids, and various phenolics. The *Moringa* plant provides a rich and rare combination of zeatin, quercetin, β-sitosterol, caffeoylquinic acid, and kaempferol. In addition to its compelling water-purifying powers and high nutritional value, *M. oleifera* is reported to have medicinal value. Various parts of this plant such as the leaves, roots, seed, bark, fruit, flowers, and immature pods act as cardiac and circulatory stimulants; possess antitumor, antipyretic, antiepileptic, anti-inflammatory, antiulcer, antispasmodic, diuretic, antihypertensive, cholesterol-lowering, antioxidant, antidiabetic, hepatoprotective, antibacterial, and antifungal activities. They are being employed for the treatment of different ailments in the indigenous system of medicine, particularly in South Asia (reviewed in Anwar *et al.*, 2007).

The enthusiasm for the health benefits of *M. oleifera* is in dire contrast with the scarcity of strong experimental and clinical evidence supporting them. Reported studies in experimental animals and humans, although limited in number and variable in design, seem concordant in their support for the potential of *M. oleifera* leaves to prevent or treat chronic hyperglycemia and dyslipidemia, as symptoms of diabetes and CVD risk (reviewed in Mbikay, 2012).

Drumstick leaves chutney, a recipe from Sairam (1998)

Ingredients

A handful of drumstick leaves
1 onion
1/2 inch piece of ginger
1 tbsp of tamarind pulp
1 tbsp of coconut oil
Salt to taste
Green chilies to taste

Preparation

Grind a handful of drumstick leaves along with one onion, half inch piece of ginger, and one teaspoon of tamarind pulp into a very fine paste along with salt and green chilies to taste. Slightly fry with coconut oil. It is excellent with basmati rice.

Potato–drumstick curry, a recipe from Sairam (1998)

Ingredients

1 onion
1 tbsp ghee
1 tbsp curry powder
8 small potatoes
A handful of drumstick leaves
Salt to tatse
Water

Preparation

Roast thinly diced onion in a little *ghee* and fry small, whole potatoes along with it. Add curry powder to taste and allow cooking with the necessary amount of water. Add salt to taste. When the potatoes are tender, add the chopped drumstick leaves, lower the fire, and stir well. It is excellent with *chappati*s.

REFERENCES

Anwar F., Latif S., Ashraf M., Gilani A.H. *Moringa oleifera*: A food plant with multiple medicinal uses. *Phytother. Res.* **2007**, *21*(1), 17–25.

Khalsa K.P.S., Tierra M. *The Way of Ayurvedic Herbs*. Motilal Banarsidass Publishers, Delhi, India, **2010**, 185–186.

Mbikay M. Therapeutic potential of *Moringa oleifera* leaves in chronic hyperglycemia and dyslipidemia: A review. *Front. Pharmacol.* **2012**, *3*(24), 1–12.

Nadkarni K.M. *Indian Materia Medica*, vol. 1. Popular Prakashan, Mumbai, India, **2002**, 811–816.

Paranjipe P. *Indian Medicinal Plants. Forgotten Healers*. Chaukhamba Sanskrit Pratishthan, Delhi, India, **2005**, 244–245.

Sairam T.V. *Home Remedies*, vol. I. Penguin Books, Delhi, India, **1998**, 203–211.

Shih M.-C., Chang C.-M., Kang S.-M., Tsai M.-L. Effect of different parts (leaf, stem and stalk) and seasons (summer and winter) on the chemical compositions and antioxidant activity of *Moringa oleifera. Int. J. Mol. Sci.* **2011**, *12*, 6077–6088.

Thurber M.D., Fahey J.W. Adoption of *Moringa oleifera* to combat under-nutrition viewed through the lens of the "Diffusion of innovations" theory. *Ecol. Food Nutr.* **2009**, *48*(3), 212–225.

Venkat P., Gangadharan G.G., Lakshmithathachar M.A., Alwar M.A. (eds.) *Bhojanakutuhalam*. I-AIM, Bangalore, India, **2012**, 109.

Warrier P.K., Nambiar V.P.K., Ramankutty C. (eds.) *Indian Medicinal Plants*, vol. 4. Orient Longman, Chennai, India, **1995**, 59–64.

10.2.9 Hogweed

Scientific name: *Boerhaavia diffusa* L.
Family: Nyctaginaceae
Hindi: Sant, Gadahpurna
Sanskrit: Punarnava
Tamil: Mukkarattai kirai

10.2.9.1 Origin and Distribution

This wild weed is quite common in roadsides and wastelands throughout the warmer parts of India and Pakistan, growing abundant especially during the rainy season up to an altitude of 2000 m. It is cultivated in some parts of West Bengal (Mahesh *et al.*, 2012). The herb distribution is very wide: we can find it in the tropical and subtropical parts of Asia, Africa, the United States, and the Pacific Islands.

10.2.9.2 Features

It is a perennial herb with many procumbent branches, "generally spreading on ground or partly ascending" (Paranjipe, 2005). Leaves are simple, opposite, and ovate or oblong or roundish; flowers are small in irregular clusters, pink or red in color (Dastur, 1962; Warrier *et al.*, 1994).

10.2.9.3 Properties and Uses in Ayurvedic System of Medicine

The Sanskrit name *punarnava* means "one which renews the body" (Paranjipe, 2005): *Charaka* reports that the herb promotes longevity. It is one of the best *rasayanas* with an excellent rejuvenating action on the urinary system. It is a general tonic that helps to take out the excessive fluids, cleanses the body, reopens the channels, and properly nourishes them. It works well in case of general body swelling, anemia of *kapha* type, congested cardiac conditions, edema, cough, and rheumatic swollen joints. In small doses, it is an appetizer and a mild laxative.

> Sushruta mentions its use in snake-poisoning and rat-bite infection
>
> **Nadkarni, 2002**

Parts used: Leaves, roots
Tastes: Pungent, bitter, sweet, astringent
Quality: Dry, light
Potency: Hot
***Dosha* effect:** VK–
Actions: Appetizer, laxative, general tonic, rejuvenating, expectorant

10.2.9.4 Traditional Uses and Folklore

An extensive review on *Boerhaavia diffusa* traditional uses are reported in Mishra *et al.* (2014).

- In the Unani system of medicine, the leaves are used as appetizers and to cure ophthalmia and joint pain. In Siddha medicine, it is used for diseases related to alteration of *kapha* and *vata* (Yoganarasimhan and Chelladurai, 2000).
- The herb has been used as a home remedy since ancient times. The powder of the roots is beneficial in asthma treatment (Bakhru, 1993). The infusion is a mild laxative and febrifuge. The juice extracted from the fresh leaves and the decoction are both an effective diuretic useful in case of dropsy and edema. The juice is also recommended for liver problems and jaundice (Nadkarni, 2002). The root decoction or infusion is used as laxative.
- The tribal communities of Chittoor district in Andhra Pradesh use the leaf decoction to cure the first stages of leprosy (Vedavathy, 1997).

Precautions: The formulations containing the whole plant or aerial part of *B. diffusa* should be taken with precautions. *Boerhaavia diffusa* extracts demonstrated several activities such as diuretic, vasodilatory, immunomodulatory, angiotensin-converting enzyme (ACE) inhibitory, and anticonvulsant. Therefore, it can be devised that the products containing *B. diffusa* as the main ingredient can have potential for interaction with medications having similar or opposing activities.

10.2.9.5 In the Kitchen

The cooked leaves are used in India as a side dish with a tonic action. The daily intake promotes general health and strengthens the stomach with a rejuvenating action. Its use as vegetable is very common in the West Bengal and Assam tribal communities. In mild cases of dropsy, the fresh leaves are boiled, salted, and eaten along with *chappati*s (Nadkarni, 2002).

10.2.9.6 Chemical Constituents and Uses as Functional Food

Boerhaavia diffusa contains various categories of secondary metabolites: flavonoid glycosides, steroids (ecdysteroid), alkaloids, phenolic and lignan glycosides, and a series of unique nontoxic rotenoids and dehydrorotenoids called boeravinones, which are used as markers for the plant standardization (Bairwa *et al.*, 2014). Boeravinones have been proposed as putative active constituents for the anticancer, spasmolytic, and anti-inflammatory activities exhibited by *B. diffusa* extracts (Ahmed-Belkacem *et al.*, 2007; Bairwa *et al.*, 2013a, 2013b; Do *et al.*, 2013).

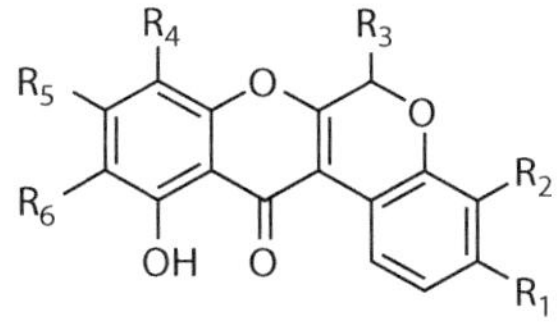

R_1=H, OH; R_2=H, OH; R_3=OH, OCH_3; R_4=H, OH; R_5=OH, OCH_3; R_6=H, CH_3

Boeravinones (dehydrorotenoids)

Boerhaavia diffusa is a good source of nutritional supplements with its 15 amino acids (6 essential) in the whole plant and 14 amino acids (7 essential) in the roots along with isopalmitate acetate, behenic acid, arachidic acid (6.3%), and saturated fatty acids (38%) (Miralles *et al.*, 1988). The presence of vitamins C, B_3, and B_2 (44.80, 97.00, and 22.00 mg, respectively) along with calcium (174.09 mg) in the roots is also reported (Ujowundu *et al.*, 2008). Based on the above studies, this use can be a validated claim.

Although leaves mostly accumulate flavonoids, the roots are reported to contain rotenoids, lignans, and alkaloids of known pharmacological activities (the roots are the official source of the drug *punarnava*), which also endowed with countereffects and drug interactions (see *Precautions* in Section 10.2.9.4).

In Ayurveda, *B. diffusa* has been classified as *rasayana* herb that possesses properties such as preventing antiaging, reestablishing youth, strengthening life and brain power, and preventing diseases, all of which imply that it provides hepatoprotection and immunomodulation (Govindarajan *et al.*, 2005).

In the current scenario of plant-based medicinal products, *B. diffusa* can prove to be an effective and affordable commodity for hepatoprotection, diuresis, and immunomodulation, but it still has not been able to claim its position in herbal market, as limited clinical studies support the traditional uses. An extensive review on *B. diffusa* pharmacological activities of plant extracts is reported in Mishra *et al.* (2014).

REFERENCES

Ahmed-Belkacem A., Macalou S., Borrelli F. *et al.* Nonprenylated rotenoids, a new class of potent breast cancer resistance protein inhibitors. *J. Med. Chem.* **2007**, *50*, 1933–1938.

Bairwa K., Singh I.N., Roy S.K., Grover J., Srivastava A., Jachak S.M. Correction to rotenoids from *Boerhaavia diffusa* as potential anti-inflammatory agents. *J. Nat. Prod.* **2013a**, *76*, 2364.

Bairwa K., Singh I.N., Roy S.K., Grover J., Srivastava A., Jachak S.M. Rotenoids from *Boerhaavia diffusa* as potential anti-inflammatory agents. *J. Nat. Prod.* **2013b**, *76*, 1393–1398.

Bairwa K., Srivastava A., Jachak S.M. Quantitative analysis of Boeravinones in the roots of *Boerhaavia diffusa* by UPLC/PDA. *Phytochem Anal.* **2014**. doi:10.1002/pca.2509.

Bakhru H.K. *Herbs That Heal.* Orient Paperbacks, Delhi, India, **1993**, 102–103.

Do T.M.L., Truong A.V., Pinnock T.G. *et al.* New rotenoids and coumaronochromonoids from the aerial part of *Boerhaavia erecta. Chem. Pharm. Bull.* **2013**, *61*, 624–630.

Govindarajan R., Vijayakumar M., Pushpangadan P. Antioxidant approach to disease management and the role of "Rasayana" herbs of Ayurveda. *J. Ethnopharm.* **2005**, *99*(2), 165–178.

Mahesh A.R., Harish K., Ranganath M.K., Raviraj A.D. Detail study on *Boerhaavia diffusa* plant for its medicinal importance—A review. *Res. J. Pharm. Sci.* **2012**, *1*(1), 28–36.

Miralles J., Noba K., Ba A.T., Gaydou E.M., Kornprobs J.-M. Chemotaxonomy in Nyctagynaceae family: Sterols and fatty acids from the leaves of three *Boerhaavia* species. *Biochem. Syst. Ecol.* **1988**, *16*(5), 475–478.

Mishra S., Aeri V., Gaur P.K., Jachak S.M. Phytochemical, therapeutic, and ethnopharmacological overview for a traditionally important herb: *Boerhavia diffusa* Linn. *Biomed. Res. Int.* **2014**, 808302.

Nadkarni K.M. *Indian Materia Medica*, vol. 1. Popular Prakashan, Mumbai, India, **2002**, 202–207.

Paranjipe P. *Indian Medicinal Plants. Forgotten Healers.* Chaukhamba Sanskrit Pratishthan, Delhi, India, **2005**, 209–211.

Ujowundu C.O., Igwe C.U., Enemor V.H.A., Nwaoguand L.A., Okafor O.E. Nutritive and anti-nutritive properties of *Boerhavia diffusa* and *Commelina nudiflora* leaves. *Pak. J. Nutr.* **2008**, *7*(1), 90–92.

Vedavathy S., Mrudula V., Sudhakar A. *Tribal Medicine of Chittor District. A.P. (India).* Herbal Folklore Research Centre, Tirupati, India, **1997**, 37–38.

Warrier P.K., Nambiar V.P.K., Ramankutty C. (eds.) *Indian Medicinal Plants*, vol. 1. Orient Longman, Chennai, India, **1994**, 281–286.

Yoganarasimhan S.N., Chelladurai V. *Medicinal Plants of India*, vol. 2. Cyber Media, Bangalore, India, **2000**, 80.

10.2.10 Nettle

Scientific name: *Urtica dioica* L.
Family: Urticaceae
Hindi: Bichu, chicru
Sanskrit: Vrishikali
Tamil: Perunkanchori

10.2.10.1 Origin and Distribution

Nettle is native to Europe, common in North and South America, West Siberia, China, the northwest Himalayas in India, Tibet, Bhutan, and Nepal.

The botanical name comes from the Latin words *urere* ("burn") and *tactus* ("touch") in relation to the hairs of the stinging nettle. The species name *dioica* comes from the Greek word *oikya*, meaning "house": the nettle has two houses as female and male flowers are on separate plants. The ancient Egyptians and Romans already appreciated the medicinal properties of the plant (Upton, 2013). The European folklore describes the herb as an aphrodisiac and a protection against evil spirits (Di Massimo and Di Massimo, 2005).

The largest percentage of stinging nettle herb is wild harvested. Cultivated supplies are available, however, and quantities continue to increase. The primary producers of stinging nettle include eastern Germany and middle-eastern European countries (Bombardelli and Morazzoni, 1997). Collection of stinging nettle herbs should be performed in select areas, avoiding locations where agribusiness and industrial runoffs occur as the plant is known to accumulate heavy metals in the leaves. After seeding, stinging nettle leaves lose their vitality quickly, so it is recommended to harvest stinging nettle before seeding or immediately after seeding and before the leaves wilt. Harvesting, storage, and drying should be performed accurately to avoid browning due to oxidation.

10.2.10.2 Features

It is a robust erect herb with lanceolate or ovate leaves. Both stems and leaves are covered with bristly stinging hairs. Flowers are greenish in axillary cymes (Joshi and Joshi, 2001).

10.2.10.3 Properties and Uses in Ayurvedic System of Medicine

It is used in sciatica and rheumatism, both externally and internally. The cold herbal tea is diuretic and anti-inflammatory; it is suggested in case of nephritis and cystitis. The warm tea is useful for asthma and for eliminating mucus from the lungs and colon (Tierra, 1999).

Parts used: Tender leaves
Tastes: Astringent, pungent, slightly bitter
Quality: Unctuous
Potency: Cold
***Dosha* effect:** PK–, V+
Actions: Astringent, hemostatic, diuretic, galactagogue, expectorant, tonic, nutritive

10.2.10.4 Traditional Uses and Folklore

It is believed in Tibetan Buddhist history that the yogi Melaralpa survived for many years only with a soup of nettle leaves during his long meditation (Dekhang, 2008). An extensive review on traditional uses is reported in Upton (2013).

- It is a common home remedy for kidney problems and hemorrhages in India and Nepal (Nadkarmi, 2002). The herbal tea is used in case of asthma, urinary complaints, nephritis, and cystitis (Tierra, 1980).
- Since ancient times, people have taken advantage of this sting by flailing arthritic or paralytic limbs with fresh stinging nettle to stimulate circulation and bring warmth to the joints and extremities in a treatment known as "urtication" (Green, 1824). This practice of urtication or rubefaction became a standard in folk medicine as a remedy for arthritis, rheumatism, and muscular paralysis, and is perhaps the most ancient medicinal use of stinging nettle.
- A diuretic effect of stinging nettle herb was documented in the medical literature of the first half of the nineteenth century.
- It is also prescribed as food in case of anemia (Di Massimo and Di Massimo, 2005; Tierra, 1980). In Tibetan medicine, it is used as a vegetable to enhance digestion and to restore the seven *dhatu*s (Dawa, 1999).

Precautions: If eaten in excess, it can be a mild laxative (Nadkarni, 2002). Caution has been advised during the first 3 months of pregnancy (Mills, 1991).

10.2.10.5 In the Kitchen

The tender leaves can be steamed and fried to prepare a variety of delicious dishes, with a tonic action. We can eat nettle alone or with other greens. It is excellent with *ricotta* cheese. Nettle tender leaves can also be dried, powdered, and mixed with Himalayan salt to be used as an aromatic salt. The daily use of herbal tea is effective against low energy and fatigue (Tierra, 1999). A delicious soup can be prepared with nettle tender leaves, ginger, carrots, and quinoa or millet.

10.2.10.6 Chemical Constituents and Uses as Functional Food

The leaves of the stinging nettle contain a fairly wide variety of chemical constituents. The compounds responsible for the stinging/burning action of the hairs on the leaves of *Urtica dioica* are acetylcholine, histamine, 5-hydroxytryptamine (serotonin), and small amounts of leukotrienes (Czarnetzki *et al.*, 1990). Minor amounts of carotenoids and fatty acids are present. Stinging nettle contains 5%–6% protein in fresh plants and 23%–24% protein in dry plants; 70% of these quantities are digestible proteins (Bombardelli and Morazzoni, 1997). All essential aminoacids are present. Stinging nettle leaves are rich in vitamins such as C (20–60 mg/100 g of dry material), the B group, and K (0.16–0.64 mg), and minerals such as calcium (853–1050 mg/100 g), iron (2–200 mg/100 g of dry material), magnesium (175 mg/100 g), phosphorus (50–265 mg/100 g), potassium (532–613 mg/100 g), and sodium (16–58 mg/100 g) (Frank *et al.*, 1998).

More than 20 phenolic structures have been identified in *U. dioica* extracts, the most abundant being chlorogenic acids, a characteristic caffeoylmalic acid, followed by the flavonoids kaempferol, isorhamnetin, and quercetin, and their 3-rutinosides and 3-glycosides (Orčić *et al.*, 2014). Pinelli *et al.* (2008) reported chlorogenic and caffeoylmalic acid concentrations to be 71.5% of the total phenolics in cultivated and 76.5% in wild samples of stinging nettle leaves.

Dried herb is processed for teas, tablets, and capsules. Freeze-dried herbs are processed mainly for use in capsules. Preparations from fresh plant material include juice, homeopathic products, and liquid extracts (Bombardelli and Morazzoni, 1997). Dried nettle leaves (often with stem and seed) are widely available for use as bulk herbal teas or teabags both alone and in combination with other herbs.

Various stinging nettle herb preparations have been studied experimentally, but for treatment effects in humans, only stinging nettle tea or juice, stew, and proprietary extracts have been evaluated.

Traditionally, the herb is primarily recommended as an adjuvant treatment of rheumatic conditions and lower urinary tract infections, and as a nutritional tonic; more recently, the fresh freeze-dried leaves have been used for the treatment of allergies. The root is used to reduce complaints associated with benign prostate hyperplasia.

There are currently no formal clinical studies proving the efficacy of any stinging nettle preparation as an antirheumatic. There has been limited scientific investigation on the use of external application of stinging nettle (urtication) in arthritic conditions, although extensive folk use is reported. There is evidence that stinging nettle herb interferes with the arachidonic acid metabolism and the production of pro-inflammatory cytokines, and also with cell adhesion molecules that facilitate both the migration of cells into the joints and the attachment of synovium to the bone and cartilage. It seems likely that the anti-inflammatory effect of stinging nettle herb is linked to lipophilic compounds.

Topical activity is related to the histamine content of stinging nettle and the counterirritant effect caused by the sting resulting in local irritation that blocks the ability of afferent sensory nerves to carry pain signal from the area of urtication (e.g., arthritic joint). This is the same activity attributed to topical capsaicin cream. Its diuretic effect has been demonstrated *in vivo*.

Use of stinging nettle herb as a nutritional tonic has not been subject to clinical or pharmacological study and can only partially assessed based on its constituent profile. In addition to the historical data, a survey regarding the use of stinging nettle herb by professional members of the American Herbalists Guild (AHG, 2004) was conducted, and the findings from the 26 respondents are that the nettle seeds are used to increase energy, as a kidney support, and as an anti-inflammatory. The stinging nettle leaves are used internally as a tonic food primarily in Europe and the United States. For this purpose, it can be used as a steamed green, as a puréed ingredient in soups, as a decocted tea, or in solid dosage forms (pills, tablets, capsules). Stinging nettle is frequently used for the treatment of anemia and convalescing patients as well as a "green drink" for general purposes. An extensive review on the pharmacology of nettle preparations is reported in Upton (2013).

REFERENCES

AHG (American Herbalists Guild). AHG professional member survey: *Urtica dioica* leaf. **2004**. Unpublished. American Herbal Pharmacopeia, 5 p. Report no: UD2004.

Bombardelli E., Morazzoni P. *Urtica dioica* L. *Fitoterapia* **1997**, *68*, 387–402.

Czarnetzki B.M., Thiele T., Rosenbach T. Immunoreactive leukotrienes in nettle plants (*Urtica urens*). *Intl. Arch. Allergy Immunol.* **1990**, *91*, 43–46.

Dawa M.R. *A Clear Mirror of Tibetan Medicinal Plants*, vol. 2. Men-Tse-Khang, Dharamsala, India, **2009**, 150.

Dekhang T.D. A handbook of Tibetan medicinal plants. Men Tze Khang, Dharamsala, India, **2008**, 197.

Di Massimo S., Di Massimo M. *Planta medica. Le erbe officinali tra scienza e tradizione.* Quaderni dell'Ambiente, Provincia di Pesaro e Urbino, vol 19, **2005**, 107–109.

Frank B., Bohn I., Uehleke B. Urtica. Volume 3. in, Blaschek W., Hänsel R., Keller K., Reichling J., Rmipler H., Schneider (Hrsg.) G. (eds.), Hagers Handbuch. Springer, Berlin, Germany. **1998**, 710–736.

Joshi K.K., Joshi S.D. *Genetic Heritage of Medicinal Plants of Nepal Himalayas*. Buddha Academic Publishers & Distributors, Kathmandu, Nepal, **2001**, 169–170.

Mills S. *Out of the Earth: The Essential Book of Herbal Medicine*. Viking Arcana, New York, **1991**, 677.

Nadkarni K.M. *Indian Materia Medica*, vol. 1. Popular Prakashan, Mumbai, India, **2002**, 1258.

Orčić D., Francišković M., Bekvalac K. *et al.* Quantitative determination of plant phenolics in *Urtica dioica* extracts by high-performance liquid chromatography coupled with tandem mass spectrometric detection. *Food Chem.* **2014**, *143*, 48–53.

Pinelli P., Ieri F., Bignolini P. *et al.* Extraction and HPLC analysis of phenolic compounds in leaves, stalks, and textile fibers of *Urtica dioica* L. *J. Agric. Food Chem.* **2008**, *56*, 9127–9132.

Tierra M. *Planetary Herbology*. Motilal Banarsidass Publishers, Delhi, India, **1999**, 333–334.

Tierra M. *The Way of Herbs*. Pocket Books, New York, **1980**, 167.

Upton R. Stinging nettles leaf (*Urtica dioica* L.): Extraordinary vegetable medicine. *J. Herbal Med.*, **2013**, *3*, 9–38.

10.2.11 Okra

Scientific name: *Abelmoschus esculentus* (L.) Moench
Family: Malvaceae
Hindi: Bendi, Bhindi
Sanskrit: Bhenda, Tindisha, Pitali, Gandhamula
Tamil: Vendakkay

10.2.11.1 Origin and Distribution

The English synonym of okra is lady's finger due to the shape of the fruit. It is also called *edible hibiscus* (Morningstar and Desai, 1999). There are many different hypotheses about the geographical origin of okra, with supporters of South Asia, Ethiopia, and West Africa. The plant grows well in all tropical, subtropical, and warm temperate countries.

A Spanish Moor who visited Egypt in 1216 described the use of the young pods as food. The presence of okra in Brazil was recorded in 1658: it was probably introduced to the Americas from Africa during the slave trade. By the year 1748, the plant cultivation was well established in Virginia. In 1806, we have the first mention of different cultivars in the southern United States.

It is widely cultivated in India up to 1200 m. It has been recently introduced in Tuscany and Puglia, Italy.

10.2.11.2 Features

It is an annual erect herb covered with the hair throughout, with leaves simple, alternate, scabrous, and palmately three to five lobed. The flowers are large, yellow with a purple center; the fruits are long and capsular, with many rounded seeds (Warrier *et al.*, 1994).

10.2.11.3 Properties and Uses in Ayurvedic System of Medicine

Cooked to enhance its moistening qualities, okra is well tolerated by all *dosha*s. Recommended for *pitta*, it has a neutral effect on *kapha* and very mildly increases *vata* (Morningstar and Desai, 1999). The bland viscid mucilage has a soothing effect, facilitates elimination, and helps prevent constipation.

It is an excellent food for problems related to high *pitta*, difficult, painful or burning urination, fevers, stomach ulcers, diarrhea and other intestinal disorders, spastic colon, diverticulitis, and constipation (Nibodhi, 2010).

The *Bhojanakutuhalam* refers that *Bhenda* imparts taste (Venkat *et al.*, 2012).

Parts used: Fruits, leaves, seeds
Tastes: Sweet, astringent
Quality: Light
Potency: Cold
***Dosha* effect:** VKP=
Actions: Emollient, cooling, aphrodisiac, stomachic, demulcent, diuretic, tonic

10.2.11.4 Traditional Uses and Folklore

In chronic dysentery, the fruit is very effective, and it is generally given in the form of soup. The decoction is administrated in gonorrheal cystitis and in other conditions where there is difficulty in micturition (Nadkarni, 2002).

10.2.11.5 In the Kitchen

In the Middle East countries, okra fruit is usually cooked in a thick stew made with vegetables and meat. It is also very popular in India, Pakistan, Malaysia, Japan, the Philippines, Caribbean islands, Brazil, and the southern part of the United States.

Usually, the unripe mucilaginous fruit is fried or cooked with curry. It is excellent with onion, tomatoes, and lemon. To retain the nutrients and self-digesting enzymes, it is better to cook the okra as little as possible. Sometimes, the young small fruits are eaten raw or used to prepare a pickle.

The entire plant is edible: leaves can be cooked or eaten raw in salads. The seeds can be roasted and ground to form a caffeine-free substitute for coffee.

Okra remains in good condition only for around 2 days at room temperature. It can be stored in the fridge for 4–5 days. It is very important to select only the small, unripe, dark green-colored, and fresh fruits. It is necessary to identify and discard the overgrown okra, as even one piece of overgrown okra in the curry will spoil the taste.

10.2.11.6 Chemical Constituents and Uses as Functional Food

Okra is a popular health food due to its high content in soluble fiber. It also contains high amount of vitamin C and folate, and is a good source of calcium and potassium (USDA Nutrient Database). Mature seeds are used for oil production and, when ground, as a substitute for coffee. Various plant parts are also used as a thickening agent in confectionaries.

Different varieties of okra contain an average of 22% crude fat, 31% crude fiber, and 27% crude protein. Okra seed is rich in both protein and fat with high lysine content. Linoleic acid is the most abundant unsaturated fatty acid (up to 50%) and palmitic acid the most abundant saturated fatty acid (up to 35%) (Jarret *et al.*, 2011).

Structural characterization of all major cell wall polysaccharides originating from okra pods was described. The sequential extraction of okra cell wall material yielded fractions rich in galactose, rhamnose, and galacturonic acid in the ratio of 1.3:1:1.3. The degree of acetylation is relatively high, whereas the degree of methyl esterification is relatively low (Sengkhamparn *et al.*, 2009).

In Asian medicine, the fruit of the okra plant, *Abelmoschus esculentus* (L.) Moench, is used as a mucilaginous food additive against gastric irritative and inflammative diseases. Use of herbal mucilages in the treatment of irritated mucous membranes has always been well appreciated and is understood as a bioadhesive interaction between exogenous polysaccharides and gastrointestinal mucins. Acidic polysaccharides and glycoproteins were isolated from okra mucilage (Bhat and Tharanathan, 1985), which showed to have effect on *Helicobacter pylori*. The antiadhesive activity is therefore due to the blocking capacity of specific *Helicobacter* surface receptors that coordinate the interaction between the host and the bacterium (Wittschier *et al.*, 2007). Neither of the active fractions showed inhibitory effects on bacterial growth *in vitro*. The antiadhesive qualities of okra were assumed to be due to a combination of glycoproteins and highly acidic sugar compounds making up a complex three-dimensional structure that is fully developed only in the fresh juice of the fruit (Lengsfeld *et al.*, 2004).

The action of a pectic rhamnogalacturonan obtained by hot buffer extraction of okra pods (okra RG-I) on melanoma cell growth and survival *in vitro* was evaluated. The cell cycle was arrested in G2/M, and apoptosis was induced. The expression of *N*-cadherin and α5 integrin subunit was reduced, and that of the multifunctional carbohydrate-binding protein, Gal-3, at the cell membrane increased. These findings suggest that okra RG-I induces apoptosis in melanoma cells by interacting with Gal-3 (Vayssade *et al.*, 2010).

Quercetin glucoside (quercitrin) and an unidentified flavonoid were detected in *A. esculentus* growing in sub-Saharan countries and popular especially during the rainy season in Nigeria and many tropical West, Central, and Eastern African countries. Due to its potent antioxidant/radical-scavenging activities, it is a good contributor to the antioxidant status and disease chemoprevention of people in these countries (Atawodi *et al.*, 2009).

A natural, moisturizing alcohol-based hand rub (ABHR) containing okra polysaccharide was formulated to reduce the dryness caused by traditional hand-cleansing products. The ABHR developed also reduced infectious disease transmission (Kanlayavattanakul *et al.*, 2012).

Okra curry, a recipe by Vaidya Kalyan Chakravarty

Ingredients

300 g okra
1 tomato or the juice of a half lemon
3 green chilies (slit)
2 tsp of black gram
1 tsp of mustard seeds

1 tsp of cumin seeds
Few curry leaves
4 tsp of ghee

Preparation

Wash okra well and spread them on a dry cloth to dry. Keep them aside for few minutes or wipe dry, eliminating the overgrown pieces. Tender okras have pointed tail ends. Take the okra and bend the pointed tail end with thumb; if it breaks easily, the okra is tender. Remove the top portion of the tender okras and cut them into even-sized pieces. Then take them into a bowl. Heat a pan and add ghee to it. When ghee is hot, add black gram and fry stirring. When the seeds start changing color, add mustard seeds and fry till they start crackling. Then add cumin seeds and fry till seeds crackle fully. Now add green chilies and fry. Then add curry leaves and mix well. To the fried seasoning, add the okra pieces and mix well. Add turmeric powder and mix well. Now cook covered on low flame for 1 minute. After 1 minute, remove the lid and gently mix the vegetable. Again cook covered on low flame for a minute. Add tomatoes pieces or lemon juice (the sour taste will help to remove the mucilage). Repeat the process of cooking covered for 1 minute and stirring afterward till the vegetable is cooked. Finally, add the salt, so the quantity of the vegetable will reduce and look dry, and the mucilage will vanish completely. By that time, the vegetable will be cooked. Transfer the curry to a serving bowl and serve hot with hot plain rice or *roti*.

Note: Green chili is an option only for *kapha*. Never add water to the curry as the mucilage will increase; for this reason, the okras need to be washed but nicely dried.

REFERENCES

Atawodi S.E., Atawodi J.C., Idakwo G.A. *et al.* Polyphenol composition and antioxidant potential of *Hibiscus esculentus* L. fruit cultivated in Nigeria. *J. Med. Food.* **2009**, *12*(6), 1316–1320.

Bhat U.R., Tharanathan R.N. Fractionation of okra mucilage and structural investigation of an acidic polysaccharide. *Carbohydr. Res.* **1985**, *148*, 143–147.

Jarret R.L., Wang M.L., Levy I.J. Seed oil and fatty acid content in okra (*Abelmoschus esculentus*) and related species. *J. Agric. Food Chem.* **2011**, *59*, 4019–4024.

Kanlayavattanakul M., Rodchuea C., Lourith N. Moisturizing effect of alcohol-based hand rub containing okra polysaccharide. *Int. J. Cosmet. Sci.* **2012**, *34*(3), 280–283.

Lengsfeld C., Titgemeyer F., Faller G., Hensel A. Glycosylated compounds from okra inhibit adhesion of *Helicobacter pylori* to human gastric mucosa. *J. Agric. Food Chem.* **2004**, *52*(6), 1495–1503.

Morningstar A., Desai U. *The Ayurvedic Cookbook*. Motilal Banarsidass Publishers, Delhi, India, **1999**, 150.

Nadkarni K.M. *Indian Materia Medica*, vol. 1. Popular Prakashan, Mumbai, India, **2002**, 1–2.

Nibodhi. *Annapurna's Prashad*. Mata Amritanandamayi Mission Trust, Amritapuri, India, **2010** (2nd edn.), 104.

Sengkhamparn N., Verhoef R., Schols H.A., Sajjaanantakul T., Voragen A.G. Characterisation of cell wall polysaccharides from okra (*Abelmoschus esculentus* L. Moench). *Carbohydr. Res.* **2009**, *344*(14), 1824–1832.

USDA Nutrient Database. http://ndb.nal.usda.gov/ndb/search/list

Vayssade M., Sengkhamparn N., Verhoef R. *et al.* Antiproliferative and proapoptotic actions of okra pectin on B16F10 melanoma cells. *Phytother. Res.* **2010**, *24*(7), 982–989.

Venkat P., Gangadharan G.G., Lakshmithathachar M.A., Alwar M.A. (eds.) *Bhojanakutuhalam.* I-AIM, Bangalore, India, **2012**, 137.

Warrier P.K., Nambiar V.P.K., Ramankutty C. (eds.) *Indian Medicinal Plants*, vol. 1. Orient Longman, Chennai, India, **1994**, 1–3.

Wittschier N., Lengsfeld C., Vorthems S. *et al.* Large molecules as anti-adhesive compounds against pathogens. *J. Pharm. Pharmacol.* **2007**, *59*(6), 1777–1786.

10.2.12 Onion

Scientific name: *Allium cepa* L.
Family: Liliaceae
Hindi: Pyaj
Sanskrit: Palanduh
Tamil: Venkayam

10.2.12.1 Origin and Distribution

Allium cepa is unknown in the wild but has been cultivated since ancient times in the Middle East and India. It is one of the oldest cultivated herbs, probably indigenous to Central Asia, possibly to the Iran–Pakistan area. It was worshipped by the ancient Egyptians for its spherical shape and concentric rings, the symbol of eternal life. Onion was depicted on Egyptian tombs as early as 3200 BC, and it was mentioned in the *Garuda Purana* and by the great Indian sage Atreya as a very useful food with a lot of medicinal properties. The Jews loved onion so much that they built a city which they named after it in 173 BC (Bakhru, 1993). In the Middle Ages, it was such an important food in Europe that people would pay their rent with onions and used to give them as gifts.

Today, it is cultivated on a global scale in 170 countries. The 10 top producers are China, India, the United States, Egypt, Iran, Turkey, Pakistan, Brazil, Russia, and the Republic of Korea.

10.2.12.2 Features

It is a biennial herb with an aromatic, fleshy, underground bulb; leaves are long, linear, and hollow; flowers are white, in globular umbels (Warrier *et al.*, 1994). All parts produce a strong odor when crushed. There are a lot of varieties different in color, shape, and aroma.

10.2.12.3 Properties and Uses in Ayurvedic System of Medicine

Ayurveda considers onion a multipurpose heart medicine. Onions are good for diabetes and for reducing phlegm and promoting expectoration. According to *Charaka Samhita* and *Susruta Samhita*, it is a great giver of strength. It increases *agni* and it is recommended for diseases due to vitiation of *vata*. Yogi Bhajan mentioned countless times the therapeutically importance of onion, garlic, and ginger combination, called trinity roots (Khalsa and Tierra, 2010).

Onion increases immunity; combats cold, flue, and general debility; eliminates parasites; and helps remove heavy metal toxins. It alleviates blood and lymph stagnation and counters sexual weakness or debility (Nibodhi, 2010). It imparts taste and treats vomiting (Venkat, 2012). Raw onion is diuretic and antiseptic.

Parts used: Bulb
Tastes: Pungent (raw), sweet (cooked)
Quality: Heavy, unctuous
Potency: Hot
***Dosha* effect:** VK–, P+ (cooked); K–, VP+ (raw)
Actions: Diaphoretic, tonic, aphrodisiac, diuretic

10.2.12.4 Traditional Uses and Folklore

- Equal amounts of onion juice and honey should be mixed, and three to four teaspoons of this mixture should be taken daily for cough (Bakhru, 1992).
- The mixed juice of onion and bitter melon cures severe indigestion.
- A *kichadi* prepared with onion, fenugreek, and coconut increases breast milk.
- Mixed with common salt, onions are a home remedy for colic and scurvy. Cooked with vinegar, they are given in jaundice, spleen enlargement, and dyspepsia. In malarial fever, they are eaten twice a day with two or three black pepper with remarkable relief. Onions eaten with jaggery stimulate the growth of children. A recipe made with roasted onions mixed with cumin, sugar candy, and cow's ghee is a home remedy for piles (Nadkarni, 2002).
- The Unani system of medicine refers to onion as dry and hot, with stomachic, aphrodisiac, and diuretic action. Use of onion is recommended for jaundice, dog bites, and piles (Nadkarni, 2002).

Precautions: Excess intake of onion is reported to cause coughing or flatulence (Sairam, 1998). Onion is *rajasic*, as aphrodisiac is second only to garlic, so in India it is usually avoided by monks and during meditation and yoga practices.

10.2.12.5 In the Kitchen

Onions can be cooked in many different ways, baked, boiled, braised, fried, roasted, sautéed, or eaten raw in salads. In India, they are used as a thickening agent for curries. Coated with gram flour and fried, they are a popular snack called *bhajiya*. In Tamil Nadu raw thinly diced onions are mixed with yogurt, salt and pepper as a side dish for *biryani* rice.

Like garlic, onions are classified as *rajasic* and *tamasic*, which means that they increase passion and ignorance. Therefore, they are avoided in the *brahmana*-style cooking of India, because they are unfit to offer to the deity. They are also forbidden to yogis and monks.

10.2.12.6 Chemical Constituents and Uses as Functional Food

Onion (*A. cepa* L.) species are found across a wide range of latitudes and altitudes in Europe, Asia, North America, and Africa. The world onion production is around 44 million tons, making it the second most important horticultural crop after tomatoes. Because of their storage characteristics and durability for shipping, onions have always been traded more widely than most vegetables. Onions are versatile and are often used as an ingredient in many dishes and are accepted by almost all traditions and cultures.

Onions are rich in two chemical groups that have perceived benefits to human health: the flavonoids and the alk(en)yl cysteine sulfoxides. Two flavonoid subgroups are found in onion: the anthocyanins, which impart a red/purple color to some varieties, and flavanols such as quercetin and its derivatives responsible for the yellow and brown skins of many other varieties (Lanzotti, 2006; Rose *et al.*, 2005). Onion bulbs are among the richest sources of dietary flavonoids ranging from more than 400 mg to 2 g/kg of fresh weight. Quercetin and its glycosides (with a dominance in glucosides) are the most abundant flavonoids present in onion (reviewed in Slimestad *et al.*, 2007, and Griffiths *et al.*, 2002).

Processing may decrease the flavonoid content as a result of preparation and/or leakage from the vegetable/fruit. However, it can increase the extractability from the matrix resulting in a higher apparent content of flavonoids. Deglycosylation can occur due to enzymes released from disrupted plant tissue and, after consumption, due to β-glucosidases of the consumer's body and those of microbial origin (Nemek and Piskula, 2007).

Fisetin (3,7,3′,4′-tetrahydroxyflavone) belongs to the flavonol subgroup of flavonoids with quercetin, myricetin, and kaempferol, and is found in several fruits and vegetables, including strawberries, apples, persimmons, and onions. Fisetin is showing promise as a useful natural agent against cancer and has been evaluated for its potential inhibitory role against cancer in several *in vitro* and *in vivo* studies (Khan *et al.*, 2013; Syed *et al.*, 2013). Dihydroquercetin (taxifolin), also present in onions, shows therapeutic promises in the major disease states such as cancer and cardiovascular and liver diseases (Weidmann, 2012).

Epidemiological studies beginning in the 1990s have reported that the intake of quercetin is inversely related to CVD. More recent work using hypertensive animals and humans indicates a decrease in blood pressure after quercetin supplementation. A number of proposed mechanisms may be responsible for the observed blood pressure decrease such as antioxidant effects, inhibition of ACE activity, and improved endothelium-dependent and endothelium-independent function. The majority of these mechanisms have been identified using animal models treated with quercetin, and relatively few have been corroborated in human studies (Larson *et al.*, 2012).

Several recent studies have revealed that phytochemicals, including curcumin (yellow pigment in turmeric), resveratrol (phytoalexin in grapes), quercetin (in onions and others), and ITCs (preferentially present in cruciferous vegetables, such as broccoli and cabbage), are remarkable regulators of protein quality control systems, suggesting that their physiological and biological functions are exerted, at least in part, through activation of such unique mechanisms (Murakami, 2013).

Most onion cultivars contain about 89% water, 4% sugar, 1% protein, 2% fiber, and 0.1% fat. They are high in vitamin C, vitamin B_6, and folic acid, and are a good source of dietary fiber. They are very low in fats and in sodium, and with an energy value of 166 kJ (40 kcal) per 100 g serving, they can contribute their flavor to savory dishes without raising the caloric content to any great extent (USDA Nutrient Database).

REFERENCES

Bakhru H.K. *Herbs That Heal.* Orient Paperbacks, Delhi, India, **1992**, 134.

Bakhru H.K. *Herbs That Heal.* Orient Paperbacks, Delhi, India, **1993**, 132–136.

Griffiths G., Trueman L., Crowther T., Thomas B., Smith B. Onions—A global benefit to health. *Phytother Res.* **2002**, *16*(7), 603–615.

Khalsa K.P.S., Tierra M. *The Way of Ayurvedic Herbs.* Motilal Banarsidass Publishers, Delhi, India, **2010**, 60–62.

Khan N., Syed D.N., Ahmad N., Mukhtar H. Fisetin: A dietary antioxidant for health promotion. *Antioxid. Redox Signal.* **2013**, *19*(2), 151–162.

Lanzotti V. The analysis of onion and garlic. *J. Chromatogr. A* **2006**, *1112*(1/2), 3–22.

Larson A.J., Symons J.D., Jalili T. Therapeutic potential of quercetin to decrease blood pressure: Review of efficacy and mechanisms. *Adv. Nutr.* **2012**, *3*(1), 39–46.

Murakami A. Modulation of protein quality control systems by food phytochemicals. *J. Clin. Biochem. Nutr.* **2013**, *52*(3), 215–227.

Nadkarni K.M. *Indian Materia Medica*, vol. 1. Popular Prakashan, Mumbai, India, **2002** 63–64.

Nemek K., Piskula M.K. Food content, processing, absorption and metabolism of onion flavonoids. *Crit. Rev. Food Sci. Nutr.* **2007**, *47*(4), 397–409.

Nibodhi. *Annapurna's Prashad.* Mata Amritanandamayo Mission Trust, Amritapuri, India, **2010** (2nd ed.), 105.

Rose P., Whiteman M., Moore P.K., Zhu Y.Z. Bioactive *S*-alk(en)yl cysteine sulfoxide metabolites in the genus *Allium*: the chemistry of potential therapeutic agents. *Nat. Prod. Rep.* **2005**, *22*(3), 351–368.

Sairam T.V. *Home Remedies*, vol. 1. Penguin Books, Delhi, India, **1998**, 9–15.

Slimestad R., Fossen T., Vågen I.M. Onions: A source of unique dietary flavonoids. *J. Agric. Food Chem.* **2007**, *55*, 10067–10080.

Syed D.N., Adhami V.M., Khan M.I., Mukhtar H. Inhibition of Akt/mTOR signaling by the dietary flavonoid fisetin. *Anticancer Agents Med. Chem.* **2013**, *13*(7), 995–1001.

USDA Nutrient Database. http://ndb.nal.usda.gov/.

Venkat P., Gangadharan G.G., Lakshmithathachar M.A., Alwar M.A. (eds.) *Bhojanakutuhalam*. I-AIM, Bangalore, **2012**, 99.

Warrier P.K., Nambiar V.P.K., Ramankutty C. (eds.) *Indian Medicinal Plants*, vol. 1. Orient Longman, Chennai, India, **1994**, 88–92.

Weidmann A.E. Dihydroquercetin: More than just an impurity? *Eur. J. Pharmacol.* **2012**, *684*(1–3), 19–26.

10.2.13 Prickly Chaff Flower

Scientific name: *Achyranthes aspera* L.
Family: Amaranthaceae
Hindi: Latjira
Sanskrit: Apamarga
Tamil: Shiru-kadaladi, Nayurivi

10.2.13.1 Origin and Distribution

It is a weed probably indigenous to Southeast Asia and/or Africa, naturalized in tropical and subtropical areas. It is an invasive species in Tanzania, Uganda, and Kenya. In India, it is common on roadsides and wastelands up to an altitude of 900–1000 m. There are many references in the *Veda*s to a plant that "sweep[s] away chronic diseases and every curse" (*Atharva Veda* IV. 18.7). It is one of the 21 plants used during the Hindu rituals offered to Ganesha. The Sanskrit name *apamarga* "means that which keeps away the *doshas*" (Paranjipe, 2005).

10.2.13.2 Features

It is an erect annual or perennial hairy herb of about 1 m in height with few thick opposite leaves and numerous flowers greenish white, born on slender spikes about 45 cm long. Seeds are subcylindric and reddish brown (Riazudin and Siddiq, 2009; Warrier *et al.*, 1994). Fruits are adhering to animals and clothes (Vedavathy *et al.*, 1997). In Sanskrit, it is also called *mayuraka*, as the spikes are similar to the peacock head (Paranjipe, 2005).

> As once when time began the Gods with thee expelled the Asuras
>
> **Atharva Veda IV. 19.4**

10.2.13.3 Properties and Uses in Ayurvedic System of Medicine

Whole plant is used in the treatment of disease of head and ear (Udayan and Balachandran, 2009). The herb is used mainly in case of asthma and cough: it helps to expel phlegm and have tranquilizing properties. It is also recommended in case of hemorrhoids, urinary diseases, liver problems, and indigestion. It can be eaten as a general tonic in case of anemia (Warrier *et al.*, 1994). The *Charaka Samhita* reports the diuretic and anthelmintic properties (Sharma and Bagwan Dash, 1976).

The decoction of the whole plant is suggested to reduce fever due to *kapha*. The ash of the roots is an antacid, and it is given with honey in case of cough and asthma (Dastur, 1962; Nadkarni, 2002; Sairam, 1998). The seeds rubbed with rice water are recommended for bleeding piles (Nadkarni, 2002; Paranjpe, 2005).

The white and red varieties have different properties: the white one is laxative and carminative. It is used in case of pain in the abdomen and blood and heart diseases. It is recommended for *vata*. The red variety is emetic, constipating, and recommended for *vata* and *kapha* (Riazudin and Siddiq, 2009).

Parts used: Seeds, roots, shoots
Tastes: Pungent, bitter
Quality: Light, dry and sharp
Potency: Cold
***Dosha* effect:** White variety V–, red variety VK–
Actions: Expectorant, depurative, anti-inflammatory, diuretic, laxative, carminative, digestive, tranquilizing

10.2.13.4 Traditional Uses and Folklore

- The crushed plant boiled in water is a traditional remedy for pneumonia. The decoction is useful in rheumatism and arthritis (Mitaliya *et al.*, 2003). The Raipur district's tribes reduce the whole plant to powder, mix it with wheat flour, and prepare *chappati*s, to treat leucorrhea, diabetes, and internal heat (Jain *et al.*, 2003).
- The decoction of powdered leaves with honey is prescribed in the early stages of diarrhea.
- The tribal and rural communities of the Bhavnagar district in Gujarat use the juice of the leaves to treat piles, jaundice, cough, asthma, bronchitis, and fever (Bhatt *et al.*, 2002).
- The tribal community of the Chittoor district in Andra Pradesh uses the root paste mixed in water as a remedy for stomach pain. A tablespoon of seed powder boiled in milk is given daily to cure urinary troubles (Vedavathy, 1997).
- The local application of the plant paste is a common home remedy for bites of poisonous snakes, scorpions, and mad dogs. It is also used for scabies and other skin diseases. The twigs and the roots are a traditional toothbrush. They are considered effective in case of pyorrhea and other gum troubles.

Precautions: It should be avoided during pregnancy. The juice in large doses can cause abortion (Dastur, 1962). The fresh roots are used by Santals tribes to cause abortion (Pal and Jain, 1998). The root decoction is a traditional contraceptive in the Bhavnagar district (Bhatt *et al.*, 2002).

10.2.13.5 In the Kitchen

The seeds and leaves are cooked and eaten by the hermits who are living isolated in the forests. In Tamil Nadu, rural people believe that they can skip a meal just eating the seeds as they do not feel hungry. Due to this anorexient property, they can be eaten in case of obesity (Paranjipe, 2005). The leaves are cooked like spinach with other wild herbs and have an excellent taste. The seeds are often cooked with milk (Manandar, 2002): they are considered a good brain tonic (Nadkarni, 2002).

10.2.13.6 Chemical Constituents and Uses as Functional Food

Preliminary phytochemical screening of the various extracts of *Achyranthes aspera* leaves revealed the presence of cardiac glycosides, reducing sugars, and alkaloids (Jurry *et al*., 2013). Tannins were the most abundant. Terpenoids, reducing sugars, and cardiac glycosides were prominently found in sequential extracts of stems. In leaves, flavonoids and terpenoids were found to be present in moderate amounts (Narayan and Kumar, 2014; Sharma *et al*., 2013; Srivastav *et al*., 2011). Three oleanolic acid glycosides and a cyclic fatty acid-like compound were isolated from the seeds (Chauhan *et al*., 2002; Rameshwar and Akito, 2007).

In the absence of the identified active molecules, global gene expression analyses helped deciphering the mode of action of plants with recognized folkloric uses. Gene expression analyses were performed using whole human genome microarrays, by exposing human pancreatic cancer cell line MiaPaCa-2 (CRL-1420) to 34 μg/mL of *A. aspera* leaf extract for 24, 48, and 72 hours. The results showed that the leaf extract affected the pathways of immune response, metabolism, development, gene expression regulation, cell adhesion, cystic fibrosis transmembrane conductance regulation, and chemotaxis. Disease biomarker enrichment analysis identified the leaf extract to regulate genes involved in vasculitis inflammation of blood vessels. The outcome of disease network analysis supports the medicinal use of *A. aspera* to stop bleeding, as a cure for pancreatic cancer, as an antiarthritic medication, and so forth (Subbarayan *et al*., 2013).

Treatment with an *A. aspera* ethanolic extract significantly reduced blood glucose levels in alloxan-induced diabetic mice. *Achyranthes aspera* extract prevented lipid peroxidation as gauged by thiobarbituric acid reactive substances and hydroperoxides, and also increased the activity of catalase and reduced NO levels in alloxan-induced diabetic mice (Talukder *et al*., 2012).

Ethanolic extracts of *A. aspera* seeds significantly suppressed the increase in body, retroperitoneal adipose tissue, liver weights, and serum parameters such as total cholesterol, total triglyceride, and LDL cholesterol level. The delay in the intestinal absorption of dietary fat in high-fat diet-treated mice by inhibiting pancreatic amylase and lipase activity was also assessed. *Achyranthes aspera* seed may prevent obesity by reducing the excess accumulation of body fat and changing the serum lipid profile (Rani *et al*., 2012). An extensive review on the pharmacological activity of *A. aspera* extracts is reported in Srivastav *et al*. (2011).

REFERENCES

Bhatt D.C., Patel N.K., Mitaliya K.D., Ant H.M. Herbal magic by contact therapy among tribals and rurals of Gujarat, in Singh V., Jain A.P. (eds.), *Ethnobotany and Medicinal Plants of India and Nepal*, vol. 1. Scientific Publishers, Jhodpur, India, **2003**, 306.

Chauhan A.S., Rawat G.S., Singh C.P. Phytochemical study of Achyranthes aspera Linn. *Asian J. Chem.* **2002**, 14(2), 1059–1061.

Dastur J.F. *Medicinal Plants of India and Pakistan*. Taraporevala Sons and Co., Mumbai, India, **1962**, 7.

Jain P.C., Jain S.P., Singh S.C., Singh J., Sushil K. Ethno-medico-botanical survey of Raipur District, Chattisgarh State, in Singh V., Jain A.P. (eds.), *Ethnobotany and Medicinal Plants of India and Nepal*, vol. 1. Scientific Publishers, Jhodpur, India, **2003**, 266.

Jurry, R., Gupta, S., Mishra, P. Preliminary phytochemical screening of *Achyranthes aspera* L. *Asian J. Chem. Environ. Res.* **2013**, *6*(1/2), 105–110.

Manandar N.P. *Plants and People of Nepal*. Timber Press, Portland, OR, **2002**.

Mitaliya K.D., Bhatt, D.C., Patel J.N., Patel, S.K. Observations of plants used for joint disorders by rural and tribal people of Gujarath, in Singh V., Jain A.P. (eds.), *Ethnobotany and Medicinal Plants of India and Nepal*. Scientific Publishers, Jhodpur, India, **2003**, *1*, 213.

Nadkarni K.M. *Indian Materia Medica*. Popular Prakashan, Mumbai, India, **2002**, *1*, 21–22.

Narayan C., Kumar A. Antineoplastic and immunomodulatory effect of polyphenolic components of *Achyranthes aspera* (PCA) extract on urethane induced lung cancer in vivo. *Mol. Biol. Rep.* **2014**, *41*(1), 179–191.

Pal D.C., Jain S.K. *Tribal Medicine*. Naya Prokash, Kolkata, India, **1998**, 49.

Paranjipe P. *Indian Medicinal Plants. Forgotten Healers*. Chaukhamba Sanskrit Pratishthan, Delhi, India, **2005**, 12–13.

Rameshwar, R.D., Akito, N. Three oleanoliic acid glycosides from medicinally important seeds of Achyranthes aspera. *Nat. Prod. Commun.* **2007**, 2, 727–730.

Rani N., Sharma S.K., Vasudeva N. Assessment of antiobesity potential of *Achyranthes aspera* Linn. seed. *Evid. Based Complement. Alternat. Med.* **2012**, 715912.

Riazudin A.S., Siddiq M. *Vedic Plants. Medicinal and Other Uses*. Chaukhamba Orientalia, Varanasi, India, **2009**.

Sairam T.V. *Home Remedies,* vol. I. Penguin Books, Delhi, India, **1998**, 8–16.

Sharma R.K., Bagwan Dash. *Charaka Samhita*. Chowkhamba Sanskrit Series, Varanasi, India, **1976**, Sutrasthana A-4.

Sharma V., Agarwal A., Chaudahary U., Singh M. Phytochemical investigation of various extracts of leaves and stems of *Achyranthes aspera* Linn. *Intl. J. Pharm. Pharm. Sci.* **2013**, *5*(1), 317–320.

Srivastav S., Singh P., Mishra G., Jha K.K., Khosa R.L. *Achyranthes aspera*—An important medicinal plant: A review. *J. Nat. Prod. Plant Resour.* **2011**, *1*(1), 1–14.

Subbarayan P.R., Sarkar M., Nathanson L. *et al.* In vitro global gene expression analyses support the ethnopharmacological use of *Achyranthes aspera. Evid. Based Complement. Alternat. Med.* **2013**, Article ID 471739, 13pp.

Talukder F.Z., Khan K.A., Uddin R., Jahan N., Alam M.A. In vitro free radical scavenging and anti-hyperglycemic activities of *Achyranthes aspera* extract in alloxan-induced diabetic mice. *Drug Discov Ther.* **2012**, *6*(6), 298–305.

Udayan P.S., Balachandran I. *Medicinal Plants of Arya Vaidyasala Herb Garden*. Centre for Medicinal Plants Research, Arya Vaidya Sala, Kottakal, India, **2009**, 25.

Vedavathy S., Mrudula V., Sudhakar A. *Tribal Medicine of Chittor District, A.P. (India).* Herbal Folklore Research Centre, Tirupati, India, **1997**, 16–17.

Warrier P.K., Nambiar V.P.K., Ramankutty C. (eds.) *Indian Medicinal Plants*. Orient Longman, Chennai, India, **1994**, *1*, 39.

10.2.14 Pumpkin

Scientific name: *Cucurbita maxima* Duchesne ex Lam.
Family: Cucurbitaceae
Hindi: Kaddu
Sanskrit: Pitakusmandah
Tamil: Pusinikkay

10.2.14.1 Origin and Distribution

Pumpkins, like other squash, are probably native to North America. Pumpkin-related seeds dating between 7000 and 5500 BC were found in Mexico. Pumpkins are widely grown in many countries; the biggest international producers are the United States, Canada, Mexico, India, and China.

The word *pumpkin* originates from the Greek word *pepon*, meaning "large melon." Pumpkin in Nepal symbolizes a son, so women cultivate it near the houses and are not allowed to cut it into two when it is whole. It is considered sacred and often used during Hindu rituals (Majupuria and Joshi, 1997).

10.2.14.2 Features

It is a trailing annual plant with somewhat prickly or hairy stem; simple, alternate, lobed leaves with deep sinus at the base; large yellow flowers; and fleshy fruits (Warrier *et al.*, 1994). The fruits have a thick shell containing the seeds and pulp.

10.2.14.3 Properties and Uses in Ayurvedic System of Medicine

Pumpkin, well cooked and spicy, is an excellent tonic for the nervous system, recommended by Ayurvedic physicians through autumn as it will build up *ojas* and help to keep the body healthy during the winter. It is beneficial for the three *dosha*s.

The *Bhojanakutuhalam* reports that among fruits growing on creepers, pumpkin is the best one. It cures urinary obstruction and several renal calculitis, reduces fluid quantity in urine and feces, mitigates thirst, and nourishes weakened limbs; it is aphrodisiac and very sweet, and treats loss of taste. Tender pumpkin alleviates *pitta* and aggravates *kapha* in its intermediate state of maturity. The complete mature fruit is light, hot in potency, cordial, and wholesome for those suffering from diseases of semen. It is alkaline, stimulates the digestive fire, cleanses the bladder, and alleviates all *dosha*s (Venkat *et al.*, 2012).

The seeds have a healing effect on the liver, gall bladder, colon, spleen, pancreas, and stomach (Nibodhi, 2010).

Parts used: Fruit, seeds, leaves
Taste: Sweet
Quality: Heavy
Potency: Hot
***Dosha* effect:** VPK= (in excess the seeds can aggravate *vata*)
Actions: The fruits are refrigerant, emollient, diuretic, sedative, and tonic. The seeds are anthelmintic, diuretic, and nervine tonic.

10.2.14.4 Traditional Uses and Folklore

The seeds are good for treating painful gout and for removing parasites and worms (Nibodhi, 2010). To remove tapeworms, they are given at bedtime with sugar, followed on the next morning by a dose of castor oil. The dried pulp in the form of confection is a home remedy for hemorrhages from the pulmonary organs (Nadkarni, 2002).

10.2.14.5 In the Kitchen

Most parts of the pumpkin are edible: the fleshy shell, the seeds, the leaves, and even the flowers. There are a lot of recipes around the world: the ripe pulp can be boiled, baked, steamed, or roasted as a salty side dish or cooked with sugar as a dessert. In North America, it is a common typical autumn dish, mashed or as soup and purées. Pumpkin pie is a traditional part of thanksgiving meals.

A very famous sweet delicacy from the Middle East is the *halawa yaqtin*. In India, pumpkin is cooked with ghee, sugar, and spices in a dish called *kadu ka halwa*, and it is also an ingredient of the soup called *sambar*, a typical side dish of the South India. In Italy, it is an ingredient of a savory stuffing for *ravioli*.

Pumpkin leaves are a popular vegetable in the western and central regions of Kenya.

In the southwestern United States and Mexico, pumpkin flowers are used to garnish dishes or are fried in oil. Pumpkin seeds are a popular and healthy snack in many countries.

10.2.14.6 Chemical Constituents and Uses as Functional Food

Pumpkin is an annual vine; it is cultivated from sea level to high altitudes in different parts of the world. Three pumpkin species, *Cucurbita pepo* L., *Cucurbita maxima* Duchesne, and *Cucurbita moschata* Duchesne, represent economically important species cultivated worldwide and have high production yields.

Most parts of the pumpkin are edible, from the fleshy shell to the seeds. They are cooked and consumed in many ways.

Pumpkins have considerable variation in nutrient contents depending on the cultivation environment, species, or part. The fruit contains polysaccharides, pectin, carotenoids, proteins, and peptides. It has a good β-carotene content and a moderate content of carbohydrates, vitamins, and minerals (Caili *et al.*, 2006; Yadav *et al.*, 2010).

Polysaccharides, including protein-bound polysaccharides, are the bioactive materials of pumpkin fruit. Various hypoglycemic polysaccharides have been characterized from the fruit pulps (Jun *et al.*, 2006).

In various reports, the pumpkin exhibited acute hypoglycemic activity (blood sugar lowering) in temporarily hyperglycemic rabbits, in alloxan-induced diabetic rabbits (Quanhong *et al.*, 2005), and in type 2 diabetic patients. Pumpkin has the hypoglycemic activity like a standard drug (tolbutamide) in healthy animals with temporary hyperglycemia and in mild diabetic animals, but not in severe diabetic animals (reviewed in Yadav *et al.*, 2010).

D-*chiro*-Inositol was identified in pumpkin (especially in *Cucurbita ficifolia*), and this compound has been considered as an insulin action mediator (insulin sensitizer) (Xia and Wang, 2006).

Local healers recommend the ingestion of crude aqueous extract of pumpkin fruits for the treatment of type 2 diabetes or noninsulin-dependent diabetes mellitus (Kwon *et al.*, 2007). By considering all these facts, it can be concluded that pumpkin has potential antidiabetic properties, which may suggest the inclusion of this plant in antidiabetic regimens to treat human diabetes.

Pumpkin seeds are of nutritional and medicinal value. They contain all nine essential amino acids. The histidine, leucine, and valine contents are higher than the other essential amino acids. *Cucurbita pepo* contains high amounts of arginine.

The major fatty acids in pumpkin are palmitic (C16:0), stearic (C18:0), oleic (C18:1), and linoleic (C18:2) acids (Kim *et al.*, 2012). Pumpkin seeds contain high amount of β-sitosterol (more than 240 mg/kg). β-Sitosterol might have broad biological effects, including lowering cholesterol, estrogenic activity, and anticarcinogenic activity (Barakat and Mahmoud, 2011; Piironen *et al.*, 2000).

Diets high in pumpkin seeds have also been associated with lower risk of gastric, breast, lung, and colorectal cancers. There are also potential health benefits, including anticarcinogenic effects, to be gained from the various carotenoid pigments found in pumpkin seed oil. The carotenoids from pumpkin fruits have been linked to the prevention of prostate cancer (Jian *et al.*, 2005). It is suggested that the administration of pumpkin seed oil and saw palmetto oil are clinically safe and may be effective as complementary and alternative medicine treatments for benign prostatic hyperplasia (Hong *et al.*, 2009).

Pumpkin soup

Ingredients

1 large pumpkin
1 1/2 cups water
1 tsp of ghee or olive oil
1/2 cup chopped ginger
1 tsp of cumin
1 tsp of curry powder
1/2 tsp of turmeric powder
Aromatic salt to salt
1 cup of milk
Nutmeg for garnish

Preparation

Preheat oven to 200°C. Wash the pumpkin, cut in half, and remove seeds. Place the pumpkin halves face down in a baking pan. Pour in the water and cover the pan with foil. Bake for 30 minutes. Let the pumpkin cool and then remove the pulp. Place it into a bowl. Heat a soup pot over medium heat and add the oil or ghee. Add ginger and cumin. Add three cups of warm water; add the aromatic salt as for your taste. Add the curry and turmeric powder and three cups of pumpkin pulp. Bring to a boil. Reduce the heat and simmer for about 10 minutes. Using a food processor, purée the soup to a smooth and creamy consistency. Return the soup to the pot, add the milk, and reheat if necessary. Add some black pepper and garnish with a sprinkle of nutmeg. A vegan version can be made without milk or with soya milk.

This recipe is an excellent *tridoshic rasayana* suggested in the evening during the autumn season.

REFERENCES

Barakat L.A., Mahmoud R.H. The antiatherogenic, renal protective and immunomodulatory effects of purslane, pumpkin and flax seeds on hypercholesterolemic rats. *N. Am. J. Med. Sci.* **2011**, *3*(9), 411–417.

Caili F., Huan S., Quanhong L. A review on pharmacological activities and utilization technologies of pumpkin. *Plant Foods Hum. Nutr.* **2006**, *61*(2), 73–80.

Hong H., Kim C.S., Maeng S. Effects of pumpkin seed oil and saw palmetto oil in Korean men with symptomatic benign prostatic hyperplasia. *Nutr. Res. Pract.* **2009**, *3*(4), 323–327.

Jian L., Du C.J., Lee A.H. *et al.* Do dietary lycopene and other carotenoids protect against prostate cancer? *Int. J. Cancer* **2005**, *113*, 1010–1014.

Jun H.I., Lee C.H., Song G.S. *et al.* Characterization of the pectic polysaccharides from pumpkin peel. *Food Sci. Tech.* **2006**, *39*, 554–561.

Kim M.Y., Kim E.J., Kim Y.N., Choi C., Lee B.H. Comparison of the chemical compositions and nutritive values of various pumpkin (Cucurbitaceae) species and parts. *Nutr. Res. Pract.* **2012**, *6*(1), 21–27.

Kwon Y.I., Apostolidis E., Kim Y.C., Shetty K. Health benefits of traditional corn, beans, and pumpkin: In vitro studies for hyperglycemia and hypertension management. *J. Med. Food* **2007**, *10*, 266–275.

Majupuria T.C., Joshi D.P. *Religious and Useful Plants of Nepal and India.* Gupta M.(ed.), Publishers, Lashkar, India, **1997**, 243.

Nadkarni K.M. *Indian Materia Medica*, vol. 1. Popular Prakashan, Mumbai, India, **2002**, 407–408.

Nibodhi. *Annapurna's Prashad.* Mata Amritanandamayi Mission Trust, Amritapuri, India, **2010** (2nd edn.), 132.

Piironen V., Lindsay D.G., Miettinen T.A., Toivo J., Lampi A.M. Plant sterols: Biosynthesis, biological function and their importance to human nutrition. *J. Sci. Food Agric.* **2000**, *80*, 939–966.

Quanhong L.I., Caili F., Yukui R. *et al.* Effects of protein-bound polysaccharide isolated from pumpkin on insulin in diabetic rats. *Plant Food Hum. Nutr.* **2005**, *60*, 13–16.

Venkat P., Gangadharan G.G., Lakshmithathachar M.A., Alwar M.A. (eds.) *Bhojanakutuhalam.* I-AIM, Bangalore, India, **2012**, 125.

Warrier P.K., Nambiar V.P.K., Ramankutty C. (eds.) *Indian Medicinal Plants*, vol. 2. Orient Longman, Chennai, India, **1994**, 238–240.

Xia T., Wang Q. D-chiro-Inositol found in *Cucurbita ficifolia* (Cucurbitaceae) fruit extracts plays the hypoglycaemic role in streptozocin-diabetic rats. *J. Pharm. Pharmacol.* **2006**, *58*, 1527–1532.

Yadav M., Jain S., Tomar R., Prasad G.B.K.S., Yadav H. Medicinal and biological potential of pumpkin: An updated review. *Nutr. Res. Reviews* **2010**, *23*, 184–190.

10.2.15 Spinach

Scientific name: *Spinacia oleracea* L.
Family: Amaranthaceae
Hindi: Palak
Sanskrit: Palankya
Tamil: Pasalai keeray

10.2.15.1 Origin and Distribution

Spinach was probably first cultivated in ancient Persia (modern Iran and neighboring countries) about 2000 years ago. Arab traders carried spinach into India, where it is abundantly grown today. The plant was then introduced into ancient China via Nepal around AD 647. In AD 827, the Saracens introduced spinach to Sicily and the Mediterranean area. From there, the plant spread to other parts of the world (Bakhru, 1994). From the tenth century, we find various records about the plant properties in medical and agricultural Arab treatises. During the twelfth century, the great Arab agronomist Ibn al-'Awwam called it the *captain of leafy greens*. By the fourteenth century, the spinach became very popular in Europe as it appears in the early spring, when other vegetables are still scarce. Catherine de' Medici from Florence insisted it to be served at every meal, so even today dishes made with spinach are known as *Florentine*.

10.2.15.2 Features

The plant is a cold season annual crop with a round, smooth, and delicate stem. The leaves are succulent, fleshy, and dark green in color. The bottom of the leaf is shiny, with thick veins running across (Bakhru, 1994).

10.2.15.3 Properties and Uses in Ayurvedic System of Medicine

Spinach has been used in Ayurvedic medicine for centuries. The leaves and the tender stem, when boiled and seasoned, are an excellent cooling, nutritious, and demulcent dish (Nadkarni, 2002). In the *Bhojanakutuhalam*, we can read that the herbaceous part is slightly pungent, sweet, wholesome, and cold. It treats hemorrhagic tendencies and is highly nourishing (Venkat *et al.*, 2012). It is excellent for the lungs; it helps soothe the mucus membranes and alleviates fever, burning sensations in the lungs, and dry cough (Nibodhi, 2010). Spinach can increase *vata dosha*, so it is recommended to cook it with spices, ginger, garlic, and milk.

Parts used: Leaves, stem
Tastes: Sweet, astringent, bitter, pungent depending on the variety and method of preparation
Quality: Dry
Potency: Cold
***Dosha* effect:** K–, VP+
Actions: Laxative, demulcent, refrigerant

10.2.15.4 Traditional Uses and Folklore

Spinach is an excellent food remedy for constipation, prevention of night blindness, acidosis, and respiratory disorders (Bakhru, 1994).

Precautions: It should not be eaten in cases of prostate enlargement, gallstones or kidney stones, liver disorders, or arthritic conditions (Nibodhi, 2010). Consumption of raw spinach should be avoided as it can lead to problems such as arthritis, rheumatism, and gout due to presence of oxalic acid in it. It is recommended to boil the leaves before consuming it. It is also recommended to buy only organic farming spinach as the plant is much affected by pesticide use.

10.2.15.5 In the Kitchen

Spinach leaves are juicy, so it is better to cook it at a moderate heat without additional water. It is recommended to consume it completely once it is cooked rather than eating the stored vegetable.

A very tasty and healthy recipe is the soup, easy to digest and excellent for constipation. It is also effective in weight reduction due to its property of pacifying *kapha dosha*.

10.2.15.6 Chemical Constituents and Uses as Functional Food

Spinach has a high nutritional value and is extremely rich in antioxidants, especially when fresh, steamed, or quickly boiled. It is a rich source of vitamins A, C, E, K, B_2, B_6, and B_3 (niacin), and minerals such as magnesium, manganese iron, calcium, potassium, phosphorus, zinc, selenium, and copper (USDA Nutrient Database). It is especially high in lutein (a carotenoid).

Polyglutamyl folate (vitamin B_9 or folic acid) is a vital constituent of cells, and spinach is a good source of folic acid. Boiling spinach can more than halve the level of folate left in the spinach, whereas microwaving does not affect folate content. Fresh spinach loses much of its nutritional value with storage of more than a few days. Although refrigeration slows this effect to about 8 days, spinach will lose most of its folate and carotenoid content, so for longer storage, it is blanched and frozen, cooked and frozen.

The calcium as well as iron content is partially vanified by the presence of oxalates that bind these minerals, avoiding their absorption. The calcium in spinach is the least bioavailable of calcium sources. By comparison, the body can absorb about half of the calcium present in broccoli, yet only around 5% of the calcium in spinach.

Broccoli and spinach had the highest amount of free phenolics (81 and 80 mg/100 g of sample, respectively), expressed as milligrams of gallic acid equivalents per 100 g of fresh weight of the edible part of vegetables. The total antioxidant activity of 100 g spinach is equivalent to 737 mg of vitamin C. With this evaluation, spinach ranked first with the highest phenolics antioxidant index, which was calculated by dividing the phenolic content by the median effective concentration (EC_{50}) of the total antioxidant activity after subtracting the contribution from vitamin C (Chu *et al.*, 2002).

Among different dietary vegetables, spinach had the highest antiproliferative capacity with the lowest EC_{50} on the growth of HepG2 human liver cancer.

Purified monogalactosyl diacylglycerol (MGDG), a major glycoglycerolipid from spinach, selectively inhibits the activities of mammalian replicative DNA polymerases (α, δ, and ε). Galactolipids are a class of compounds widely found in the plant kingdom, including edible plants, and are an important part of the cell membranes. Galactolipids in plants consists mainly of MGDGs and digalactosyldiacylglycerols containing one or two saturated and/or unsaturated fatty acids linked to the glycerol moiety. Several galactolipids have been shown to possess *in vitro* and/or *in vivo* antitumor-promoting activity and anti-inflammatory activity. However, the function of MGDG following ingestion is not well known (Akasaka *et al.*, 2013; Maeda *et al.*, 2013).

The chemopreventive effects of dietary spinach on microRNA (miRNA or miR) changes in rat colon tumors induced by a cooked meat carcinogen has been recently described (Parasramka *et al.*, 2012), by influencing many biological pathways. miRNAs are small, noncoding RNAs that act at the posttranscriptional level to degrade or repress translation from target mRNAs. They are implicated in several human pathologies, including cancer, and are influenced by lifestyle factors such as cigarette smoke exposure and the intake of dietary chemopreventive agents. Although the precise mechanisms await further study, the current investigation provides further support for research at the interface of epigenetics, diet, and cancer prevention.

Spinach soup

Ingredients

1 onion
2 tomatoes
1 bundle of spinach
1/2 tsp of ginger paste
1/4 tsp of turmeric powder
4 garlic flakes (finely minced)
1/2 cup of milk
1/2 tsp of ghee
Aromatic salt to taste

Preparation

After washing carefully the spinach, cut the onion, tomatoes, and spinach. Boil the chopped onion and tomatoes for about 10 minutes. Add the spinach and the ginger paste and cook on a low fire for 10 minutes. Blend together into a smooth pulp. Heat the ghee and fry the minced garlic for a minute. Add the pulp, salt, and turmeric, and bring to boil. Add milk and cook again for few minutes on low heat.

Instead of milk you can add small pieces of paneer fried with few cumin seeds and a tsp of turmeric to have a delicious side dish to eat with rice or chapattis. A vegan option can be made with tofu and oil instead of paneer and ghee.

REFERENCES

Akasaka H., Sasaki R., Yoshida K. *et al.* Monogalactosyl diacylglycerol, a replicative DNA polymerase inhibitor, from spinach enhances the anti-cell proliferation effect of gemcitabine in human pancreatic cancer cells. *Biochim. Biophys. Acta* **2013**, *1830*(3), 2517–2525.

Bakhru H.K. *Foods That Heal*. Orient Paperbacks, Delhi, India, **1994**, 145–147.
Chu Y.F., Sun J., Wu X., Liu R.H. Antioxidant and antiproliferative activities of common vegetables. *J. Agric. Food Chem.* **2002**, *50*(23), 6910–6916.
Maeda N., Kokai Y., Hada T., Yoshida H., Mizushina Y. Oral administration of monogalactosyl diacylglycerol from spinach inhibits colon tumor growth in mice. *Exp. Ther. Med.* **2013**, *5*(1), 17–22.
Nadkarni K.M. *Indian Materia Medica*, vol. 1. Popular Prakashan, Mumbai, India, **2002**, 1164–1165.
Nibodhi. *Annapurna's Prashad*. Mata Amritanandamayi Mission Trust, Amritapuri, India, **2010** (2nd edn.), 108–109.
Parasramka M.A., Dashwood W.M., Wang R. *et al.* MicroRNA profiling of carcinogen-induced rat colon tumors and the influence of dietary spinach. *Mol. Nutr. Food Res.* **2012**, *56*, 1259–1269.
USDA Nutrient Database. http://ndb.nal.usda.gov/.
Venkat P., Gangadharan G.G., Lakshmithathachar M.A., Alwar M.A. (eds.) *Bhojanakutuhalam*. I-AIM, Bangalore, India, **2012**, 109.

10.2.16 TOMATO

Scientific name: *Lycopersicum esculentum* Mill.
Family: Solanaceae
Hindi: Tamatar
Sanskrit: Raktamaci
Tamil: Takkali

10.2.16.1 Origin and Distribution

The botanical name *Lycopersicum* means "wolf peach," and it is related to the German legends: the nightshade fruits were believed to be used by witches and sorcerers in potions to transform themselves into werewolves. The fruit was called by the Aztecs *xitomatl*, and from this name derived the majority of western European names.

The plant is probably native to Mexico, where it was already cultivated and used as food by Aztecs by 500 BC. During the colonization, the Spanish spread the cultivation throughout their colonies in the Caribbean and in the Philippines. They also brought the tomato to Europe, where it grew easily in the Mediterranean climates.

Pietro Andrea Mattioli, an Italian physician and botanist, described for the first time tomato as a food in a script written in 1544. After 10 years, he named it as *pomi d'oro*, or golden apple. Until the late seventeenth or the early eighteenth century, the fruit was sought mainly for its beauty and used in many areas solely as a tabletop decoration. After 1860, the use of tomato as food became very popular and the cultivation was spread throughout the entire world, with hundreds of different varieties. At present, tomato holds the second place among the vegetables produced in the world after potato (Bakhru, 1994). It is widely cultivated in Europe, India, China, Malaysia, Indonesia, the Philippines, Africa (Central, East, and West), the United States, tropical America, and the Caribbean. The largest producer is China, followed by the United States and India. In Europe, there are some areas that grow high-quality tomatoes with a Protected Geographical Status: *Pomodoro di Pachino* from Sicily, *Pomodoro San Marzano* from Agro Sarnese Nocerino, *Pomodoro del Piennolo del Vesuvio* from the Vesuvius area, and *Tomaten von der Insel Reichenau* from Reichenau Island in Germany are some of these special varieties.

10.2.16.2 Features

Lycopersicum esculentum is an unarmed herbaceous spreading pubescent annual plant with leaves unevenly pinnate, curled, acute, or acuminate, yellow flowers, fruits globose, red or yellowish red on ripening (Warrier *et al.*, 1995).

10.2.16.3 Properties and Uses in Ayurvedic System of Medicine

The tomato is an excellent appetizer, helpful in digestion, with curative effects in low appetite and stomach ache. It is recommended in case of piles, jaundice, weakness, and constipation. When well cooked, it is useful for people having excessive *vata* and *kapha*. Raw tomatoes heat the stomach and intestines, whereas cooked tomatoes clear heat from the body and help detoxify the blood (Nibodhi, 2010).

As the postdigestive taste is sour, in excess it can aggravate *pitta* and *kapha*. *Vata* is aggravated by peel and seeds, so it is always better to remove them (Morningstar and Desai, 1999).

Ayurveda recommends eating tomatoes only when fully ripe. Tomatoes are invaluable in case of tendency of biliousness, as they promote the flow of bile; they are also an excellent corrective for kidneys; a gentle, natural stimulant; and very helpful in washing away the poison that contaminates our systems (Nadkarni, 2002).

Parts used: Fruits
Tastes: Sweet, sour, astringent
Quality: Light
Potency: Hot
***Dosha* effect:** The raw tomato increases all *dosha*s; the cooked one reduces *vata* and *kapha*.
Actions: Emollient, carminative, digestive, depurative, intestinal antiseptic, liver and kidney stimulant and tonic

10.2.16.4 Traditional Uses and Folklore

- The tomato is recommended as food for diabetic patients. One or two ripe tomatoes taken in the early morning as breakfast for few months help to reduce bodyweight (Bakhru, 1994).
- Raw tomato sprinkled with salt and dry ginger increases appetite.
- The soup cures chronic constipation.

Precautions: Tomatoes should be avoided in case of preexisting acidity, gout, sciatica, arthritis, kidney stones, and enlarged prostate.

10.2.16.5 In the Kitchen

The tomato can be consumed raw or as an ingredient in many dishes, soups, sauces, and drinks. It is used extensively in the Mediterranean cuisines as a key ingredient in pizza and pasta sauces. Although it is botanically a fruit, it is considered a vegetable for culinary purposes. It can be preserved dry or as pickle. It is not recommended to refrigerate as this can harm the flavor.

10.2.16.6 Chemical Constituents and Uses as Functional Food

The most abundant carotenoid in tomato is lycopene, followed by phytoene, phytofluene, *z*-carotene, γ-carotene, β-carotene, neurosporene, and lutein, which are known antioxidants, acting synergistically. The antioxidant activity of carotenoids is based on their singlet oxygen-quenching activity. Tomatoes are the most concentrated food source of lycopene, although apricots, guava, watermelon, papaya, and pink grapefruit are also significant sources. Lycopene levels in the blood were higher after eating cooked tomatoes than after eating raw tomatoes or drinking tomato juice. This suggests that lycopene in cooked tomatoes such as tomato sauce or paste may be more readily absorbed by the body. Being lycopene a fat-soluble pigment, its absorption by the intestines is higher if consumed with a small amount of oil or fat (e.g., salad oil or cheese on pizza). Lycopene is generally stable to processing when present in the plant tissue matrix and less when isolated as pure compound. Lycopene is also available in soft gelatin capsules and liquid supplements (Kachik *et al.*, 2002).

Recently, lycopene has been studied in relation to its potential health effects. Although promising data from epidemiological as well as cell culture and animal studies suggest that lycopene and the consumption of lycopene-containing foods may affect cancer or CVD risk, more clinical trial data are needed to support this hypothesis (reviewed in Story *et al.*, 2010).

Lycopene metabolism has been studied, concluding that its biological activities can be mediated by apo-10′-lycopenoids formed by endogenous cleavage by carotenoid cleavage oxygenases (Blanche and Xiang-Dong, 2014; Wang, 2012).

REFERENCES

Bakhru H.K. *Foods That Heal*. Orient Paperbacks, Delhi, India, **1994**, 148–151.

Blanche C.I., Xiang-Dong W. Non-alcoholic steatohepatitis and hepatocellular carcinoma: Implications for lycopene intervention. *Nutrients* **2014**, *6*, 124–162.

Khachik F., Carvalho L., Bernstein P.S., Muir G.J., Zhao D.Y., Katz N.B. Chemistry, distribution, and metabolism of tomato carotenoids and their impact on human health. *Exp. Biol. Med. (Maywood)* **2002**, *227*(10), 845–851.

Morningstar A., Desai U. *The Ayurvedic Cookbook*. Motilal Banarsidass Publishers, Delhi, India, **1999**.

Nadkarni K.M. *Indian Materia Medica*, vol. 1. Popular Prakashan, Mumbai, India, **2002**, 756–758.

Nibodhi. *Annapurna's Prashad*. Mata Amritanandamayi Mission Trust, Amritapuri, India, **2010** (2nd edn.), 111–112.

Story E.N., Kopec R.E., Schwartz S.J., Harris G.K. An update on the health effects of tomato lycopene. *Annu. Rev. Food Sci. Technol.* **2010**, *1*, 189–210.

Wang X.D. Lycopene metabolism and its biological significance. *Am. J. Clin. Nutr.* **2012**, *96*(5), 1214S–1222S.

Warrier P.K., Nambiar V.P.K., Ramankutty C. (eds.) *Indian Medicinal Plants*, vol. 3. Orient Longman, Chennai, India, **1995**, 354–357.

10.3 NUTS AND SEEDS

10.3.1 Properties and Uses in Ayurvedic System of Medicine

Nuts should not be taken in large amounts as they are quite heavy to digest. The high oil content can disturb *pitta* and *kapha*, but it is recommended for *vata*; therefore, nuts are excellent for *vata* constitution, good in small quantity for *pitta*, and less beneficial for *kapha*. Generally, all nuts and seeds need to be chewed very well, and the kind of preparation can change the effect in the different constitutions. Lightly roasting makes them easier to digest, helpful also in releasing more nutrients and preserving them longer (Smith, 2011).

Almost all nuts and seeds are nutritive, strengthening, rejuvenating, warming, and sweet in taste. They are considered a *sattvic* food that builds *ojas*, perfect in a yogic diet (Nibodhi, 2010).

Ayurveda recommends taking nuts daily, but in small quantity, to rejuvenate the body and to increase the memory.

Precautions: Nuts and seeds are extremely susceptible to pesticide accumulation, so it is necessary to buy only organic products. They need to be fresh and well preserved. Overcooking can aggravate *kapha* (Smith, 2011).

10.3.2 Chemical Constituents and Uses as Functional Food

Like other foods, nuts have a wide variety of cultural connections to the areas where they grow and to the people who live there or eat them. History, symbolism, and legends reveal the ancient tradition of nuts and how they are related to the lives of our ancestors. Archeological excavations in eastern Turkey have discovered the existence of a nonmigratory society whose economy centered on harvesting nuts. This shows that nuts have been a staple in the human diet since the beginning of history. Moreover, since ancient times, nuts have been used for their medicinal properties (Casas-Agustench *et al.*, 2011).

Nuts are nutrient-dense foods with complex matrices rich in unsaturated fatty acids and other bioactive compounds, such as L-arginine, fiber, minerals (magnesium, potassium, and copper), vitamins (vitamins E and B_6, folic acid, and niacin), phytosterols, and polyphenols. Phytosterols in nuts range from 95–280 mg/100 g. α- and β-Carotene, β-cryptoxanthin, lutein, and zeaxanthin are found in mg/100 g amounts in some nuts but at 1–3 mg/100 g in pistachios and none at all in Brazil nuts, macadamia nuts, and peanuts. Walnuts are particularly rich in total phenols (including phenolic acids, flavonoids, and stilbenes) with 1625 mg gallic acid equivalents per 100 g. The stilbene resveratrol is found in peanuts and pistachios at 84 and 115 μg/100 g, respectively. The flavonoid content of nuts as provided in the USDA Database for the Flavonoid Content of Selected Foods lists totals in pecans at 34 mg/100 g, almonds at 15 mg/100 g, and pistachios and hazelnuts at 12 mg/100 g. Proanthocyanidins are found in almonds, cashews, hazelnuts, pecans, pistachios, peanuts, and walnuts, with concentrations varying from 9 to 494 mg/100 g (Chen and Blumberg, 2008). Nut phytochemicals have been associated with numerous bioactivities known to affect the initiation and progression of several pathogenic processes.

In general, nuts are energy dense and provide 23.4–26.8 kJ/g of food with a high fat content (45%–75% of weight), but mostly unsatured fat. Walnuts are especially rich in linoleic acid, α-linolenic acid, and PUFAs (Vadivel *et al.*, 2012).

Frequent nut consumption has been connected to decreased incidences of several chronic diseases, including hypertension, hypercholesterolemia, coronary heart diseases, and obesity (Ros *et al.*, 2010); hence, frequent nut consumption is recommendable to improve the health status. Although nuts have a high-fat content, most of them consist of unsaturated fats, which are poorly absorbed and actually induce energy expenditure by thermogenesis. Further, because of the high-energy density and protein and fiber content, nuts increase satiety.

The pistachio is a nutrient-dense nut with a heart-healthy fatty acid profile as well as protein, dietary fiber, potassium, magnesium, vitamin K, and γ-tocopherol. Its unique green and purple kernel color is a result of its lutein and anthocyanin content. Among nuts, pistachios contain the highest levels of potassium, γ-tocopherol, vitamin K, phytosterols, and xanthophyll carotenoids. Five published randomized cardiovascular trials have shown that pistachios promote heart-healthy blood lipid profiles. Exploratory clinical studies suggest that pistachios help to maintain healthy antioxidant and anti-inflammatory activities, glycemic control, and endothelial function (Dreher, 2012).

REFERENCES

Casas-Agustench P., Salas-Huetos A., Salas-Salvadó J. Mediterranean nuts: Origins, ancient medicinal benefits and symbolism. *Public Health Nutr.* **2011**, *14*(12A), 2296–2301.

Chen C.Y., Blumberg J.B. Phytochemical composition of nuts. *Asia Pac. J. Clin. Nutr.* **2008**, *17*(Suppl 1), 329–332.

Dreher M.L. Pistachio nuts: Composition and potential health benefits. *Nutr Rev.* **2012**, *70*(4), 234–240.

Nibodhi. *Annapurna's Prasad.* Mata Amritanandamayi Mission Trust, Amritapuri, India, **2010**, 126.

Ros E., Tapsell L.C., Sabaté J. Nuts and berries for heart health. *Curr. Atheroscler. Rep.* **2010**, *12*(6), 397–406.

Smith V.A. *Ayurvedic Nutrition.* Motilal Banarsidass Publishers, Delhi, India, **2011**, 126–128.

Vadivel V., Kunyanga C.N., Biesalski H.K. Health benefits of nut consumption with special reference to body weight control. *Nutrition* **2012**, *28*(11/12), 1089–1097.

10.4 NUTS

10.4.1 Almond

Scientific name: *Prunus amygdalus* Batsch
Family: Rosaceae
Hindi: Badam
Sanskrit: Vatadah
Tamil: Badamkottai

10.4.1.1 Origin and Distribution

Almond is considered to be one of the earliest domesticated tree nuts as it appears in the Early Bronze Age (3000–2000 BC). Wild almond species are toxic, whereas domesticated almonds are not: this is probably due to a genetic mutation created by early farmers, "at first unintentionally in the garbage heaps, and later intentionally in their orchards" (Diamond, 1997).

The plant is native to the Mediterranean climate region of the Middle East and was spread by humans in ancient times along the shores of the Mediterranean into North Africa and southern Europe. It has been more recently transported to the United States. It is now cultivated in Kashmir (India), Baluchistan (Pakistan), Afghanistan, Iran, Morocco, southern Africa, Cyprus, Turkey, Spain, Portugal, Italy, Australia, and California (the United States). The United States are actually the largest producer in the word. According to Ayurvedic scholars, the best variety is from Iran.

The Bible describes the almond as one of the best fruits, a symbol of the inner divine light. The way to the immortality is under the feet of an almond tree. Christian paintings often include almonds encircling Jesus and the Virgin Mary, as a symbol of their aura (Brosse, 1985).

For the ancient Greeks, the almond was a symbol of love and truth. They believed in its supernatural power (Sairam, 1998). Therefore, the Romans called it the *Greek nut* (Bakhru, 1994).

In the Greek mythology, the almond is the seed spread by Zeus, from which was born a hermaphroditic deity named Agdistis. The other gods castrated him, and from his blood an almond tree was grown. One of the fruits impregnated the virgin Nana, who conceived Attis. The early flowering of the almond tree was for the Greeks, a symbol of a virginal love (Brosse, 1985).

10.4.1.2 Features

It is a small tree, 5–6 m high, with bright green leaves and pale pink or white flowers. There are several varieties of almond, but broadly it can be classified into two: bitter and sweet. The bitter variety is poisonous and is used only for external use and for producing almond oil, perfumes, and cosmetics (Warrier *et al.*, 1995). According to Hebraic tradition, the road of Aaron bore sweet almonds on one side and bitter on the other: if the Israelites followed the Lord, the sweet almonds would be ripe and edible, but if they were to forsake the path of the Lord, the bitter almonds would predominate.

10.4.1.3 Properties and Uses in Ayurvedic System of Medicine

Almond is a *rasayana*, an effective health-building food, for both body and mind. Per the Ayurvedic concept, dry fruits increase *ojas*, the vital energy, and out of all dry fruits, almond is a king, being highly nutritious. It suppresses *pitta* and *vata*; it is nutritive for the brain and improves eyesight. The Jain monks of Karnataka mention the relation between consumption of almond and longevity (Puri, 2003).

Parts used: Kernel, oil
Tastes: Sweet, bitter
Potency: Hot
***Dosha* effect:** VP–, K+ (in excess)
Actions: Revitalizing, general tonic, laxative, nutritive, aphrodisiac

10.4.1.4 Traditional Uses and Folklore

- The best method to digest almonds is to soak the kernel in water overnight. In the morning, separate the outer seed coat and mash the white kernel in a blender to get a thick milky paste (Puri, 2003). This paste is called almond butter, an excellent tonic especially for the elderly.
- As a brain tonic, steep in water or orange or lime juice three parts almond, two parts walnut, and two parts pine kernel until soft and make a cream in a blender. Almonds mixed with figs are laxative and relieve pain in the bowels (Nadkarni, 2002).
- For constipation, grind separately five almonds and five dried dates and mix with 10 teaspoons of honey. Take three teaspoons of this mixture twice a day (Sairam, 1998).
- To enhance their assimilation, almonds are taken in combination with other spices such as black pepper, turmeric, and cardamom (Sairam, 1998).

Precautions: It is always better to remove the outer skin as it heats and can create problems during the summer.

10.4.1.5 In the Kitchen

Almond seed kernel is often adulterated with that of apricot, but the two have a different shape and size, so they can be easily differentiated.

Almond seed kernel contains heat-sensitive compounds, so it should not be heated while making any of its preparation (Puri, 2003). Almond should always be blanched in hot water as the skin is indigestible (Nadkarni, 2002).

Almonds are used to prepare delicious desserts all over the world. In India, the nuts are the base ingredients of *pasanda*-style curries. The recipes including almonds effectively stimulate a sluggish digestive tract (Morningstar *et al.*, 1999).

10.4.1.6 Chemical Constituents and Uses as Functional Food

The seed is composed of triglycerides (45%–50%), proteins (20%–25%), mineral salts (2%–3%), cellulose fiber (4%–5%), mono- and disaccharides (5%–6%), and vitamins (A, B_1, B_2, B_3, B_6). It also contains an enzymatic complex, including emulsin (β-glucosidase), amylase, and oxynitrilase.

The seed of the wild form contains the glycoside amygdalin, which upon the action of β-glucosidase is decomposed into the essential oil benzaldehyde, the extremely toxic cyanide, and two molecules of glucose (Moetel *et al.*, 1981).

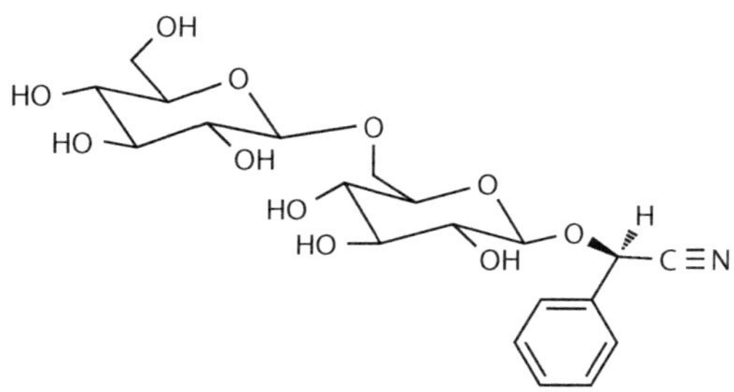

Amydgalin

A pale yellow, sweetish oil is obtained from seeds by pressure. Bitter almonds contain the cyanogenic glycoside amygdalin, but the oil does not contain this substance (Dicenta *et al.*, 2002). Therefore, both sweet and bitter almonds are indicated by Pharmacopeia for oil production. The oil has a slight odor and a nutty taste. The chief constituents of the lipidic fraction are oleic (75% of the total lipids), linoleic (20%), palmitic (3%), stearic (1.5%), and linolenic acid (0.5%) (Sathe *et al.*, 2008). The fatty acid profile along with the presence of other nutrients (phytosterols) is predictive of the LDL-lowering benefits of almonds (reviewed in Berryman *et al.*, 2011). Phytosterols act as an antagonist for the absorption of dietary cholesterol through competition for incorporation into mixed micelles in the gut (Calpe-Berdiel *et al.*, 2009).

Among nuts, almonds have the highest fiber content and are considered a good source of dietary fiber, providing approximately 12% of the daily recommended amount of fiber per serving. Although almonds are recognized for their unique fatty acid profile, approximately 15% of their energy is protein, making almonds a good protein source.

REFERENCES

Bakhru H.K. *Foods That Heal*. Orient Paperbacks, Delhi, India, **1994**, 181–184.

Berryman C.E., Preston A.G., Karmally W., Deckelbaum R.J., Kris-Etherton P.M. Effects of almond consumption on the reduction of LDL-cholesterol: A discussion of potential mechanisms and future research directions. *Nutr. Rev.* **2011**, *69*(4), 171–185.

Brosse J. *Les arbres de France*. Librarie Plon, Paris, France, **1985** (Italian version Ed. Studio Tesi, 1991, 123–126).

Calpe-Berdiel L., Escolà-Gil J.C., Blanco-Vaca F. New insights into the molecular actions of plant sterols and stanols in cholesterol metabolism. *Atherosclerosis* **2009**, *203*, 18–31.

Diamond J. *Guns, Germs and Steel. The Fates of Human Societies*. W.W. Norton & Company, New York, **1997** (Italian version, Einaudi ed., 89–90).

Dicenta F., Martinez-Gomez P., Grane N. *et al.* Relationship between cyanogenic compounds in kernels, leaves, and roots of sweet and bitter kernelled almonds. *J. Agric. Food Chem.* **2002**, *50*, 2149–2152.

Moetel C.G., Ames M.M., Kovach J.S. A pharmacologic and toxicological study of amygdalin. *J. Am. Med. Ass.* **1981**, *245*, 591–594.

Morningstar A., Desai U. *The Ayurvedic Cookbook*. Motilal Banarsidass Publishers, Delhi, India, **1999**, 227.

Nadkarni K.M. *Indian Materia Medica*, vol. 1. Popular Prakashan, Mumbai, India, **2002**, 1010–1013.

Puri H.S. *Rasayana. Ayurvedic Herbs for Longevity and Rejuvination*. Taylor & Francis, New York, **2003**, 59, 63.

Sairam T.V. *Home Remedies*, vol. I. Penguin Books, Delhi, India, **1998**, 224–230.

Sathe S.K., Seeram N.P., Kshirsagar H.H., Heber D., Lapsley K.A. Fatty acid composition of California-grown almonds. *J. Food Sci.* **2008**, *73*, C607–C614.

Warrier P.K., Nambiar V.P.K., Ramankutty C. (eds.) *Indian Medicinal Plants*, vol. 4. Orient Longman, Chennai, India, **1995**, 363–365.

10.4.2 Coconut

Scientific name: *Cocos nucifera* L.
Family: Arecaceae
Hindi: Nariyal-ka-ped
Sanskrit: Narikelah
Tamil: Thengai

10.4.2.1 Origin and Distribution

The botanical name *Cocos* comes from the sixteenth-century Portuguese *coco*, meaning "head" or "skull," due to the shape of the coconut shell that resembles a face with its three small holes similar to the eyes. In Bengal, it is said that because the coconut has three eyes, it will never fall on a person head (Rajaiah, 2010). As God Shiva has also three eyes, coconut is considered a symbol of the Lord (Gupta, 2001). The fruit is a very auspicious essential part of every Hindu ritual all over India. It is customary to break a coconut in front of a deity at the start of something important (a travel, a business, etc.) as a way to surrender the ego represented by the nut (Rajaiah, 2010).

> Coconut adds beauty and prosperity to a house.
>
> ***Matsya Purana*, 255, 24**

The *Veda*s describe the plant as *kalpavriksha*, a wish-granting heavenly tree, brought to the earth by Vishnu. It is also called *Karpagavalli*, which means giver of health, energy, strength, longevity, and peace (Bakhru, 1994). In the Philippines, the coconut is commonly called the *tree of life* as it provides not only a nutritious food, but everything we need in life, such as medicine, fuel, furniture, and even a shelter.

There is a debate about the exact place of origin of coconut. Many authorities suggest an Indo-Pacific origin either around Melanesia and Malaysia or the Indian Ocean, whereas others see the origin in northwestern South America. The nut can float on the high sea for long periods of time, covering a very long distance, and still germinate when in the dry land. The cultivation spread across the tropics, especially throughout India, the Maldives, the Philippines, Sri Lanka, Indonesia, Australia, and Central and South America (Mexico, Brazil). In Africa, the largest coconut-producing countries are Mozambique, Tanzania, and Ghana.

The Philippines are the world's largest producer of coconuts. In India, the southern states account for almost 92% of the total production in the country.

Coconut is the national tree of the Maldives and is considered the most important plant in the country.

10.4.2.2 Features

It is a straight, 15–25 m tall, unbranched palm bearing a crown of large leaves that are pinnate, 2–5 m long, narrow, and tapering. The fruits are green or yellowish fibrous drupes with one seed oval or spherical with a hard endocarp and oily white endosperm and sweet milky or watery fluid in the large cavity (Warrier *et al.*, 1994).

10.4.2.3 Properties and Uses in Ayurvedic System of Medicine

The *Charaka Samhita* in the Sutrasthana A-27 categorizes coconut as a bulk promoting herb (Paranjipe, 2005). The *Nighantus* divides the coconut into three types: tender coconut, half-mature coconut, and fully mature coconut. The tender coconut water is cooling and lubricates the dryness caused by *ama* in the gastrointestinal tract. The half-mature coconut is said to be the most nutritious. The fully mature coconut is heavy to digest and can build up toxic *ama*. For this reason, it is not advised for people with low *agni*. The best way to eat it is as chutney, combined with healthful ingredients such as roasted chickpeas, curry leaves, and mustard seeds.

The *Bhojanakutuhalam* reports that "Partially ripened coconut treats thirst and depletion of *dhatu*s but is very difficult to digest. Madhunarikela (sweet coconut) treats burning sensation, thirst, pain and vitiation of *pitta*, promotes strength, nourishment and excellent luster, causes increase in virility and imparts taste" (Venkat *et al.*, 2012). It also treats vitiation of *vata* and strengthens muscles, the cardiovascular system, and the seven body tissues but causes disorders of *ama* and *kapha* (Acharya, 1980, Sutrasthana A-46). Coconut cools *sadhaka pitta*, which is associated with emotions, so it has also an antistress action. It can also restore emotional stability in menopausal women.

Coconut oil is greatly appreciated in Ayurveda for the cooling effect. It is recommended for *pitta* constitutions for massage. It is also very beneficial for hair growth, skin diseases, and wound healing. As cooking oil, it enhances fat metabolism and is particularly popular in South India where the coconut tree grows abundantly.

Parts used: Kernel, oil
Taste: Sweet
Quality: Unctuous, heavy
Potency: Cold
***Dosha* effect:** VP–, K+
Actions: Cooling, appetizer, aphrodisiac, laxative, tonic, diuretic

10.4.2.4 Traditional Uses and Folklore

- In India, coconut is considered an ideal food traditionally recommended to every woman during pregnancy as it provides all the essential ingredients for the child (Sairam, 1998).
- Plenty use of tender coconut water is a traditional home remedy for stomach ulcer and vomiting. The tender kernel is highly beneficial for indigestion and colitis (Bakhru, 1994). Combined with cinnamon, cardamom, ginger, cloves, cumin, coriander, and turmeric, coconut heals the digestive system and promotes better metabolism.
- Use of coconut in the daily diet prevents intestinal parasites.

10.4.2.5 In the Kitchen

In Sri Lanka, coconuts are the main source of dietary fat. Not only the pulp is very nutritious, but also the tender coconut water is the world's safest natural soft drink: there are a number of stories of island and coastal people surviving months of drought only by drinking it.

Milk can be extracted by blending the grated pulp with some hot water and squeezing and straining it (Aggarwal and Yost, 2011). The milk can be added to tomato and lentil soup with a cooling effect. In South India, grated coconut or coconut milk are frequently added to curries. Every day people eat chutney made with salt, chilies, and mustard seeds as a side dish served with the typical breakfast. The pulp is widely used in a variety of sweets and desserts. Due to the cooling properties, coconut is recommended during the summer.

Coconut flour has also been developed for use in baking to combat malnutrition.

10.4.2.6 Chemical Constituents and Uses as Functional Food

Coconut water contains sugars (4%), mainly glucose, fructose, and sucrose, whereas other components include malic and citric acids, proteins (0.1%), fats (0.1%), amino acids, vitamins (particularly vitamin E), and mineral salts (0.4%), of which the most abundant is potassium, followed by calcium and phosphorus, whereas iron is also relatively abundant (Khan *et al.*, 2003).

The fresh pulp of the seed is composed of lipids (about 35%) and sugars (about 10%), whereas in the dried copra, the content of lipids is higher (about 70%) (Santoso *et al.*, 1996).

An oil is extracted from the seed pulp, which has a low level of unsaturated fatty acids, forming an ivory-white, aromatic solid mass below a temperature of 25°C. The main component is a substance known as cocoin, a mixture of various glycerides, of which the most abundant contains lauric acid (about 50%). Other fatty acids include myristic, palmitic, caprylic, caproic, stearic, oleic, and linoleic acids.

The woody shell of the coconut is maximally composed of cellulose (53%) and lignin (33%).

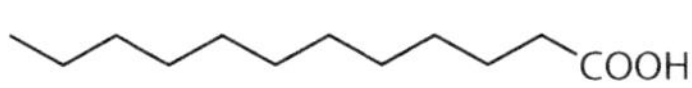

Lauric acid

Coconut water is an isotonic liquid with refreshing, rehydrating, and antioxidant properties, which can also be administered intravenously.

The composition of saturated fatty acids in coconut oil, which is unique in the plant kingdom, helps to prevent heart diseases and the hardening of arteries (Sindhurani and Rajamohan, 1998). The high content in lauric acid, a compound that is abundant also in maternal milk, makes the oil easily digestible. In addition, this fatty acid strengthens the immune system and protects against viral, bacterial, and fungal infections.

The action of coconut lipids on the liver results in a higher metabolic rate and an increased production of lipoproteins, thus improving triglyceride secretion from the liver and preventing liver steatosis (Nevin and Rajamohan, 2004; Trinidad *et al.*, 2004). Coconut lipids also induce a higher production of biliary salts, involved in the digestion of fat.

The oil exerts beneficial effects on various organs and reduces the risk of diseases such as cancer, osteoporosis, and diabetes. Medium-chain fatty acids, which are a typical component of the oil, promote the consumption of the body's energy reserves and are therefore useful in the treatment of obesity or hypothyroidism.

REFERENCES

Acharya J.T. (ed.) *Sushruta Samhita*. Chaukhambha Orientalia, Varanasi, India, **1980**.

Aggarwal B.B., Yost D. *Healing Spices*. Sterling Publishers, New York, **2011** (Italian version, Armenia, 134–140).

Bakhru H.K. *Foods That Heal*. Orient Paperbacks, Delhi, India, **1994**, 185–189.

Gupta S.M. *Plant Myths and Traditions in India*. Munshiram Manoharlal Publishers, Delhi, India, **2001**, 15–16.

Khan M.N., Muti-Ur-Rehman, Khan K.W. A study of chemical composition of *Cocos nucifera* L. (coconut) water and its usefulness as rehydration fluid. *Pak. J. Bot.* **2003**, *35*, 925–930.

Nevin K.G., Rajamohan R. Beneficial effects of virgin coconut oil on lipid parameters and in vitro LDL oxidation. *Clin. Biochem.* **2004**, *37*, 830–835.

Paranjipe P. *Indian Medicinal Plants. Forgotten Healers*. Chaukhamba Sanskrit Pratishthan, Delhi, **2005** (2nd publ.), 183–184.

Rajaiah R. *How the Banana Goes to Heaven. And Other Secrets of Health from the Indian Kitchen*. Westland, Chennai, India, **2010**, 177–187.

Sairam T.V. *Home Remedies*, vol. 1. Penguin Books, Delhi, India, **1998**, 70–74.

Santoso U., Kubo K., Ota T., Tadokoro T., Maekawa A. Nutrient composition of kopyor coconuts (*Cocos nucifera* L.). *Food Chem.* **1996**, *57*, 299–304.

Sindhurani J.A., Rajamohan T. Effect of dietary fiber from coconut kernel (*Cocos nucifera*) on cholesterol metabolism. *J. Clin. Biochem. Nutr.* **1998**, *24*, 125–132.

Trinidad T.P., Loyola A.S., Mallillin A.C. *et al.* The cholesterol-lowering effect of coconut flakes in humans with moderately raised serum cholesterol. *J. Med. Food* **2004**, *7*, 136–140.

Venkat P., Gangadharan G.G., Lakshmithathachar M.A., Alwar M.A. (eds.) *Bhojanakutuhalam*. I-AIM, Bangalore, India, **2012**, 142–143.

Warrier P.K., Nambiar V.P.K., Ramankutty C. (eds.) *Indian Medicinal Plants*, vol. 2. Orient Longman, Chennai, India, **1994**, 146–152.

10.4.3 Walnut

Scientific name: *Juglans regia* L.
Family: Juglandaceae
Hindi: Akhor
Sanskrit: Aksotah
Tamil: Akrottu

10.4.3.1 Origin and Distribution

The tree is native to the mountain ranges of Central Asia, extending from the Balkans to the Himalayas and southwest China. In these countries, there is a great genetic diversity, in particular ancestral forms with lateral fruiting. This ancient variety, called Persian nut, was introduced in Macedonia and Greece by Alexander the Great and was spread in southern Europe and North Africa by Romans. During its migration to western Europe, the ancient variety became a large tree with terminal fruiting. The tree was introduced in the Americas in the seventeenth century by English colonists. Now walnut is cultivated in France, Serbia, Greece, Bulgaria, Romania, Hungary, China, California, and Chile.

In India, it is found wild in the temperate Himalayas: it is largely cultivated in Kashmir, Tibet, and Afghanistan (Nadkarni, 2002). It is a sacred plant dear to Lakshmi, the goddess of wealth. The nut symbolizes longevity, and it is an important offering during the Hindu rituals (Majupuria and Joshi, 1997). In the European folklore, the fruit is a symbol of prosperity. As aphrodisiac, it is also part of many traditional rituals related to the marriage (Brosse, 1985).

The word "nut" derives from the Latin *nux*, with a suggested derivation from *nox* (night) owing to the dark juice of the nut, which was used as a natural dye. In ancient Rome, the fruits were called *nuces juglandes* and the tree *juglans* or *Jovi glans*, the nut of Jove (Brosse, 1985). *Juglans regia* means the "royal nut of Jove."

The inscriptions of ancient Chaldeans (2000 BC) reveal the existence of walnut groves within the famed Hanging Gardens of Babylon, and there is evidence of walnut consumption dating from the same era on carved *stelae* containing the Code of Hammurabi, in a section devoted to food. We can find references to walnut groves also in the Old Testament (Song of Solomon 6:11).

The walnut appears in Greek mythology in the story of Carya, a Lakonian maiden with whom the god Dionysus fell in love. When she died, Dionysus transformed her into a walnut tree and the goddess Artemis ordered to build a temple in her memory. The temple's columns, sculpted in wood in the form of young women, were called caryatides or nymphs of the walnut tree.

10.4.3.2 Features

It is a large deciduous tree with leaves alternate, imparipinnate; flowers are yellowish green; fruits are drupes, 5 cm long with leathery exocarp, woody, wrinkled, hard endocarp enclosing four-lobed, corrugated, oily, and edible seeds (Warrier *et al.*, 1995).

10.4.3.3 Properties and Uses in Ayurvedic System of Medicine

Walnut is mentioned in all Ayurvedic scriptures as an excellent nervine tonic and aphrodisiac, very effective in *vata*-related diseases. The properties are similar to those of almond, but walnut is a more potent aphrodisiac (Morningstar and Desai, 1999). It is a tonic for both male and female reproductive systems (Nibodhi, 2010).

Parts used: Seeds, oil
Taste: Sweet
Quality: Heavy, unctous
Potency: Hot
***Dosha* effect:** V–, KP+
Actions: Nutritive, aphrodisiac, laxative, tonic, carminative

10.4.3.4 Traditional Uses and Folklore

- Many home remedies and Ayurvedic medicines are prepared with leaves, bark, and fruits.
- The paste of the fruits is effective in reducing rheumatic pain and swelling.
- The leaves and the bark are used for cleansing the teeth and for strengthening the gums.
- The roasted fruits serve to relieve the dry cough.
- The oil extracted from the fruits is laxative and, if massaged to the scalp, induces sound sleep (Paranjipe, 2005).

10.4.3.5 In the Kitchen

Walnuts are extremely perishable and should be stored in an airtight container and placed in the refrigerator. In this way, they can be kept for about 6 months. Walnuts can be chopped and added to granola, salads, vegetables, fruits, and dessert.

10.4.3.6 Chemical Constituents and Uses as Functional Food

According to the food composition database published by the US Department of Agriculture, 100 g of walnuts contains 15.2 g protein, 65.2 g fat, and 6.7 g dietary fiber. Whereas most nuts are high in monounsaturated fatty acids, walnuts are composed largely of PUFAs (47.2 g), especially α-linolenic acid (18:3*n*-3; 9.1 g) and linoleic acid (18:2*n*-6; 38.1 g) (Li *et al.*, 2007).

Several reviews have described possible explanations and mechanisms by which this unique fatty acid profile can beneficially affect cholesterol concentrations and other CVD risk factors (Ros and Mataix, 2006). Whether by simply replacing saturated fatty acids in the diet or by explicitly acting on various pathways in CVD development, walnuts appear to have the potential to beneficially affect CVD risk. Feeding trials have shown cholesterol-lowering trends in walnut-enriched diets compared with control diets (reviewed in Feldman, 2002). Many qualitative reviews have also suggested beneficial effects of walnuts on cholesterol concentrations.

A recent meta-analysis study found significant improvements in lipid profiles with high walnut consumption compared with various control diets. Walnuts may also have potential benefits on oxidative stress and inflammatory markers. Despite their high fat content, walnuts do not appear to adversely affect body weight (reviewed in Banel and Hu, 2009).

Spaghetti with walnut sauce (serves: 4)

Ingredients

300 g spaghetti
2 cloves garlic, peeled (or a piece of ginger)
1 teacup walnut pieces
1 cup parsley leaves, firmly packed
2 tbsp of milk cream
1/2 cup ricotta cheese
1 tbsp of olive oil
2 tsp of salt
Salt and freshly ground pepper
Parmigiano-Reggiano

Preparation

In the bowl of a food processor, chop and blend all the ingredients. Boil the spaghetti in a large pot of salted water and drain, reserving half cup of the cooking water. Add some of the water to the sauce and mix with the spaghetti. Sprinkle with Parmigiano-Reggiano (or gomasio) and chopped parsley (or basil) and serve very hot.

REFERENCES

Banel D.K., Hu F.B. Effects of walnut consumption on blood lipids and other cardiovascular risk factors: A meta-analysis and systematic review. *Am. J. Clin. Nutr.* **2009**, *90*, 56–63.

Brosse J. *Les arbres de France*. Librarie Plon, Paris, France, **1985** (Italian version Ed. Studio Tesi, 1991, 147–152).

Feldman E.B. The scientific evidence for a beneficial health relationship between walnuts and coronary heart disease. *J. Nutr.* **2002**, *132*, 1062S–1101S.

Li L., Tsao R., Yang R., Kramer J.K., Hernandez M. Fatty acid profiles, tocopherol contents, and antioxidant activities of heartnut (*Juglans ailanthifolia* var. *cordiformis*) and Persian walnut (*Juglans regia* L.). *J. Agric. Food Chem.* **2007**, *55*, 1164–1169.

Majupuria T.C., Joshi D.P. *Religious and Useful Plants of Nepal and India.* Gupta M.(ed.), Publishers, Lashkar, India, **1997**, 243.

Morningstar A., Desai U. *The Ayurvedic Cookbook.* Motilal Banarsidass Publishers, Delhi, India, **1999**, 227.

Nadkarni K.M. *Indian Materia Medica.* Popular Prakashan, Mumbai, India, **2002**, 709.

Nibodhi. *Annapurna Prasad*. Mata Amritanandayi Mission Trust, Amritapuri, India, **2010** (2nd ed.), 134.

Paranjipe P. *Indian Medicinal Plants. Forgotten Healers*. Chaukhamba Sanskrit Pratishthan, Delhi, India, **2005**, 5–6.

Ros E., Mataix J. Fatty acid composition of nuts: Implications for cardiovascular health. *Br. J. Nutr.* **2006**, *96*(Suppl 2), S29–S35.

Warrier P.K., Nambiar V.P.K., Ramankutty C. (eds.) *Indian Medicinal Plants*, vol. 3. Orient Longman, Chennai, India, **1995**, 264–267.

Section IV

Conclusion

11 Crossing Communities and Environments
Rethinking Diets from an Intercultural Perspective

Marina Calloni

CONTENTS

Concluding a book is not easy, mainly when different cultures, knowledge, and disciplines have been challenged. However, let me try to make some comments, starting from the focus of the book, that is, how plants can be employed as both aliments and adjuvants, able to improve health and the quality of life, comparing the Indian Ayurvedic tradition with the Mediterranean diet.

In this context, the main point I would like to stress concerns not only the *organic* compatibility between different dietetic traditions and foods but also the necessity to understand any civilization as a continuous dynamic development, which can be enriched only through a mutual and interactive learning process, as the examples of the use of plants have showed fruitfully and peacefully over centuries.

11.1 EMPOWERING HEALTHY FOOD AS A MEDICINE

In the fifth century BC, during the golden age of the Greek *polis*, Hippocrates, the father of Western medicine, pointed out that food is a medicine and that the medicine is food, stressing the intrinsic dialectic that characterizes any nourishment: the quality of food is the core of health and vice versa. Nutrition can be thus conceived as a daily process of self-caring, and self-healing in relation to the choice that an individual can operate about appropriate aliments, a healthy conduct of life, and a well-balanced diet within a given community.

Hippocrates's comprehension—which is similar to some Oriental conceptions—indicates that, since antiquity, incessant connections and exchanges of knowledge have characterized the interaction between Western and Eastern cultures. One of the most important common aspects that have characterized and made compatible the comparison between Western and Oriental traditions about the philosophy of plants and aliments is without any doubt a reflection upon the best nutritional lifestyle that human beings have to adopt to achieve longevity and psychophysical well-being, depending on personal preferences or material conditions.

The analysis of Western and Oriental ancient testimonies and codes—as discussed in Chapter 1—put in evidence that one of the central and common issues discussed in philosophical and scientific debates on the use of plants was the search toward the best conduct of life a human being should have, in light of the preservation of health and the obviation of a disease risk. This question was crucial mainly because medicaments were at that time *natural*, meaning *not industrialized* or *pharmacologically and chemically reproduced*.

The debate on the *nature* of plants and alimentary stuff was thus polarized on the definition of opposite qualities that determined their substance and consequent effects. Oppositions were conceived in both absolute and relational terms, that is, in itself, or depending on the relationship between the interested individual and the object. The judgment about the balsamic or maleficent influences that eatable or drinkable aliments had on themselves or as a component of a mixture was thus based on the individuation of *compatible* discrepancies, which could have made possible the definition of an aliment as good or bad, healthy or harmful, material or spiritual, local or universal, high or low. Nature was pictured as based on tensions, dichotomies, and contraries, as well as on similarities, correspondences, and harmony. Consequently, human nature and food also were believed to be composed by elements, which were intrinsically contradictory and at the same time compatible because they were integrable. For instance, a certain food can be dangerous for some people but good for others. As Lucretius (1924) argued in his work *De Rerum Naturae*, in which he tried to unveil poetically the "nature of things," "What is food to one man is bitter poison to others." Therefore, a nutritional element is always essentially dialectic and antinomic. It contains in itself opposite forces, which provoke different effects in relation to diverse people and situations.

Historically, food refers to wars and conquests because of the necessity that humans have to find nutrition to survive as well as to a sense of hospitality and peace because of the pleasure of consuming common meals together with other fellow beings. Sociologically, food denotes both poverty and richness. Philosophically, it signifies vices and virtues. Religiously, it implies prohibitions and precepts. Politically, it indicates famine and social policies. Culturally, it represents both conflicts and interrelationships between different civilizations. Scientifically, it infers morbidity and wholesomeness. Food is thus a controversial picture that portraits ambivalently the physical and moral vulnerability and precariousness of human nature, where people have to deal with the common and final destiny of death. Food is thus the concrete metaphor of a surrogated compromise in search of a healthy and enduring longevity.

11.2 MAKING COMPATIBLE DIVERSITIES AND POLARITIES

The scrutiny of the *qualities* of drinkable and eatable things, that is, aliments, has always been central to any civilization. The *nature* of nourishment is thus a constitutive part of a broader search toward the *roots* of the world in a cosmic perspective, which implies the examination of the common bases that determine objects and living beings, the flora and the fauna.

This kind of examination was at the basis of the ancient Greek and Indian philosophies. Both traditions tried to individuate in four or five opposite but integrating constituents the recurrent substance that regulates the nature as well as the human characters. The universal and basic roots, commonly identified, were air, fire, earth, and water for the physical Greek philosophers plus the ether for the Indian thinkers.

The understanding of nutriment reproduces the same ambivalent dynamics that determines the structure of nature in the tension between materiality and spirituality, low and high. Nutrients have in fact material shapes as well as spiritual connotations, if understood from a creationist viewpoint.

From a religious perspective, food is a factual vehicle that connects transcendentally human beings with a god or gods. From this perspective, nutritional and remedial elements have divine origin like living creatures, such as the flora and fauna. Therefore, diet is not only a means, which permits a daily survival or a mystic experience, but also a matter of well-being, pleasure, and longevity, depending on a well-balanced way of life. Furthermore, nourishment—plants in particular more than the eating of meat, which is controversial under many aspects—contains a cross-cultural significance and a dialogical potential, when it is conceived in a comparative way.

The daily practice of nourishment implies overlapping learning processes for the share-stakeholders who are involved in the action. It implies the knowledge that an individual has about himself/herself, the comprehension of the environment he/she lives in, the culture of the community he/she shares with other people, the scientific and medical reasons that indicate the grades of health or disease. In sum, food is what permits discovery of *the variegate and ambivalent nature of things* and the variable relationships (affinities or dissimilarities) they have with human beings, dependent on the digestion, idiosyncrasies, allergies, intolerances, or preferences an individual has as well as on the different cultural, environmental, and scientific contexts that an interested person experiences.

All these questions have been at the core of the chapters collected in this volume, which aims at comparing two ancient traditions that apparently relate to different conceptions of food, nutrition, philosophy of mind, and conception of the body, which have been taken as elements of distinction and separation between Western and Oriental cultures.

On the contrary, one of the main challenges of writing this book is the attempt to support a cultural and scientific compatibility between the principles of the Ayurvedic tradition and the practices of the Mediterranean diet, showing a common understanding about the knowledge and use of plants. Next to differences, many similarities have been stressed, so that common cultural roots have become more evident than they were in the past, symbolizing preexisting exchanges that modernity, incessant processes of industrialization, and devastating forms of colonization have denied and refused for centuries.

The building of a cross-border bridge could facilitate the reconnection of traditions, which had common cultural roots but whose dialogue was interrupted due to the expansion of different kinds of socioeconomic development and political power. An innovative interaction between overlapping and fusional alimentary traditions can contribute to rethinking intercultural praxes, aimed at increasing individual well-being and collective prosperity across the globe, improving the quality of life of individuals and populations.

Multidisciplinary reflections upon similarities and potential—which have lied at the core of this book and are related to the substance, processing, and use of plants as medicaments and aliments—contribute to improving the strength of intercultural perspectives, considering nutrition as a matter of both self-care and social policies. The issue of a healthy and well-balanced diet has in fact become a central question to any state and supranational institution not only to struggle against shortage and poverty in countries in development/transition, but because of the necessity to increase campaigns against malnutrition, obesity, and related diseases in Western countries.

Sustainable diets and the respect of biodiversity have thus become an important element for the support of a *green economy* and the development of related strategies, so that the Food and Agriculture Organization (FAO) of the United Nations is encouraging its member states to adopt appropriate policies. For this reason, nutrition education is playing a crescent therapeutic role in the contrast not only to famine but also to cancer and diseases in Western countries (Hawkes, 2013), starting from a better knowledge of the healing power of plants.

Because a collective reflection upon the quality and the quantity of food has become a key issue in international public discourse and social policies, a stronger cross-border, multidisciplinary, comparative, and holistic approach to the analysis of the Ayurvedic tradition and the Mediterranean nutrition civilization can help to reach a new global sensitivity toward the boundaries that differentiate a state of health from illness.

11.3 BORDER CROSSING GLOBAL FOOD

In this global age, connoted by the redefinition of previous geopolitical borders, the circulation of people and goods has exponentially increased with respect to the past, when ideological conflicts and armed barriers divided Western societies one from the other. Consequently, the availability and commercialization of varying qualities and quantities of food have also increased. Globalization can be thus seen as a consequence of the crisis of industrial societies, the end of mass production in the West, the collapse of socialist regimes, and the rising of emergent markets in the east of the world, which by contrast has put in evidence the necessity to ideate a *green economy* and new environmental strategies to face ecological disasters and global warming. Different kinds of food, vegetables, and aliments previously considered *exotic* have become easily accessible, merchandizable, and popular in various parts of the world.

However, the circulation of food is not a modern or postindustrial novelty. *De facto*, although with different intensities and modalities, food in antiquity has always been the example of an incessant interchange between populations and the core of any civilization. Archeological excavations in cities and graves have showed that populations consumed some kind of nourishment, which was not local but imported from other lands. Moreover, in modernity, plants and vegetables that have become examples of national identities were originally native in other places, for example, tomatoes that connote typical Italian nourishment such as pizza but originated in the American continent. The same applies to potatoes, which evoke German dishes but are also native to the new continent discovered at the end of the fifteenth century. This verification confirms the theory—proposed by the philosopher Karl Popper—that a totally closed society has never existed: all societies are open. Diets that accommodate nutritional products coming from other cultures are the best examples and practical explanations that since archaic times the quality of life has been reached and enriched by an intensive collaboration between civilizations and countries.

As has happened in the past, food—and plants and vegetables in particular—can be meant as a sensitive cross-border barometer that measures epochal changes. Marking transformations in collective uses, representations, and commercialization, food signifies deep changes in mentalities, fashions, and daily life in contemporary societies all over the world. Socioeconomic transformations—which have brought postcapitalistic societies in the West and neo-industrial communities in the East together with new cultural trends—have induced new collective imaginaries about environmental risks and new imagination of the world (*Weltanschauungen*), able to orient the sense of daily life and consequently the meaning of nutrition.

The new ecological wave, the interest for the safeguard of the planet against global warming, the mass mobilizations for the right to food in various countries against the power of multinational companies that control financially common resources, a new generation of rights devoted to nonhuman animals and future generations, a different sensitivity toward the quality of life, and a new understanding about human development have contributed to change a general approach to nutrition, which is no longer meant as a mere instrument for the survival and a merchandise serially produced but as a matter of relationships between different components, that is, individuals, communities, and the *universe*.

11.4 REDEFINING NUTRITION IN A HOLISTIC AND COSMIC PERSPECTIVE

Global trends in healthy eating, experiments in *fusional cooking*, and new dietetic knowledge about the properties of plants as aliments and medicaments (i.e., the aim of this book) imply innovative methods in health therapies, new habits in daily life, and innovative intercultural understandings upon dietetic traditions. In analogy to antiquity, food and plants have become more understood in relational terms, that is, what are their nutritional impacts and antidisease effects on the life of people in light of the respect for the environment. Food becomes once again a *medicine*, as Hippocrates argued.

These tendencies induce deep changes in attitudes, daily life conducts, and sustenance. These new cultural inclinations cannot be thus modes or evanescent exotic fashions, as articles in popular magazines and advertisements in mass media could let appear. These recent approaches to diet—which are grounded on the longstanding inheritance of preexisting cultures—signify an imaginative interest in a better understanding of the *nature of things* that we eat and become parts of our bones and body. In an age when *natural* elements are technologically transformed (like the genetically modified organisms), an implicit countereffect of this process induces the intention to understand intrinsically the *substance* that composes aliments and the main components that determine their nutritional, dietetic, and healing efficacy.

Ancient traditions acquire a new significance when they are tackled with those challenges that present global societies have to face in terms of environmental concern, economic crisis, and political conflicts. This present situation induces a sort of *neo-cosmic* approach in the understanding of human beings as fellow of a broader universe as well as part of an anthropized and manipulated nature. The new global interest for traditional diet, plants, and lifestyle thus stresses the necessity to rethinking nutrition in a "holistic" perspective, where food becomes the mean that interconnects a human being—considered as an individual with proper qualities and deficiencies—with both specific alimentary components and the natural/urban context he/she lives in (Calloni, 2013). Diet becomes strictly related to the discipline of daily life and the practice of ethical principles—both religious and secular—a person believes in.

Thanks to the contribution of different viewpoints, the present collection compared in a prismatic and multifaceted way the Ayurvedic diet with the Mediterranean nutrition tradition, finding affinities because of a common aim they have, which is the achievement of a well-balanced way of living/eating and the accomplishment of a healthy longevity.

As mentioned above, the Ayurvedic tradition and the Mediterranean diet have different cultural and geographic origins, although thanks to continuous interchanges they have developed reciprocal influences over centuries. Differences thus become compatible, because of mutual effects in the past and thanks to comparative analyses in the present.

For example, one of the main differences between the two traditions lies in a differing approach to subjectivity. Namely, the Ayurvedic philosophy focuses on the interaction between the nature of the individual—who can be summarized in the tripartition between *vata*, *pitta*, and *kapha*—and the qualities/essence connoting any element. Ayurveda thus tries to highlight whether aliments and people are reciprocally companionable or in antagonism. The encounter can happen only between *contraries* and not similarities, so that *vata*, *pitta*, and *kapha* people have to eat different stuffs depending on their nature and habits. Only a meeting between opposites can lead to a joint beneficial influence, whereas the sameness among elements implies a mutually exclusive dynamic. It means that the association and assumption of similar elements can bring to pathologies and diseases.

Differently, the Mediterranean perspective mostly refers to environmental conditions and a shared communitarian context of life. The Mediterranean regime is thus based on a balanced diet, derived from the consumption and cooking of products, which grow naturally in a specific land and are rich in vitamins and proteins, beneficial in general for all consumers. People eat what they have at disposal.

11.5 LEARNING THE LESSON OF THE AYURVEDIC TRADITION AND THE MEDITERRANEAN DIET

Despite differences about the significance of eating and what kind of nourishment has to be consumed in relation to his/her own nature, both the Ayurvedic tradition and the Mediterranean diet share many convictions and basic presuppositions. Among others, they support a common idea that an individual is part of a natural environment and fellow of a broader cosmos. This macro principle—which is grounded on philosophical and cultural assumptions—can be revealed and applied in the micrologia of food knowledge and consumption, thanks to a productive encounter between appetites and desires, and prohibitions and intolerances.

Indeed, both Ayurvedic and Mediterranean traditions assume as constitutive the relationship that an individual has with the surrounding environment, the affiliation to other human beings, and the ability the individual has to "live with concern for and in relation to animals, plants, and the world of nature" (Nussbaum, 2003), as the philosopher Martha Nussbaum argued in her *Decalogue*, related to the theory of the functioning and empowerment of human capabilities, developed together with Amartya Sen. The quality of food (besides the necessary quantity to survive) is thus central to the improvement of individual health, the quality of a decent life, and a collective well-being, which is in the perspective of a *flourishing life* and the recognition of principles of social justice (Sen, 2009), living in dignity.

Rethinking the relationship between individuals, environment, and food induces in people overlapping and multifaceted forms of knowledge about themselves, the surrounding existential context, and the processed and manipulated nature of food and the effects it has on health. This learning process contributes to develop a hermeneutic approach to health, as a matter of decodification and interpretations of signs, which denote pathologies. The relationship between a physician and a patient becomes thus based on a dialog, where signals, symptoms, and reactions have to be soon decrypted to avoid diseases and death. Hans-Georg Gadamer's book *The Enigma of Health: The Art of Healing in a Scientific Age* (1996) provides a good example in this regard.

Starting from his experience and practice of natural medicine, Gadamer noticed the increasing different attitude that society has to denote and face the issues of health, illness, death, and implicitly nutrition. Gadamer thinks that the knowledge of health is more enigmatic and puzzling than that of disease and illness. Therefore, it is not totally clear what "to feel healthy" means. This reflection can be extended to food knowledge, that is, "what does it mean to feel healthy, thanks to a well-balanced diet?" This question can support a more conscious and sensitive concept of health and well-being against any rhetoric of a *healthiest* society, based on the commercialization, consumption, and commodification of lifestyles, which could avoid any personal responsibility toward the significance of nutriment as both a personal and a collective matter.

Another lesson that we have learned from the comparison between the Ayurvedic and Mediterranean traditions is a common notion of wisdom that refers to a stratified concept of experience, judgment, and practical reason (*phronesis* in Greek), in search of a well-balanced and moderated life conduct. In our case, it means that the involved person has to acquire an experiential or scientific ability to know his or her own qualities together with the properties of nutriment, considered in itself and as a combined whole. We commonly remember a quotation by Aristotle, who, in his treaty on ethics and the search toward a good life and happiness, indicated as a precept for moderation the following faculty: “It is best to rise from life as from a banquet, neither thirsty nor drunken.” This assertion can also be found in Ayurvedic rules.

Phronesis, that is, practical reason and the faculty of judgment, thus refers in both the Ayurvedic and Mediterranean traditions to one’s ability to find an equilibrium in life. Rethinking nowadays wisdom from the perspective of food (as a curative and nourishing mean) means the necessity to redefine the relationship between individuals, environments, global societies, and the *cosmos*, interconnecting micro-imagines (the context of our daily life) with macro-visions (the perspective of a borderless infinitum).

In sum, the interchange of aliments is a constitutive element that has connoted any civilization and promoted a productive dialog between different cultures with the aim at creating both an individual and a common prosperity and well-being. However, if we want to be nourished, we have to feed the planet and to nourish food, flora and fauna as well, bringing the world to a new vitality and natality. Food can be a medicine, only if we acknowledge and respect it together with the environment and the *cosmos* at large, as the Ayurvedic and Mediterranean diets have taught us for centuries.

REFERENCES

Calloni M. Street food on the move: A socio-philosophical approach. *J. Sci. Food Agr.* **2013**, *93*(14), 3406–3413.

Gadamer H.-G. *The Enigma of Health. The Art of Healing in a Scientific Age*. Stanford University Press, Stanford, **1996**.

Hawkes C. *Promoting Healthy Diets through Nutrition Education and Changes in the Food Environmental International Reviews of Actions and Their Effectiveness*, FAO, Rome, Italy, **2013**.

Hippocrates. *The Hippocratic Writings*. Penguin, London, **1984**.

Lucretius. *On the Nature of Things*. Harvard University Press, Cambridge, MA, **1924**.

Nussbaum M. Capabilities as fundamental entitlements: Sen and social justice. *Feminist Economics* **2003**, *9*(2/3), 33–59.

Sen A. *The Idea of Justice*. Belknap Press, Cambridge, MA, **2009**.

Acknowledgments

We thank all the authors, without whose contribution this book would have not been possible.

We thank Professor Gerry Bodeker's guidance and encouragement, which were very useful during the initial stages of the preparation of this work.

We thank Dr. Kishor Patwardhan and Dr. Subrahmanya, who reviewed the Ayurveda-related chapters. We are grateful to Laura Salaris, who checked some parts of the book, making important improvements to language and style. We thank our friends who are living in Italian country sides—Alessandra, Isabella, Luciano, Piergiorgio, and Stefania—for their contribution to the photographs of plants. Our thanks go to the Foundation for Revitalisation of Local Health Traditions, India, for permitting us to use the photographs of the date, mango, and ginger from their medicinal plant database.

Luisella Verotta personally thanks Alessandra and Maria for their useful comments and psychological support and the colleagues at the department of chemistry of the University of Milan for the helpful suggestions and encouragement. Maria Pia thanks her sister Elisabetta for her precious support during the revision of proofs and web research. We thank Marco Bianchi for his approval to use some of his food recipes in this book.

We gratefully acknowledge the University of Milan, EXPO 2015 Scientific Committee and Centre for Orientative Education (COE) for giving us the opportunity to include the book presentation in various events related to the international exhibition.

We thank Professor Roland Hardman (editor of *Traditional Herbal Medicines for Modern Times*) for his enthusiastic appreciation of the proposal to prepare this book. We gratefully acknowledge the staff at CRC Press, Taylor & Francis Group, John Sulzycki and Cheryl Wolf, for their contribution in concretizing the project and Jill Jurgensen for her support in tutoring and expert editorial attention. We greatly appreciate the technical skill and precious suggestions of Ed Curtis and Vasuki and Visvanathan at Lumina Datamatics Ltd., who were responsible for the layout and production of the book.

Glossary

abhipretha: intent
abhishyanda: a state of derangement of *kapha* in the body
adana: depleting phase
adhikarana: substratum
adhyasana: eating soon after a meal
Agastya: famous Vedic seer, said to have brought the Vedic teachings to South India; one of the founders of Siddha medicine
agni: fire as a cosmic principle of transformation—it is one Hindu deity. In the body, it is the digestive fire, the enzymatic function, and the base for a proper assimilation
agnivardhana: increasing digestive "fire"
ahara: food
ahita: inappropriate
ama: undigested food, which will become toxic. Accumulation of *ama* is due to improper metabolic functioning
amatrasana: eating inappropriate quantities
amrita: nectar of immortality
anala: digestion and metabolism
annapana vidhi: section on instructions of consumption of food and drinks
annarakasha vidhi: section on food safety
annasvarupa vijnaniya: section on classification of foods and their qualities
apana vayu: downward moving *prana* governing elimination and reproduction
apathya: unfavorable food
aristha: medicated wine
arsas: hemorrhoids
artha: means
ashwins: the divine twin healers, portrayed as horsemen. The horse is a symbol of *prana*. Their constellation marks the rejuvenative power at the beginning of the zodiac
***Astanga Samgraha*:** one of the three major classical texts of Ayurveda
***Atharva Veda*:** last of the four Vedas
Ayurveda: science of life and longevity
ayus: life or longevity
balam: strength
balya: strengthening
basmati: a light variety of rice, very easy to digest, recommended in Ayurveda during the summer. The whole variety is more nutritious, but heavier to digest, and it is recommended during the cold season
bhaishajya kalpana: pharmacy
bhoutika: relating to *panchamahabhuta*

bhuta: element
bhutagni paka: digestion/metabolism at elemental levels
brahmacarya: celibacy
brahman: the Absolute, the Divine Reality
brmhana: nourishing
***Charaka Samhita*:** one of the three major classical texts of Ayurveda
chakras: mystic energy centers of the subtle body, symbolized by various wheels in the Veda
***Chandogyopanishad*:** one of the philosophical texts
chappati: Indian flat bread, made with whole wheat flour, without yeast
chikitsa: therapy
dal: any split bean used in Indian cooking. It is also the name of the soups made from split beans
dasa vidha pariksha: 10-fold examination
deepana: increasing digestive process
desha: place
Dhanvantari: according to a mythological tale, Dhanwantari came out from the ocean churned by gods and demons, with a pot containing *amrita*, the nectar of immortality. He is considered a form of Vishnu, the Ayurveda deity of healing and rejuvenation. He is also known as Kashiraj Divodas, the king of Varanasi. He wrote 13 texts on various aspects of Ayurveda; unfortunately no more available
dharma: natural law, principle of consciousness or code of honor
dhatu: tissue element. There are seven *dhatu*s in the body: *rasa* (plasma), *rakta* (blood), *mamsa* (muscle), *meda* (fat), *majja* (nerve), and *shukra* (reproductive)
dhatuposhana: nourishment of tissues
dhatvagnipaka: tissue-specific metabolism
dhirghayu: long life
dinacharya: daily regimen
dosha: biological humor. There are three *dosha*s in our body and their balance is a key to health
doshabhediya: section on *doshas*
drava dravya vijnaniya: section on liquids
dravidians: ancient people of South India
dravya: material
dravyabhediya: section on materials
dravyagunavijnana: Ayurvedic pharmacology
dukha: sorrow
dushyam: seven body tissues
ghee: clarified butter, symbol of the clarified state of the mind it has therapeutical use in Ayurveda
gomasio: roasted and powdered sesame seeds
grahani: diseases of duodenum
guna: quality
haridra: turmeric
hingu: asafetida

hita: appropriate/good
hitayu: sustainable happiness in life
hridroga: heart disease
hridya: cordial
Indra: the king of the Vedic deities
Indus Valley Civilization: ancient civilization of the third millennium BC, possibly of Vedic origin. It was the largest of the ancient civilizations, being larger than the whole Middle Eastern realm of Egypt, Syria, and Mesopotamia combined
jaggery: brown Indian sweetener
jala: water as a cosmic principle
jangama: animal material
jatharagni paka: gross digestion at gastrointestinal tract
jvara: fever
jvaraghna: reduces fever
kala: time/season
kamala: jaundice
kapha: (abbr. K) means literally "that which binds," as it is the principle that provides cohesion in the body. It is one of the three *dosha*s, with the qualities of water (cold, clear, soft, and unctuous) and earth (solid, heavy, and grounded)
karma: action
kasa: cough
katu: pungent
kesara: saffron
kichadi: Ayurvedic recipe made with basmati rice and split yellow *mung dal*, part of rejuvenation diet
kostha: gastrointestinal system
krimighna: anthelmintic
Krishna: Vishnu *avatara* of yoga and devotion. He is famous for longevity as he is said to have lived 125 years
kusthaghna: reduces skin diseases
laghu: light
laksa: lac
Lakshmi: goddess of preservation, devotion, and wealth; married to Vishnu
langhana: depletion
lassi: yogurt shake
loka: macrocosm
madhura: sweet
Mahabharata: the great Hindu epic of Lord Krishna and his era, talking about the war that marked the end of the Vedic age
mala: waste material of the body
mandagni: metabolic "fire" that is dull
manjista: *Rubia cordifolia*
mantra: sacred sounds, syllables, and chants
matrasitiya: section on food quantities and intake

Mohenjo-Daro: another name for Indus Valley civilization, after one of its main cities
mung dal: split mung beans
nama: nomenclature
nasya: nasal administration of therapy
nidra: sleep
Nighantus: *materia medica*/lexicons
nimbu pani: a very common drink all over India made from lemon, salt, and water
nitya rasayana: ethical practices for rejuvenation.
ojas: spiritual vitality, vital essence of *kapha dosha*
oudbhida: plant
oushadha: medicine
oushadhi: herbaceous plants
pachana: increasing catabolism
padartha vijnana: Ayurvedic physics
panchakarma: five purificatory and rejuvenating procedures
panchamahabhuta: five elements—air, fire, water, earth, and ether
paneer: fresh cheese made adding lemon juice or yogurt to the boiling milk
parampara: tradition
Parvati: the daughter of Himalaya married with Shiva, the cosmic feminine energy
pathya ahara: ***pathya***, favorable food
phala: effect
pitta: (abbr. P) means literally "that which eats or cooks." It is one of the three *doshas*, related to transformation
pittahara: *Pitta* reducing
pliha: spleen disorders
poshana: nourishment
prabhava: special effect
prakriti: constitution of an individual
prakruti: genetic and phenetic constitution
prana: the life force and link to the body, mind, and soul; in the Veda, it is identified with Indra
pranayama: control, development, and transformation of *prana*
prasanna atma: contented state of self
prasannamana: contented state of mind
prasannendriya: contented state of sense organs
prthvi: Earth as a cosmic principle
Puranas: mythological texts of later Hinduism
purusha: microcosm
rajas: excitable mental state
rajasic: with *rajas* quality
rakta: blood
raktasodhaka: blood purifying
Ramayana: the epic story of Rama, an *avatara* of Lord Vishnu
rasa: taste
rasabhediya: section on tastes

rasayana: rejuvenation
rasayana cikitsa: rejuvenation therapy
***Rig Veda*:** oldest Vedic texts
rishi: Vedic seers
ritucharya: seasonal regimen
ruksa: dry
rupa: shape
sadravya cikitsa: treatment with drugs
samadhatu: equilibrium state of body tissues
samadosha: equilibrium state of *dosha*
samagni: equilibrium state of digestive processes; metabolic "fire" that is balanced
samamalakriya: equilibrium state of excretory functions in the body
samanya: general attributes
samasana: *balanced eating*
samavaya: total nature
samgrahas: treatise
Samhitas: classical Ayurvedic books
samskara: processing
Saraswati: goddess of wisdom, mother of the Vedas, married to Brahma
satmyam: habituation
sattva: disposition of the mind, which endows the individual with a very deep spiritual vibration
sattvic: pure mental state
shakti: the cosmic feminine energy of consciousness
Shiva: pure consciousness, the great lord of immortality; as *Rudra*, Shiva is the divine doctor, the foremost of all physicians in the *Rig Veda*. His mantras hold a great healing and rejuvenative power
snigdha guna: unctuous
sodhana: purificatory procedures
sothahara: reduces edema
srotas: the channel systems
srotosodhana: purification of the channels of nourishment
sukha: happiness
sukhayu: happy life
Sushruta: ancient Ayurvedic author
***Sushruta Samhita*:** one of the major three classical texts of Ayurveda
sutra: axiom used in Vedic teaching
svasa: dyspnoes
svasthya: a state of perfect health
tamas: lethargic state of mind
tamasic: with *tamas* qualities
tantra: yogic teachings on Shiva and Shakti as universal power
tiksna: penetrating
tiksnagni: metabolic "fire" that is sharp
tikta: bitter
tofu: soybean cheese used occasionally in American Ayurvedic cuisine instead of *panir*

tridosha: three humors
tridoshic: substances that calm all three *doshas*
triphala: Ayurvedic herbal preparation with rejuvenating properties
tristambha: three pillars of life
Upanishads: last portion of the Vedas, revealing the inner spiritual meanings of the *mantras*
upaya: stratagem
***upayoga vyavastha*:** consumption practices
usna virya: hot potency
usna virya: hot quality
Vagbhatta: ancient Ayurvedic author
vaidya: Ayurvedic doctor
vamana: emesis
vata (abbr. V): means literally "that which moves." It is one of the three *dosha*s, related to movement and destruction. In the *Rig Veda*, it was associated with Indra
vaya: age
Vayu: Vedic god of the wind, air as a cosmic principle
veda: spiritual knowledge
vedanta: spiritual system behind the Vedas
Vedas: ancient spiritual texts of India
vidya: way of knowledge
vikriti: alteration of *prakriti*, disease
vipaka: postdigestive effect
virecana: purgation
viruddhanna vijnaniya: section on incompatibilities of food
virya: potency
visamagni: metabolic "fire" that is irregular
visamasana: irregular eating
visarga: replenishing
visesa: specific attributes
Vishnu: the great Lord of preserving, maintaining the universe
vyakhyas: commentaries
yashtimadhu: licorice
yoga: spiritual practices to bring union with the Absolute
yuktivyapasraya: one of the three ways of disease management in Ayurveda using logical method

Appendix

SUGGESTED READINGS AND NOTES

Indian Classical Texts

Acharya J.T. (ed.) *Sushruta Samhita*. Chaukhambha Orientalia, Varanasi, India, **1980**.

Devi Chand M.A. (trans.) *The Atharva Veda*. Munshiram Manoharlal Publishers, Delhi, India, **1980**.

Gupta M. (ed.) *Vishnu Purana*. Gita Press, Gorakhpur, India, **1952**.

Hoernle A.F.R. *The Bower Manuscript*. Govt. Press, Kolkata, India, **1893**.

Kunte A.M. *et al.* (eds.) *Ashtanga Hridayam*. Chaukhambha Orientalia, Varanasi, India, **1982**.

Ray P., Gupta H. *Charaka Samhita: A Scientific Synopsis*. Indian National Science Academy, Delhi, India, **1965**.

Ray P., Gupta H., Roy M. *Sushruta Samhita: A Scientific Synopsis*. Indian National Science Academy, Delhi, India, **1980**.

Shankar D., Ved D.K., Gangadharan G.G., Lakshmithathachar M.A., Alwar M.A., Shankar R., Ananth M.A. (eds.) *Ksemakutuhalam*. I-AIM, Bangalore, India, **2009**.

Sharma R.K., Bagwan Dash (trans.) *Agnivesa, Charaka Samhita,* Sutrasthana (Text with English translation). Krishnadas Academy, Varanasi, India, **2009**.

Sharma R.K., Bagwan Dash (trans.) *Charaka Samhita,* Chikitsasthana. Chaukhamba Sanskrit Series office, Varanasi, India, **1976**.

Sharma S. (ed.) *Chhandogya Upanishad*. Hariyana Sahitya Sansthan, Rohtak, India, **1983**.

Shastri K.N. (ed.) *Charaka Samhita*. Chaukhambha Vidyabhavan, Varanasi, India, **1970**.

Shastri R. (ed.) *Bhagavata Purana*. Sri Pauranik Karyalaya, Varanasi, India, **1989**.

Shastri R. *Yajurveda*. Chaukhamba Sanskrit Pratishthan, Varanasi, India, **1992**.

Trivedi K.P. (ed.) *Dhanvantari Vanaushadhi Visheshank* (Text with Hindi translation). Dhanvantari Karyalaya, Vijaygarh, India, **1961**.

Valiathan M.S. *The Legacy of Charaka*. Orient Longman, Chennai, India, **2003**.

Venkat P., Gangadharan G.G., Lakshmithathachar M.A., Alwar M.A. (eds.) *Bhojanakutuhalam*. I-AIM, Bangalore, India, **2012**.

Historical Texts

Allchin F.R., Allchin B. *Rise of Civilization in India and Pakistan*. Cambridge University Press, Cambridge, **1982**.

Arnold T., Guillaume A. (eds.) *The Legacy of Islam*. University Press, Oxford, **1931**.

Banerji G.N. *Hellenism in Ancient India*. Munshiram Manoharlal Publishers, Delhi, India, **1961**.

Diamond J. *Guns, Germs and Steel. The Fates of Human Societies*. W.W. Norton & Company, New York, **1997**.

Filliozat J. Ayurveda and foreign contacts. *Bull. Inst. Hist. Med.*, annual vol., **1981**.

Frawley D. *Gods, Sages and Kings. Vedic secrets of Ancient Civilization*. Motilal Banarsidass Publishers, New Delhi, India, **2010** (reprint).

Jaggi O.P. *Scientists of Ancient India*. Atma Ram & Sons, Delhi, India, **1966**.

Jayne W.A. *Healing Gods of Ancient Civilizations*. University Books, New York, **1925**.

McCrindle J.W. *Ancient India as Described in Classical Literature*. Philo Press, Amsterdam, the Netherlands, **1901**.

Nath C.D. Origin and development of medicine in ancient India. *Indian J. Hist. Sci.* May **1970**.
Pandey V., Pandey A. A study of Bower Manuscript. *Bull. Inst. Hist. Med.* January **1988**.
Ranade S., Deshpande R. *History and Philosophy of Ayurveda.* Chaukhamba Sanskrit Pratishtan, Delhi, India, **2006**.
Reddy D.V.S.R. Health and medicine in Buddhist times. *Indian J. Hist. Med.* December **1962**.
Sharma P.V. *Indian Medicine in Classical Age*. Chaukhamba Pratishthan, Varanasi, India, **1972**.
Wujastyk D. *The Roots of Ayurveda.* Penguin Books, Delhi, India, **1998**.

Ayurveda and Nutrition

Aggarwal B.B., Yost D. *Healing Spices.* Sterling Publishers, New York, **2011**.
Bakhru H.K. *Foods That Heal.* Orient Paperbacks, New Delhi, India, **1994**.
Ballentine R. *Diet and Nutrition: A Holistic Approach.* Himalayan International Institute, Honesdale, PA, **1978**.
Chakravarti A.K. Diet and disease, some cultural aspects of food use in India, in Noble A.G., Dutta A.K. (eds.), *India: Cultural Patterns and Processes.* Westview Press, Boulder, CO, **1982**.
Dwarakanath C. Digestion and metabolism in Ayurveda. *Indian J. Hist. Sci.* May **1968**.
Gupta L.P. *Biogenic Secrets of Food in Ayurveda.* Chaukamba Sanskrit Pratishthan, Delhi, India, **2011**.
Narain, S. (ed.) *First Food. A Taste of India's Biodiversity.* Centre for Science and Environment, Delhi, India, **2013**.
Nibodhi. *Annapurna's Prashad.* Mata Amritanandamayi Mission Trust, Amritapuri, India, **2010** (2nd edn.).
Smith A. *Ayurvedic Nutrition.* Motilal Banarsidass Publishers, New Delhi, India, **2011**.
Tripati B. *Pathyapathyanirnaya* (Text with Hindi translation). Chaukhambha Sanskrit Pratishthan, Delhi, India, **1998**.
Yeshi T. *Dietary Guidance for the Prevention of Pulmonary Tuberculosis.* Men-Tsee-Khang Clinical Research Department, Dharamsala, India, **2011**.

Western Dietetic Classical Texts

Brothwell D., Brothwell P. *Food in Antiquity: A Survey of the Diet of Early Peoples.* Thames & Hudson, London, **1969**.
Cornford F.M. *Plato's Cosmology.* Routledge & Kegan Paul, London, **1937**.
Edelstein L. The dietetics of antiquity, in Temkin O., Temkin C.L. (eds.), *Edelstein, Ancient Medicine: Selected Papers of Ludwig Edelstein.* Johns Hopkins University Press, Baltimore, MD, **1967**.
Flower B., Rosenbaun E. (trans.) *Apicius: The Roman Cookery Book.* George G. Harrap, London, **1980**.
Gardner Wilkinson J. *A Popular Account of the Ancient Egyptians.* J. Murray, London, **1854**.
Goold G.P. *Hippocrates. The Loeb Classical Library*, vol. 4. Harvard University Press, Cambridge, MA, **1992**.
Gremen M.D. *Diseases in the Ancient Greek World.* Johns Hopkins University Press, Baltimore, MD, **1989**.
Harington J. *The School of Salernum: Regimen Sanitatis Salerni.* Ente Provinciale per il Turismo, Salerno, **1957** (orig. publ. 1607).
Jackson R. *Doctors and Diseases in the Roman Empire.* British Museum Publishers, London, **1988**.
Keys A., Keys M. *How to Eat Well and Stay Well, the Mediterranean Way.* Doubleday, New York, **1975**.

Lyons A.S., Petrucelli R.J. *Medicine. An Illustrated History*. Abradale Press, New York, **1987**.
Manniche L. *An Ancient Egyptian Herbal*. The American University in Cairo Press, Cairo, Egypt, **2006**.
Panayotatou A.C. *L'Hygiene chez les anciens Grecs*. Vigot Freres Editeurs, Paris, France, **1923**.
Shorey P. (trans.) *Plato in Twelve Volumes,* vols. 5 & 6. Harvard University Press, Cambridge, MA, **1969**.
Tannahill R. *Food in History*. Eyre Methuen, London, **1973**.
Taylor A. *A Commentary on Plato's Timaeus*. Oxford University Press, Oxford, **1928**.

Scientific and Divulgative Texts on Natural Products, Phytopharmaceuticals, Nutrition, Ethnobotany, Ethnomedicine, and Ayurveda

Bagwan Dash. *Materia Medica of Indo-Tibetan Medicine*. Classic India Publishers, Delhi, India, **1989**.
Benzie I.F.F. Wacthel-Galor S. (eds.) *Herbal Medicine: Biomolecular and Clinical Aspects*. CRC Press, Boca Raton, FL, **2011** (2nd edn.).
Bhattacharjee, S.K. *Handbook of Aromatic Plants*. Pointer Publishers, Jaipur, India, **2000**.
Bianchi A., Cozzi L., Speciale A. (eds.) *Forest Medicine. Traditional Medicines, A Resource to Discover*. Centre for Orientative Education (COE), Milan, Italy, **2003**.
Bisset N.G. *Herbal Drugs and Phytopharmaceuticals—A Handbook for Practice on a Scientific Basis*. CRC Press, Tokyo, Japan, **1994**.
Brigelius-Flohé R., Joost H.G. (eds.) *Nutritional Genomics*. Wiley-VCH Verlag GmbH, Weinheim, Germany, **2006**.
Chopra R.N., Chopra I.C., Varma B.S. *Supplement to Glossary of Indian Medical Plants*. Publications and Information Directorate, CSIR, Delhi, India, **1969**.
Chopra R.N., Nayar S.L., Chopra I.C. *Glossary of Indian Medicinal Plants*. Council of Scientific and Industrial Research, Delhi, India, **1956**.
Codex Alimentarius Commission. *Guidelines for the Use of Nutrition and Health Claims*. Codex Committee on Food Labelling, Rome, Italy, CAC/GL 2-1985, as modified **2013**.
Codex Alimentarius Commission. *Guidelines on Nutrition Labelling*. Codex Committee on Food Labelling, Rome, Italy, CAC/GL 2-1985, as modified **2013**.
Colonna S., Folco G., Marangoni F. *I cibi della salute. Le basi chimiche di una corretta alimentazione*. Springer-Verlag Italia, Milan, Italy, **2013**.
Consumers International. Recommendations for the UN high-level summit on non-communicable diseases (NCDs). **2011**.
Consumers International. Recommendations towards a global convention to protect and promote healthy diets. July **2014.** http://www.consumersinternational.org/media/745830/ci%20proposed%20outcomes%20for%20un%20ncd%20hlm%20sep%2011.pdf.
Dans A., Ng N., Varghese C., Shyong Tai E., Firestone R., Bonita R. The rise of chronic non-communicable diseases in southeast Asia: Time for action. *Lancet* **2011**, *337*, 680–689. http://wphna.org/wp-content/uploads/2014/07/2014-03_CI-WO_Framework_Convention_draft.pdf.
Dastur J.F. *Medicinal Plants of India and Pakistan*. Taraporevala Sons, Mumbai, India, **1962**.
Dawa M.R. *A Clear Mirror of Tibetan Medicinal Plants*, vol. 1. Tibet Domani, Rome, Italy, **1999**.
Dawa M.R. *A Clear Mirror of Tibetan Medicinal Plants*, vol. 2. Men-Tse-Khang, Dharamsala, India, **2009**.
FAO (Food and Agriculture Organization). *Sustainable Diets and Biodiversity: Directions and Solutions for Policy Research and Action*. FAO, Rome, Italy, **2012**.

Foster G.M. An introduction to the ethnomedicine, in Bannerman R.H., Burton J., Wen-Chieh, C. (eds.), *Traditional Medicine and Healthcare Coverage*. World Health Organization, Geneva, Switzerland, **1983**.

Frawley D. *Soma in Yoga and Ayurveda. The Power of Rejuvenation and Immortality*. Motilal Banarsidass Publishers, New Delhi, India, **2013**.

Frawley D., Lad V. *The Yoga of Herbs*. Motilal Banarsidass Publishers, New Delhi, India, **1994**.

Gregori D., Halpern G. Multifunctional food and traditional ingredients A competitive marriage. *Intl. J. Food Sci. Nutr.* **2012**, *63*(S1).

Kapoor L.D. *Handbook of Ayurvedic Medicinal Plants*. CRC Press, New York, **1989**.

Kastorini C.-M., Milionis H.J., Esposito K., Giugliano D., Goudevenos J.A., Panagiotakos D.B. The effect of Mediterranean diet on metabolic syndrome and its components: A meta-analysis of 50 studies and 534,906 individuals. *J. Am. Coll. Cardiol.* **2011**, *57*(11), 1299–1313.

Keys A. *Seven Countries: A Multivariate Analysis of Death and Coronary Heart Disease*. Harvard University Press, Cambridge, MA, **1980**.

Keys A., Keys M. *Eat Well and Stay Well, the Mediterranean Way*. Doubleday, New York, **1975** (1st edn.).

Khalsa K.P.S., Tierra M. *The Way of Ayurvedic Herbs*. Motilal Banarsidass Publishers, New Delhi, India, **2010**.

Khare C.P., Katiyar C.K. (eds) *The Modern Ayurveda. Milestones beyond the Classical Age*. CRC Press, Boca Raton, FL, **2012**.

Koyama A., Houston D.K., Simonsick E.M., Lee J.S., Ayonayon H.N., Shahar D.R., Rosano C., Satterfield S., Yaffe K. Association between the Mediterranean diet and cognitive decline in a biracial population. *J. Gerontol. A Biol. Sci. Med. Sci.* **2014**, pii, glu097.

Kurup, P.N.V., Ramadas, V.N.K., Joshi, P. *Hand Book of Medicinal Plants*. Central Council for Research in Ayurveda and Siddha, New Delhi, India, **1979**.

Leslie C. (ed.) *Asian Medical Systems: A Comparative Study*. Motilal Banarsidass Publishers, New Delhi, India, **1998**.

Malhotra S.C. (ed.) *Pharmacognostical Investigations of Certain Medicinal Plants and Comp. Formulations used in Ayurveda & Siddha*. Central Council for Research in Ayurveda and Siddha, New Delhi, India, **1996**.

Mishra L.C. *Scientific Basis for Ayurvedic Therapies*. CRC Press, New York, **2003**.

Misra A., Singhal N., Khurana L. Obesity, the metabolic syndrome, and type 2 diabetes in developing countries: Role of dietary fats and oils. *J. Am. Coll. Nutr.* **2010**, *29*(3), 289S–301S.

Morton J.F. *Fruits of Warm Climates*. J.F. Morton, Miami, FL, **1987**.

Morton J.F. *Herbs and Spices*. Golden Press, New York, **1976**.

Nadkarni K.M. *Indian Materia Medica*. Popular Prakashan, Mumbai, India, **2002**.

Narayanan A.N., Morandi A. (eds.) *An Integrated View of Health and Well-Being. Bridging Indian and Western Knowledge*. Springer, Berlin, Germany, **2013**.

Ng M., Fleming T., Robinson M., Blake Thomson B.A., Nicholas Graetz B.S., Christopher Margono B.S., Erin C. Mullany B.A. *et al.* Global, regional, and national prevalence of overweight and obesity in children and adults during 1980–2013: A systematic analysis for the Global Burden of Disease Study 2013. *Lancet* **2014**, *384*(9945), 766–781. doi:10.1016/S0140-6736(14)60460-8.

Ogden C.L., Carroll M.D., Kit B.K., Flegal K.M. Prevalence of childhood and adult obesity in the United States, 2011–2012. *JAMA* **2014**, *311*(8), 806–814. doi:10.1001/jama.2014.732.

Paranjipe P. *Indian Medicinal Plants. Forgotten Healers*. Chaukhamba Sanskrit Pratishthan, Delhi, India, **2005** (2nd edn.).

Phillips H. *History of Cultivated Vegetables: Comprising Their Botanical, Medicinal, Edible, and Chemical Qualities; Natural History; and Relation to Art, Science, and Commerce*. Henry Colburn, London, **2010** (2nd edn.).

Pieroni A., Price L.L. (eds.) *Eating and Healing. Traditional Food as Medicine*. The Haworth Press, New York, **2006**.

Pollan M. *Cooked: A Natural History of Transformation*. Penguin Press, New York, **2013**.

Pollan M. *The Omnivore's Dilemma: A Natural History of Four Meals*. Penguin Press, New York, **2006**.

Pruthi J.S. *Spices and Condiments*. National Book Trust, New Delhi, India, **1998** (5th edn.).

Puri H.S. *Rasayana. Ayurvedic Herbs for Longevity and Rejuvenation*. Taylor & Francis, London, **2003**.

Rastogi R.P., Mehrotra B.N. *Compendium of Indian Medicinal Plants*. Publications and Information Directorate, CSIR, New Delhi, India, **1993** (reprinted edition).

Riazudin A.S., Mohammed S. *Vedic Plants. Medicinal and Other Uses*. Chaukhamha Orientalia, Varanasi, India, **2009**.

Sairam, T.V. *Home Remedies*, 2 vols. Penguins Books, Delhi, India, **1998**.

Sharma P.V. *Fruits and vegetables in Ancient India*. Chaukhamba Orientalia, Varanasi, India, **2009**.

Sharma P.V. *Introduction to Dravaguna (Indian Pharmacology)*. Chaukhambha Orientalia, Varanasi, India, **1976**.

Sivarajan V.V., Balachandran I. *Ayurvedic Drugs and Their Plant Sources*. Oxford and IBH, Delhi, India, **1994**.

Svoboda R.E. *Ayurveda Life, Health and Longevity*. Penguin Books, Delhi, India, **1993**.

Warrier P.K., Nambiar V.P.K., Raman Kutty C. *Indian Medicinal Plants*, vol. 1–5. Orient Longman, Chennai, India, **1993–1996**.

Webb G.P. *Dietary Supplements and Functional Foods*. Wiley-Blackwell, **2011** (2nd edn.) 328pp.

Willcox D.C., Scapagnini G., Willcox B.J. Healthy aging diets other than the Mediterranean: A focus on the Okinawan diet. *Mech. Ageing Dev.* **2014**, *136–137*, 148–162.

World Health Organization. *Medicinal Plants in the South Pacific*. WHO Regional Publications, Western Pacific Series, Series 19, Manila, the Philippines, **1998**.

Yoganarasimhan S.N., Chelladurai V. *Medicinals Plants of India*, 2 vols. Cyber Media, Bangalore, India, **2000**.

Zimmer H.M. *Hindu Medicine*. Johns Hopkins University Press, Baltimore, MD, **1948**.

Zysk K.G. *Medicine in the Veda. Religious healing in the Veda*. Motilal Banarsidass Publishers, New Delhi, India, **2009** (reprint).

Websites: Phytochemistry, Nutrition, and Dietetics Journals; Integrative and Complementary Medicine Journals; and Botanical Websites

Dr. Duke's Phytochemical and Ethnobotanical Databases. http://www.ars-grin.gov/duke/plants.html.

Ethnobotanical databases: KEW gardens. http://www.kew.org.

Evidence-based information about complementary and alternative medicine including dietary supplements. http://www.naturalstandard.com.

Home of the electronic version of "A Modern Herbal" by Maud Grieve, originally published in 1931. http://www.botanical.com/.

Integrative and Complementary Medicine journals. http://ip-science.thomsonreuters.com/cgi-bin/jrnlst/jlresults.cgi?PC=D&SC=OI.

Missouri Botanical Garden. http://www.missouribotanicalgarden.org.

Nutrition journals. https://metranet.londonmet.ac.uk/services/sas/library-services/subject-help/subjects/nutrition/home.cfm.

PubMed. It comprises more than 24 million citations for biomedical literature from MEDLINE, life science journals, and online books. http://www.ncbi.nlm.nih.gov/pubmed.

SciFinder (The database from Chemical Abstracts). It currently reports more than 43 million chemical substances and 22 millions citations. https://scifinder.cas.org/.

Systematic Reviews in Ayurveda. http://www.systematicreviewinayurveda.org/index.php?option=com_content&view=article&id=61&Itemid=66.

Journals

Agricultural and Food Chemistry (ACS). http://pubs.acs.org/journal/jafcau.
Ancient Science of Life. http://www.ancientscienceoflife.org/.
European Journal of Clinical Nutrition. http://www.nature.com/ejcn/index.html.
Fitoterapia (Elsevier). http://www.journals.elsevier.com/fitoterapia/.
Food Chemistry (Elsevier). http://www.journals.elsevier.com/food-chemistry/.
Journal of Ayurveda and Integrative Medicine. http://www.jaim.in/.
Journal of Natural Products (ACS). http://pubs.acs.org/journal/jnprdf.
Nutrition (Elsevier). http://www.journals.elsevier.com/nutrition/.
Phytochemistry (Elsevier). http://www.journals.elsevier.com/phytochemistry/.
Planta Medica (Thieme). https://www.thieme.de/de/planta-medica/journal-information-4809.htm.

Food Recipes and Blogs

Banchek L. *Cooking for Life: Ayurvedic Recipes for Good Food and Good Health*. Harmony Books, New York, **1989**.
Bianchi M. *Le ricette dei magnifici 20*. Ponte alle Grazie, Adriano Salani Editore, Milano, Italia, **2011**.
Bianchi M. *Un anno in cucina con Marco Bianchi*. Ponte alle Grazie, Adriano Salani Editore, Milano, Italia, **2012**.
Giri J., Jain P. *Sukham ayu. Cooking at Home with Ayurvedic Insights*. KARE, Mulshi Lake, Pune, India, **2013**.
Johari H. *The Healing Cuisine: India's Art of Ayurvedic Cooking*. Healing Art Press, Rochester, VT, **1994**.
Lad U., Lad V. *Ayurvedic Cooking for Self-Healing*. Ayurvedic Press, Albuquerque, NM, **1994**.
Morningstar A. *Ayurvedic Cooking for All. Familiar Food Prepared with Ayurvedic Principles*. New Age Books, Delhi, India, **2002**.
Morningstar A., Desai U. *The Ayurvedic Cookbook*. Motilal Banarsidass Publishers, New Delhi, India, **1999**.
Prakash O. *Food and Drinks in Ancient India*. Munshiram Manoharlal Publishers, Delhi, India, **1961**.
Rajiaiah R. *How the Banana Goes to Heaven*. Venkat Towers, Chennai, India, **2010**.
Ranade S. *Ayurvedic Nutrition & Cooking*. Chaukhamba Sanskrit Pratishthan, Delhi, India, **2007**.
Tiwari M. *Ayurveda. A Life of Balance: The Complete Guide to Ayurvedic Nutrition and Body Types with Recipes*. Healing Art Press, Rochester, VT, **1995**.
http://www.foodnews.org/fulllist.php.
http://www.giallozafferano.it/.
http://www.joyfulbelly.com.
http://nutritiondata.self.com/.
http://www.thecookingacademy.co.uk.
http://www.whfoods.com/index.php.

Organizations and Projects

Ayurveda Maharishi. http://www.ayurvedamaharishi.net.
Biology Web Directory. http://www.biologydir.com.
The Blue Zones. http://www.bluezones.com/.
CIFOR (Centre for International Forestry Research). http://www.cifor.org.
CIKS (Center Indian Knowledge Systems). http://www.ciks.org/.

COE (Centre for Orientative Education). http://www.coe.org.
COMPAS. http://www.compasnet.org.
European Heart Network: Diet, Physical Activity and Cardiovascular Disease Prevention in Europe, **2011**. http://www.ehnheart.org/publications/publications/publicatdisease-prevention.html.
European Project. AnThocyanin and polyphenol bioactives for Health Enhancement through Nutritional Advancement (ATHENA). http://www.athena-flora.eu/.
FAO (Food and Agriculture Organization). http://www.fao.org/.
FRLHT (Foundation for Revitalisation of Local Health Traditions), an NGO based in Bangalore whose mission is to revitalize Indian Medical Heritage. It has established comprehensive databases on Indian medicinal plants, a herbarium, and research laboratories. It also has a 100-bed Ayurvedic research hospital that systematically documents the effectiveness of treatment of patients. It has recently attained the status of a State Private University, legislated by the Government of Karnataka. The university is called the Institute of Trans-disciplinary Health Sciences & Technology (ITD-HST). http://www.frlht.org/.
Global Diversity Foundation. http://www.global-diversity.org.
IASO (International Association for the Study of Obesity). UN summit—Obesity and nutrition outcomes, **2011**.
IRD (Institute de recherche pour le dévéloppement). http://en.ird.fr/.
ISE (International Society of Ethnobiology). http://www.ethnobiology.net.
Isht University. http://www.ihstuniversity.org/.
Italian School of Ayurvedic Medicine. http://www.ayurvedicpoint.it.
IUCN (International Union for Conservation of Nature). http://www.iucn.org.
The Moli-sani project. http://www.moli-sani.org/.
NTBG (National Tropical Botanical Garden) database. http://ntbg.org/plants/.
NuGO (Nutrigenomics Organization). http://www.nugo.org.
SIFOR (Smallholder Innovation for Resilience). http://www.bioculturalheritage.org.
The Seven Country Study. http://sevencountriesstudy.com/.
UN (United Nations). Report of the United Nations conference on sustainable development. Rio de Janeiro, Brazil, June 20–22, **2012**, A/Conf.216/16.
UN DESA (United Nations Department of Economic and Social Affairs). Realising the future we want for all—Report to the Secretary General, United Nations System Task team on the Post-2015 UN Development Agenda, New York, June **2012**.
UN General Assembly. Political declaration of the high-level meeting of the General Assembly on the prevention and control of non-communicable diseases, Resolution 66/2 adopted by the United Nations General Assembly, September 19, **2011**.
University of Milan, Department of Chemistry. http://www.unimi.it; http://www.chimica.unimi.it.
WCRF (World Cancer Research Fund) International. WCRF International Food Policy Framework for Healthy Diets: Nourishing. **2014**. http://www.wcrf.org/int/policy/nourishing-framework/.
WHO (World Health Organization). Factsheet No. 311. Obesity and overweight. March **2013**.
WHO. Global strategy on diet, physical activity and health. 57th World Health Assembly, WHA 57.17, May 22, **2004**.
WHO. Set of recommendations on the marketing of foods and non-alcoholic beverages to children. WHA 63.14, May 21, **2010**.
WHO/Europe. Vienna declaration on nutrition and non-communicable diseases in the context of health 2020. WHO, Geneva, Switzerland, **2013**.

Index

Note: Locators followed by "*f*" and "*t*" denote figures and tables in the text

C

D

E

F

O

P

Q

R

For Product Safety Concerns and Information please contact our EU representative GPSR@taylorandfrancis.com Taylor & Francis Verlag GmbH, Kaufingerstraße 24, 80331 München, Germany

Batch number: 10726532

Printed by Printforce, the Netherlands